£7.99
(10)
Hist 2

Alexander Del Mar

A History of the Precious Metals from the Earliest to the Present

Elibron Classics
www.elibron.com

Elibron Classics series.

ISBN 1-4021-7302-4 (paperback)
ISBN 1-4021-3163-1 (hardcover)

This Elibron Classics Replica Edition is an unabridged facsimile of the edition published in 1902 by the Cambridge Encyclopedia Company, New York.

A HISTORY OF THE PRECIOUS METALS.

BY THE SAME AUTHOR:

A HISTORY OF MONEY IN ANCIENT STATES,

FROM THE EARLIEST PERIODS TO THE DARK AGES,

COMPRISING

CHINA, JAPAN, INDIA, ARIANA, BACTRIA, CABUL, AFGHANISTAN, EGYPT, PERSIA, ASSYRIA, BABYLON, PALESTINE, ABORIGINAL EUROPE, GREECE, GREEK COLONIES, CARTHAGE, ETRURIA AND ROME.

8vo, pp. 400; cloth and gold, $3.

MONEY AND CIVILIZATION:

A HISTORY OF THE

MONETARY LAWS AND SYSTEMS OF VARIOUS STATES SINCE THE DARK AGES, AND THEIR INFLUENCE UPON CIVILIZATION.

8vo, pp. 475; cloth and gold, $3.

A HISTORY OF MONETARY SYSTEMS;

A RECORD OF ACTUAL EXPERIMENTS IN MONEY MADE BY VARIOUS STATES OF THE WORLD,

AS DRAWN FROM THEIR

Statutes, Customs, Treaties, Mining Regulations, Jurisprudence, History, Archæology, Coins, Nummulary Systems, and other sources of information.

8vo, pp. 450; cloth and gold, $2.

HISTORY OF MONEY IN AMERICA;

FROM THE SPANISH CONQUEST TO THE FOUNDATION OF THE AMERICAN CONSTITUTION.

8vo, pp. 200; cloth, $1.50

HISTORY OF MONETARY CRIMES.

8vo, pp. 110; illustrated; cloth, 75 cents.

THE SCIENCE OF MONEY;

OR THE

PRINCIPLES DEDUCED FROM ITS HISTORY.

Third edition, 8vo, pp. 226; cloth, $1.

CAMBRIDGE PRESS, NEW YORK.

A HISTORY OF

THE PRECIOUS METALS

FROM THE EARLIEST TIMES TO THE PRESENT

BY

ALEX. DEL MAR, M. E.

FORMERLY DIRECTOR OF THE U. S. BUREAU OF STATISTICS, MINING COMMISSIONER TO THE U. S. MONETARY COMMISSION; AUTHOR OF "A HISTORY OF MONETARY SYSTEMS," "ANCIENT BRITAIN," "THE MIDDLE AGES REVISITED," ETC., ETC.

SECOND EDITION—REVISED

NEW YORK
PUBLISHED BY THE
CAMBRIDGE ENCYCLOPEDIA COMPANY
1902

HISTORY OF THE PRECIOUS METALS

LIST OF CHAPTERS.

PUBLISHERS' ADVERTISEMENT.

Del Mar's "History of the Precious Metals" was begun in 1858, laid aside in 1862, when the author published his first work on Money, resumed in 1869, completed in 1879, and published during the following year by Messrs. Geo. Bell & Sons, of London. The merits of the work were at once recognized by the literary press, portions of it were translated and republished with approbation by the *Journal des Economistes* and other Continental journals, and the edition was soon exhausted. But until within late years the author's occupation as a Mining Engineer took him so far afield that he found it impracticable to revise the work for a second publication: a task which he always expressed himself anxious to accomplish.

The plan of the present work embraces a history of the precious metals in each important country by itself, from the earliest times to the present. After this, the subject is treated from a general standpoint; the produce of the world at various epochs is brought together, the consumption in the arts and for coinage is shown, together with the resulting stocks on hand in the form of coins and bullion. This is the framing of the work: the body of it is filled with the most interesting and in some instances very important historical material. The history of the Veneti of Pontus, an ancient race of miners who settled in Greece and North Germany, and afterwards in Italy at a very remote period, is here brought to light for the first time; the Tschudic remains found in the Siberian tundras; and the Roman mining laws of the Peak of Derby, which are still enforced, are among the strikingly original subjects treated in this volume. The South American Revolution of 1732, and the system of contract-labour which prevails to-day in the Mysore and British South African gold mines, to the disadvantage of Australia, California, Colorado, and the free mining States generally, are also new lights shed upon a little understood, but highly important, subject.

While, as the author shows, the search for gold has carried the torch of civilization and the claims of Christianity to the remotest regions, it has also extended the area and prolonged the establishment of slavery. Of the twenty thousand million dollars worth of the precious metals produced since the Discovery of America, fully one-half has been wrung from the blood and tears of conquered and enslaved races. The value at which this crime-stained metal has entered the exchanges of the world keeps down the value of the portion produced by free labour; so that as yet, the latter is sold to the mints at less than its average cost. In other words, thus far, and perhaps for generations yet to come, the precious metals gained by free-labour, cost in the long run more to produce than they are worth; a conclusion long familiar to practical mining men, but sounding strangely to those who ignore the hazards, expense, and losses of gold and silver mining.

Among the graphic portions are the details of Albuquerque's sack of Muscat; the story of the Paraguayan Missions; the tragedy of Lonira, as told by Humboldt; the Coolgardie Butchery; the silver sidewalks of San José; the auction sales of golden graves in Chiriqui and of golden loot in Pekin; the conquest of Siberia by Yermak, and the storming of Seringapatam by the British.

The Cost of Production theory, which has been so widely entertained since the publication of Adam Smith's celebrated treatise, is shown by our author to have no footing in respect of the precious metals, whose production has been largely due to conquest and slavery. "Is there a price of blood," he asks; "Is there a price of anguish, of life, of death, of the extinction of races and of their inheritance of experience, invention, law, religion, and the moral code"? Another theory drawn from the same prolific source of economical error: That the volume of money obeys the scale of prices and the demands of commerce, rather than that the latter obey the former, also engages our author's attention.

The scope of the volume is enormous; the details full of interest. A Roman silver ingot found within the Tower of London, in 1777, is proved by its weight to be a talent, the origin of the thaler or dollar, and therefore the oldest dollar now extant. The cession to the British of the three provinces of Bengal, with a population of 25 millions, and a net revenue of 20 million dollars, was made in 1765, from a throne which consisted of a dining-table and arm-chair in Lord Clive's tent. The plunder of Japan in the 17th, of India in the 18th, and of China in the 19th century, events intimately connected with the history of the precious metals, are replete with graphic incidents which hold the attention while they profoundly instruct the reader.

A HISTORY OF THE PRECIOUS METALS.

PREFACE.

The favourable reception accorded to the author's previous work on the Precious Metals, the entire edition having long since been disposed of, has induced him to resumė the subject, although upon a somewhat different plan. In some portions of the former work the search for the precious metals was treated by countries, whilst in others it was treated with reference to the physical devastation which it had entailed, as well as the gambling, insanity and crime which had everywhere marked its course. In the present work these subjects are omitted from view and the historical method is followed throughout.

As since the preparation of the first edition the author has, in his capacity of Engineer, visited most of the localities described, he feels increased confidence in the accuracy of his descriptions. The dangers with which his authorities invested some of the gold regions in distant countries disappeared upon closer acquaintance with them; while the method of working the mines, described by others, proved to be faulty or inaccurate.

Although, as will be seen when this portion of the subject comes to be dealt with, money is not the principal use to which the precious metals are devoted, it is by far the most important one. It is a peculiarity of money that it cannot with propriety be treated in its functional capacity apart from other money, because, unlike physical measures, its function is affected by numbers. To increase or diminish the number of yardsticks or pound weights would have no effect upon the measurement of length or weight; whilst to increase or diminish the number of coins or bank notes would have a very decided influence upon the measurement of value. This is a principle which has not been left to modern discovery, for it will be discerned in the most ancient works which have been spared to us by time and proscription. In its relation to value, money therefore means all the money of a given epoch, or else all the money within a given jurisdiction.

To admit that it is the quantity of the precious metals coined and in circulation which influences their value and not what it may have cost to produce them, is to admit that value is a *relation* and not an *attribute* of commodities; it is to admit that money is a measure or measurer of value; that like other measures it is a creation of Law; and that to work equitably it should, like other measures, be defined and limited by law with precision.

Such admissions will never wittingly be made by those who profit by the existing lax laws of money or by the abuses which have grown up under them. Yet they are being continually made unwittingly. When California threatened to greatly increase the quantity of gold money, Holland and Belgium demonetised gold; France was on the point of following suit; while England gravely listened to Maclaren and Cobden, who advised a recourse to silver money and barter. When the Comstock Lode of Nevada threatened to greatly increase the quantity of silver money, England, by her Mint Act of April 4, 1870, deprived her Queen of the power to remonetise silver; Germany and France followed her lead, and in the course of a comparatively few years the stampede to gold was completed.

In neither of these great movements was the least consideration accorded to the interests of mankind, of the people at large, of the masses, to which both you and I, dear reader, or else the majority of our decendants must necessarily belong. Their object was primarily to arrest a threatened Rise of Prices having a Governmental basis (gold or silver coins) long enough to convert it into a Rise of Prices having an Individual basis (private bank notes).

Similar admissions concerning the dependance of value upon Quantity (*e.g.* the Numbers of coins), are tacitly made whenever the expected course of prices is viewed through the movements of money, or of the precious metals. The apprehensions excited by the exportation of these metals would have no basis at all were their value and its influence upon trade and speculation due to the cost of their production. That such apprehensions are valid is an unconscious admission that value is due to Quantity; and no amount of sophistry can permanently remove this conviction from the intelligent mind.

Yet its place is not yet assured. Three thousand years of contention on this subject—a contention which dates backward to the very origin of metallic money—evince at once the tremendous results that hang upon its decision, and the facility with which a simple truth has been made to baffle the popular comprehension.

So long as this state of affairs continues, so long as Individuals in place of Government retain control of the Monetary Measure, there can be no real Religion, there can be no real Liberty, there can be no real National Life. The bases of Religion are Love and Fraternity. There can be no Fraternity whilst an Unjust Measure is permitted to introduce discontent and strife into all the transactions of social life. The basis of Liberty is Justice. There can be no Justice whilst an Unjust Measure continues to nullify the lessons of wisdom and experience. The basis of National Life is Political Equality. There can be no Equality so long as an Unjust Measure continues to rob the many for the benefit of the few. The control of Weights and Measures, including Money and the materials of which its symbols are made, is a prerogative and a necessary prerogative of National Life. It has been so held by all the Courts of Judicature, by all the exponents of Law, from Plato to Paulus, from Paulus to Grimaudet, from Grimaudet to the Mixt Money case, and from the latter to the decisions of the Supreme Court of the United States.

The Silver Question is settled: it is to be hoped, forever; but the Gold Question remains; and in our rapid and strenuous life the Gold Question involves considerations which not only seriously affect the future of this nation, but of mankind at large.

Under these circumstances the time seems to be approaching when the Money Question will have to be debated upon much higher grounds than any which have hitherto been advanced in the contention. It will have to be argued not by appeals to ignorance and passion, but by appeals to Religion and Patriotism; to Fraternity and to Equity.

To approach so profound a subject without the advantages of the world's knowledge and experience which are embodied in a History of the Precious Metals, would be the height of temerity: a consideration which the author ventures to believe is a sufficient apology for the publication of the present work.

BIBLIOGRAPHY.

The following List of Books is to be read in connection with the lists published in the author's previous works. The numbers at the foot of each title are the press marks of the British Museum Library. When several editions or translations of the same work are given, the object is to help the student to find authority in distant libraries for the statements in the text.

ACADEMIES. Giornale degli Economista. Bologna, 8vo. *In progress*. PP.1424. ayb.
——— Institute International de Statistique. Rome, 8vo. *In progress*. Ac. 2404.
ADAMS (Brooks). The Law of Civilization and Decay. London, 1895.
AGRICOLA. De Re Metallica. Basil, 1657.
AGUERO ESCALANO (Gaspar de). Gazophilacium regium Perubicum. (Gazofilacio real del reyno de Peru). Lat. and Spanish. Madrid, 1647, fol. 501, g. 8.
ALDROVANDI (Ulysses). Musæum Metallicum. Bononiæ, 1648, fol.
ALLEN and TEXEIRA. Monnaies d'Or, Suevé-Lusitaniennes, Extrait de la Revue Numismatique, Nouvelle Series, tome X, 1865, by Edwardo Augusto Allen and Henrique Nunes Texeira. Pamphlet, Oporto, 1865, 8vo. 7755. bb.
ALLEN (Grant). Evolution of the Idea of God.
AL-MAKKARI (Ahmed-ibn-Mahommed). Selections from his History of the Mahometan dynasties. Trans. from Arabian to English by Pascual de Gayangos. London, 1840–3, 2 vols, 4to. 452. m. 1–2.
AMERICA. Recopilacion de Leyes de las Indias. Madrid, 1774, 4 tom. fol.
ANCHARANO. See *Petrus de Ancharano*.
ANDERSON (A. Hay). Notes of a Journey to the auriferous quartz regions of Southern India. Edinburgh, 1880.
———Ophir, or the Indian gold mines. London, 1880.
ANDERSON (James). History of Commerce. London, 1787, 4 vols, 4to.
ANNESLEY (Arthur), *Earl of Anglesey*. Memoirs. London, 1693, 8vo. 1415. a. 17.
———State of the Government and Kingdom. London, 1694. E. 1973. (8.)
ANONYMOUS. Theory of Money, in connection with some of the prominent doctrines of Political Economy, by a Scotch Banker. Edinburgh, 1868.
———Della Moneta. Milano, 1769, 4to. (A valuable work.)
———Observations on the Scheme of Mr. Law in France and of Sir H. Mackworth in Great Britain. London, 1720, fol.
ARRIAN. See *Mac Crindle* and see *Chinnock*.
ASHLEY (Wm. James). English Economic History and Theory. Second ed. London, 1892, 8vo. 2240. aa. 1.
ATKINSON (Edwin T.) Economic Geology of the Hill District, North-West provinces of India. Allahabad, 1877.
ATKINSON (Thomas Witlam). Bronze relics found in an auriferous sand in Siberia. Qt. Jour, Geolo. Soc. XVI, 241.
———Oriental and Western Siberia. London, 1858.
AXERIO (Giulio). Mineral industry of Italy. Jour. Inst. C. Eng., XLII.
BACON (Francis) *Lord Verulam*. Works. London, 1874, 14 vols, 8vo. 2044. f.
BALL (V.) Mineral Resources of India and their Development. Jour. Soc. Arts, XXX.
———Manual of the Geology of India. London, 1881, 8vo. 484. g. 7.
BALUZE (Etienne). Capitulaires des Rois François. See tom. II, 174, for the Edictum Pistense or Piste, which provided for recoinages in ten cities at the ratio of 12 for 1; also for Weights and Measures of the period. Paris, 1779, 8vo. 708. a. 12.
BANKS (W.) Guide to Lewes, England. Lewes, 8vo, 1887, brochure.

Barba (Albaro Alonso). Arte de los Metales, en que se enseñar el verdadero beneficio de los de Oro y Plata con azogue, el apartar unos de otros. Madrid, 1729, 4to.
———The same. An earlier edition. Madrid, 1640, 4to.
———The same. A later edition. Con el tratado de las Antiguas Minas de España. Lima, 1817.
———The same, trans. into English by the Earl of Sandwich. London, 1740, 12mo.
Barbacciani-Fedeli (Ranieri). Decisione dell' Illus. sig. avvocato R. B–F. . . . nella Piscien. Nullit. Contract. tra il. . . . Cavaliere Antonio Cecchi Toldi e i Signori Alientari. (Upon the Nullification of Contracts of Sale when affected by changes in Money.) Pescia, 1829, 8vo. 898. h. 2. (6.)
Barbon (Nicholas). A discourse concerning coining the new money lighter. In answer to Mr. Locke. London, 1696, 8vo. 1139. c. 7. (1.)
———Decus and Intamen; or our New Money, as coined, etc. A reply to Sir R. Temple and Dr. B. London, 1696, 8vo. 1139. c. 25.
Bartolus de Saxoferrato. See *Savigny*.
Bazinghen. Weights, Measures and Monies. (The portion devoted to England is evidently borrowed from Lowndes and Fleetwood. As an universal Cambist this work, though much earlier than Dr. Kelly's, is almost as complete.) 603.g.12.
Beattie (James) *Rev.* The Nature and Immutability of Truth. London, 1812, 8vo.
Beccaria (C.) Del disordine e de remedii delle Moneta nello Stato de Milano nel 1762. By the Marquis Cesare Beccaria Bonesana, of Milan. Lucca, 1762, 8vo, pp. 50. (A short treatise on the disordered coinages of Milan.) 140. a. 26.
———Political Economy. (Published in the Scrittori Classici Italiani, wherein Beccaria is ranked as one of the three ablest economists of his age.) See *Italiani Scrittori*.
Bellefonds (Linant de) L'Etbaye, pays habite par les Arabes Bicharich. Paris, 1868.
Belluga (Joannis de). De mutatione monetarum.
Benites (Cantos). (A Spanish Numismatic work.)
Bergius (John H. L.) Catalogue of Books on Finance. Cameralisten Bibliothek. Nuremburg, 1762, 8vo. 8206. b. 36
Biaeus (Jacobus). La France Metallique. Paris, 1636, fol.
———Numismata Imperatorum Romanorum.
Bigandet (Paul A.) *Bishop of Ramatha*. The life or legend of Gaudama, the Buddha of the Burmese. Third edition. London, 1878, 2 vols, 8vo. 2318. f. 7
Bircherod (Thomas Broderus). Specimen Antiquæ Rei Monetariæ Danarum, etc. Halfinæ, 1701, 4to. 603. d. 20. (2.)
Bizot (Pierre). Histoire Metallique de la Republique de Hollande. Nouvelle ed. Amsterdam, 1688, 8vo.
Blanchard. Essay on the Three meanings of the word "Denier."
Blanford (W. T.) Mineral Resources of India. Jour. Soc. Arts, xxi.
Boden (John). The Republic. First published in French, 1577, and afterwards in Latin, 1586. Translated into English by Knolles. (Boden opposes all changes in the value of money as affecting contracts.)
Boguliubsky (I.) Ocherk Amurskago kraia. St. Petersburg, 1877.
Botero (Giovanni). A treatise concerning the causes of the magnificence and greatness of Cities. Trans. into English by Robert Peterson. London, 1606, 8vo.
———Relations of the most famous kingdoms and commonwealths throughout the world. Trans. into English by R. I. London, 1630. (A valuable work.)
Boureau Deslands (Andre François). Dying Merrily. Trans. from the French of "Reflexions sur le grande hommes qui sont mort en plaisantant," by T. W. London, 1745, 12mo. 12330. bbb. 34.
Bouteroue (Claude). Recherches curieuses des Monoyes de France. Paris, 1666, fol.
Bowring (John). The Kingdom and People of Siam. London, 1857.
Bracton (Henry de). Laws and Customs of England. Edited by Sir Travers Twiss. London, 1878. 6 vols, 8vo. 2063–68.
Brandis (Johannes). Das Münz-Mass und Gewichtswesen in Vorderasien, bis auf Alexander den Grossen. Berlin, 1866, 8vo. 7756. dd. 3.
Brisson (Barnabe), *Pres. Par.*, *Paris*. De Verborum. Lugduni, 1559, fol. 5207. g. i.
———Trois Discours. Incluye "Du Commencement de l' An." Paris, 1609, fol.
Bullarium. Commentaries. 484. g. 7.
Burachkov (P.) Coins of the Greek Colonies on the Black Sea. Odessa, 1884, 4to.

BURNES (A.) Travels into Bokhara. London, 1834.
BURTON (Richard F.) *Sir*. The Gold Mines of Midian. London, 1878.
CADAMOSTO. Delle navigationi et viaggi. Venice, 1563.
CÆSAR (*Sir* Julius). Catalogue of his Manuscripts. London, 1757, 4to. 11903.
CAJIGAL (Ricardo). Las minas de Oro existentes en las márgenes del Sil. Santander, 1877.
CALVERT (John). Gold Rocks of Great Britain and Ireland, London, 1853, 8vo.
CAMERON (V. L.) *Lieut.* Across Africa. London, 1877.
CARDAN (J.) Trattato di numeri e misure. Edition, 1555.
CASTELNAU (François de). Expedition dans les parties centrales de l' Amerique du Sud. Paris, 1850–7.
CASTRO (Manuel F. de). Las minas de Oro de la isla de Cuba. Habana, 1884.
CAZENOVE (John). Considerations on the accumulations of Capital and its effects on Profits. London, 1822. Pamph. 8229. aaaa. 21. (12.)
CHABAS (F. J.) Determination metrique, etc. Chalon-sur-Saone, 1867, 8vo.
———Le Calendrier Egyptienne. Chalon, 1890, 8vo. 7702. dd. 20.
———Etude sur l'Antiquite. Paris et Chalon, 1873, 8vo. 2258. e. 4.
———Egyptian Obelisks. Paris, 1878, 8vo. 7702. a. 33.
———Inscriptions des Mines d'Or. Chalon, 1862, 4to. 7704. L. 16, (3.)
CHARLEVOIX (P. F. X. de) *Jesuit.* Histoire et Description Generale du Japon. Paris, 1736, 2 tom. 4to.
CHINNOCK (Edward James). Arrian's Anabasis of Alexander the Great. London, 1884, 8vo. 9026. ff. 18.
CHOPIN (René), *Bishop of Argos.* Traité du domain de la couronne de France. Paris, 1662, 5 tom. fol. 710. m. 3.
CLEVELAND (Duchess of). Memoirs. London, 1709, 8vo. G. 13786. (1.)
———Case of the Duchess of Cleveland. 816. m. 10. (59.)
———The gracious answer of the Countess of Castelemaine. London, 1668, 8vo.
———The arraignment, trial and conviction of Robert Fielding (who married the Duchess of Cleveland). London, 1708, fol. 515. l. 3. (14.)
———Memoir of Barbara, Duchess of Cleveland, by Geo. S. Steinman. (Privately Printed). Oxford, 1871–8, 8vo. 10825. dd. 11.
———The Duchess of Cleveland's Memorial. London, 1707, sheet fol. 1872. a. (158.)
———The case of Barbara, late Duchess of Cleveland. London, 1715, 12mo.
———A faithful account of the Examination of R. Fielding (who married the Duchess of Cleveland). London, 1706, sheet fol. 1851. c. (38.)
COHEN (Henry). Monnaies de la Republique Romaine (Medailles consulaires.) Paris, 1857, 4to. 77. 56. e. 7.
COIGNARD. Essai sur les Monnois; ou Reflexions sur le rapport entre l' argent et les denrées. Paris, 1746.
COSSA (Luigi). An introduction to the study of Political Economy. Trans. by Louis Dyer, M. A., of Balliol College. London, 1893 8vo. 2020. e.
COVARRUVIAS (Diego). Veterum numismatum collatio, cum his quæ modo expenduntur, publica et regia authoritate percusa. Salmanticæ, 1562, fol. 7755.c.c.1. (The same work is included in the Collection of Monetary Essays, by Budelius.)
CRAWFORD (John). History of the Indian Archipelago. Edinburgh, 1820.
CREVIER (J.B.L.) Histoire des Empereurs Romains. Amsterdam, 1750, 10tom. 12mo.
CUNNINGHAM (*Sir* Alex.) The Stupa of Bharhat; a Buddhist monument ornamented with numerous sculptures, illustrative of Buddhist legend and history in the Third Century, B. C. See pp. viii, and 143. London, 1879, 4to. 7708. g. 21.
———Book of Indian Æras; with tables for calculating Indian dates. Calcutta, 1883, 8vo. 8562. bb. 12.
———Ladak: Physical, Statistical and Historical. London, 1854.

DARIEN. History of "Caledonia," or the Scots Colony in Darien. By a Gentleman lately arrived. London, 1699, 8vo Tract. 636. B.
DAUBREE (Aug.) Distribution de l' or dans la plaine du Rhin. Ann. des Mines, X.
DEBOMBOURG (Georges). Gallia aurifera. Etude sur les alluvions aurifères de la France. Lyons, 1868.
DEL MAR (Alex.) Precious Metals Product of the World. Annual circulars in sheets, 1884 to the present time. London and New York, folio.

DEL MAR (Alex.) Progress of the Demonetisation of Silver; a Memorandum on the dates of the limitation, suspension or abrogation of the privilege formerly accorded to Individuals to have full legal-tender silver coins struck for them by Government free of seigniorage; commonly called "Private Coinage," or "Free Coinage." Printed as Table A to Mr. Henry Huck's Gibbs' (Lord Aldenham's) "Colloquy on Currency." Third edition. London, 1894.

——The Ratio of Value between Gold and Silver in the Occident: England, 1560–1816; France, 1560–1803; Spain, 1475–1890; Portugal, 1510–1880; Germany, 1521–1871; the Netherlands, 1492–1885. Printed as Table B to Lord Aldenham's edition of 1894.

——Letter written at the request of the Royal Commission on the Depression of Trade and Industry: the Earl of Iddlesleigh (formerly Sir Stafford Northcote), Chairman. Third Report of the Commission. London, 1886, folio.

——Reports on the Geology, Production, Present Condition and Prospects of the following Mines and Mining Districts: The Comstock Mines of Virginia City; the Providence, Pittsburg, Empire, Maryland, and other gold mines of Grass Valley; the Dutch Flat and Forest Hill gold mines; the placer and quartz mines of Placer and El Dorado Counties; the Park Canal and Mining Company's ditches and gold mines in El Dorado County; the Hathaway and other hydraulic gold mines of Nevada County; the Tin mines of San Bernardino County; the New Albany, Divoll-Bonanza and other quartz gold mines of Toulomne County; the Tellurides of Tuttletown; the gravel gold mines of the Mohave River; the Plymouth, Keystone, De Witt, and other quartz gold mines of Amador County; the Black Hawk Gold Mines of San Bernardino County; the Copper Mines of Calaveras County; the Ivanpah mines of San Bernardino County; the Vanderbilt, Geddes-and-Bertrand, Hillside and other Silver mines of Eureka, Nevada; the Jennie A. and other Silver mines of White Pine; the Maryland, Day and other Silver mines of Pioche; the Great Western, Vanderbilt and other Silver mines of Esmeralda County; the Borax deposites of Esmeralda County; the Ontario, Christy and other Silver mines of Utah; the Centennial, Toughnut, Goodenough, and other Silver mines of Tombstone, Arizona; the Copper mines of New Jersey; the Gold mines of the Adirondacks; the Gold mines of Salisbury, North Carolina; the Gravel mines of Minhas Geraes, Brazil; the Tin mines of St. Austell, Cornwall; the Iron ores of Senegal, Africa; the Gold mines of the Rio Sil, in Spain; the Gold mines of the Carpathian Mountains, Hungary, etc. Brochures, printed variously in San Francisco, New York, London, Paris and Madrid, 1874 to 1895.

——A History of the Precious Metals from the Earliest Times to the Present. London, 1st ed., Geo. Bell & Sons; 1880, 8vo, pp. 400. Second ed., New York, Cambridge Press, 1902, 8vo, pp 500.

——Special Report, as Mining Commissioner, to the United States Monetary Commission on the present and prospective Production of Silver in the United States, particularly from the Comstock Lode. Washington, Government Press, Feb. 24, 1877, Senate Report, 703, 44th Congress, 2nd Session.

——Minutes of evidence given before the Hon. Alex. Del Mar, acting on behalf of the United States Monetary Commission, concerning the Production of Gold and Silver in the United States. Washington, 1877, Appendix L, Senate Report 703, 44th Congress, 2nd Session.

——World's Production of Gold and Silver, 1492–1877; Relative value of Gold and Silver, 1700–1877; World's Population and Stock of Precious Metals, 1492–1877; Demonetisation of Silver in Germany; Payment of the French War Indemnity of 1871; Movement of the Precious Metals to India, 1835–1877; Standard Coins of the United States, 1792–1873; Coinage of the United States, 1793–1876; Standard Coins of Great Britain and Ireland; Monetary System of Austro-Hungary; Monetary System of China. A series of Essays prepared for the United States Monetary Commission and printed as a portion of their Report. Washington, 1877. Senate Report 703, 44th Congress, 2nd Session.

DEL MAR (Manuel). Individual Resource, or Self-Help. An autobiography. London, 1858, MS.

——Historia de Mejico. New York, 1828, 8vo.

DEMOLOMBE. Cours de Code Civile.

DENNIS (George). Cities and Cemeteries of Etruria. London, 1878, 2 vols, 8vo.

DE ROSSI. Inscriptiones Christianæ urbis Romæ, septimo sæculo. Paris, 1855.
DESMAILINS (Charles.) See *Molinaeus*.
DOUGLAS (Robert) *Rev*. Darkest England's Epiphany. London, 1891, 8vo.
DU HALDE (Pierre) *Father*. The General History of China. Trans. from the French. London, 1736. 4 vols, 8vo.
DUMOULIN (Charles). See *Molinaeus*. DURANTON. Cours de Droit Français.
DUTENS (Louis). The Origin of Discoveries. Trans. from the French. London, 1769, 8vo. 1135, i. 1.
DUVAL JAMERAY. (See *Jameray*.) DUVERGIER. Traité du Pret. Paris, 8vo.

ECONOMIC PAMPHLETS. Gassiot, on Monetary Panics, 1867; "Overend, Gurney and Co., or, the Knack of Muddling through Millions," 1866; Sillar, on Usury, 1867; Carpenter, Perils of Policy-Holders, 1860; Alex. Del Mar, in N. Y. Social Science Review, National Bank Act of 1864; Faucher, Precious Metals, 1852; Torrens, Conveyancing by Registration of Title, 1859.
ENGLAND. Ancient Laws and Institutes; comprising laws enacted under the "Anglo-Saxon" Kings, from Athelbriht to Cnut; those of Edward Confessor, William I., and Henry I., being in English. London, fol., 1840.
——State Papers (Domestic) Interregnum XVI. (Court and Times of Charles I.) Edited by Wm. Douglas Hamilton. 2076. c. d.

FAUCHER (Leon). Recherches sur l' Or et sur l' Argent, considérés comme étalons de la valeur. Paris, 8vo, 1843.
——Researches concerning the Precious Metals. Trans. by Thompson Hankey, Jr., Governor of the Bank of England. London, 1852, 8vo. 8226. c. 73.
FERGUSSON (James). Tree and Serpent Worship in India, in the first and fourth centuries. London, 1850, 8vo. 2032. g.
FLAMMARION (Camille). Lumen et Uranie. Paris, 1880, 8vo.
FLORES. Medalas de España. FOLKES (Martin). Coins. London, 1750, 8vo.
FOSTER (Clement le Neve). A Journey up the Orinoco to the Caratal gold-field. Raliegh's El Dorado. Illustrated. See Foster's Travels I, ii.
FULLARTON (John). Regulation of Currencies, being an Examination of the Principles on which it is proposed to restrict, within certain fixed limits, the future issues on Credit of the Bank of England. London, 1844.
FUZIER-HERMAN (Edouard). Repertoire generale alphabetique du Droit Français. Paris, 1889, 5 tom. 4to. *In progress*. 5424. ee. 6.

GALIANI (F.) Della Moneta. Napoli, 1750, 4to.
GAMBOA (Francisco Xavier de). Commentaries on the Mining Ordinances of Spain. Trans. from the Spanish by R. Heathfield, Barrister at Law. London, 1830, 2 vols, 8vo. 1127. k. 27.
GERMANY, Die Deutche Münzfrage. (The German Curreney Question.) Rinteln, 1861. Pamph. 8225. bb. 38.
GMELIN (S. G.) Journey through Siberia. Gottingen, 1751–2.
GOLDAST VON HEMINGSFELD (Melchoir). Catholica Rei nummariæ. Frankfort, 1652.
GOBET (Nicolas), Les Anciens Minéralogistes du royaume de France. Paris, 1779.
GRANT (C. Mitchell). The gold mines of Oriental Siberia. San Francisco, *Kosmas*.
GRIMAUDET (François). Des Monnoyes. Paris, 1576, 8vo. (Another ed. 1586.).
——The Law of Payment. New York, Cambridge Press, 1900, 8vo.
GRONOVIUS (James). Thesaurus Antiquitatum Græcarum. See tom. XI, on Money.
GRUCHOT (J. A.) Die Lehre von der Zahlung des Geldschuld. (Concerning the Payment of Money debts.) Berlin, 1871. 8vo. 8226. eee. 40.

HAKLUYT (Richard). Collection of Voyages. London, 1810, fol.
HAMILTON (R.) Money and Value: An Inquiry into the Means and Ends of Economic Production; with Appendix on the Depreciation of Silver and Indian Currency. London, 1878, 8vo.
HAMILTON (William J.) Researches in Asia Minor, Pontus and Armenia, with some account of their antiquities and geology. London, 1842.
HANSEN (Peter Ancreas). Tablettes de la Lune. 1857.
HARTT (Ch. Fred.) Geology and Physical Geography of Brazil. Boston, 1870.
——The Gold Mines of Brazil. Engineering and Mining Journal, vol. VIII.

HAWKINS (Edward). Silver Coins of England. Third edition, with alterations and additions by R. L. Kenyon. London, 1887, 8vo. 2032. d.

HAWKINS (Christopher). Gold found in the tin stream works of Cornwall. Trans. Royal Geol. Soc. of Cornwall. See vol. I., p. 235.

HEEREN (A. H. L.) Historical Researches into the Politics, Intercourse and Trade of the principal Nations of Antiquity. London, 1846.

HEISS (Alois). Description des Monnais Antiques de l' Espagne. Paris, 1870, 4to.

———Descripcion general de las Monedas Hispano-Cristianos. Madrid, 1865–9; 3 vols, 4to. (The best work on the subject of Spanish moneys yet met with.)

HEYLYN (Peter). Mikrokosmos; a Little Description of the Great World. London, 5th edition, 1631, 4to.

HILDRETH (R.) Japan as it Was and Is. Boston, 1855.

HOTMAN (François). De re numaria. Paris, 1585, 2 parts, 8vo.

HUBBARD. On the Currency. London, 1843, 8vo.

HULTSCH (Frederick). Griechische und Romische Metrologie. Berlin, 1862, 8vo.

HUMBOLDT (Alex. von) *Baron*. Personal Narrative of Travels to the Equinoxial regions of America. London, 1852.

———The Fluctuations of Gold. New York, Cambridge Press, 1900, 8vo.

———Essai politique sur le royaume de la Nouvelle Espagne. Paris, 3 tom. 8vo, 1825.

HUSSEY (Robert). Essay on Ancient Weights, Money and Measures. Oxford, 1836, 8vo.

IMBERT (Jean) Enchiridion; ou Brief Recueil du droict escript; gardé et observé, ou abrogé en France. Edition augmenté. Paris, 1603, 4to. 5402. cc. 2.

INDIAN GOVERNMENT. Note on the Report of Mr. Brough Smyth, on the Gold Mines of the Wynaad. London, 1880, fol.

ITALIANI. Scrittori Classici. Florence, 1751, 25 vols, 8vo. 2020. a.

ITALY. Diritto Civile Italiana. Napoli, 1888, 5 vols, 8vo. 5359. eeee.

JACKSON (James Grey). An account of the Empire of Morocco and the districts of the Suse and Tifilet. London, 3rd edition, 1814.

JACOB (William). An Historical Inquiry into the Production and Consumption of the Precious Metals. Philadelphia, Carey and Lea, 1832, 8vo.

JACOLLIOT (Louis). Voyages aux Indes. . . . Le pays aux perles. Paris, 8vo.

JAMERAY DUVAL (Valentin). Numismata Cimelii Caesarii Regii Austriaci, Vindobonensis; quorum rariora iconismis, cetera catologis exhibitor, etc. Vindobonæ, 1755, fol. 2pt. 7756. h. 17.

JAMES (H.) Essays on Money, Exchanges, and Political Economy. London, 1820.

JENSEN (Adolf). Aus Briefen Adolf Jensen's mit einem Vorworte des Empfängers. Berlin, 1879, 8vo. 10920. ee. 13.

JEVONS (W. S.) Investigations in Currency and Finance, edited by Prof. Foxwell. London, 1884.

———Money and the Mechanism of Exchange. London, 1878.

———A Serious Fall in the Value of Gold ascertained, and its social effects set forth. London, 1863.

JOBERT (Louis) *Jesuit*. The Knowledge of Medals, 1697, 8vo. 602. c. 5.

JOHN, OF NIKIOS. Chronique de Jean, eveque de Nikiou, texte Ethiopien, publie et traduit par N. Zotenberg. Paris, 1883, 4to 755. k. 18.

This is an epitome of General History from the Creation of the World to the Conquest of Egypt by the Arabs, compiled by a christian monk, John Nabaddar, priest of Nikios, who lived during the last quarter of the seventh century. The work was originally written in Greek, then, in 1602, translated into Coptic, and (recently) into French. There are two copies of the Ethiopian text extant, one each in the Bibliotheque Nationale and the British Museum Library. The latter is in excellent condition, written on clear parchment and bound in thick wooden boards, covered with leather. The earlier portions of the work consist of Biblical tradition and Greek fable; the Roman history, though greatly coloured by secretarian hatred, is often fresh and interesting; the contemporaneous history, chiefly Egyptian and Arabian, is of historical value.

JUSTIN (N. de). "De Moneta," Jena, 1757. (For writing this treatise on Money, de Justin was imprisoned by the King of Prussia. No copy of it is in the British Museum Library.)

KEARY (C. F.) Catalogue of English Coins. London, 1887, 8vo.

———A guide to the study of English coins by the late William Henfrey. New and revised edition by C. F. Keary. London, 1885, 8vo. (This work should be entitled: "An attempt to prove that the Kings of England neither debased, degraded, nor altered the value of their coins.")

KER (David). The Mineral Wealth of Central Asia. Geog. Mag., II.
KŒHNE (Bernard) *Baron von.* Description du Musée de fue le Prince B. Katschoubey. St. Petersburg, 1857, 2 vols, 4to. 7705. cc. 11. 12.

LABBE (Phillippe) Bibliotheca nummaria. (Hebræis, Græcis, Romanis.) Three editions, 1675, 1686, 1692, 4to. The shelf number of the last one is 278. i. 4.
——— Concordia Chronologia. . . . technicam et historicam. Paris, 1670.
Phillippe Labbe, a Jesuit of Bourges, born 1607, was a compiler, rather than an author. His principal work was a "Collection of the Councils," 1672, in 18 vols. Oblivion has kindly absorbed his pious but tedious labours in the field of chronology.
LAMAS (Manuel de). Spanish Numismatic Author.
LANDRIN (H.) De l' or, de son etat dans la nature, de son exploitation, de la métallurgie, de son usage et de son influence en économie politique. Paris, 1851.
———Traité de l' or. Paris, 1863.
L'ANONYME D' ACQUITAINE. Chronique.
LAROMBIERE. Theorie et practique des obligations.
LANSDELL (Henry) *Rev.* Through Siberia. London, 1882, 8vo.
LASTAMOSA. Spanish Numismatic Author.
LAUDERDALE (*Earl of*). Inquiry into the Nature and Origin of Public Wealth, and into the means and causes of its Increase. Edinburgh, 1804, 8vo.
LAUR (P.) De la production des métaux précieux en Californie. Paris. 1862, 8vo.
LAVOIX (Henri). Catalogue des Monnais Musulmanes de la Bibliotheque Nationale. Paris, 1887, 3 vols, 8vo.
LAW (John), *of Lauriston.* Money and Trade considered; with a proposal for supplying the Nation with Money. Edinburgh, 1705, 4to.
LEAKE (William M.) *Lieut.-Col.* Travels in Northern Greece. London, 1835, 8vo.
———Topography of Athens. London, 1821, 8vo.
LEAKE (Stephen Martin). Historical Account of English money. London, 1726, 8vo.
(An arid little book, purely numismatic.)
LEBON (Gus.) The Crown.
LEGALE (F. O.) Codice di Commercio Ottomano. Versione Italiana, 1851, 8vo.
LEITCH (John). Scientific System of Mythology. Trans. from the German of C. O. Mueller, 1844, 8vo.
LENORMANT (François). Essai sur l' organization politique et économique de la monnaie dans l' antiquité. Paris, 1863, 8vo.
———La Monnai dans l' antiquité. Paris, 1878, 3 tomes, 8vo.
LE PESANT (Pierre) *Sieur de Bois-Guillebert.* Le Detail de la France. (Included in the "Economistes financiers du XVIII Siécle," ed. by E. Daire. 1843, 8vo.
———Le Detail de la France, sous le regne de Louis XIV. Rouen, 1707, 12mo.
LESLIE (T. E. Cliffe). Essays in Political and Moral Philosophy. London, 1879, 8vo.
LETTRÉ (E.) Dictionaire de la Langue Française. Paris, 1869, 4 tom., 4to. (See Titles: "Or," "Noblesse d'or"; "Argent"; "Monnaie," etc.) 2217. c.
LEVASSEUR (E.) La question de l' or. Les mines de Californie et d' Australie. Les Ancienne mines d' or et d' argent. Paris, 1858, 8vo.
LIGHTFOOT (J. B.) *D.D.* The Apostolic Fathers. London, 1889, 6 vols, 8vo.
LIPENIUS (Martin). Bibliotheca Juridicæ. Catalogue Desk.
LOFTIE. London, by W. J. Loftie. London, 1887, 8vo.
LONGPERIER (Adrienne de). Œuvres. Paris, 7 vols, 8vo.
LORIA (Achille). La rendita fondiaria e la sua elisione naturale. Milano, 1880. 8vo.
———La teoria economica della constituzione politica. Torino, 1886, 8vo.
———Analisi della proprietà, capitalista, etc. Torino, 1887, 2 vols, 8vo.
LYALL (Sir Alfred). Asiatic Studies, religious and social. London, 1882, 8vo.

MAC CRINDLE (J. W.) Commerce of the Erythræn Sea. Trans. from Arrian. London, 1879, 8vo. 10026. f. 14
———The Invasion of India by Alexander the Great, collated from Arrian, Q. Curtius, Diodorus, Plutarch, Justin and others. Westminster, 1893, 8vo. Illus.
MAC CULLOCH (J. R.) Treatise on Metallic and Paper Money and Banks. Edinburgh, 1858.
MACKENNA (B. Vicuña). La edad de Oro en Chile. Santiago, 1881, 8vo.
MAGENS (Nicholas). *Merchant.* Essay on Insurances. London, 1775, 2 vols, 4to.
MALEVILLE (Jacques de) *Marquis.* Analyse raisonné de la discussion du Code Civile au Conseil d'Etat. Paris, 1807, 4 vols, 8vo. 1128. c. 2.

MARCHEVILLE. La Rapport entre l' or et l' argent au temps de Saint Louis.
MARCUS (L.) Essai sur le commerce que les anciens faisaient de l'Or, avec le Soudan. Journal Asiatique, 2nd series, III.
MARIANA (Juan de) *Jesuit*. Copy of a late Decree of the Sorbonne condemning the opinion concerning the murdering of princes maintained by the Jesuits and amongst the rest by J. Mariana (in his work entitled "Del Rey y la institucion real.") London, 1610, 4to. 4091. d.
MARIANA *Padre*. De Ponderis et Mensuris.
MARTELLO (Tullio). La Moneta. (With an Appendix on the theory of Value.) Firenze.
MAWE (John). Travels in the Interior of Brazil, particularly in the Gold and Diamond districts. London, 1812, 8vo.
MEMORIAL NUMISMATICO. (Spanish Numismatic work, published in Barcelona.)
MENANT. Les cylindres Orientaux á La Haye. Paris, 1879, 8vo.
MEREDITH (Evan Powell). The Prophet of Nazareth. London, 1865, 8vo. 4226.cc.
MERLIN (C. L. W.) Catalogue of Greek, Byzantine and Crusaders' Coins. London, 1859, pamph, 8vo. 7757. c. 10. (2.)
MERLIN (Philippe Antoine) *Count*. Annotations des cinq Codes. Paris, 1826, 4to.
———Table generale alphabetique et raisonné des matieres contenues dans le Repertoire de Jurisprudence. Paris, 1827, 18 tomes, 4to. 497. f. 1–18.
———The same work, continued by L. Rondonneau. Paris, 1829, 4to. 497. e. 10.
———Recueil Alphabetique des questions de droit. Paris, 1827–30, 4to. 5425.gg.
MIGNÉ (Jacques Paul) *Abbé*. Patrologiæ Cursus Completus. Paris, 1857, 4to.
———Advantages of dealing at the Ateliers Catholiques and of investing in the Abbé Migné's Loan. Paris, 1863, 4to. 1220. m.
MIONNET (Theodore E.) Description des Médailles Antiques, Greques et Romaines. Paris, 1807–37, 16 tomes, 8vo. 602. f. 1–16.
———De la rareté et du prix des médailles romaines. 3rd edition. Paris, 1847, 2 tomes, 8vo. 7755. c. 4.
MOLINAEUS (Carolus). Tractatus commerciorum et usurarum redituumque pecunia constitutorum et monetarum. Parisiis, 1546, fol. 502. g. 5.
———Abus des petites dates, reservations, preventions, annates, et autres usurpations et exactions de la Cour de Rome, contre les edicts et ordonnances des roys de France. Lyon, 1564, 4to. 5051. c. 11.
———De Mutatione Monetarum. Coloniæ Agrippinæ, 1591, 4to. (This learned and celebrated Essay commences on p. 475 of Budelius, "De Monetis." Cap. 100 was cited by the British Bench in the case of the "Mixt Monies."
———Coutumes de la Prevoste et vicomté de Paris. Paris, 1691, 12mo. 5402. e. 1.
MOMMSEN (Thedor). Ueber den Verfall des Romischen Münzwesen in der Kaiserzeit. Dresden, 1857, 8vo.
———History of Rome. Eng. Trans. by Rev. William P. Dickson. London, 1868. 4 vols, 8vo. (A poor translation.)
———De l'organization financiere chez les Romains, traduit par Albert Vigie. Paris, 1888, 8vo.
MONA. Manual de Contadores. (Spanish Numismatic work. Not in the British Museum Library.)
MONEY. Conférence Monetaire International. Paris, 1881, 3 vols, fol.
———Mystery of Money explained and illustrated by the Monetary History of England, from the Norman Conquest to the present time. 2nd ed., London, 1863.
MONGEZ (A.) Considerations Generales sur les Monnaies.
MONTALVO. See *Ordoñez Montalvo.*
MONTEMAYOR Y CORDOVA DE CUENCA (Juan Francisco de). Resumen de Reales Cedulas para las Indias. Mexico, 1678, fol. 6784. h. 5.
MONTESQUIOU-FERENSAC (Anne Pierre) *Marquis de*. Court reflections sur l'Emission de deux milliards d'assignats. Paris, 1793, Brochure.
MONTPEREUX (Dubois de). Voyage autour de Caucase, Crimée, etc. Paris, 1839–43, 6 tomes, 8vo. 790. e. 4–6.
MULENUS (Joannes). Numismata Danorum. Halfinæ, 1670, 8vo. 603. d. 2.
MUNRO (Henry S.) Mineral Wealth of Japan. Engineering and Mining Jour. XXII.
MURRAY (A. S.) A History of Greek Sculpture. London, 1885, 2 vols, 8vo. 2031.
MYRI (Wernerus Christophorus). De Re Nummaria Vet. N. T. in J. Harduinum (Jean Hardouin). Holmstadii, 1711, 4to. 813. c. 15.

NASMITH (David.) Outlines of Roman History, including a translation of the Twelve Tables. London, 1890, 8vo. 9041. i. 12.
NERI (P.) Osservazione sopra el prezzo legale delle Monete. (Observations on the legal value of money.) By Pompeo Neri. Florence, 1751. A new edition, 1803.
———Documenti Annessi alle osservazioni sopra il prezzo legale delle monete. (An Appendix to the former work.)
———Delle Monete commercio. See also *Italiani Scrittori*.
NICHOLSON (J. Shield). Treatise on Money, and Essay on Present Monetary Problems. London, 1888.
———Science of Exchanges, 4th edition. London, 1873, 8vo.
NICHOLSON. English History.
NIKIOU (*Bishop of*.) See *John of Nikiou*, *Nikios*, or *Nikious*.
NORTH (*Sir* Dudley). Discourses upon Trade. London, 1691, 4to. 8223. b. 5.
———Considerations upon the East India Trade. London, 1701, 8vo. 1139. g. 3.
NORTH (*Hon.* Roger). Life of Sir Dudley North. London, 1744, 4to. 708. f.1.(2.)

O'BRIEN (G.) Gold and Silver. A Treatise by Geo. O'Brien. London, 1884, 8vo.
OLMEDO Y LAMA. Ordenanzas de Minera. Mexico, fol., 1873.
ORDOÑEZ MONTALVO (Juan). Arte ó Novo modo de Beneficiar los Metales de Oro y Plata, y de Plata con ley de Oro, por azogue. Mexico, 1758, 4to. 7104. aa.(3.)
ORESMII (Nicolai). De Origine et jure nec non et de Mutationibus Monetarum. This is a tract of 23 chapters in 28 pages: a model of brevity.
ORTELIUS (Abraham). Dearum Dearumque Capita ex antiquis Numismatibus collecta, etc., 1699, fol. Included in vol. VII of "Græcarum Thesaurus." q. v.
ORTOLAN (J. L. E.) History of the Roman Law. London, 1871, 8vo. 2228. c. 13.

PARK (Mungo). Travels in Africa. London, 1800, 8vo.
PATIN (Charles.) Travels. Translated from the French. London, 1697, 12mo.
———Historia de las Medallas. Madrid, 1771, 8vo. Translated from the Latin.
PAUCTON (Alexis J. P.) Metrologie. Paris, 1780, 4to. 8531. e. 22.
PAUSANIAS. Description of Greece (circ. A. D. 174.) Trans. from the Greek, by Thomas Taylor, *the Platonist*. London, 1794, 3 vols, 8vo.
PELLERIN (Joseph). Recueil des Medailles du Rois. Paris, 1762–68, 9 tome, 4to.
PENKA (Carl). Die Herkunft der Arier. Wien, 1886, 8vo. 10007. h. 28.
PENNA (Lucas de). *Jurisconsult of Naples*. Placita principal, sea constitutiones Regni Neapolitani, cum glossis. 1534, 4to. This work is cited by Grimaudet to the effect that money is intended for public use, not for private advantage.
PETRUS DE ANCHARANO. Rubrice juxta ordinem decretatium, etc. This work supports the doctrine that money is for public use, not for private advantage. Rome, 1474, fol. 5061. e. 4.
PEUCHET (Jacques). Statistique élémentaire de la France. Paris, 1805.
PHILLIPS (Fabian) *Royalist*. The Reforming Registry: or a representation of the many mischiefs and inconveniences of public registries, etc, London, 1671, 4to.
PHILLIPS (John). The mining and metallurgy of Gold and Silver. London, 1867, 8vo.
PINKERTON (John). Collection of voyages and travels. London, 1808–14.
POCOCKE (E.) India in Greece. London, 1852, 8vo. (A suggestive work.)
POIS (Antoine de). Discours sur les Medailles et graveures antiques. Paris, 1579, 4to.
POULLAIN (Henri), Traitez des Monnoyes. Paris, 1621, 12mo. Another ed., 1709.
———De priz de nostre Escu. Paris, 1613, 8vo. 1139. k. 13.
———Response a l'advis de M. Denys Godefroy. Paris, 1613.
PREJEVALSKY (N.) Mongolia, the Tangut country, and the solitudes of Northern Thibet. Trans. by E. Delmar Morgan. London, 1876.

QUADRADO (José). España. Barcelona, 1885, 8vo. 2060.
QUINONES (Dr. Ivan de). Explicacion de unas Monedas de Oro de los Emperadores Romanos que se han hallado en el Puerto de Guadarrama. Madrid, 1620. Pamph., 8vo. (A tract of 100 pages, useful for its mention of old Spanish authors on Money and the Precious Metals.)

RALEIGH (Walter) *Sir*. The Discovery of the . . . empire of Guiana. London, Hakluyt Soc., 1848.
RAVALOVICH (Artur). Les Finances de la Russie, 1887–9. Paris, 1889, 8vo.

RICHTHOFEN (F. von) *Baron*. Die Metall-Produktion Californiens und der augrenzenden Länder. Gotha, 1864.
ROBERTS (Lewes). Merchants' Map of Commerce. London, 1638, folio. Another edition, 1677; another, 1700. 522. m. 8. 20.
———The Treasure of Traffic, or Discourse of Foreign Trade. London (Reprint by the Political Economy Club), 1856, 8vo. 2240. f. 1.
RODOCANACHI (E.) Les Corporations Ouvrieres à Rome, depuis la Chute de l' Empire Romain. Paris, 2 tom., 4to,
ROE (*Sir* Thomas). Speech in Parliament relating to the merchants. 712. m. 1.(1–83)
ROME. Il Muratori. Essay in the Archæological Journal. P. P. 3557. ca.
———The News. (An English weekly scientific Journal). P. P. 9260. a.
———Memoria Numismatiche. P. P. 1925. d.
———Institut International de Statistique. Ac. 2404.
———Corpus Juris Civilis. Leipsig, 1720, 2 tom., 4to, 499. b. 1. 2.
ROME DE L' ISLE (Jean Baptiste Louis de). Metrologie. Paris, 1789, 4to.
ROUGE (Jacques de) *Vicomte*. Monnais nouvelles de Nomes d' Egypt. Paris, 1882. 8vo. (A pamphlet on the Roman coins of the Egyptian nomes, or military districts.)
RUGGIERO (Ettore). Dizionario Epigrafico di Antichita Romane. Roma, (Loreto Pasqualucci), 1886, 8vo. *In progress*. 760.
RUSHFORTH (G. Mc. N.) Latin Historical Inscriptions. Oxford, 1893, 8vo.
RYMER. Fœdera. (For "Turneys" or "Tournies," see V., 113.)

SABATIER. Production de l' or, de l' argent, et du cuivre, chez les Anciennes, et les hotels monetaires Romains et Byzantines. Paris, 8vo.
SABAU Y DUMAS (Tomás). Descripcion de los terrenos auríferos de Granada, y observaciones imparciales sobre su explotacion y beneficio. Madrid, 1851.
SAIGAY (Jacques Frederic). Traité de Metrologie. Paris, 1834, 12mo.
SAINT CHAMANS (August de) *Viscount*. Nouvel essai sur la Richesse des nations. Paris, 1824, 8vo.
SAVIGNY (Frederich Carl von). A Treatise on the Conflict of Laws, (in English), with an Appendix containing the Treatises, (in Latin) of Bartolus de Saxoferrato, C. Molinæus, Paul Noet, and Ulrich Huber. Edinburgh, 1880, 8vo, 2nd ed.
———An Epitome and Analysis of Savigny's Treatise on Obligations, in the Roman Law, by Archibald Brown, Barrister at Law. London, 1872, 8vo. 5206. cc.
SAVOT (Louis). Discours sur les Medailles Antiques. . . . estoient Monnoyes: de leur matiere, poids, prix, etc. Paris, 1627, 4to. 602. h. 6.
SAYCE (Archibald H.), Member of the Society of Biblical Archæology. Ancient Empires of the East. Herodotus, I–III. London, 1883, 8vo. 2280. e.
———Principles of Comparative Philology, 3rd ed. London, 1880, 8vo. 2260. a.
SAY (Léon). Turgot. Trans. by G. Masson. London, 1888, 212 pp., 8vo.
This is not only an excellent biography of Turgot, but at the same time a treatise on the economic questions which agitated France during the eighteenth century.
SCALIGER (Joseph Juste). Constantini Imp. Byzantini numismatis argentei exposite duplex, etc. Leyden, 1604, 4to. 602. c. 16.
———De Nummaria antiquorum dissertatio. 1616, 8vo.; 1697, fol. 602. c. 10. 2068. d.
———De Emendatione Temporum. Coloniæ Allobrogum Geneva, 1629, fol.
SCHUMBERGER (G.) Lead Moneys of the Holy Land. Paris, 1878, 8vo.
SCHMIDT (Gustavus). Civil Law of Spain and Mexico. New Orleans, 1851, 8vo.
SEISMIT-DODA (F.) Delle Condizioni Finanziarie del Regno. A financial speech delivered July 26, 1867, by Frederico Seismit-Doda, Firenze, 1867.
SENIOR (N. W.) Political Economy. London, 1850, 8vo.
SESTINI (Domenico) *Abbé*. Descriptio Numorum veterum. Lipsiæ, 1796, 4to.
SIMONIN (L.) La vie souterraine, ou les Mines et les Mineurs. Paris, 1867, 8vo.
SINCLAIR (*Sir* John). History of the British Revenue. London, 1820, 3 vols, 8vo.
SITZGSBER (Christ). Transactions in der Munchen Akademi. See Vol. I, p. 68.
SLINGSBY (*Sir* Henry). Diary of London, 1836, 8vo. 1202. k. 25.
———Trial of. London, 1836, 8vo. E. 753. (5.)
SMITH (Charles Roach). Dictionary of Roman Coins. London, 1889, 4to. 2259. d.
———Dictionary of Greek and Roman Antiquities. Several editions.
SNELLING (Thomas). A view of the gold coins and coinage of England, from Henry III. to the present time. London, 1763, fol. 7756. h. 4. (1.)

SNELLING (Thomas.) A view of the Copper coins, etc. London, 1766, 4to.
———Miscellaneous views of the coins struck (at the Mint). London, 1769, 4to.
———View of the coins at this time current throughout Europe. London, 1766, 8vo.
SOLORZANO (Juan de). Politica Indiana. Madrid, 1647, fol. (There are numerous other editions, from 1629 to 1776, both in Latin and Spanish, some by Juan and some by Pereyra de Solorzano. "Indiana" relates to America.) 521. l. 11.
SOUTHEY (Robert). History of Brazil. London, 1810–19, 4to. C. 61. k. 1.
SPEED (Joannes). Sigilla et monetæ. Londra, 1601, folio.
SQUIER (E. G.) Nicaragua: its people, scenery, monuments, resources, condition and proposed canal. New York, 1860, 8vo.
STANOSA (*Don* Vincentio Juan de la). Museo de las Medalas desconocidas Españolas. Oscæ, 1645, 4to.
STAUNTON (*Sir* G. T.) A selection from the Penal Code of China. (Ta tsing leu lee.) London, 1810, 4to.
STENERSEN (L. B.) Myntfundet. Christiania, 1881, 4to, with plates of coins.
———Om et Myntfund. Christiania, 8vo, 1889. Brochure, with plates.
STEUART (*Sir* J.) Principles of Money, applied to the present state of the coin of Bengal. 2nd edition. London, 1767.
———Inquiry into the Principles of Political Economy, being an Essay on the Science of Domestic Policy in Free Nations, in which are particularly considered Population, Agriculture, Trade, Industry, Money, Coins, Interest, Circulation, Banks, Exchange, Public Credit, and Taxes. London, 1767, 2 vols, 8vo.
STRAHLENBERG (Phillip Johan von). Northern Europe and Asia. London, 1736, 4to. Trans. into English. 149. b. 14.
STUBBS (W.) Registrum Sacrum Angliorum. (A work on Roman-British coins.)
SUESS (E.) Ueber die zukumft des Goldes. Vienna, 1877.

TAPPING (Thomas). High Peak Mineral Customs. London, 1851, 12mo, 1383. f.
———Derbyshire Mining Customs. London, 1851, 16mo. 1380. b. 11.
TARRASSENKO-OTRESCHKOFF (Narcès). De l'or et de l'argent. Paris, 1856.
TAVERNIER (John Baptista). Six Voyages. London, 1678.
TEGOBORSKI (Louis de). Essai sur les consequences éventuelles de la découverte des gîtes aurifères en Californie et en Australie. Paris, 1853.
TEYLOR. Vorhandeilingen entgogeren door Teylor's Teveede Genootschap.
THOMSEN (C. J.) Monnais des Moyen Age. Copenhagen, 3 vols, 8vo.
TILL (William). The Roman Denarius and English Penny. London, 1837, 12mo.
TOD (James) *Lieut.-Col.* Annals and Antiquities of Rajasthan. Second ed. Madras, 1873, 2 vols, 8vo.
TORRENS (R.) *Lieut.-Col.* Principles and Practical Operation of Sir Robert Peel's Act of 1844 explained and defended. Second ed., with additional chapters on Money, the Gold discoveries, International Exchange, etc. London, 1857, 8vo.
TOWNSHEND (Heywood) *M. P.* Historical Collections: or an exact Account of the Proceedings of the Four last Parliaments of Queen Elizabeth. London, 1680.
TROPLONG (M.) *Conseiller à la Cour de Cassation.* Le Droit Civil expliqué suivant l' ordre des articles du Code. Paris, 1845, etc., 32 tomes, 8vo. See tomes XIV, and XV, on "Paiments.") 5403. e. 1. 2.
———Du Pret. Paris, 1847, 8vo.
TULLIER (Charles Bonaventure.) See *Troplong*, vols. XIV and XV, articles on Money Payments.

VANDALE (A.) De Consecratione Principium, 1678, fol.
VAN DER BORGHT (R.) Statistiche Studien über die Bewährhrung der Actiengesellschaften. Academia Fredericana. 1883, 8vo Ac. 2320
VAN DER CHIJS (P. O.) De Munten der Frankische en Deutsch Nederlandsche, vorsten door P. O. Van der Chijs. Haarlem, 1866, 2 vols, 4to. (A very complete work on early Frankish and Dutch coins.)
VANDERLINT (Jacob). Money answers all things. London, 1734, 8vo. 104. g. 10.
VASCO. Essay on Moneys, 1772. See *Italiana Scrittori.*
VELASQUEZ Congeturas sobre las medalas de los reyes Godos. Malaga, 1759.
VENEZUELA. Noticia de estado que han tenido y tienen estas missiones de Capucinas de la provincia de Caracas, desde el año, 1658, etc. 4565. f. 17.

VERGIL (Polydore.) Three Books of P. V.'s English History. Camden Society, XXIX. London, 1844, 4to. Ac. 8113. 29

———De Rerum Inventoribus. Trans. by Thomas Langley, Canon of Winchester. A reprint of the 1603 ed. Agathynian Club, II. New York, 1867, 8vo.

VERTUE (George). Medals, coins, etc, (principally of the Commonwealth and Charles II.). Impressions from the works of J. Simon. London, 1753, 4to. 2032.g.

VOIGHT (Moritz). Die XII Tafeln. 1883, 8vo. 5254. c.

VOLNEY (C. F.) *Count.* Recherches Nouvelles sur l'histoire Ancienne. Paris. 2nd edition, 1825, 8 tom., 8vo. Tom. VI, *Chronologie.* 12235. k. 2.

VOLTAIRE (Arouet de). Siecle de Louis XIV. Paris, 1895, 8vo. 9771. aa. 18.

VON TSCHUDI (*Dr.* J. J.) Travels in Peru, 1838–42, Trans. from the German by Thomasini Ross. London, 1847, 8vo.

WADDINGTON (Wm. Henry). Fastes des Provinces Asiatiques de l' Empire Romaine. *In progress.* Paris, 1872, 8vo. 9041. cc. 3.

WALKINS. Roman Cheshire. Liverpool, 1886, 4to.

WALSH (R. H.) Elementary Treatise on Metallic Currency. Dublin, 1853.

WARD (H. G.) *Sir.* Mexico in 1827. London, 1828, 2 vols, 8vo.

WATERUS DE HEMINGBURGH. See his Preface, p. xlv, re Base Coins. 7707. o. 50.

WATT (P.). Theory and Practice of Joint Stock Banking. Edinburgh, 1836, 8vo.

WEBSTER (R.) Principles of Monetary Legislation. London, 1874.

WERDENHAGEN (Johann Angelius von). De Rebus publicis Hanseaticis. Tractatus generalis. Frankfort, 1641, fol.

WEX (Joseph). Metrologie grecque et romaine. Traduite de l' allemand par P. Monet. Paris, 1886, 12mo.

WHITELAW (T. N.) Just Money. Glasgow, 1886, 8vo. (One of the few works which marks the influence of a monetary system upon the equity of exchange. It is marred by the advocacy of an absurd paper scheme.)

WHITWORTH (*Sir* Charles). Inquiry into Prices. London, 1768, 8vo.

WILLIAMS (Charles). The Seven Ages of England; or its advancement in Arts, Literature and Science. London, 1836, 8vo.

WILLIAMS (S. Wells). The Middle Kingdom (China). London, 1848, 8vo.

WILSON (Horace H.) A glossary of judicial and revenue terms . . . relating to . . . British India. London, 1855, 4to. 2115. c.

———Observations on Lieut. Burnes' Collection of Bactrian coins and other coins, by H. H. W. and Mr. T. Prinsep. London, 1839. 602. b. 25.

———Arriana Antiqua. London, 1841, 4to. 811. l. 12.

WINCHEID (Bernard). Diritto delle Pandette. Torino, 1887, 8vo. *In progress.*

———Lehrbuch des Pandekten rechts. Frankfurt, 1891, 8vo. 5254. c. 13.

WITWATERSRAND. *Chamber of Mines.* Johannesburg. Monthly Reports of Gold Production. Nov., 1893, to the present time.

WORDSWORTH (John) *Bishop of Salisbury.* Fragments and Specimens of early Latin. London, 1874, 8vo. 2320. g.

WRIGHT (Wm.) *Mem. Soc. Biblical Archæology.* Empire of the Hittites. Second ed. London, 1886, 8vo. 2378. d.

YULE (Henry). The book of Ser Marco Polo, the Venetian. London, 1875.

HISTORY OF THE PRECIOUS METALS

CHAPTER I.

INDIA.

Gold, the pioneer of conquest and civilization—The only metal furnished by nature ready for use—The earliest metal known to man—Its use for coinage—Method of extracting it—Copper more ancient than silver—Silver the latest of the precious metals—Placer and vein mining—Earliest gold coins, the rama-tankahs of India—No copper or silver mining before the invention of iron—Example of the Mexicans and Peruvians—The archaic Indians and the Bactrians used gold rings for money—The small change currency was cowrie shells—The Iron Age in India is marked by the Mahabharata wars—Iron first supplied the people with arms—Not yet used in mining—The perturbations which had been occasioned by the irregular supplies of gold, caused the Buddhist conquerors to forbid its use—Fable of the gold ants—Mines of India mentioned by ancient authors: Herodotus, Strabo, Pliny—Geology of India—Auriferous deposits—Placer mines—Ancient use of mercury—Conditions of mining under the Brahmins—Buddhists—Bramo-buddhists—Moslems—British—Mysore mines.

DESIRE for the precious metals, rather than geographical researches or military conquest, is the principal motive which has led to the dominion of the earth by civilized races. Gold has invariably invited commerce, invasion has followed commerce, and permanent occupation has completed the process. This order of events is to be observed in the remotest past; it is to be seen in operation to-day. The same motive that led Scipio to Spain and Cæsar to Gaul, impelled Columbus to Cathay, Cortes to Mexico and Pizarro to Peru. It led Clive to the conquest and Hastings to the plunder of Bengal; at the present moment it is impelling a host of hardy adventurers into Africa and Alaska, before whom the native tribes will surely melt away, leaving their conquerors to lay the foundations of new empires.

Of the three coining metals, gold is the only one which nature furnishes to man, ready for use, in more than minute quantity; and this she only does in alluvions, "washings," placers, or beds of auriferous gravel. Masses of native silver or copper are not unknown to the miner, but they are rare. On the other hand, native gold is or was to be found in every country which possessed a range of mountains;

for it is by the erosion of mountains, most of which contain more or less auriferous rocks, that gold is collected by nature throughout lengthy ages, into alluvions, or placers. Hence gold must have been the earliest metal known to man; and its use, first for purposes of ornament and afterwards as an universal, intertribal and international article of barter, (not money,) long preceded the like use of silver or copper.

This deduction is supported by the most ancient monuments of the Orient, by the oldest scriptures extant, and by the rock-carvings, place-names, and other philological and epigraphic remains of Oriental antiquity. Gold is mentioned or found in all extant scriptures; and it is mentioned more often and familiarly than any other metal. The Rig Veda Sanhita alludes to "purses" of gold coins.[1] It is unfortunate for the supposed antiquity of this scripture—we shall presently see why—that it also mentions silver coins,[2] The Mahabharata epic and the Code of Manu are both replete with allusions to gold and silver money, whilst a Chinese author, referred to by the Fathers Dentrecollis and Du Halde, cites "ancient books" which affirm that in the reign of Yu, both "gold, silver and copper coins" were employed in China. The reign of this hero is supposed to have begun with the second year of the Third Jovian Cycle, corresponding to our B. C. 2218. I have myself possessed bell-shaped bronze coins purporting to be of a dynasty scarcely less remote, stamped with the legend "Good for gold," words which convey the implication that gold coins were previously in use for money.[3] Similar bell-shaped bronze coins are now in the great collection at Paris. But the ages assigned to them are doubtful, because the periods and dates attributed to the heroes by whom they are supposed to have been issued, are astrological. As this reasoning would demolish not only Yu and Sung, but also many other personages in whose existence and æras the Western World is accustomed to place reliance, I have hith-

[1] Strictly speaking a coin is a piece of stamped money, struck by a cuneus, or punch; but as money was cast long before it was struck, and as our language possesses no distinctive name for a piece of cast money, the word coin will be used throughout this book to include all kinds of metallic money, whether cast, struck, or milled.

[2] Gen, Alex. Cunningham, in his "Book of Indian Eras," ed. 1883, p. 31, gives reasons for believing that the Code of Manu in its present form is certainly posterior to Buddhism. The twelve-spoked wheel and the twelve-felloed wheel of the Sun, in the Rig Veda, together with other evidences, prove that that work also, that is, in its present form, is post-Buddhic, meaning, of course, later than the first Buddhic period.

[3] The coins of Sung mentioned in the text were cast. Del Mar's "History of Money in Ancient States," p. 21.

erto preferred rather to accept than to doubt them. For reasons connected with the use of iron in quartz mining, I can accept them no longer.

An absurd theory, which of late years has acquired some currency, contends that copper was the earliest metal used for coins; then silver; and lastly, gold; that is to say, in the order of their present relative value; the hypothesis being that man has been continually progressing from imperfect to more perfect media of exchange and that gold metal or gold coins are the best and consequently the latest of such media. Nothing could be farther from the truth. It is a mere tissue of absurdity.

It can be proved by a thousand evidences that gold was used for money, ages before either copper or silver. It can also be demonstrated that although originally the best substance out of which money can be made, it is now one of the worst. Although it is so widely diffused by nature that it may be found in almost any kind of metamorphic rock or earth; or in any of the drift, washings, or sediment from such rocks, yet it is usually deposited ab initio in quartz and that, too, in such fine particles that it is difficult and expensive to collect. Hence when the superficial "placers" or natural washings which contain the coarsest gold are exhausted, the supply of the metal which has now to be obtained from the rocks become unstable, spasmodic, or irregular; and these are vital objections to its unrestricted use for coins. Another serious objection is due to its indestructibility, a fact which by increasing its volume (as money) often lowers its value throughout long periods and impairs its exactness as a measure of the value of other things. This was the cause of the Anti-gold crusade of 1850.

The Hindus, whom we affect to despise, were better informed on this subject. Their astrology, which in our Western conceit is only a subject for contempt, evinced a profounder knowledge of mining than our newly born political economy. Their Four Ages of the World were first the Creda-joga, or Golden Age; second, Treda-joga, or Silver Age; third, Touvabar-joga, or Bronze Age; and fourth, Cali-joga, or Iron Age. But in fact neither of these systems are correct. There is no such harmony or symetry in the order of the discovery or use of the metals.

The ancient method of extracting gold was the same as that which is still practiced, excepting perhaps as to the use of mercury for amalgamation: and even as to this art, there are not wanting evidences that it was well known to the East Indians, the Phœnicians, the Persians, and the Greeks.[4] The old and new methods of obtaining gold differ chiefly in details, not in principle. The gold was anciently separated from the gravel, sand, or pulverized rock which contained it, by means of water and its own superior gravity: and such is still the method in vogue. The mechanical power in ancient times was the labour of slaves and captives, and this was quite as cheap and perhaps cheaper than is now machinery, with either steam or water power.

Copper was probably the next metal to gold which man learnt to extract and reduce.[5] It is often found in a native state; it is very widely diffused; it is quite abundant. The present state of archæological research does not enable us to locate all the principal copper mines of the ancients, nor to compute the quantity of metal which they yielded: but it is believed that both the Hindus and Chinese commanded supplies sufficiently ample to warrant them in making coins of this metal at a period which may be fixed approximately at about twelve centuries before our æra. From that time down to the Roman period, the vicissitudes of mining and of wars, which altered the possession of mines, were such that copper was often too scarce in one or another country to warrant its sole use for coins: and other expedients had to be resorted to. But on the whole, it maintained its ground longer and more satisfactorily everywhere than either gold or silver. Indeed, until the opening of the Saxon silver mines, in the middle ages, copper, throughout the entire civilized world, was much more truly and essentially the material of money, than either of the other metals, or indeed any other substance. Viewed from this point, its history as one of the precious metals, has not received the attention which its importance deserves. Copper was one of the greatest articles of commerce with the Phœnicians, who derived a large supply from the mines of Nubia, that at one time supplied the whole of

[4] Beneath the ancient mining-mill for pulverizing quartz—and often beneath the modern mill—is placed a covered box into which the *mercury*, laden with gold caught from the "pulp" of the mill, is made to flow. It is into this box that dishonest workmen are wont to dip their fingers, to steal the precious amalgam. It seems more than likely that the ancients, in order to intimidate thieves, feigned that the metal was sacred to Mercury, and thus gave it the name of that god, whom mythologists have commonly identified with Bacchus, Osiris, Buddha, Ies Chrishna, etc. Should this conjecture prove to be well founded, it will account for the fable attached to most of these deities, that whilst yet an infant he was confined in a box and that the rash mortal who opened it was immediately blinded or struck with madness. Cf. voc. Eurипilus and Ericthonius in Bell's "Pantheon," I, 122, 292, from Pausanias. In modern mills the mercury box is under lock and key.

[5] Æsculanus, or Æs was the father of the god Argention; "because copper is older than silver." Noël, Dic. Fable.

the western world. They combined with it the tin obtained from the islands of Cyprus and Britain, to make the bronze of commerce. Copper was used by some of the northern nations of Europe in the fabrication of weapons, at a period and under circumstances when steel or iron appeared to be more precious than gold. This has been illustrated in Denmark, by the opening of tumuli of very remote ages, from which have been collected knives, daggers, swords and implements of industry, many of which are preserved in the national museum at Copenhagen. Some of these implements have blades of gold and edges of iron. Some of them are formed of copper, with edges of iron. The profuse application of copper and gold, when contrasted with the parsimony evident in the expenditure of iron, seems to prove that the former were much more common products than the latter.[6]

Silver is rarely found in a native state, and then only in comparatively small quantities. It is commonly mingled with other metals and minerals, forming a compound known as ore and presenting none of those attributes, either to the eye-sight, or the other senses, which distinguish it when freed from its matrices. Most of the ores of silver are difficult to reduce; a fact that bespeaks a long familiarity with the art of smelting, before silver could have become as amenable to treatment as copper. It is therefore deemed quite safe to regard this one as the last of the three great coining metals which came into use. With regard to its diffusion, silver was originally as widely spread as gold, that is to say, it occurred in nearly all the metamorphic and some of the primitive rocks and (when in situ) nearly always in conjunction with gold. Its susceptibility to oxidation prevented its further diffusion.

Gold remains unaltered and uninfluenced by the action of the elements, it is often carried away long distances from its original place of occurrence by the breaking down of the rocks which contain it, and their formation anew elsewhere, either as other rocks, or as placers of gravel and sand; whilst silver is only to be found in the places of its original occurrence. A single gold belt, as the Mother Lode of California, may be the parent of innumerable gold districts and Poor men's mines, some of them far removed from the lode. Silver is soon driven into new mineral combinations, or dissipated and lost to view. A silver belt, though it be as rich as the Comstock Lode of Nevada,

[6] Scientific Press, August 7, 1880. In the early portion of 1897 native copper workings of great antiquity were discovered about twenty miles from the head of Gitche Gumee, the Big Sea-Water of Indian legend and the Lake Superior of to-day. Del Mar's "History of Money in America," p. 29.

produces no progeny. Hence silver is only to be obtained by quartz-mining, or reef-mining, which requires the aid of capital, association and machinery. This fact has had no little influence upon the history of money: and of mankind.

Veins of metalliferous quartz are usually imbedded in hard rock, such as trap, slate, or gneiss; and these veins are for the most part narrow and expensive to extract, on account of the great extent of excavation needed in order to secure a comparatively small quantity of the vein. Thus, if the passages are made high enough for a man to walk through erect and wide enough for two men to pass each other, or to permit the swing of a hammer they must be seven by four feet in dimensions. If the quartz vein is but one foot wide (a high average) it follows that twenty-eight cubic feet of rock have to be excavated, in order to secure seven cubic feet of quartz, to say nothing of shafts, winzes, adits, cross-cuts, pumping and other dead work. Now seven cubic feet of quartz will weigh about half a ton, and if the quartz goes £2 sterling or $10 to the ton weight (a high average) the produce is £1 sterling against the cost of excavating and conveying to grass twenty-eight cubic feet, or more, of worthless stone. When this work has to be done, as the ancients did it, without the use of explosives, it becomes exceedingly toilsome and expensive: and only the richest class of quartz veins—of which the number was exceedingly limited—and these only for a short distance downward, paid to work. If we go back to a period before the use of steel tools, it may be safely concluded that the opening of a quartz mine was practically impossible. The only mines that could have been extensively worked were gold placers, or alluvions.

The oldest coins extant are probably the ram-tenkis or rama-tankahs of Brahminical India. They are made of electrum, a mixture, in this case probably a natural mixture, of gold and silver. Their precise æra cannot be determined, but they are probably later than the First Buddha and yet older than any coins of the Occident. The reader will find this question treated at length in my works on Money, a subject to which the comparative antiquity of coins more properly belongs. I am here concerned less with money than with the precious metals, of which money has usually been made.

Although we have no copper coins or moneys of the Occident earlier than the Etruscan or Roman ingots attributed by Pliny to Servius Tullius, we have bronze weapons and implements several centuries older than the æra asigned to that monarch; whilst the general use of copper and bronze, certainly in the Orient, and perhaps in the

Occident, may be safely carried back at least a century and probably several centuries earlier than the appearance of silver. As for the evidences of Egyptian archæology and of the archæological remains of Assyria and other countries whose æras and chronology have been built upon the supposed antiquity of Egyptian monuments, their alleged dates rest upon the fragmentary and doubtful evidences ascribed to Manetho, as reported by various conflicting ecclesiastical authorities; a discussion of which would open up topics having no connection with the present work. It is enough to say that in the opinion of many eminent Egyptologists, the chronology built upon Manetho's alleged list of alleged kings is far from being satisfactory. The æra which we are now about to adduce has a much more solid foundation.

The concurrent testimonies of both Indian, Assyrian, Babylonian and Greek tradition, as well as its own etymology, fix the discovery of iron—that is to say the invention of smelting iron ore and of making iron and steel, or the escape of the invention from the temples, which before that time they may have known and kept secret—at a period not earlier than the fifteenth century B.C. At all events, this is the highest date that can be assigned to it, for it was a rule and a necessity of sacred tradition and mythology to date its heroes and their exploits as far back as contemporary credibility would permit, both in order that they might not be ante-dated by like heroes and exploits in the traditions and mythologies of other nations, as well to procure for them all the veneration that is accorded to superior antiquity. This æra—1406 B.C.—was that of the apotheosis of Brahma, (1406,) of Ies Chrishna or the first Buddha, (1394,) of Nin-Ies, or Nineis, (1406,) of Belus, (1390,) of the Dionysian Jasius and the Ten Dactyles of Mount Ida, (1406,) and of Osiris, (1350.) All these heroes were credited with the invention of iron, and there is nowhere in history or mythology any mention of iron of a date which can positively be fixed prior to their epochs.

Until this date is overthrown, not hypothetically, but positively, it may be safely assumed that it marks not only the æra of iron, but also that of quartz-mining; whether of gold, copper, or silver. I am well aware that gold can be picked out of quartz crevices with a boar's tooth and that a bronze tool can be made of sufficient hardness to cut (with difficulty) some of the softer and rarer kinds of silver-bearing and copper-bearing rocks; but it will require something more than mere literary evidences to prove that quartz or rock-mining, upon a scale sufficiently ample and permanent to supply the states of antiquity with gold and silver money, or with bronze weapons, tools and utensils, could have been conducted without the use of iron and steel.[7] The Mexicans and Peruvians had no iron, consequently they had neither silver nor copper, except in such comparatively small quantities as nature furnishes of these metals in a native state. Their principal metal was gold, which they got from placers. When these placers were exhausted and their produce was absorbed in the arts, the native mining industry and the use of gold coins came to an end. For money, they were afterwards obliged to use cacao beans and a few pieces of tin, which metal, like placer-gold, can be got by washing, or "streaming" surface deposits. An instructive chapter on this subject can be found in the author's "History of Money in America."

The testimony gathered by Sir John Lubbock in his "Prehistoric Times," is very decisive concerning the antiquity of coins. "No inscriptions or coins have yet been found in any antiquities which can be referred to the Bronze Age." (p. 7.) The reason is that neither "inscriptions," meaning inscriptions on stone, nor "coins," which involve the use of a "die," can practically be cut without steel tools. "The entire absence of coins and of inscriptions on the bronze finds is very remarkable." (p. 16.) On the contrary, I should regard it as still more remarkable if a "coin" or epigraphic inscription could be traced to an age when it was practically impossible to produce either of them. "Mr. Ramsauer, for many years Director of the Salt mines at Hallstadt, near Salzburg in Austria, has discovered an extensive cemetery belonging to this transitional period," (between the Bronze and Iron Ages.) "He has opened no less than 980 graves, evidently of those who even at that early period worked the salt mines which are still so celebrated." (At that early period the "mines" may have consisted of little more than a salt-spring, or a salt-marsh, the efflorescence of the great deposit beneath.) "Another interesting point in the Hallstadt bronze, as in that of the true Bronze Age, is the absence of silver, lead and zinc, excepting of course, as mere impurities in the bronze. This is the more significant, inasmuch as the presence not only of the tin itself, but also of glass, amber, and ivory in the graves indicates the existence of an extensive commerce. Moreover, as Mortet well pointed out, the absence of silver cannot be accidental, because the bronze of Hallstadt contains no lead and the absence of lead entails that of silver, since the latter could not, at least in Europe, be obtained without the former. In the above 980 graves 525 were of bodies buried in the ordinary manner. These contained gold

[7] The bearing of this conclusion upon the pretended antiquity of the Egyptian, Assyrian and Chaldean monuments, is a subject with which we have no present concern.

ornaments, 6; bronze ornaments, 1471; bronze vessels, 3; bronze weapons, 18; iron weapons, 161; other objects of iron, 33; amber ornaments, 165; glass ornaments, 38; pottery, 334; stone, 57. The remaining 453 graves contained burnt corpses." In these remaining graves the proportion of iron to bronze objects was nearly double.

The Massagetæ made use (probably for weapons) of both gold and bronze; they used neither iron nor silver. (Herod. Clio., CCXV.) The Greek and Roman writers were accustomed to speak of all other nations with so much contempt that the significance of this passage will be lost upon the reader who may not be aware that the Massagetæ formed a portion of the great Scythic commonwealth, which, at the period alluded to by Herodotus, was far more populous and wealthy than both Greece and Rome combined. (Encyc. Brit., art. "Khozares.")

The Scythians possessed gold, but neither silver nor brass. (Herod. Mel., LXXI.) Some such condition of affairs as this must have existed generally throughout the Orient previous to the invention of iron—that is to say, before the fifteenth to the thirteenth centuries before our æra—a date that corresponds roughly with that of the Mahabharata wars. The Indians and perhaps also the Mongolians may have used gold coins or baugs for money; but they certainly had no money of silver or copper. The small change currency of both peoples consisted of cowrie shells, which are mentioned in writings of the highest antiquity.

The fitful supplies of gold from the placers of the Orient and the great perturbations of prices which must have been occasioned when there were no adequate supplies of silver or copper to complement the gold, are deductions which derive corroboration from an otherwise inexplicable ordinance of the Vinaya. This is an interdict against using the precious metals, traces of which will be found wherever the Buddhic religion penetrated, namely, in Laddak, Tibet, Ceylon, China, Corea, Malayia, (the "Golden Chersonesus" of the Greek writers,) and even in Babylonia, Greece and Italy. Two instances, one of the earliest and one of the latest enforcements of this interdict, will be given in this chapter. The reader will find a number of them mentioned elsewhere.

"The Katiari (Getæ and other Scythians of the time of Darius" (and probably also those of a much earlier date) "regarded gold as sacred. Two of the sons of their founder, Targitaus, Son of God, were burnt by touching sacred gold, but the youngest was unharmed and by this sign he became their ruler. They have a tradition that if the person in whose custody this (sacred) gold remains, sleeps in the open air during the time of their annual festival, he dies before the end of the year." (Herod. Mel., v-vii.)

"Oppert (1880) states that the searching and digging for precious metals (in Corea) is strictly forbidden to the natives and that this prohibition is so rigorously kept up that transgressors are threatened with capital punishment." [8]

During the first Buddhic period, that is to say, some time between the fourteenth and ninth centuries before our æra, copper and afterwards silver appears to have been first produced from the native mines of India. This inference is derived from the archæological finds of bronze implements, which become far more plentiful after the Iron Age than before it, and from the apparent epoch of the Behat and Coimbatore silver coins, which, in the opinion of Prinsep, Wilson, Thomas, and Marsden, are more ancient than the Greek invasions of India.[9] Quintus Curtius, in his life of Alexander the Great, informs us that Omphis, one of the native kings, presented the conqueror, among other things, with "coined silver to the amount of eighty talents." [10]

The use of silver money in India at this early period could hardly have been general. It must have been confined to those nations or districts where Brahminism or Brahmo-Buddhism prevailed. Among the Buddhists proper the interdict with reference to the use of the precious metals seems to have been always religiously observed. Gorgus, the miner, who accompanied Alexander the Great, remarked that the subjects of Sopeithes were so little acquainted with mining and smelting gold and silver as not to know the value of their own resources and that "the Musoi (Buddhists) made no use of gold or silver, although they had mines of both of these metals." Pausanias has a similar remark. The use here alluded to is undoubtedly money; and the remark will apply as well to many parts of the Orient to-day as it did then. The reason for this abstention is to be found in the venerated ordinances ascribed to Buddha, or Ies Chrishna.

As for the story of the gold-seeking ants told by Herodotus and repeated by Nearchus (in Strabo), by Pliny and other ancient writers

[8] See Pliny's Nat. Hist., vi, xxxi, 7; Moorcroft's Travels; Treveck's Travels; Cunningham's "Laddak," (Little Tibet), p. 232; V. Ball, "Geological Survey of India," part III, Econ. Geol., ed. 1881, p. 213: Lock on "Gold," pp. 303, 360; Appleton's Encyc., art. "Buddhism," iv, 68; and the authorities mentioned elsewhere throughout the present work.

[9] "Hist. of Money in Ancient States," p. 67. [10] Quintus Curtius, viii, xiii, 41.

and "explained" by numerous modern ones, it is a mere fable derived from the Mahabharata epic and sent down the stream of time, like a "sluice-robber," to gather substance as it went along.[11]

Turning now from inferences to fact, so far as any facts were known to the classical authors concerning the sources of gold in India, let us give hearing to Pliny and Strabo. Says Pliny: "The Dardæ inhabit a country the richest of all India in gold mines, while the Setæ have the most abundant mines of silver." (VI, xxii, 4.) "In the country of the Nareæ on the other side of the Capitalia Range (the highest in India) there is a very great number of mines, both of gold and silver, in which the natives work extensively." (VI, xxiii, 5.) "Just within the mouth of the River Indus there are two islands, named Chryse and Argyre, so called, I think, from the deposits of gold and silver which have been found there: for I cannot believe, what some have asserted, that the soil consists wholly of these metals." (VI, xxiii, 11.)

The countries of the Dardæ and Setæ have been conjectured to be Darjilling and Sikkim, situated upon an affluent of the Ganges, which descends from the Himalayas north of the southern borders of Great Tibet. It is through the valley formed by this affluent that winds the main road from Great Tibet by which a part of the gold and silver produced near the confines of that country finds its way to the valleys of India. During the eighteenth century of our æra the annual value of the trade by this route, including the precious metals, silk, tea, knives, paper, horses and other commodities, was only 30,000 rupees, say £3,000. On the other hand, Dr. Campbell, in a recent report, says that the gold imported from Tibet through Nepal, Darjilling and Sikkim, amounted to the value of two lakhs (£20,000) per annum. (Lock, 342.) What it amounted to in the days of Megasthenes, or even of Pliny, we have no means of determining. Mac Crindle ("Invasion of India by Alexander the Great," 1893, p. 342) places the Dards in Cashmir on the Upper Indus; in other words, Tibet. Lock (461) regards the Derdai of Megasthenes, the Dardæ of Pliny, the Daradrai of Ptolemy and the Daradas of Sanscrit literature, as the Dards of to-day and their land the plateau of Chajatal, in Great Tibet. "The gold fields are carefully watched by the L'hassa authorities: a gold commissioner called Sar-pun (Sar=seer: pun=gold)

[11] Sluice-robbers are lumps or balls of pipe-clay, found in placer mines, to which loose pieces of gold tenaciously adhere so that in the ordinary process of sluicing, they pass down into the debris, gathering grains at every step, and wasting them at last.

superintends the whole of them and each field has a separate master.[12] Any individual is allowed to dig, provided he pays the annual tax of one sarshoo weight of gold, which is about half a tola or two-fifths of an ounce (say eight dollars) . . . In Tibet gold is sacred to the Grand Lama. . . The same Sar-pun makes the round of the Tibetan-gold-fields once a year to collect the taxes. . . Nearly the whole of the gold collected in Western Tibet finds its way to Gar-tokh (tokh, or thok, means a mine) and ultimately, through the Kumauni merchants, to Hindustan." The annual product is estimated at about £8,000.

As for the Argyre and Chryse of Pliny, these, if indeed they had any foundation in fact, were insignificant sources of the precious metals, whose importance was magnified by imperfect information and a vast distance.

Lock (333), who is anxious to locate the Ophir of King Solomon, which he thinks must have been in a distant country that also produced "ivory, apes and peacocks," (all of which, by the way, were produced in Babylon) has discovered the Narae of Pliny in the Nairs of Malabar: but as Megasthenes is not known to have dwelt in or visited this district, and Pliny expressly says that the Narae dwelt on the other side of Capitalia, the highest mountain of India "mons altissimas Indicorum Capitalia," our author in this respect is probably mistaken.[13] Moreover Nairs is not the name of a nation or tribe, but of one of the castes of Brahminism, a caste which is not known to have existed so long ago as the time of Pliny and which, even if it did exist, is not likely to have been recognised or distinguished by the Western writers of that period.

Says Strabo: "It is reported by Gorgus, the miner of Alexander, that in the mountains of the territory of Sopeithes . . . are situated valuable mines of gold and silver (xv, i, 30). The people of Musicanus[14] . . . make no use of gold nor silver, although they have mines of these metals (xv, i, 34). These peculiar people have no slaves"

[12] Pun was also the Phœnician name for gold. The "Land of Pun" was in eastern Egypt.

[13] There is also a district called Dardistan, which was once "the most prolific portion of the Indus" valley for placer mines. Lock, 347. On the whole, Great Tibet seems to have been the principal source of the gold reported by Megasthenes and recorded by Pliny.

[14] Homer associates the Teuckroi and Musoi, whom he says were worshippers of Dionysius, (Buddha.) Moëss, or Moses, was also one of the surnames of Buddha. The Mongol title of Chan or Khan may be detected not only in Musicanus, but also in Porti-canus, Oxy-canus, Assa-canus, etc. The Rev. Dr. Vincent, "Voyage of Nearchus," p. 129.

(xv, i. 54). Says Sandrocottus: The affluents of the Ganges "like the rivers of Iberia (Armenia) bring down gold dust, part of which is paid as a tax to the king" (xv, i, 57).

Gold occurs in beds of detrital gravel in all the mountainous and hilly regions of India, and is borne by the streams to lower levels. The native names for gold, as Pon, or Pun (Tamil) Suvarna or Soowurn (Sanscrit) and Hemma, Honna and Chinna (Kanada), Son, etc., are still borne by innumerable districts, towns, gulches and streams, in testimony of their auriferous character (Lock 318-23 and 331). But these alluvial deposits have long since been exhausted, so that nothing remains which deserves the name of pay-dirt. Quartz mines also occur in numerous districts of India, chiefly in Mysore, in the hills near Ponderpoor on the upper Christna, in Bengal and in the Punjab near Attock on the upper Indus. Some of these mines were worked by the Indians at various periods before our æra, down to that of the Moslem conquest, others by or under the Moslems down to a recent period, and again others, chiefly re-workings and extensions of ancient Indian mines in Mysore, by the English; these last being substantially the only mines which are productive at the present time. The geology of India, especially with reference to its gold-bearing rocks, has been ably treated by Mr. Ball and other writers, most of whom have been alluded to in Lock's bulky work. It is sufficient here to say that the gold occurs for the most part in quartz veins, in metamorphic and sub-metamorphic rocks, especially the latter, and that the most ancient miners treated it as we do now, with mercury. (Lock, 337.) The few notices which occur of the ancient silver or copper mines of India will be found in the "History of Money in Ancient States," p. 70, note 3, and in the first edition of the present work p. 6, n. These mines were never important.

The conditions under which gold was obtained by the earlier Brahmins of India are indicated by several circumstances. The Brahmins were the conquerors of the country; they enslaved the conquered; they did not permit them to bear arms; the use of iron or steel implements was unknown; the mines belonged to the sovereign; and gold was made sacred. When the Brahminical rule was overthrown by the Buddhists, about the 13th century before our æra, the latter forbade the mining of gold and silver. These circumstances point to slave-mining on the part of the Brahmins and to a cruelty so hideous that it was made a prime article of the Buddhist religion to shut down the mines altogether.

In the course of a few centuries the Brahmins recovered their control of India, when mining, presumably with slaves, was resumed. In the reign of Asoka, during the third century B.C., the Buddhist religion once more regained its ascendancy. Knowing its hostility to slavery, we must believe that slave-mining was again put down. When the Chinese missionary Fa-hian visited India, A.D. 400, the Buddhist temples were once more overthrown, and Brahminism, this time Buddho-Brahminism, had become the prevailing religion. After that period, downward to the Moslem conquest, slave-mining was probably resumed; though there are not wanting indications that in some parts of India, notably the Punjab, the native rulers resigned the diggings to free workmen, who, for the privilege of working them, paid a fourth of their scant produce to the government. Such at least was the proportion paid to the Sikh government in recent times. (Lock, 346-8.) In Laddak, which is north of the Punjab and more properly belongs to Cashmir or Little Tibet, the Buddhists retained their ascendancy and always refrained from mining; although at the present day this interdict is practically relaxed in favour of the Moslems, who are permitted freely to search for the forbidden metal.

To the period which preceded the Moslem conquest probably belongs that once magnificent temple of Melukote, formerly the richest in Mysore, which was built with the wealth derived from the gold mines of Chinnataghery, near Bellivetta. The slaves have perished; the temple is levelled to the dust; the religion which shared this ill-gotten wealth is obsolete; nothing remains but the gold, the cause of all this ruin. The vast spoil in gold which was captured in India at various times by the Moslems (see Chap. XXVI) appears to indicate a considerable native production of this metal; but when it is remembered that this spoil represented the accumulations of many centuries and especially when it is borne in mind that no Indian quartz mines of more than superficial depths have been discovered in modern days; and that no large "dumps" of mining stone or debris are known to exist anywhere in India, it must be concluded that the product was derived chiefly from placers and river beds.

The conditions of Moslem mining in India are but little known to us, but generally speaking they included the prerogative of mines-royal, and the practice of slavery, and mine-farming. These conditions were not changed when the British forces effected the conquest of India: the changes that have been made are of later date. In 1830 Mr. F. H. Barber testified before the Lords Committee on East Indian Affairs that there were no less than one hundred thousand slaves in Malabar, who were bought and sold like cattle, the value of an adult man ranging from 5 to 20 rupees, or twice as many sterling shillings. In some cases the only clothing of these people was a plantain leaf. Their wretched condition was suggestive of baboons rather than men. Many of them were originally the children of freemen, kidnapped from their parents and sold to the mine proprietors. Wretched as they were, the government demanded and collected a revenue from them: it taxed their puttis, or wooden trays, in which they washed the miserable pittance of gold-dust won from unwilling nature. The tax often exceeded the utmost which they could glean from the gravel. In 1831 Lieut. Nicolson made an official survey of the goldfields and mines of Malabar, and reported that at Nilambar the mines were worked by Korumba slaves, "who were subjected by the English to horrible cruelties if the gold they found was deficient in quantity" compared with what they were required to produce. This ill-gotten wealth now lies stored in Lombard Street: it calls itself "sound money;" and it demands the dominion of the earth.

The recent developement of gold-mining in Mysore has directed attention afresh to the mineral resources of India. Of those resources gold forms but a single item, and an item interesting for its future possibilities rather than for its present yield. Yet the yield is quickly mounting to a considerable figure. A complete statement of the annual gold production for the whole of India cannot be furnished. The washing of auriferous beds and river deposits takes place in almost every province. But, except in Mysore, it is a purely native industry, spasmodically conducted, on a very small scale, and in regard to which no verified statistics can be obtained. Dr. George Watt, the Reporter on Economic Products to the Government of India, has, however, been able to collect returns for eight tracts. The output for 1894 was 211,770 ounces, valued at 14⅓ million rupees.

Of this amount Mysore furnished nearly the whole. The government of that State, under its Prime Minister, Sir Sheshadri Iyer, offers facilities to mining; and the labour difficulty, although sometimes felt, exists only in a modified form. The experimental crushings by the rude native methods in 1870 yielded a fair percentage of gold. English capitalists afterwards embarked on this enterprise, but the developement of the Mysore goldfields took place slowly. In 1889 their output of gold only amounted to 4⅓ million rupees. The increasing price of gold expressed in the silver coins of India gave a great impulse to gold-mining. In 1894 the Mysore out-turn of gold had risen to 14¼ million rupees. It is still advancing, and Mysore has taken a definite place among the goldfields of the world. We shall say nothing more of the cruel means by which the metal is obtained and the wicked mint laws which have so enhanced its value and depressed the value of silver that the native is deprived of every means but death to escape the terrible effects of a drought.[15]

Of the seven other provinces and native States reported upon, Madras heads the list with a yearly production valued at Rs. 50,000. The Punjab comes next with an annual yield of Rs. 22,000. Cashmir and the Nizam's dominions also find a place. But the total yield of all India, with the exception of Mysore, only amounted in 1894 to Rs. 86,352. In 1897 the East Indian output is given as 389,779 oz.; 1898, 417,124 oz.; 1899, 447,971 oz.; while for 1900, it is estimated at 500,000 oz.; equal in value to about ten million dollars. The native methods, still followed everywhere except in Mysore, have come to the end of their capabilities. Gold washing in the Indian river-beds has gone on for ages; therefore no surface finds of any importance can now be hoped for. Indeed, gold washing is in Indian districts the last resource of the poor and thriftless classes, or an occasional industry adopted by the peasant during some few days of his agricultural year. Yet the same geological indications which attracted mining with modern machinery to Mysore are not wanting in the Central Plateau and the broken country amid which that plateau slopes down to the Gangetic valley.

[15] No words are needed to emphasize the hardship on the natives of India, to which Mr. Donald Graham alluded in his recent speech to the Manchester Chamber of Commerce. This is the hardship of being unable, owing to the mints being closed, to sell their silver ornaments at full value, as formerly. These ornaments, made chiefly from melted rupees, constitute their hoards, or savings, to be used in time of stress; and apart from any political danger which may be involved, it is grievous in the extreme that in their present direful plight, with this appalling famine—*the worst of the century*—raging, they should, owing to the closed mints, have such a loss forced upon them. Mr. Vicary Gibbs, M.P., in a letter to *The Times*, estimates that in up-country districts this loss will be 60 per cent. We have heard former members of the I. C. S. put the loss at possibly 75 per cent., and there seems some grounds for this calculation in the fact that in the Orissa famine, while ornaments of silver (the mints being open to that metal) realized full value, gold ornaments (there being no gold mints) in up-country districts realized only one-quarter of their value. The loss now entailed upon the natives is all the more grievous seeing that, as is admitted, the famine is not because of an insufficiency of food to feed the people—food-stuffs being practically unaltered in price—but because of an insufficiency of money.—London *Bi-Metallist*, April, 1900.

CHAPTER II.

CAUCASIA, COLCHIS AND PONTUS.

The Argonautic expedition—Probable origin of the story—The Veneti of Colchis and Pontus—Their proficiency in mining—Mines of the Caucasus, Olgassys and Taurus ranges—Exhaustion of the placers—Discovery of iron—Iron mines of Chalybia—Opening of the quartz mines—Gold, copper, silver, mercury—Mines of Georgia—Caves inhabited by the miners—Gold mines of Trebizond—Their importance—Electrum mines—Mines of Tamasus—Balgar-Dagh—The future of Colchis and Pontus as mining regions.

THE Argonautic expedition is evidently a Pontine story adapted and perverted by the Greek priests and poets. According to one version of the story, as we have it, Argos, the son of Danaus, built the ship Argos, in which Jason or Jasius and his companions sailed from Thessaly to obtain the golden fleece, which hung from a tree in Colchis. Other versions make Argos the son of Jupiter, the son of Agenor, the son of Phryxus, etc. After landing in various places, among them Salmyd-Essus, Jason reaches the river Phasis in Colchis, where, having been charmed by Medea against fire and steel, he seizes the golden fleece and escapes to the river Eridanus and eventually to Italy, where he consecrates the fleece to Neptune. Danaus, Jason, Agenor, Eridanus and other of these names are Pontine or Venetian, not Greek. They belong to the Venetians of the Pontus, who were sailors, fishermen and miners, and in whose country, to the present day, the gold of the Phasis and other rivers, flowing from the mountains of Caucasus, is saved by means of sheeps' fleeces in their sluice-boxes, just as we of the West now use woollen blankets for the like purpose. Steel was unknown to the Greeks of the period assigned to this story, namely, 79 years before the taking of Troy. After Ies the Sun-god and Dion-Issus, his representative, the favourite deity of the Veneti, who were a sea-faring people, was Poseidon, or Neptune, and in their mythology this god appears to have played a prominent part.[1] The story of the Argonauts seems more likely to have been originally connected with a mine-hunting expedition of the Veneti to Greece and Italy and thus to have been turned around and embellished with the romantic incidents in which Hellenic fancy has seen fit to clothe it.

The Venetoi, Benetoi, or Henetoi, are mentioned under one or other of these names by Homer, Hecatæus, Sophocles, and other very ancient authors, as occupants of the coasts of the Euxine, in Colchis, Pontus and Paphlagonia, from whence they are said to have been driven by the Assyrians (Assuroi) before the period ascribed to the Trojan war, a contest in which the Veneti are said to have taken part.

The Veneti were located by the Greek authors in Paphlagonia (now Kasta-Muni) also in Pontus on the river Am-Issus, at Cerasus (now Kerosoun) and at Trap-esus, now Trebizond: indeed they seem to have occupied the entire shores of the Euxine, from the Palus Mæotis, by Colchis and Pontus to Paphlagonia. The names of Cerasus, Trap-esus, Hyssus, Issonium, Amisenus, Amasia, and Am-Issus, are too significant, and they too closely resemble those by which the Veneti of the Baltic designated their towns, rivers, and capes in Iestia or Esthonia, to have been bestowed by any but a Gothic race. For example, Am-Issus was the ancient name of the Dutch Ems and Tam-Issus of the British Thames. Similar place-names, as Charisia, or Chrisia, Trapezeus, Hypsus, Issus, Cyparissius, Larissa, Oxiae, Lissus, and Histria mark the wanderings of the Veneti from Arcadia to the modern Venice.

The Veneti of the Euxine are characterized by Homer. They were seamen, miners, traders and breeders of horses and mules. Among their possessions were the rich Alybean silver mines (Il. II, 1045). They are also characterized as horse-breeders, in the "Hippolytus" of Euripides. Cato and Strabo mention the part which the Heneti took in the siege of Troy, whilst Cato, Pliny and Cornelius Nepos enable us to trace their removal from Paphlagonia to the Adriatic.

[1] "The prince of Mingrelia (Colchis) assumes the title of Dadian (Dar-dion?) or Master of the Sea, though he possesses not even a fishing-boat." Malte-Brun, I, 304. At Samos during the sixth century B. C. the king threw into the sea a golden ring. (Herod., Thalia, 41.) In Venice of the Adriatic, the doge or duke annually weds the sea by casting a ring into its waters. This custom is said to have been instituted by the Pope of Rome in A.D 1177. There were five rival popes at that period, viz., Alexander, Victor, Pascal, Calixtus and Innocent. The custom is of far higher antiquity.

The Veneti transmitted this custom to the Norsemen who conquered Dublin in the 8th century and to-day the Lord Mayor of that city keeps it alive by triennially casting into the sea a gilded dart.

They are also mentioned as of Paphlagonia by Pomponius Mela, A. D. 50, Arrian, A.D. 90-170, and Solinus, A.D. 230.

Upon being driven out of Pontus, and after many adventures, in which perhaps Egypt was not neglected, the Veneti made their way into Greece, and afterwards into Italy, Spain and Gaul, and as Prof. Evans suggests, perhaps also Britain.

Possessing a numerous marine, this people evidently found it advisable, when they were dislodged from Paphlagonia and elsewhere in the Levant, to steer their ships close to the Pontine coast and eventually seek a refuge in Europe. Some unexplained obstacle in Bithynia, perhaps the occupation of the strait by the Assyrians, prevented the Paphlagonian Veneti from crossing the Bosphorus; so probably leaving their fleet in the Euxine, to effect a junction with them after they crossed, they traversed Mysia and approached the Hellespont. At the siege of Troy, the services of some of them appear to have been engaged by the Trojans, for it is recorded that they lost their king Pylæmenes in that celebrated contest, and after the fall of Illium placed themselves under the leadership of Antenor, (brother of Jasus, and son of Triopas,) who, rejecting the proposed passage of the Hellespont, led them back to the Bosphorus, which they could only have crossed after defeating the enemy. This account differs from several others and especially from that of Herodotus, which makes them cross the Bosphorus and conquer the Thracians "before the Trojan war," then advance westward, probably in their shipping, to the river Peneus in Arcadia and Elis. Arrian says that they suffered greatly in wars with the Assuroi (Assyrians) and having escaped into Europe were thereafter known as Benetoi, instead of Henetoi, without alluding to any participation of theirs in the Trojan war. Other authors engage the Veneti at Troy before the main body is driven out of Pontus.

Reverting to the Venetian passage of the Bosphorus, Homer alludes to the Veneti as the Teuckroi and Musoi, and says they worshipped Dion-Issus. Herodotus alludes to one of their tribes called the Satræ, whom he says were never subdued. Strabo, who was a native of Pontus, says the Veneti were called Leuco-Syrians on account of their fair complexion. After the removal of the Veneti from Pontus, this tribe (the Satræ) inhabited the mountains of Thrace, worshipped Bacchus (or Dion-Issus) and maintained priests and priestesses called Bessi. Bury, II, 15, calls the latter the Bessi, or Satri, of Rhodope. Ovid, who lived in exile among the Getæ or Goths of the Danube, thus alludes to the Bessi:

Viva quam miserum est inter Bessosque Getasque.

Both Sophocles and Arrian allude to the Veneti in Thrace. Homer, in the passage above cited, also in the opening of Book XIII, carries them into Mœsia, which, it is well known, was inhabited by Goths, down to the Medieval Ages. Its name is derived from that of their god or prophet, Moess, or Bacchus. These references extend the Gothic occupation of the valley of the Danube, (Ister,) to a period of nearly twenty centuries, over the whole of which period ecclesiastical history has drawn an almost impenetrable veil.

In another passage Herodotus carries the Veneti from Arcadia to Illyria and there leaves them, whilst Livy and Strabo take them at once by sea from the Bosphorus to the Adriatic, which, as they must in fact have possessed many ships, and as their land-passage through western Greece would have been disputed by many enemies, seems more probable. However, in another place Herodotus alludes to the "Eneti on the Adriatic" and this phrase may be construed to mean not only the Veneti of Illyria but those also of the lagoons and islands.

Here too the place-names identify them. Polybius in his first and second books mentions the Illyrian promontory of Lissa, the city of Issa and the Venetian name of the river Po, which was Eodencus and at a later period Eridanus after the Eridanus in Pontus. Odin, Boden, or Woden, the Gothic name for their god Dion-Issus, is still commemorated in our Wednesday. There was a town named Bodincomagus, on the same river, not far from the modern Turin. One of the Illyrian provinces was called and is still called Istria: the name of the Illyrian queen was Teuta, also a Gothic name. In describing the opposite shore of Italy Polybius says: "Below all these and nearest to the Adriatic were the Venetians, *a very ancient people*, whose dress and manners greatly resembled those of the Gauls, though they used a different language. This is that nation of whom the tragic poets have recorded so many monstrous fables." The Illyrians and Venetians, who appear to have been one people, were noted for their fondness for the sea, prominence in mining, love of plunder, addiction to piracy and trade, skillfulness as mariners and the strength of their marine. Their avoidance of the use of gold money at an æra when in hierarchical states gold, as shown in my "History of Monetary Systems," was regarded a sacred metal and was systematically doubled in value by the coinage laws of the sovereign-pontiffs, is very remarkable.

The decline of the Venetians of Illyria is to be dated from A. U. 524, when, instigated by the infirm but zealous pontiff L. C. Metellus, the Romans and allies, with an immense fleet of two hundred ships, defeated Queen Teuta, occupied Corfu, destroyed the Dion-issian shrines, made great slaughter of the Venetians, almost depopulated Illyria and Istria, compelled the survivors to submit to the payment of tribute and ordered that thenceforth they should "never sail beyond Lissa with more than two galleys and those unarmed."[2]

Among the many strange coincidences which occur in comparing the circumstances of the Veneti of the Euxine and the Veneti of the Baltic, none is more striking than their persecution by the hierarchies, first of Assyria, and afterwards of Greece and Rome. In this generalization we may perhaps include the Egyptian hatred of the Hyksos. By which of the sovereigns of Assyria the Veneti of the Euxine were attacked, is uncertain, though probability points to Tiglath-pil-esar. It is, however, well known that in a later age Pope Metellus, as above related, destroyed the Veneti of Illyria, that Julius Cæsar drove the Veneti of Brittany from the Loire, and that Charlemagne and his successors, of the Latin empire, exterminated the Veneti of the Baltic.

This coincidence does not merely consist in the fact that the Veneti were attacked without provocation and pursued with vindictiveness, but in the circumstance that their assailants in every case were the lords of hierarchies, especially embittered against the Dion-issian worship, and that in each case the suppression of the Dion-issians was followed by the establishment or revival of gold-mining.

Modern archæological researches have brought to light many interesting memorials of Assyria and Chaldea, nationalties which, at one time or another, controlled the countries embraced in this chapter. But I am far from being convinced that the remote dates attributed to these remains have any nearer relation to fact than is to be discerned by the misleading lights of astrology. Baked clay tablets have been found in a mound of Mesopotamia containing, in cuneiform characters, accounts of the transactions of a merchant, many of which are said to be couched in silver shekels and ring-money and some in gold dust. If the readings and translations are to be depended upon, the use of silver for money, unless the reasons set forth in a previous chapter herein are quite overthrown, must challenge the antiquity sought to be attached to these finds. It is certainly rather remarkable that whilst the archæological department of the British Museum is asking us to believe in the use of silver money in Chaldea over six thousand years before the Christian æra, the coin department of the

[2] A similar policy was pursued by the Japanese in the 17th century, who, to prevent Europeans from repeating their attempt to forcibly evangelize their empire, restricted them to two unarmed ships a year and these to the island of Deshima.

same institution continues to insist that money was invented by the Lydians more than five thousand years later.[3] One department asserts that inscribed coins are not older than the sixth century before our æra, whilst the other is able to read a cuneiform script which it assumes remained unaltered for upwards of sixty centuries. (Cf. "Coins and Medals" by S. Lane-Poole, p. 38.) From which it would follow that inscriptions were made upon bricks during the immense period of time represented by fifty-four centuries, before the idea occurred of making similar inscriptions upon coins! Such theories seem hardly less monstrous than those complained of by Strabo, some eighteen centuries ago. The truth appears to be that coins are much older and the Chaldean ruins much more recent than the dates assigned to them respectively by these inharmonious savants.

The mines of Colchis and Pontus were numerous and some of them important. Nearly every stream that descended from the Caucasus, Olgassys and Taurus mountains brought down its contribution of gold.[4] These resources were, however, probably exhausted by the aborigines before the arrival of the Veneti. The latter, caring nothing for gold, turned their attention to the production of iron, a knowledge of which they had either brought, or acquired, from Greater Asia. The iron ore was obtained in the range which extends from Tocat in Chalybia towards the province of Trebizond where the mountains separate the basin of the Euxine from that of the Euphrates. There were also other deposites, both of iron, copper, and electrum nearer to the city of Trebizond, in the districts known as Koureh and Gumish-Khana. From these Chalybian mines we retain the name of chalybeate, which is still given to waters containing iron.

When the iron industry had advanced sufficiently to enable the Veneti to supply the Assyrians with steel weapons and tools, the latter, driving away and supplanting their benefactors, prospected and opened the quartz mines of the vicinity and from them extracted for themselves their gold, copper, silver and mercury, the latter from Mount Olgassys, or Olgasus, in Paphlagonia.

The ores were reduced at the Sandaraca Works in Pimolis on the river Halys. Says Strabo (XII, iii, 40): "The Sandaracurgium is a mountain hollowed out by vast excavations made by the miners. The mining work is always carried on at the public charge," because, as

[3] Cuneiform Texts from Babylonian Tablets in the British Museum. London, 1897; printed by order of the Trustees.

[4] Strabo, XI, passim: Pliny, xxxiii, 3: Plut., in Pompey: Appian, de Bello Mith. Procop. Bell. Persic,, besides numerous modern authorities cited by Malte-Brun.

he explains further on, it did not pay, "and the slaves are convicts. Beside the severity of the labour, the atmosphere of the mines is filled with poisonous odours, which commonly prove fatal to the workmen. The works are frequently suspended from the difficulty of obtaining labourers whose full number is two hundred, which is being continually diminished by disease and accident."

These mining operations which originated with the Veneti were afterwards extended by them to the Caucasus range, and eventually embraced the entire country from Georgia to Paphlagonia. Malte-Brun mentions gold, silver or copper mines in Georgia, Imeritia, Mingrelia (Colchis), Abassia, Kabardia, the country of the Kubashes, (who claim to be descended from the Venetians, or as they term them, "Franks,") and also various other parts of Colchis and Pontus. Both Strabo and Pallas state that in the vicinity of the Caucasian mines there are numerous caves hollowed out of the solid rock in which the miners dwelt, and which still contain vestiges of their former inhabitants. Similar caves are to be seen on the eastern coasts of Egypt, in Mauretania (Strabo, XVII, iii, 7), in that part of Spain which was inhabited by the Goths and on the coast of Nova Scotia (the Ovens) explored by the Norsemen (Veneti) of the eleventh century. Hamilton (1842) says that in the reign of Justinian the gold mines of Trebizond were so important as to become a subject of dispute between that monarch and Chosroes I. He gives a long account of the electrum mines at Gumish-Khana (Gumusthane) which produced annually 250 to 300 dr. of gold without mentioning the silver which they likewise yielded. The mines of Balgar-Dagh, on the slopes of the Taurus are said to be exceedingly rich, yielding £8 ($40) gold and £5 ($25) silver per ton, besides a large proportion of lead. The British Consul Wrench (1880) grows quite enthusiastic over the mines of Pontus and predicts that whenever the Turkish government establishes laws that will define and protect the miner's rights and effectually shield him from official exactions, this country will regain a considerable proportion of the eminence it once enjoyed as a productive mining region.

CHAPTER III.

PERSIA.

Greek perversion of Persian literature—The Persian dynasty of Geimschid—Reputed treasure of Susa—Spoil taken at Platæa—Antiquity of the Persian daric—£ *s. d.* System—Scythian invasion and pillage of Persia—Silver of Laristan—Gold of Irak—Persian mines in the time of Trajan—Unimportance of Persia as a mining country.

THE Greek writers have so thoroughly obliterated the early history of Persia that but little remains of it at the present time; and even that little is difficult to distinguish from fable. Dr. Hyde mentions an early dynasty whose most notable figure was Geimshid, the reputed discoverer of an intercalary cycle of 120 years, which is still in use. The æra of Geimshid is assigned in another work [1] to the year B.C. 703, but it is much to be feared that Geimshid is a Persian myth built upon the Hindu myth of Buddha. Between his æra and that of Cyrus the Elder, the history of Persia is a blank.

In the time of Cleomenes, king of Sparta, 491 B.C., the barbarians of Asia, this includes the Persians, are represented as being exceedingly rich in gold and silver, the Lydians with "a profusion of silver," and the Persians with so vast a hoard in Susa as made them vie in affluence with Jupiter himself. At the same time the Arcadians and Argives were "destitute of the precious metals." [2] Susa is situated on the river Kerkah (anciently the Choaspes) in the province of Kurzistan, Persia. This stream was once called the Golden Water and the kings of Persia drank of this water only; for which purpose it was bottled up and carried with them on their journeys. The origin of the custom was probably the fact that the Choaspes was an auriferous stream, from which it was deemed desirable, by means of royal appropriation, to warn off all intruders. Similar customs are recorded of other auriferous countries. [3]

At the battle of Platæa (B.C. 479) an enormous spoil was gathered by the Greeks from the camp and persons of the vanquished Persians,

[1] Del Mar's "Worship of Augustus Cæsar." [2] Herodotus in Terpsichore, 49.
[3] Consult Herodotus in Clio, 188, and the authorities in Beloe's notes.

including "scimitars of gold," presumably scimitars with gold hilts or scabbards. The Ægintae purchased some of this spoil "at the price of brass" from the Greek slaves, or helots, (who had stolen and secreted it.)[4]

Although Mionnet is of opinion that the darics of Persia are the oldest coins (familiar to the Western World) which are still extant, we know but little of the sources whence the metal was obtained of which these coins of Persia were composed. If it be answered that the source was plunder, we know of no nations rich enough in the precious metals and within reach of her armies which could have supplied Persia with the materials of coinage at so early a period. Egypt was not plundered by Cambyses until B. C. 526, nor India by Darius until B. C. 521; while the darics, according to Mionnet and the Hebrew Scriptures, are of an earlier date than either of these expeditions. These coins, whose name was possibly derived from the zodiacal sign of the Archer (Indian, *danaus*) with which they were stamped, or else from the Indian gold coin dharana, contained 129.275 English grains of fine gold, something heavier than a modern English sovereign, or American half-eagle. Each daric went for 20 silver shekels, of 83.96 grains of fine silver: and each shekel for 12 bronze or copper coins. The ratio of silver to gold was 13 for 1. Thus, the oldest monetary system, of which any succinct account is now deducible, was almost precisely the same as the £ *s. d.* system of England at the present day, barring the gratuitous and unrestricted coinage and the adjuncts of paper notes, employed in the latter. Substitute darics, shekels and cashbequis, or kasbequis, for pounds, shillings and pence—and the resemblance is almost complete. Assuming Mionnet's antiquity of the daric to be well founded, the source of the gold of which they were made—for they were very numerous—must be left to conjecture.

About B. C. 633 Persia was overrun by the Scythians; and when these fierce marauders evacuated the country, some quarter of a century later, they probably left but little gold behind them. However, it is not improbable that this raid imparted to the Persians that knowledge of the auriferous wealth of Scythia of which they made such ample use in their subsequent expeditions into that country under Darius Hystaspes, and that meanwhile they inaugurated the system of mining of which Onesicritus observed the indications. Said that writer, who was of the fourth century before our æra: "A river in Carmania brings down gold-dust; there are mines of silver, copper and minium and there are two mountains, one of which contains ar-

[4] Herodotus in Calliope, 80.

senic, the other salt.[5] The silver mines here mentioned probably comprised those of Laristan. Auriferous quartz deposits were also found near Zinjan in Irak, but nothing is known of their history or exploration, except that a Berlin engineer, who visited them in A.D. 1877, could find no pay ore. (Lock, p. 366.) The rivalry between the Persians and Greeks for the trade of the Orient is an almost unmistakeable indication that the former possessed some permanent and reliable source of the precious metals, especially of silver, without which such rivalry would have been commercially ineffectual. The silver mines of Laristan were hardly of sufficient importance to supply this want;[6] and it must therefore be supposed that the Persians were more completely and for a longer time in control of the argentiferous parts of Greece, both in Asia and Europe, than their classical rivals have chosen to admit. This silver, when exchanged for Oriental gold, may explain the source of the most ancient Persian darics.[7] The treasures of Persepolis and Pasagarda were of a later date.

In the reign of Trajan, a Roman slave, one Callidromus, who had been sent as a present to Pacorus, King of Parthia, was by him consigned for punishment to the mines, whence he afterwards escaped to Nicomedia and took refuge in a sanctuary dedicated and sacred to the Roman Emperor. Upon being interrogated by Pliny the Younger, who was then pro-prætor in Pontus, Callidromus drew forth a "small ingot of Parthian gold which he said he brought from thence, out of the mines. I have affixed my stamp to it, (wrote Pliny,) the design being a chariot drawn by four horses." During a previous pro-consulship (that of Velius Paullus), Flavius Archippus, a "philosopher," had been unjustly "condemned to the mines for forgery." He was pardoned and restored to his possessions by Trajan.[8] The location of the Pontine mines to which the philosopher was consigned is of no consequence, but it would have been interesting to learn that of the diggings whence Callidromus obtained the ingot which Pliny for-

[5] Minium was, strictly speaking, a protoxide or peroxide of lead, but as the colour was that of cinnabar, the ore of mercury, it is quite likely that minium here meant the mercury used in catching the gold of the Carmanian streams. Onesicritus is quoted in Strabo, XV, ii, 14.

[6] From the silver of Laristan were made loops or baugs of silver-wire, stamped with a monetary mark. They were called larins, and were each of about 72 grains, or much the same weight as a French franc.

[7] Adarkonim, the Hebrew plural for darics, are mentioned in I Chron., XXIX, 7; Ezra, II, 69 and VIII, 27 and Neh., VIII, 70-72. In the English version they are translated "drams," a palpable corruption.

[8] Pliny's Letters, X, xxvi, and lxvi.

warded to Trajan. There is nothing, however, upon which to rest a conjecture.

Neither the ancient Parthia nor the modern Persia are known to have possessed any extensive gold or silver mines, though the vast spoil in the precious metals captured by Alexander at Persepolis (Ishtakar) has led to a contrary inference.

CHAPTER IV.

EGYPT.

The Nile in ancient times—Its fall was greater than now, while its banks were wooded—These physical changes probably due to gold-mining—Oriental origin of the Egyptians—Social changes—Many of these due to gold-mining—Vast extent of the mining regions—Outlying districts—The great central or Bisharee region—Antiquity and wealth of these mines—Treasury of Rhampsinitus—Plunder of Cambyses—Tribute of Darius—The mines as described by Agatharchides and Diodorus—Worked by the Arabs—Explored by the moderns—Their exhaustion—Devastated aspect of the surrounding country.

IT may, perhaps, appear surprising that the Nile, which, from time immemorial, has been regarded chiefly as a great agricultural stream, should be classed among mining rivers; yet there is ample warrant for so doing. Not only this; but its history as a mining river appears to be more ancient, and—considering the geographical changes it has brought about—it is certainly more important than as an agricultural one.

In the remotest historical period the Nile, though never a rapid stream below the junction of the Blue River, had, undoubtedly a greater fall than at present. The country-rock, which appears at numerous places along its bed, did not form, as now, merely rapids, but cataracts. This their ancient names attest. The hills of Sennaar and Nubia, which now are destitute of timber and water, were once wooded; and from their flanks flowed numerous feeders, which, after enriching the soil, added their floods to the Nile. These feeders now flow underground, through the mining debris which underlies the sand of the plains.

The Bisharee or gold country was once cultivated, and populous. Much of it is now, and has been since the time of Cambyses, a desert of sand and gravel. The Nile, which now washes the Arabian side of the valley, formerly skirted the Libyan side. From lat. 27 40 N. to the sea, a portion of Egypt was at one time a morass. It was like the delta of the Po, and covered with trees. This was probably during the first æra of mining, and before the river was diked by Menes. It is now all dry land and treeless, although its delta is now far to the northward. The Nile once emptied into the sea a little below where Cairo now stands. Nor need this period have been very remote. At the present time the river runs ninety miles farther through a Delta whose vast dimensions probably owe some part of their foundation to the operations of the multitudes who for generations were employed in washing the sands of the Bisharee region for gold.

After the æra of mining which had bestrewn the plains of Lombardy with sand and gravel from the mining regions of Piedmont, the annual floods of the Po were able to cover the mining deposits with layers of loam, which, in the course of ages, became sufficiently deep to restore the ancient agricultural character of the country. Not so with the Nile, which, from Fazogle, in lat. 12 N., to the Delta, had scarcely enough fall for this purpose. Consequently, the ruin brought upon Nubia was irreparable; and its inhabitants were driven for support to the narrow valley of the Nile proper, where agriculture had to be conducted under conditions that led to the permanent enslavement of the people and to those extraordinary systems of government and religion which have exercised such a potent influence upon the destiny of the world. Next to the comprehensive study of these systems, nothing so well marks the uncertain nature of Egyptian agriculture, and the abject and defenceless character of the people who had to depend upon it for support, than the frequency of their subjection to foreign conquerors, and the number of alien dynasties to which Egypt has been compelled to submit. This furnishes a curious and instructive contrast with China, which during forty centuries has not changed the nationality of its dynasties half-a-dozen times.

After the devastation and physical ruin of the Bisharee region, and before the Egyptian Delta assumed its present vast proportions, the only considerable agricultural surfaces in Egypt were in the Fork, anciently called the Island of Meroe, and in the plains below the foothills of Nubia. The edifices and works of art discovered in Meroe, and elsewhere in Nubia and Upper Egypt, appear to have preceded thos: which have been found in the lower country.

The Egyptians and Nubians are regarded as of Indian origin; their physical appearance, their complexion, their pyramids and earliest edifices, their ecclesiastical and political institutions and the habitat of some of the plants and many of the articles of commerce found in their tombs, all point to this conclusion.

The term Nubia appears to have originated in Egypt, where Nob or Nub signifies gold, hence Nubia, the Land of Gold. The ancient inhabitants were called by the Greeks, Nobatæ. Malte-Brun regards

Nubia as the Ethiopia supra Egyptum of the ancients. It embraces the foot-hills of the mountainous ranges that constitute a great part of Abyssinia; whilst Egypt is a country of low plains bestrewn with sand. The three distinguishing characteristics of all great gold-mining countries are here apparent: the sierras, Abyssinia; the foot-hills, Nubia; the plains, Egypt; the connecting link between them all, the Nile. Analogous geographical surroundings characterize those other great mining rivers, the Po, the Rhone, the Rhine, and the Sacramento. Like Italy and California, Nubia has also its coast range. These mountains separate the Bisharee country from the Red Sea, and rise to an altitude of 6000 feet. Bish-marmak means five fingers; so does Penta-dactylos, the mining region of the Ptolemies.

Below the foot-hills of Nubia is a vast expanse of sand and gravel, known as the Bisharee, or Great Nubian Desert. This, as we shall presently see, was once the centre of the greatest gold-mining works known to the ancient world. According to Diodorus, Book iv, the Pharaohs derived from this region in gold and silver, a sum which Mr. William Jacob computes to have been equal to £6,000,000 per annum. According to the hieroglyphics in the Memnomium under the figure of the King, who is offering produce to Amun Ra, the gold and silver mines of Nubia are held to have yielded altogether 3,200,000 minæ, a sum which Rev. John Yeats (Hist. Commerce, 48) regards as equivalent to £7,000,000: but which we believe amounted to 2½ times as much. If this computation can be depended upon, the Nubian mines were as prolific as have since been those of either Italy, Spain, Brazil, Russia, Australia, or California. An able writer in Appleton's Encyclopedia says that in Lower Nubia, in or about the same latitude as the Second Cataract of the Nile, lie "vast and fertile, but neglected plains, which it is conjectured were, at some remote period, reached by the inundations of the Nile." If the inundations of the main stream could have reached these plains at any former period, they certainly can reach them now, that the bed of the river is higher than it ever was before. Marsh testifies that since the Augustan æra the river bed has risen seven feet at Thebes and nearly four feet at the Delta. The Encyclopedist must therefore be mistaken as to the cause of the desolation noticed. The fact is, that anciently Nubia was watered by numerous small streams that flowed into the Nile, and were employed to irrigate these desert plains—once the Lombardy of Africa—but which after they had been diverted by the gold miners, and their sources of supply cut off by the destruction of the forests in the foot-hills for mining timbers, fell into the condition of "washes," which now are only flooded for a brief period during the rainy season of March to May, and are quite dry for the remainder of the year. These "washes," or dried-up water-courses, are called by the Arabs "wadys." Besides these there are the underground rivers alluded to in Reclus, 278. Without surface rivers, there are, of course, no irrigating canals.

Such is Nubia to-day. Its hills have been leveled into a plain of sand and gravel; its alluvial soil has been washed into the Nile, which has transported it a distance of a thousand miles to fill up the morasses, and form the Delta of Egypt; its forests have been cut down; its rivers have been either dried up or submerged; and where the native has not abandoned it altogether, he has degenerated to the level of a savage. The Arabs call the natives "Berbers," a term equivalent to the Roman word barbarian, and possibly of the same origin. Treachery, dishonesty, drunkenness, and filth characterize the men, and vulgarity and licentiousness the women, many of whom are worked as beasts of burden, to plough the land or tow the boats on the Nile. Both sexes go naked, and the money of the country is a sort of broom-corn, called dourra. Such are the results of exploiting the country for gold.

Gold has been found in nearly every region tributary to the Nile, from the Equator to the First Cataract. The following summary of these regions affords a brief view of the information which has been collected on this subject:

Darfoor. Hon. Robert Curzon in his "Armenia," p. 120, written in 1854, says that some years previously he met at Assouan, a European from the mountains beyond Darfoor, lat. 12 N., long. 26 E., who showed him several strongly made iron-bound chests full of gold from that region. Some of the gold was in nuggets, but the most part was in the form of rings (baugs) the size of bracelets, and others of the size of large heavy finger-rings, all of pure gold. These rings were passed in Darfoor as money, and were of the same form as those used for a similar purpose by the ancient Egyptians, in fact the same as the rings found in all Gothic countries, including ancient Britain.

Kordofan. This region lies between Darfoor and the Nile. It abounds with auriferous placers, which are washed by slaves. Yet it uses an overvalued iron money. Forty or fifty years ago the production and trade in gold was monopolized by the pasha; nevertheless gold was sold clandestinely at the rate of $8 in silver for 430 grains of fine gold, equal to $17.30 per oz. Troy, or about 15 per cent. under the mint value.

Takalé District, White Nile. Sheiban, modern Seizaban, on the west bank of the White Nile, within the Takalé or Takla district, in the Kordofan country, lat. 12 N., is spoken of by Pinkerton as a placer gold country. At Luca, apparently in the same district, as well as at Sheiban, gold, probably gold dust in quills, was the only money employed.

Shoa. The Portuguese commander Albuquerque, landed on the coast of Abyssinia about the year 1510, and at once proceeded to ransack the country for gold. Amongst the means which he employed was to enslave the living in the mines, and plunder the graves of the dead. His rapacity and cruelty left so strong an impression on the natives, that up to a very recent date, 1841, and perhaps even to the present day, the use or possession of gold is strictly forbidden in the kingdom of Shoa. The monopoly of the gold trade by the sovereign of that country may have something to do with this interdict. In spite of it, however, gold in 1841 was sold surreptitiously for about nine silver dollars per ounce Troy.

Fazooglu Country on the Blue Nile. The Saukin merchants deal in gold brought from this region, which is in lat. 11 N. In October, 1838, Mahomet Ali left Cairo in a steamboat to visit this country. The mines proved to be near the confluence of the Blue Nile and the Fazangoro. After inspecting them, Mahomet left a colony to work them for wages, and laid the foundation for a town near by, to contain fifteen hundred families; but the mines failed to pay, and the town has gone to ruins. The previous productiveness was due to the fact that the mines were worked by slaves, who were paid nothing for their labour but blows.

Kaffa. Herr Camill Russ (1878) says that in the Kaffa country south of Abyssinia, lat. 7 N., gold is found so plentifully that it is not much dearer than silver coins of the same weight. The reader can believe as much of this statement as he pleases.

Sasu Country on the Takáze. Adowa is in Abyssinia, about fifty miles from the Takáze, and in lat. 14 N. Gold is one of the principal articles of the transit trade through this place, and of the export trade from Abyssinia generally. Some of this gold probably comes from the Kaffa country. Cosmos, a Greek writer who visited Ethiopia (Abyssinia) about A.D. 535, gives an account of this trade, which, in his time, appears to have centered at Axum, which was the capital. The gold came from a country called Sasu.

Meroe. Strabo, XVII, ii, 2, while describing Meroe, says "there are also mines of copper, iron, gold, and various kinds of precious stones." This reference is applicable not merely to Meroe, which is an alluvial plain containing no gold mines, but to the whole of Nubia. The mention of gold mines in connection with Meroe may have been derived from the fact that Shendy, lat. 17 N., was a mart for the gold of the Upper Nile regions. McCulloch, writing in 1838, says it is so still, the price of gold being $16 an ounce and that Sennaar on the Blue Nile, lat. 13 N., is also one, the price of gold being $12 an ounce in silver dollars.

Below the Cataracts. There are no gold mines below the cataracts says Englehardt. The first or lowest cataract is in lat. 24 N., and this, therefore, is the northern limit to the gold mines of Egypt.

Somali. The above list comprises all the gold mines in the valley of the Nile (except the Bisharee) from the 11th Parallel, which was regarded as the extreme southern limit of Egyptian authority, to the 24th Parallel, north of which no mines have been found. Besides those of the Nile Valley, there are numerous gold mines in the surrounding countries, other than those mentioned by Herodotus, which are supposed by Rawlinson to have been in the Somali country. This is on the south coast of the Gulf of Aden. At the present time there are some small lots of dust shipped from Leila and Berbera, long. 45 E. There are gold mines also in the Kaffa country previously noticed, and others in the coast range of the Red Sea. We now come to the principal gold regions of Egypt, and the greatest of all antiquity.

The Bisharee Mines. These auriferous mines are in the Bisharee country, situated in the great Nubian bend of the Nile, between lat. 20 and 22 40 N., and between long. 32 30 and 35 20 East. The quartz veins have a course N.W.-S.E. In the Pharaonic period the produce of these mines was sent down the Nile. The shipping port was Edfou, or Apollinopolis Magna, or Redesiah, in lat. 24 58. It was ten days journey N.W. from the mines. Opposite to Edfou was Bahayreh. In the Ptolemaic period the produce was transported to various ports on the Red Sea, among them Berenice, lat. 24 N., a distance of about 260 miles from the most productive mines. This place is now a mere ruin. The Bisharee country forms the foot-hills to the sierras of Abyssinia. The sierras are 12,000 to 15,000 feet, whilst the highest of the hills is about 6000 feet in altitude. The latter gradually diminish until they melt into the plains of Egypt. According to Linant de Bellefonds, there are remains of gold mines at the following places in the Bisharee region:

Oum Guereyatte; Ceiga, 22 30 N., 33 50 E.; Gebel Offene; Gebel Abdulla; Gebel Matchouchelennaye; Gebel om Cabrille; Tamilla Gebel Essewed; Gebel Tellatabd. All these are situated in the hilly country of the Cawatil Arabs, between about 21 30 N. and 32 50 and 34 20 E.

Oum Tayour; Wady Sohone; Wady Hagatte; Wady Affawe; Wady Daguena; Wady Camolit; Derehib, 21 40 N., 35 E.; and Wady Chawanib. These are in the country of the Mansour Melecat Arabs, lying east of the foregoing and west of 35 20 E. At Derehib the quartz excavations are of immense extent.

Raft; Kelle; and Absab. These are in the country of the Foukar Arabs, between 20 21 N. and about 32 30 E.

Among the water resources now traceable are the following wadys (a term applied by some Arabs to all water courses whether filled with a running stream of water or dried up): Al-aki (anc. Akita); Sohone; Hagatte; Affawe; Daguena; Camolit; and Chawanib. It seems probable that in remote times all these wadys were running streams. There are also numerous streams that, in flood-time, still swell into devastating torrents and lose themselves in the desert, to make their way underground to the Nile. Numerous wells tap these underground streams.

Next to the mines of the Altai mountains of Asia, the Bisharee mines of Egypt are among the oldest in the world; and in view of the evidently Oriental origin of the Egyptians, and the distant researches and conquests which have been made by leading nations in all ages for the acquisition of gold, it seems not at all improbable that there existed a close connection between the discovery of the Egyptian mines and the original settlement of the country by Asiatic races. At what æra this occurred cannot be determined with any approach to certainty. Scholars of the present age are no less anxious to find a place in chronology for Hesiris and Isis, than were Hecatæus and Herodotus to assign an actual æra to Hercules and Venus; and the entire current of literary and archæological research is vitiated by the unwarrantable intrusion of personages into history whose only legitimate domain lies within the pages of mythology and fable. If conjecture be admitted where dates are thus in doubt, it appears likely that the Bisharee mines were worked for placer gold so early as the 16th or 15th centuries B.C. However this may be, these mines are believed by Mariette Bey to have been worked for quartz gold so long ago as the XIIth dynasty of Egypt. From the fact, understood by every miner, that quartz is never worked so long as the placers contain the smallest practical quantity of metal, and judging from the experience of Italy, Spain, and Brazil—where extensive placer deposits were worked, as in Egypt, by the hand labour of slaves—the Bisharee placer mines were at least two hundred years old when the quartz was worked under the XIIth dynasty of the Pharoahs. Bunsen, by according 33 years to the reign of each of Manetho's kings, dates this B.C. 2466-1733. We should be disposed to lower these dates to the extent of a millenium. The next date at which these mines are believed to have been worked is under Thotmes III., of the XVIIIth dynasty; under Seti Sethos, or Sethothis, of the XIXth dynasty, and under Rames, or Rameses II., son of Seti. All these names are derived from inscriptions on the neighbouring rocks and temples, but whether they are valid, or not, is questionable.

During the XVIIIth dynasty the Egyptians also obtained gold in bags from the Soudan. About 1500 B.C. gold was produced in the land of Pun (an East Indian word for gold) on the southeastern shores of the Red Sea. About 1300 B.C. the Egyptians captured a number of gold, silver and brass vessels from the Libyan chief who invaded Egypt with his Mediterranean allies, (Prof. Petrie, in Harper's, July, 1888,) but this gold may have come from a distant part of Africa. The vast sum of "silver and gold money" which is said to have been contained in the treasury of Rhampsinitus, an Egyptian monarch whose æra is assigned by Rawlinson to the XVIIIth dynasty, points both to the working of the Bisharee mines for gold and to commerce with Spain for silver. The former are the only mines known to have been worked at that period whence any large supplies of gold could have come; the latter is the only country which at that period could have supplied any considerable quantity of silver. The disappointment of Cambyses in respect of the Egyptian spoil and the character of the treasures which he carried away from Egypt, about B. C. 526; the tribute imposed by Darius, which was only 700 talents in money, together with 7000 talents worth of corn and the produce of Mœris; as well as the annual revenues of Ptolemy Auletes, the father of Cleopatra, which were only "12,500 talents" a year; imply that the production of the precious metals upon a large scale had ceased. The country was exhausted.

In the reign of Ptolemy Philopata, B. C. 180-170, the Bisharee mines were visited by Agatharchides of Cnidus, who has left us a brief and unsatisfactory account of them. In B. C. 50 the Bisharee mines were visited by Diodorus Siculus. He says: "On the confines of Egypt and the neighbouring countries there are regions full of gold mines, whence, with the cost and pains of many labourers, much gold is dug. The soil is naturally black, but in the body of the earth there are many veins of shining white quartz, glittering with all sorts of bright metals, out of which those appointed to be overseers cause the gold to be dug by the labour of a vast multitude of people. For the kings of Egypt condemn to these mines not only notorious criminals, captives taken in war, persons accused of false dealings, and those with whom the king is offended, but also all the kindred and relatives of the convicts. These are sent to this work, either as a punishment, or that the profit and gain of the king may be increased by their labour.

"There are thus infinite numbers thrown into these mines, all bound in fetters, kept at work night and day, and so strictly surrounded that there is no possibility of their effecting an escape. They are guarded by mercenary soldiers of various barbarous nations, whose language is foreign to them and to each other, so that there are no means of forming conspiracies or of corrupting those who are set to watch them. They are kept to incessant work by the rod of the overseer, who often lashes them severely. Not the least care is taken of the bodies of these poor creatures; they have not a rag to cover their nakedness; and whoever sees them must compassionate their melancholy and deplorable condition, for though they may be sick, maimed or lame, no rest nor any intermission of labour is allowed them. Neither the weakness of old age, nor the infirmities of females, excuse any from the work, to which all are driven by blows and cudgels; until borne down by the intolerable weight of their misery, many fall dead in the midst of their insufferable labours. Deprived of all hope, these miserable creatures expect each day to be worse than the last, and long for death to end their sufferings."

During the Roman period the Bisharee mines were visited and described by Cosmos, a Greek writer, A.D. 535, but we have no account of Roman production from this source. During the Arabian period they were visited or described by Edrisi, A. D. 1099-1164; Abulfeda, king of Hamah in Syria, 1273-1331; Ibn-al Wardy, d. 1358; Macrizi, 1385; and Massandi, of the 14th or 15th century, all Arabians. These writers allude to the Bisharee country as the "land of Bega."

In recent times these mines were visited by Belonzi, Linant de Bellefonds, Bonomi, and Mahomet Ali, the last of whom, as above mentioned, reopened and worked some of them for a short time, but without success. The following extract is from Wilkinson's "Ancient Egyptians": The gold mines of Egypt, or of Ethiopia, though mentioned by Agatharchides and later writers, and worked even by the Arabian caliphs, long remained unknown. Their position was only ascertained a few years since by M. Linant de Bellefonds and M. Bonomi. They lie in the Bisharee desert, the land of Begah, or of the "Bugait" mentioned in the inscription at Axum, about seventeen or eighteen days journey to the southeastward from Derow, which is situated on the Nile, a little above Kom Ombo, the ancient Ombos. Those travellers met with the same Cufic funereal inscriptions, which, from their dates, show that the mines were worked in the years A. D. 951 and 989, the former being in the fifth year of the Caliph Al Mostakfi Billah a short time before the arrival of the Fatimites in Egypt, the latter in the fourteenth year of El Aziz, the second (fifth) of the Fatimite dynasty. The mines continued to be worked till a much later period, and were afterwards abandoned, the value of the gold barely covering the expenses; nor did Mahomet Ali, who in the present century sent to examine them and obtain specimens of the ore, find it worth while to work them for long.

The matrix is quartz: and so diligent a search did the Egyptians establish throughout the deserts of the Nile, for this precious metal, (continues Wilkinson,) that I never remember to have seen a vein of quartz in any of the primitive ranges which had not been carefully examined by their miners; certain portions having been invariably picked out from the fissures in which it lay and then broken into small fragments. The same was done in later times by the Romans and Arabians. . . . The gold mines are said by Abulfeda to be situated at El Allaga, (or Ollagee,) but Eshuranib (or Eshuanib) the principal place, is about three days journey beyond Wady Allaga, according to M. Bonomi, to whom I am indebted for the following account of the mines: "The direction of the excavations depends on that of the strata in which the ore is found; and the position of the various shafts differs accordingly. As to the manner of extracting the metal, some notion may be given by a description of the ruins at Eshuranib, the largest station, where sufficient remains exist to explain the process adopted. The principal excavation, according to M. Linant's measurement, is about 180 feet deep. In the valley near the most accessible part of the excavation are several huts, built of unhewn stone, their walls not more than breast high; perhaps the houses of the excavators or the guardians of the mine. Separated from them by the ravine or course of the torrent is a group of houses, about three hundred in number, laid out very regularly in straight lines. In those nearest the mines lived the workmen who were employed to break the quartz into small fragments, (the size of a bean,) from whose hands the pounded stone passed to the persons who ground it in stone hand-mills, similar to those now used for corn in the valley of the Nile, one of which is to be found in almost every house at these mines, either entire or broken. The quartz thus reduced to powder was washed on inclined tables, furnished with two cisterns, all built of stone. Near these inclined tables are generally found little white dumps of debris, the residue of the operation. Besides the numerous remains of houses at this station, are two large buildings, with towers at the angles, built of granite. The valley has many trees, and in a high part of the torrent-bed is a sort of island, or isolated bank, on which we found many tombstones, some written in the ancient Cufic character, very similar to those at Assouan."

Says Diodorus Siculus: "When the stone containing the gold is hard, the Egyptians soften it by the application of fire, and when it has been reduced to such a state that it yields to moderate labour, several thousands (myriads) of these unfortunate people break it with iron picks. Over the whole work presides an engineer, who views and selects the stone, and points it out to the labourers. The strongest of them, provided with iron chisels, cleave the quartz by mere force, without any attempt at skill; and in excavating the shaft below ground they follow the direction of the glistening stratum, without keeping a straight line. In order to see in these dark windings, they fasten lamps to their foreheads, having their bodies painted sometimes of one and sometimes of another colour, according to the nature of the rock; and as they cut the stone, it falls in masses on the floor, the overseer urging them to the work with commands and blows. They are followed by little boys, who take away the fragments as they fall, and carry them out into the open air. Those who are above thirty years of age are employed to pound pieces of the stone, of certain dimensions, with iron pestles in stone mortars, until reduced to the size of a bean. The whole is then transferred to women, and old men, who put it into mills arranged in a long row, two or three persons being employed at the same mill, and it is pounded until reduced to a fine powder. . . At length the masters take the stone thus ground to powder, and carry it away to undergo the final process."

The mines described by Diodorus are supposed by Mariette Bey ("Histoire ancienne d'Egypte," 1867, p. 96) to have been situated at Attaki or Allaki near the Red Sea, 120 miles back from Ras Elba, the headland midway between Berenice and Suakin. Mariette goes on to say that in the reign of Seti I., of the XIXth dynasty, wells were dug on this route, in order that the mines which had been explored at an earlier period might be re-opened. These mines are probably those which were worked by the Phœnicians, who called the country the land of Pun, or gold. Petrie is of opinion that during the XVIIIth dynasty Egypt also obtained gold-dust in bags from the Soudan: this would of course mean from the placer mines. The earliest mention of the Egyptian mines in western literature occurs in Herodotus, (Euterpe, 121,) which gives us the wonderful story of Rampsinitus and his vast treasure and how it was partly abstracted by a clever robber. The Father of History omits to inform us of what value the treasure consisted, but Diodorus Siculus, who calls the monarch Rhemphis, says that it amounted to 400 thousand talents in gold and silver. If "talent" here means anything more than a sum of five staters, the sum is excessive and incredible. Even if many centuries be deducted from the interval of time usually accorded to the XVIIIth dynasty, it is impossible that any large amount of silver could have been collected at that period, when iron was unknown and quartz-mining impracticable. The account is probably fabulous.

The extensive area covered by the gold deposites of Egypt and the great antiquity of the country, when viewed as a populated region, has led to gold-mining upon so vast and long-continued a scale, as to greatly alter the physical aspects and by consequence the social circumstances, of that favoured land. Even in Brazil, where gold-mining history did not cover a period of two centuries similar consequences are to be observed. The water courses of Minhas Geræs (that is to say, many of them) are altered by bodily diversions or collateral ditch lines; the hills have been washed into the valleys; the arable lands are strewn with gravel; and the inhabitants are reduced to poverty, nakedness and crime.

CHAPTER V.

GREECE.

Silver mines of Laurium—Antiquity—Extent—Worked by slaves—Their frequent revolts—Scheme of Xenophon—Modern re-opening of Laurium—Strange good fortune—Mines of Thasus—Their revenues—Destroyed by inundations of the sea—Gold mines of Pangæus—Reputed to have been opened by Cadmus—Re-opened by Philip—Mines of Samos—Venetian custom of marrying the sea—Story told of Polycrates—Lead coins cased with gold—Description of the Samian mines by Theophrastus—The "Pelasgians"—Mines in Albania—Dalmatia—Croatia—Bosnia—Servia—Thrace and Bulgaria—Mines of Sidherokapsa—Immense revenues formerly obtained from them by the Turks—Greek colonies in Asia.

THE mines of Laurium were situated near Cape Sunium, on the coast of Attica, about 30 miles from Athens. They stretched from Anaphlystus to Thoricus, a distance of about 8½ English miles. All of these places were fortified.[1] The ores of the mines were reduced by smelting and the smelting works were situated at or near Anaphlystus, now called Ergasteria. The mines of Laurium are referred to by some of the oldest Athenian and Roman writers whose works remain to us. These include Æschylus, Themistocles, Herodotus, Thucydides, Xenophon, Pythocles, Aristotle, Cornelius Nepos, Pausanias, and Hesychius. When first discovered, the mines of Laurium probably yielded native silver in small quantities. The architectural ruins and animal-headed gods and other remains recently found at Mycenæ and Tiryns do not at all resemble Greek remains, whilst they do resemble Phœnician ones. Dr. Dorffeld says: "The conformity between the structures of Tiryns and Byrsa (Carthage) must be looked upon as a proof that both were erected by Phœnician builders." For these reasons, we are inclined to the belief that Laurium, like Thasus, was originally opened either by the Phœnicians, or the Venetians of Pontus: a conjecture that derives corroboration from the statement in Julius Pollux that Eric-theus (a Venetian demigod or hero whose æra is fixed at B.C. 1406) first introduced coins into Greece. This archaic condition of the mines—that of working them for such native silver as appeared in the quartz—probably con-

[1] Boeckh, 279.

tinued a long while before they were attacked with steel tools and systematically opened. It was during this interval, or shortly after its close, that Pheidon of Argos, B.C. 748, struck coins in the island of Ægina, which is within sight both of Athens and Laurium, from which latter source the metal was doubtless obtained. The ore from these mines was known to the Greeks as cadmia (probably from Cadmus) and is now known as calamine. It is a silicate of zinc and, as found in the Laurium mines, contains zinc, lead, silver and some other metals. In the time of Themistocles, about B. C. 500, when these mines appear to have been very productive, they seem to have yielded to the State, which let or leased them in fee-farm to private persons, about 30 to 40 talents per annum.[2] As this was the twenty-fourth part of the produce, the latter must have amounted to between 720 and 960 talents yearly. The Attic talent was equal to about half a hundred weight: so that if a talent weight is here meant, the produce of silver amounted to between 360 and 480 cwt. At the epoch of Themistocles silver was valued by law in Greece at one-tenth of its weight in gold. Reckoned in gold the annual produce of Laurium was worth 36 to 48 cwt. of that metal. This, if coined into sovereigns of the present day, would equal about £200,000 to £233,000, or over a million dollars. The mines were worked by slaves, who, on more than one occasion, revolted from the cruel labour to which they were forced, and to which they were again forced by merciless repression.[3] In the time of Socrates and Xenophon these mines so far as the mechanical resources of that age are concerned, were practically worked out: yet the latter, in his treatise on the "Revenues of Athens," offers a scheme to re-open them with the labour of 10,000 slaves. That this scheme was actually put in practice may be inferred from the exemption of the silver mines from taxation, which appears in the following extract from the laws of Athens: "Those who do quit their own estates for those of their neighbours shall be obliged by oath to discover them in this form: 'I will fairly and honestly make known the estimate of all my possessions, except such as consist in those silver mines from which the laws exact no duties.' No one shall be compelled to exhibit his estate which lies in mines."[4] In spite of these advantages it is evident that the mines were not profitable and that as time went on it became more and more difficult to sustain the monetary circulation with coins; and yet to preserve the customary level of prices. Witness the following law: "Let no

[2] Boeckh, 417.
[3] Athenæus, VI, 104; Moyle's Xenophon, Note 23.
[4] Potter's Antiq. of Greece, I, 188.

Athenian or sojourner lend money to be exported, unless (to pay) for corn or some such commodity allowed by law."[5] We are far from wishing to ascribe the decline of the Greek power to the failure of the mines; yet it cannot be denied that at a period when metallic money alone seemed practicable, such failure must have had a tremendous effect upon the politics of the state. Sir Archibald Alison has clearly demonstrated this to be true with regard to Rome at a subsequent period. Why not in respect of Greece at an earlier one?

After the Roman Conquest of Greece, that is to say, in the time of Strabo, some of the richest of the ekvolades or refuse ores of Laurium, of which vast mounds or "dumps" lay piled up on the surface, were re-worked; a sign that some new and more economical process or mode of working had become known which rendered such re-working profitable. This conjecture is corroborated by what Strabo says of the Roman silver mines of Turdetania. Demetrius of Phalereus, an Attic orator of the third century B. C., having boasted that his countrymen worked their mines with an energy that "promised to dig up Plutus himself," he was answered by Posidonius of Apamea, a Greek philosopher of a later age, who visited the Roman works in Spain: "There," said he, "you will observe no wasted energy. The Romans are quite different from the Attic miners, whose mining work may justly come under the Homerian enigma, that 'what I have taken up I have not kept and what I secured, I threw away.' In short, the Turdetanian miners know their business, and make a good profit. They get 25 per cent. of metal from their copper ores, while from the silver mines a single person has taken as much (per day) as an Eubœan talent." Something of this was doubtless exaggeration, for Strabo says that his language was turgid enough to have been dug from the mines themselves.[6] Nevertheless Posidonius was one of the ablest men of his age and the great difference which he noticed between the mining methods of the Greeks and Romans must have had a large basis of truth.[7] Our own observation of the works which the ancient Greeks have left in Attica and the ancient Romans in Asturias, etc., strongly confirms this view. In the time of Pausanias the mines of Laurium were again abandoned; and yet a notable revival of these very ancient mines took place in recent times.

About the year 1870 heaps of ekvolades and scoria left by the ancients and imperfectly reduced by their process, enlisted the attention of French and Italian capitalists, who bought the property from

[5] Potter, I, 198. [6] Strabo, III, ii, 19.

[7] Livy, XLV, 18, alludes to the profitable character of Roman mining in Macedonia.

its Greek owners or their nominee and began the reduction of the refuse. A score or more of mining companies at once sprang into existence, and from first to last something like two millions sterling worth of silver were extracted from the refuse materials.[8] The discovery of the refuse occurred in a curious manner. Signor Serpiere, an Italian engineer, was engaged in constructing some works on the coast of Italy, when a vessel arrived from Greece in ballast. This ballast having been thrown near Serpiere's works, the latter examined it and recognized it as calamine ore. Upon ascertaining whence it came and that there was plenty more to be had, the enterprising engineer at once went to Ergasteria, surveyed the property, secured an option upon it and sold this at an enormous advance to the capitalists above alluded to. Besides this, he secured a number of subsidiary advantages, all of which together enabled him to amass a large and well deserved fortune. According to the "Miners and Smelters Magazine," vol. VI, pp. 286-322, the Phœnician slags or scoria, by which is probably meant the oldest of the refuse, yielded no silver and only $5 per ton in lead; whilst the Grecian slags yielded 6¼ per cent. of lead, and about $90 silver per ton, a proof that the Phœnicians were better smelters than their conquerors. Besides the minerals above mentioned, the Laurium deposits yielded cinnabar (the ore of mercury) and sil, a colouring material, highly esteemed by the ancients.

Thasus, an island of the Thracian coast, (written Thasso by the Greeks [9] and Thassus by Livy,) was originally colonized by Phœnicians from Tyre [10] and went by the various names of Odonis, Ogygia, Ceresis, Acte, Chryse, etc., all of which names are of Buddhic or Bacchic origin. Odonis (or Odin) and Ogygia were surnames of that deity, whilst Ceresis and Acte were surnames of Ceres, the Holy Mother, and Chryse or Chryses, the name of that priest of the Sun whose daughter became the prize of Agamemnon at Ilium. Thassus itself is probably a corruption of Iassus, for Pausanias informs us that Thassus was the son of Agenor, the brother of Europa and the leader of the Phœnicians: details that belong to the myth of Iassus. Herodotus says that he himself visited the island of Thasus, where he saw a temple to the Thasian Hercules "erected by the Phœnicians, who built Thasus while they were engaged in search of Europa, an event which happened five generations before Hercules, the son of Amphitryon, was known in Greece."[11] The "Thasian Hercules" was Iassus.

[8] U. S. Commercial Relations 1873, p. 661, and private information derived from some of the engineers connected with these enterprises.

[9] Nisard's note to Pomponius Mela, II, 2. [10] Pausanias, V, 25. [11] Euterpe, 44.

We know but little more of the early history of Thasus beyond the fact that its mines were celebrated for their yield of gold and silver; that the most productive ones were in the S. E. district between Ænyra and Cœnyra; and that the Thasians, in addition to the mines of the island, owned and worked those of Scapte Hyle (or Scaptesyla) on the Thracian main. These last in the time of Darius yielded an average annual product worth or equal to 80 talents. The mines on the island did not produce so much at this period, although at an earlier one they had annually yielded between two or three hundred talents.[12]

About 60 miles S. S. E. of Cape Sunium is the island of Siphnos, which in the time of Polycrates B.C. 580-22 and perhaps long before, was famous for its rich mines of gold and silver. "Their soil produced both gold and silver in such abundance that from a tenth part of their revenues they had a treasury at Delphos equal in value to (all) the riches which that temple possessed." [13] Under the Athenian supremacy they only paid 3600 drachmas of annual tribute. In the Roman period, time of Strabo, Siphnos was noted for its poverty: for says Pausanias, speaking of the interval, "Afterwards, their gold mines were destroyed by an inundation of the sea." [14]

Mount Pangæus is in Thrace on the river Nestus, about 200 miles W.N.W. from Constantinople. Pliny says that the gold mines in this range were opened by Cadmus: indeed it is probable that all the mines of ancient Greece were opened by the Phœnicians, or the Venetians, before they were worked by the Greeks. Philip of Macedon, about B. C. 358, being informed that in ancient times these mines had been productive, caused them to be re-opened, with the result that he obtained from them annually more than a thousand talents. It is from the gold of Pangæus that he struck his "philips," [15] whose type, during the following century, was copied so extensively by the Gauls.

The island of Samos, once called Cypar-Issa, is on the west coast of Asia Minor near the mouth of the Caystrus and ruined Ephesus. It was colonized originally by the Bacchidæ, who were presumably Phœnicians or Venetians, and who on being driven out of Samos by the Ionians, settled afterward in Samothrace. We know little of the early history of Samos. Before the time of Polycrates it was prob-

[12] Herod., Euterpe, 44; Erato, 46. Boeckh, 417, has a different reading of this passage.

[13] Thalia, 57.

[14] Phocics, II; vol. III, pp. 131, 138, 151.

[15] Diod. Sic., II, 88.

ably the custom, and an ancient one, for the island governor or king to throw each New Year day into the sea a gold ring, an observance out of which the imaginative Greeks wrought a marvellous tale.[16] The Samian mines were of gold and silver, the ores of which were reduced in works situated on the river Imbrasus. The extant gold, silver, and electrum coins of Samos are numerous. Some of those commonly attributed to Sardis, were ascribed by Sestini to Samos. Herodotus reports that Polycrates bought off the Lacedæmonians, who tried to deprive him of the island, with a subsidy of lead coins thinly cased with gold, and thus cheaply got rid of his unwelcome visitors.[17] The mines of Samos were still worked in the time of Theophrastus, about 240 B. C., for he wrote concerning them: "Those who work in these mines cannot stand upright, but are obliged to lie down either on their sides or backs: for the vein they extract runs lengthwise and is only two feet deep, though considerably more in breadth and is enclosed on every side with hard rock. From this vein the ore is obtained." [18]

Mines of gold or silver or both were worked by the so-called Pelasgians in many parts of Greece, chiefly in the mountains of Albania, Dalmatia, Croatia, Bosnia, Servia, Thrace and Bulgaria. The remains of a smelting furnace composed of colossal hewn stones, together with heaps of refuse silver ores, can still be seen in Albania, almost in sight from the houses of Corfu.[19] Similar structures and remains are said to exist in Dalmatia. In Bosnia at Slatnitza, on the road to Scopia, six miles from Traunick, the Romans worked gold mines on an extensive scale and they were probably worked by the Greeks before the Romans. There are reported to be gold mines in several mountains near Zvornick and Varech. The rivers Bosna, Verbatch, Drina, and Latchva are auriferous. Many silver mines have been worked in the neighbourhood of Rama or Prezos, Foinitca and other villages, called Sreberno, Srebernik, or Srebernitza. Cinnabar is obtained near the convent of Chressevo, and this deposit was probably

[16] Thalia, 41 and Note to Chapter II herein. See also "The Worship of Augustus Cæsar," chap. VII, sub anno B.C. 1200.

[17] Thalia, 56.

[18] De Lapidibus, c. CXIX.

[19] The following directions for reaching these mines were given to us in January, 1885, by Maj. Gen. Mackenzie: "Go due north from Corfu to Santa Quaranta by boat, then to the Gardiki Pass in the mountains, then eastward along the southern slopes to the ruins of the Pelasgian smelting works, altitude about 2500 feet above the sea." There are also mining remains and ruins at Butrinto which is on the sea shore and which drains the Pelasgian works above. All these remains are well known to the natives.

worked for mercury in very ancient times.[20] About B.C. 470 Alexander, son of Amyntas, possessed a mine near Lake Prasis and Mt. Dysia in Macedonia, which yielded him a talent per diem.[21]

In Servia there were silver mines near Nova-Berda, and (Roman) gold mines near Saphina. Ancient mines of both gold and silver, chiefly the latter, exist in other parts of Servia, but little is known of their early history. There are some twenty thousand acres of alluvions within fifty miles of Belgrade which might yet richly reward the hydraulic process. There is plenty of water with good heads and good grades for the gravel. Bulgaria also abounds in mines of the precious metals, but like most of those in the territory comprised within ancient Greece, they have fallen to ruins and their history is forgotten.

Peter Belon, who visited the mines of Sidherokapsa, in the middle of the 16th century, asserts that he found 500 or 600 furnaces at work in different parts of the mountain: that besides silver, which was the chief product, gold was obtained from pyrites: that the mines and works employed 6000 men: and that the revenue of the Turkish government from the enterprise often amounted to 30,000 ducats (say £15,000) per month.[22]

In many parts of Greece or European Turkey, where ancient mines were worked, a superstition is said to prevent the peasants from visiting them. Malte-Brun especially mentions this of the old Roman mines near Traunick, and we have ourselves noticed the same superstition in the vicinity of the Roman gold mines in the Carpathian foot-hills. This superstition is probably due to the traditions of that cruel and relentless slavery to which their forefathers were subjected by the Greek and Roman lords who once owned these mines. Valdivia, writing to the emperor Charles V., declared that every castellano of gold from Peru cost a measure of human blood and tears.[23] What was the cost of gold to the ancient Romans or the still more ancient Greeks it would be hard to say: but a human life for every ounce would probably be well within the mark.

There are reasons for believing that the interdict of the precious metals ascribed to Lycurgus of Sparta was followed in some of the other Greek states. We know that to a certain extent this policy was imitated in the Greek Colonies, for Aristotle explicitly informs us that Clazomenæ (probably during the sixth century B.C.) struck iron money, whilst Aristides of Hadrianopolis and Hesychius of Alexan-

[20] Cf. Malte-Brun, IV, 118.

[21] Terpsichore, 17.

[22] Lock, on "Gold."

[23] Mackenna, "Libro de Oro."

dria both allude to the sidareos or iron money used in Byzantium and other portions of Greece and the Greek Colonies during the Peloponnesian war.[24] The scant or fitful supplies of the precious metals, or the apprehension that the use of them for money was attended with danger to the state, or to popular liberty, whatever the reason was that occasioned the interdict or the supplanting of these metals in Greece, appears to have stimulated their production in the colonies; for it is at corresponding periods that we first hear those turgid accounts of the auriferous wealth of Asia Minor which Greek literature has transmitted to all time.

The gold dust which descends from Mount Tmolus is mentioned by Herodotus in Clio, XCIII. This related to the Pactolus, about 60 miles east of Smyrna, which Strabo says was the source of Crœsus's wealth. Its importance as an auriferous stream has no doubt been vastly exaggerated. At the time of Herodotus the houses of Sardis, the capital of Lydia, situated on the Pactolus, were built of rushes; a sure sign of poverty. In the time of Strabo, and probably long before, the Pactolus had ceased to yield any gold.[25] About one hundred miles east of the Pactolus stood the mountain village of Celænæ, near which rises the river Meander, in the Taurus range. In the time of Xerxes, king of Persia, B.C. 485-65, there dwelt in Celænæ a Lydian named Pythius, son of Atys, next to Xerxes himself, the richest of all mankind, his wealth consisting of no less than "two thousand talents of silver and four millions, wanting only seven thousand, of the gold staters of Darius," all of which, beside his slaves and farms, he freely offered to Xerxes, who, (of course,) not only declined the princely gift, but gave Pythius the seven thousand staters necessary to make up the even four millions.[26] Those who are fond of the marvellous will find plenty of this sort of thing in Herodotus and the modern commentators, Larcher, Jacob, Rawlinson, Lock, etc. Practical experience of mining admonishes us to give these tales but scant credence.

We rather suspect that with regard to the Greek interdicts of the precious metals and the use of nummeraries, we are dealing with Greek fiction, rather than Greek history. After the Greeks had been taught the arts of civilized life, which the Pelasgians had fetched from Pontus and India, after the former had adopted the institutes and religion of the latter, they turned upon their teachers, drove them from the country and usurped both their possessions, and their history. They

[24] Aristid. Orat. Platon., II, 145, ed. Jebb; Hesychius, v. Sidareoi.

[25] Strabo, XIII, iv, 5.

[26] Polymnia, 27.

changed the names of their gods, they erected new temples from the cyclopean stones of the old, they modified the laws, renamed the months and perverted the history of their benefactors; so that little of it is left to posterity beyond what has been recovered through conjecture. The principal trade of the Phœnicians, as carriers between Asia and Europe, was spices, perfumes, ivory, jewelry and tissues, (from east to west,) and the precious metals gold, silver and bronze, from west to east. But whilst the Phœnicians were both navigators and miners, they were also Buddhists; and as such they never used the precious metals for money. They used iron and copper and perhaps bronze, but neither gold nor silver. Hence the iron moneys of Lycurgus (the myth) and of Clazomenæ and Byzantium may all have been Phœnician institutions perverted by Greek fancy and floated down the stream of time, to the bewilderment of antiquarian research and the confusion of history.

CHAPTER VI.

ITALY.

Auriferous streams—Bodencus (Po)—Rhenus—Tiber—Ancient mining camps—Rome—Bologna—Turin—The Tyrrhenians, or Etruscans—Their gold and silver coins—Principal mining regions—Piedmont, for gold; Sardinia, for silver; Temesa, for copper—The Salassi—Their conquest by Augustus—Ischia—Interdict of mining by the Roman Senate—Its motive—Its Buddhic or Bacchic origin—Enacted in Ancient India, China, Babylon, Sparta, and other states—Its probable æra—Mines of Aquileia—Blunders of Strabo—His auriferous Aquileia probably identical with Turin—The Italian agricultural skeleton—Devastation of Lombardy wrought by placer-mining—Necessity of artificial irrigation—Accumulation of gravel in the valley of the Po—Raising of its bed—Dykes—Floods—Present condition of gold-mining in Italy—A losing industry—Plundering more profitable than mining.

THE Durias Major (Doria Baltea) Durias Minor, Sessites, and other Alpine affluents of the Bodencus (Eridanus, Padus, or Po) and their valleys, were anciently auriferous; so were the Renus, Chiana, Arno and Tiber.[1] We have ourselves "panned" gold from the sands of the upper Tiber. From evidences which we observed in following the course of that river upward, from the ancient works that still exist in Rome to Mount Albanus and from the results of the dredging operations recently conducted in the river near the Capitol, we are rather inclined to the belief that Rome itself was originally a mining camp. It may also be conjectured that Bononia (Bologna) was originally founded in connection with the mines of Mount Albanus.[2] Turin was undoubtedly a mining camp. The earliest settlers upon the Italian streams, of whom any account has come down to us, were the Tyrrhenians, whom the Romans of a subsequent age called Tusci or Etrurians.[3] Among the cities built by this nation were Ar-

[1] Rhenus. The Italians call this river, the Reno. The Rhenus of the Menapians is now called the Rhine.

[2] Bononia. The Italians call this city Bologna. The Bononia of the Veneti, in Picardy, is now called Boulogne. The significance of these place-names has a bearing upon the history of the precious metals. They imply that, like America in the 16th century and Africa recently, Europe was first settled by races who were in search of gold. The gold veins of Mount Albanus were the sources of the auriferous particles won from the bed of the Tiber. Bononia and Albanus are names evidently bestowed by the ancient Veneti.

[3] Tyrrhenia probably is from Tyr, the Tuscan Mars, or god of war.

retium, Clusium, Cosa, Cortona, Fæsulæ, Graviscæ, Luna, Pisæ, Perugia, Populonia, Telamon, Turino, Veii, Venetia, Vetulonia, Villanova, Vollaterræ and Volscinii. We have Etruscan gold coins of Graviscæ, and Populonia, Etruscan silver coins of Fæsulæ and Populonia and Etruscan bronze coins of all these places. After the river washings and placers of Italy were more or less exhausted of their geological accumulations of gold, vein mines were opened by the Tyrrhenians, both of gold, silver and copper. Gold quartz was found chiefly in the Piedmont country, below the Pennine and Cottian Alps; silver was obtained chiefly from Sardinia; whilst copper mines were opened in Calori, Montegazza, in numerous places between Lucca and Volterra, in Temesa, Enna, (Castro Giovanni,) in Sicily and other localities. The Salassi dwelt at the source of the Doria Baltea, which rises in the foothills of the great St. Bernard. Says Strabo: (IV, vi, 7:) "The country of the Salassi contains gold mines, of which formerly in the days of their power, they were masters, as well as of the Alpine Passes. The river Doria afforded them great facilities for obtaining the metal, by supplying them with a water-head for washing the gold; so that they actually emptied the main bed by the numerous ditches they employed to draw the water to distant places. This operation, though profitable to miners, injured the agriculturists below, as it deprived them of irrigation from a river which, by its altitude, was capable of watering their plains. This gave rise to frequent wars between these tribes, of whom, when they fell beneath the arms of Rome, the Salassi lost both their gold works and their country. However, as they continued to hold the mountains (and therefore the river-heads) they sold water to the (Roman) lessees or contractors of the gold mines, (publicani,) with whom they had continual disputes, due to the avidity of the latter. Hence the Roman military commanders of the district always had a pretext for making war. Until recently, the Salassi, whether at peace or war with the Romans, found frequent opportunities, through their control of the passes, to inflict serious damage upon their enemies. They exacted from the followers of Decimus Brutus, on his flight from Mutina, (about B. C. 43,) a drachm, (denarius,) per man. Messala, likewise, having taken up his winter quarters in their vicinity, (about B. C. 27,) was obliged to purchase from them both timber for fuel and elm-wood for javelins. On another occasion they plundered (a portion of) Cæsar's (military) treasure, by rolling down huge masses of rock upon the soldiers, satirically pretending that they were building roads and bridges. Afterwards Augustus completely overthrew them and transported them to Eporedia (Ivrea) a Roman colony which had been planted as a bulwark against the Salassi, although its inhabitants had been able to do but little against them until the tribe was destroyed.[4] The numbers of the Salassi were 36,000; besides 8,000 men capable of bearing arms. Terentius Varro, the general who defeated them, sold them all as captives at public auction.[5] Three thousand Romans sent out by Augustus founded the city of Augustus (Aouste) on the site of Varro's camp. Now the country, even to the summit of the mountains, is at peace."

Strabo, (v, i, 12,) also mentions the placer mines of Vercelli. This camp was on the river Sessites, about 30 miles S. W. of Eporedia. He says: "The mines are not worked now so diligently, because not equally profitable with those of Transalpine Celtica and Iberia;" that is to say, of Gaul and Spain. They must have been productive during the interval between the Civil Wars and the Empire, for Pliny says (XXXIII, xxi, 12): "The publicani were forbidden by a censorial law to employ more than 5,000 men" in the mines. "Men," or "hominum," here means slaves. The text does not enable us to say whether each, or all, of the publicans (lessees) were forbidden to employ more than "quinque millia," but we take it that *each* is meant: and that there were several, perhaps many, publicani. Strabo alludes (v, iv, 9), to the "once productive gold mines" in the volcanic island of Pithecussæ (Ischia) outside the Bay of Naples. They were probably of comparatively little importance.

Pliny in two places (III, xxiv, 5 and xxiii, 21) alludes to an ancient Senatus Consultum which forbade mining for the precious metals in Italy. This, of course, was during the Commonwealth and more specifically during the fourth and third centuries before our æra, whilst the nummulary system of money continued to preserve its integrity.

As this interdict of mining is of great importance in the history of the precious metals, it may be of advantage in this place to allude to its origin and bearing more fully than has hitherto been done. There are numerous examples of laws and customs forbidding mining for the precious metals, both in Asia, Africa, Europe and America. The earliest of them all seems to be the Buddhic interdict, which is to be found, probably in altered form, in the Hindu or Brahmo-Buddhic

[4] Eporedia was on the Durias Major, about 30 miles from Bodincomagus on the Padus. The Augustan trophy which commemorated the conquest of the mountain tribes is dated Trib. XV, which, according to Ruggiero, agrees with A. U. 745-6. What a lesson for the Boers of South Africa!

[5] The slavery of the Salassi was to last for twenty years. Dio.

code of the Vinaya, containing rules for the discipline of the priests, although one of its parts, called Sila, (to learn,) refers to the moral duties of laymen. In the Vinaya the Sramanas (sense-tamers) are bound to observe 250 ordinances. Of these, ten are essential and constitute what may be termed the Ten Commandments of Brahmo-Buddhism. Of these commandments the last one relates to our subject. It is expressly forbidden "to receive precious metals." This includes both gold and silver. The ordinance appears to be of high antiquity: for traces of its observance are to be seen in all those states in which the Buddhic, (or Bacchic,) religion found lodgment. A similar ordinance appears in the ancient laws of China, Perak, Corea and Japan, in the laws of Sparta, (tempo Lycurgus,) in those of Babylon, and, as before remarked, in those of Italy, during the Roman Commonwealth. It was probably brought from Asia into Europe by the Phœnicians or Veneti. A like ordinance also appears in the observances of many of the African and American tribes, as will be seen on referring to another part of the present work, where the subject is dealt with more at length. But we take it, that among these instances of the interdict, some of them were derived from the Buddhic law, while others were of comparatively recent date and were due to the remembrance of cruelties exercised toward mining-slaves by modern nations; for example, the Spaniards, Portuguese and Dutch, in America. Similar cruelties were no doubt the cause of the original Buddhic interdict: but we are here endeavouring to distinguish between the abandonments of mining which were due to the ancient Buddhic interdict and the abandonments due to more recent circumstances. That the Buddhic religion found its way at a remote date into Europe is attested by innumerable circumstances, such as the use of Buddhic æras, place-names, the names of gods (*e. g.* Thammuz and Nissus, in the Hebrew scriptures,) the names of the months and days, the sacred symbols of the Phœnicians and Venetians; the explicit statements of Herodotus, Pliny and other ancient writers; the observance of Buddhic institutes; (such as the interdiction of ecclesiastical privileges, social caste and slavery;) the preservation of Buddhic rites and festivals; the appearance of Buddhic symbols on ancient monuments; and other matters mentioned in the researches of modern writers. Hence there is reason to believe that the interdictions enforced by Lycurgus and by the Romans of the Commonwealth, no matter what fanciful reasons have since been assigned for them, either by ancient or modern writers, were due in fact to superstitious veneration for that Buddhic or Bacchic religion which originated in

India and fonnd its way westward many centuries before the christian æra and centuries even before that second "incarnation" of Buddha, which the Cingalese are made to assign to the sixth century before our æra.

Nothing less than a law of high antiquity and clothed with sacerdotal authority will serve to explain the numerous instances of the voluntary abandonment of mining in very ancient times, which appear in other parts of this work; and nothing less than religious veneration can explain the aversion to mining for the precious metals which still prevails in Buddhic States, for example, in Ladak, Ceylon, China, Perak and Corea. Such being the case, and remembering Pliny's remark (III, xxiv, 5) that Italy is naturally rich in the precious metals, it is not too much to regard the ancient Senatus Consultum, in reference to mining for the precious metals, as having had a sacerdotal origin. Should this conclusion be verified, the interdict of mining mentioned by Pliny must be assigned to the earliest period of Rome, nay, it may even be carried backward to the æra of the Tyrrhenians or Etruscans, from whom Rome obtained much of its civilization and religion. On the other hand, should it be due to any other cause than rcligion, the Senatorial interdict will most likely have to be assigned to the period of the nummulary system of money.

Says Strabo (IV, vi, 12): "Polybius tells us that in his time, (about B.C. 140,) the gold mines were so rich about Aquileia, but especially in the countries of the Taurici Norici, that if you dug but two feet below the surface you found gold and that the diggings (generally) were not deeper than fifteen feet. In some instances the gold was found pure, in pellets about the size of a bean and which diminished in the fire only about one-eighth; in other instances, though requiring more fusion, the gold was still very profitable. Certain Italians aiding the Barbarians in working (the mines,) in the space of two months the value of gold was diminished throughout the whole of Italy by one-third."

The fragments of Polybius which still survive do not contain this passage, so that it cannot be verified by comparison with the original. However, there are several errors of fact in the quotation. There are no gold mines near the Aquileia of Strabo's time, a place situated near the mouth of the river Taurus, in Venetia. This stream has its source in the Carnicæan Alps, which separate its valley from Noricum (Corinthia). The country of the Taurici Norici was drained by this stream. It is destitute of any remains of gold mines of more importance than river-washings. However, there were Roman gold placers on the affluents of the Drave; and still further to the north-eastward there are extensive openings in the Carpathian mountains which were originally worked by the Iesiges Metanastes and afterwards by the Greeks and Romans. The writer has personally examined some of these openings and studied the geoiogy of the country, without being able to find in either of them any distinct corroboration of Strabo's statements.

As Polybius (II, 2) locates the Taurini (not Taurici) in the foot-hills of the Alps above the plains of Piedmont, it is included in the auriferous country already described. In case the Taurini were meant, (in place of the Taurici,) then the Aquileia of Strabo was the mining camp successively called Colonia Julia and Augusta Taurinorum, now the great city of Turin. Placer-gold, such as Strabo mentions of Aquileia, does not require to be submitted to fusion. It is impossible that pure placer-gold, such as he describes, should lose an eighth by fire. In B.C. 218 and in 206 the value of gold (to silver) was fixed by the law of Rome at 1 for 10; and this continued until B.C. 84, when Sylla lowered it to 1 for 9. At these rates it was bought and coined by the mints. This interval covers the period alluded to by Strabo, during which time it was just as impossible for the value of gold in silver to have diminished throughout the whole of Italy by one-third. as it would be to-day, through the discovery or working of any number of gold mines, however rich. If by "the value of gold" is meant the general level of prices, it should not be forgotten that the period alluded to by Strabo was that of the Second Punic War, when Hannibal was in Italy, in possession of the entire mining region of Piedmont and when great fluctuations of prices occurred, both from military, currency, and other causes, not connected with the production of gold. It is quite time that history and Strabo's oft-repeated error should part company.

No one who has examined the auriferous region of Piedmont and sailed down the great river of Italy can doubt that the former was mined and the latter deranged, by a people more ancient than the Romans. These could only have been the Tyrrhenians. But, if as we have herein ventured to surmise, the religion of this ancient people forbade the receiving of precious metals, how is such an hypothesis to be reconciled with their great activity in mining? Precisely as it is reconciled to-day in Tibet. The religions of this country are Dalai-Lamaism in Great Tibet and Brahmo-Buddhism in Little Tibet. Both of these religions are corrupted forms of Buddhism. In the last named region the inhabitants refuse to mine for gold: in the former, they evince no such aversion. The religion of the Basena, or Tyrrhenians, was also a corrupted Buddhism; for, with numerous undoubtedly Buddhic elements, it included caste and ecclesiastical privilege, both of which institutes were unknown to Buddhism proper. When the Romans conquered Tyrrhenia they adopted these and many other institutes of the conquered. When afterwards the Romans overthrew their own monarchy and erected a Republic in its place, they swept away privilege and caste, and reverted to the purer form of Buddhism, or Bacchism. This view would account for the Senatorial interdict of the precious metals.[6]

It must not be supposed that the Romans began their accumulations of the precious metals by mining for them. In fact they began by plundering them. It was not until after they had gathered all the spoil which the surrounding states of Italy afforded them, that they fell to mining; indeed, it was not until after the spoils of Tarentum had found their way to the jewelers, the temples and the grave, that gold-ming became systematic. But mining is slow work compared with plundering; and the Romans soon discovered the superior productiveness of the latter. In another portion of the present work the reader will be able to trace the measure of their success.

Formerly the government worked some of the mines, as those of Alagna in the valley of the Sesia, where there is a group of interesting veins; but gold mining in Italy has generally been left to private enterprise. According to the mining laws of Upper Italy, quartz mines, being considered *res nullius*, (belonging to nobody) are the object of concessions, granted gratis by the government, to those willing to take them, precedence being given to the discoverer. More than 60 localities are known where operations have been undertaken, generally on government concessions, but a greater part of these have been abandoned as unprofitable. There remain some 20 or 25 concessions in force; although the number of mines actually at work does not exceed 8 or 10. Alluvions of auriferous sands belong in Piedmont to the owner of the soil, a provision of some antiquity and one that was probably designed to save the agricultural bottoms from being covered with mining debris.

Allusion has been made to the deranged regimen of the river of

[6] For authorities on Etruscan mines and coins, consult Calori, in London Anthropological Review, (paper by J. Barnard Davis, 1863 to 1870); Montegazza, 1867 to 1872; Blacas' Mommsen, II, 213, 234, 372, 375, 379, 389; Dickinson's Mommsen; Burton's Castellien of Istria, Anthropologia, pp. 59 and 376; Denis, or Dennis, on Etruria; Isaac Taylor; R. S. Charnock; Livy, V, 33 and 54.

Italy. In the plains of Lombardy, ancient grape-vines have been found, buried in sand and clinging to elm-trees, many feet beneath the surface of the soil. In other places vast heaps of bricks, denoting the former level of the plains, have been unearthed many feet below the present surface. From the Dora Baltea to Ciglione, and from Milan to Pavia, there are enormous tracts of gravel. The valleys of the Sesia, Tesino and other affluents of the Po, are covered with gravel wastes, brought down in ancient times from the gold mines. Lower down, what the traveller sees is the loamy covering which human industry has spread over the entire plain of Lombardy. What he fails to see is a substratum of gravel, that drains the rain-water from the roots of plants and necessitates continual irrigation by artificial means. This gravel, from the gold mines of antiquity, is the skeleton that still lingers in the Italian agricultural closet. In the main river itself the accumulations of gravel are so vast that in some places the bed has been raised fifteen or twenty feet above the adjoining plains and the stream has had to be sustained by dykes of the most solid construction. In flood time the river hangs suspended over the cities and cultivated lands of this fair region, threatening both with destruction.

The present condition of mining in Italy can be disposed of very briefly. In 1869 there were twelve quartz gold mines operating in the province of Novara and two in Alessandria, besides four others that had been recently worked and abandoned. The fourteen going mines and their reduction works employed 722 artisans and labourers, at an annual cost of 302,421 lire, or francs. The total value of the gold produced was 236,331 lire; so that, as in many other gold-mining districts, the gold obtained cost, for labour alone, more than it was worth; to say nothing of the outlay for plant and other expenses. In 1877 the value of the annual product of gold, in Italy, was 257,400 lire; in 1884, 475,170 lire;[7] in 1900, 900,000 lire. Most of the mines are owned by English companies, which are managed, (or mismanaged,) in London.

[7] The Statistico Italiano for 1884 adds 590,000 lire worth of gold from amalgamating establishments; but this appears to be the same gold previously credited to ores. The value of the silver product of the year is given at 2,238,000 lire. In 1900 it was about 5,000,000 lire, coining value.

CHAPTER VII.

SPAIN AND GAUL.

Theories of Homer and Strabo regarding the discovery of Erythia, Iberia, or Spain —Its æra—Mines of the Guadalquiver—Conquest of Spain by the Carthaginians—Gold gravel mines—Water-wheels—Hydraulic tunnels—Silver mines of Turdetania —Tin—Gold placer mining distinguished the earlier, while silver mining marked the later Carthaginian epochs—Influence upon Carthage—Bebulo—New Carthage—Carteia—Capture of Saguntum—Second Punic War—Conquest of Spain by the Romans—Renewed attack upon its mines—Sacrifice of slave life—River washings, gravel-banks and quartz mines—Ancient gold mines of Asturias—Sierra Nevada—List of mines—Mode of draining deep quartz mines—Mode of working the gravel mines—Description of Las Medulas—Slaves—Hyraulic methods—Sluicing—Booming —Devastation described by Pliny—Modern mining in Spain—Mines of Gaul—River washings—Rhone—Isere—Hautes Alpes—Garonne—Finistere—Silver mines of the Narbonnais—Opulence of the Arverni—Mode of displaying it—Roman mints at Lyons—Coinage ratio—Gaul comparatively poor in mines—Tribute imposed by Julius Cæsar—His plunder of Lusitania and Gaul—Plunder of the Roman Treasury.

STRABO declares that the expedition of Ulysses was directed to Iberia and that the Odyssey and Iliad of Homer, though both are fabulous, were based upon facts gleaned from actual voyages of the Phœnicians to that country. In the same category of fables he places "the wanderings of Antenor and the Veneti." The last sentence appears to be a mere play upon the words Phœnician and Venetian. So far as can be ascertained, the Venetians, who had been driven from the Euxine to the Ægean and Adriatic, made voyages and left traces all along the Mediteranean, as far as the mouth of the Ebrus, which river they ascended, in search of the gold deposits of Asturias, Leon and Galicia. At a later period their voyages embraced the entire coasts of Spain. Some of the places they founded and named were credited by the Greeks to the Phœnicians (sometimes to the Phocæns), and among the "fables" built upon the assumed discoveries of the latter is here classed the only fact which is to be discerned in the whole matter, namely, the voyages of the Venetians.

"The Phocæns are the people who discovered the Adriatic and Tyrrhenian Seas and Iberia and Tartessus" says Homer. "I repeat," says Strabo, "that the Phœnicians were the discoverers (of those countries) for they possessed the better part of Iberia and Libya before the time of Homer, and continued master of those places until their Empire was overthrown by the Romans." If we regard the Venetians and Phœnicians as one people, all this hair-splitting disappears; and one they almost certainly were. For the purposes of the present work it indeed is of little consequence whether the Phœnicians or the Venetians discovered the mines of Spain and Gaul: so few details of their achievements have Greek and Roman perversion permitted to survive. As to the claim made on behalf of the Phocæns we consign it to the company of Geryon and his wonderful oxen.[1]

We learn from Stesichorus that the Guadalquiver, called by the Romans, Bætis, was known to earlier times as "The silver-bedded Tart-Essus," and that the country through which it flowed, indeed the entire Peninsular, was called Erythia. Tart-Essus is spelled by Eratosthenes as Tart-Essis. It may be a corruption of Taat-Essis. Erythia is probably a corruption of Eric-thia. Acquaintance by the Venetians or Phœnicians with the iron and copper mines of Rio Tinto, and the silver mines of Guadalquiver, betokens some earlier acquaintance with the gold alluvions of Spain.[2] In fact the gravel of all the rivers of Iberia and Lusitania, for they are all auriferous, must have been washed for gold by some people from Asia Minor not long after the period assigned to Troja Capta.

In the fifth century before our æra Spain passed into the military possession of the Carthaginians, from whom, through Greek or Roman channels, we gain details of the mines. Herodotus, Aristotle, Posidonius, Polybius, Strabo, Pliny and others have each contributed something to the subject.

Posidonius, B. C. 135–50, informs us that anciently in Spain silver was discovered accidently. The forests took fire and melted some of the ore in which the precious metal was imbedded. This is not at all improbable, for it has happened in other countries. Discours-

[1] Herodotus, I, 163, says that the Phocæns discovered Iberia.

[2] The Rio Tinto mines were worked by the Phœnicians, and afterwards by the Romans, Moslems, and Spaniards. In 1873 they were sold by the Spanish government to a British Company, who now annually extract one million tons of ores, two-thirds of which are quarried and the remainder taken from under-ground. About one-fourth of the whole product is exported, chiefly to England and the remainder treated at the mines. From the latter were produced in 1881 ten thousand tons of metallic copper, which was sent to England, and seven thousand tons of iron, chiefly to the United States. The mines and works employ 10,500 men, and a vast amount of machinery, including 25 locomotives, and six miles of precipitating tanks, etc. London Mining Journal, May 13, 1882.

ing on the diligence of the miners, he applies to them the remark of Demetrius of Phalaris, who, speaking of the silver mines of Attica, (Laurium) said the men there dug with as much energy as if they thought they could dig up Plutus himself.[3] But he regarded the Roman miners as much more efficient than the Greeks. The men who dug in Laurium were slaves and so were those who dug in Iberia. Their energy was not due to any thought of digging up Plutus, but to fear of the lash, which their cruel masters held over them. It was the Roman skill which guided them, that Posidonius contrasts with Greek ignorance or indifference. "In Turdetania," says Posidonius, "the streams are drained by means of Egyptian screws, which empty the water into deep and tortuous tunnels. These tunnels are cut into the auriferous gravel-banks so that the water may break them down." The writer saw one of these water-tunnels, still existing in the anciently worked gravel-bank of Las Medulas, in Leon.

The river Minho which rises in the mountains of Asturias and flows through Portugal to the sea, was the centre of very extensive mining operations on the part of the Romans. It drains the great placers of Gallicia, Asturias, Leon and Portugal, and is the river whose debris, according to Pliny, had shoaled the harbours and altered the sea coasts of Spain. Its principal affluents are the Sil, Duerna and Dourro, upon or near which are situated not only the alluvions above referred to, but also numerous quartz mines which were worked by the ancients and are now water-logged and abandoned to the bats and the snakes.

The Portuguese name of Minho and the Spanish name of Minio are from the Latin minium, or cinnabar, a red ore, from which is extracted the vermillion of the toilet-table and the mercury or quicksilver of the sluice-box. Technically, it is the red sulphuret of mercury, to which the Phœnicians or Venetians gave the name of sinople, sinoper, or sinopite, from Sinope, in Pontus, near which place they evidently had found deposits of this ore. This word has survived to the present time as a synonym for red lead. The river Minho was evidently given its name either because its waters were of a reddish colour, or else because deposits of cinnabar were found in the districts which it drained, both of which circumstances characterize the Minho. There was anciently a river Minio, now the Mignone, in Etruria, whose name was doubtless derived from similar circumstances.

The country drained by the Minho and its affluents bears marks

[3] Strabo, III, ii, 9.

of having been mined by successive races of people, Phœnicians, Carthaginians, Greeks, Romans, Arabs and Christians, each of whom had their own peculiar modes of working, so that the attentive observer is able to distinguish them. The difference between the Roman workings and those which preceded or followed them can be learnt with ease: but it is often difficult to decide concerning prior workings, whether they are Greek, Carthaginian or Phœnician. The presence of reptiles in the old quartz workings, which the writer examined near Ponferrada, rendered inspection difficult, but he is inclined to the opinion that these mines were worked for gold before the Roman period, by either the Carthaginians or Phœnicians, and possibly by both nations. On the other hand, the earlier marks upon the alluvions are effaced; those which remain being clearly Roman, medieval and modern. The ditch-line in slate-rock from Mount Teleno, the drift works on the Duerna, the gravel banks of Las Medulas and the booming apparatus on the Sil, are clearly Roman; the remaining works may be Arabian.

After breaking down the bank by the means above alluded to, the gravel was run into sluice-boxes and treated in the ordinary way. The dump was "boomed" off with the waters of an artificial lake. In the silver mines of Turdetania "one person has taken as much as an Eubœan talent;" meaning probably so much per day. If "talent" here means not a weight but a sum of money, it contained in Attica, in the time of Posidonius, 7620 grains of silver, equal in value, when coined, to 5 gold staters; each of the latter containing about the same quantity of fine metal as a modern guinea, or $5 piece: in short, $25 worth of silver. If a talent weight is meant, it equalled half a hundred weight of silver, or say double as much as $25 worth. In neither case is the statement incredible, but it proves nothing concerning the richness of the mines; for the poorest mines may contain lumps of native silver and the richest ones none at all. The copper ore of the Turdetanians, continues Posidonius, was one-fourth pure. Tin, he continues, is produced in Artabria (Gallicia) as well as in the Cassiterides. Thus far we can follow our author. But when he tell us that in Gallacia "the earth is powdered with silver, and tin and white gold, (that is, gold mixed with silver,) the gravel (containing it) having been brought down by the rivers," we prefer to draw the line.[4] Neither in Gallacia nor any where else was the earth ever powdered with silver, or tin, or white gold; nor have the rivers borne gravel containing electrum.

[4] Posidonius, in Strabo, III, ii, 9.

In the time of the Socratian dialogue, imputed to Æschines, a work of probably the fifth century before our æra, Carthage used a numerary money; after its conquest of Iberia, its mercenary troops were paid in gold coins. These circumstances indicate that Carthaginian mining in Spain was for a lengthy period chiefly placer mining. The conquest of this easily found gold "had bewitched her people, as it afterwards bewitched the Romans and as that of America at a later period bewitched the Spaniards. Nothing was thought of but rich mines, sudden wealth and the conquest of the world. . . . In B. C. 146, when Carthage is said to have been sacked and demolished by Scipio Africanus the Younger, all the silver found in the city (no gold at all is mentioned) amounted to a trifle less than two tons in weight, the net results of an empire, which, after having attained the ripe age of six centuries, obliterated itself forever through a mistake of monetary policy." [5]

In B. C. 238, Hamilcar Barca, without authority from his government, conducted a marauding expedition through Spain and sought to justify his conduct with the Carthaginian authorities by contending that the extension of their arms into the interior was necessary in order to make good the loss of the mines in Sicily, Sardinia, and Corsica; an argument that discloses the great importance which was now attached to the silver mines and the chagrin felt for the loss of those in the Italian islands. Strabo says that Hamilcar found that the Turdetanians used silver goblets and casks. If the remark is confined to the Turdetanian mine-owners we may credit the silver goblets, but the silver casks were probably flasks, or else they existed only in Strabo's imagination. However, it is safe to believe that Hamilcar seized them all. In 237 Hamilcar died, leaving to his son Hannibal, then but eight or nine years of age, the heritage of an undying hatred to Rome. In 228 the Roman frontier in the Narbonnais and Spain had advanced south-westward to the Iberus. Hannibal now strenghened himself in Spain by sacking the native cities and capturing the mines. Among others was a silver mine called Bebulo, which is said to have supplied him with 300 pounds weight of silver daily. This mine is now known to be the same with the modern Guadalcanal, in Cordova. There can be little doubt that the means employed for obtaining the silver was slavery. The inhabitants of the conquered cities were thrust into the mines and compelled to produce silver under the penalty of death, or torture. It was at this period that the silver mines near New Carthage were

[5] "History of Money in Ancient States," first ed., p. 175.

discovered and the city, or mining-camp, of that name was founded by Asdrubal, the brother-in-law of Hannibal. Asdrubal died in 220 and in the same year Hannibal stormed, sacked, and put to the sword, Carteia, a wealthy city and capital of the Olcades. This city was in the bay of Calpe (Gibraltar) nearly where stands the modern San Roque. It was founded by the Phœnicians or Venetians under the name of Tartessus[6] or as renamed by the Greeks, Heraclea; refounded by the Carthaginians as Carteia; and subsequently occupied by the Olcades, a native, or else a mixed race. Nearly two centuries after Hannibal's exploit Carteia was plundered by Julius Cæsar. The writer possesses four bronze coins, exhumed about the year 1830 from the ruins of this city, part of which ruins now lie under the waters of the bay. One has the legend "Irippo;" another, "Carteia;" one is two much worn to decipher; and one is of the mintage of Cæsar, about B. C. 50.

In the following year, B. C. 219, Hannibal captured and sacked Saguntum, now Murviedro, a Greek colonial city of Spain, sending the spoil captured here and elsewhere to Carthage, there to appease the authorities, win the populace, purchase supplies, and engage fresh troops for further operations. However, the capture of Saguntum afforded the Romans a pretext for hostilities; and in 218 began that memorable contest, the Second Punic War, which ended in 207 with the Roman conquest of Spain and in 202 with the downfall of Carthage.

Spain was to the ancients what Mexico and Central and South America became in later ages to the Spaniards; the Dorado, the richest mining country in the world, the place where, after the country was plundered, the metals gold and silver were found in the greatest abundance, and where they could be procured by the forced labour of captives and slaves. The fate of its aboriginal inhabitants, the subsequent struggles among the leading nations for the mastery of its precious metals, the destruction of its forests and diversion of its water-courses for the purposes of the mines, the consequent exposure of its soil to drought and devastation, the neglect of agriculture in the absorbing and senseless pursuit of gold and silver and the resulting poverty and backwardness of its population—both aboriginal and colonial—all these can be read by the nearer pictures which are accessible to us of Mexico and Peru.

At the period of the Roman conquest of Spain, the metallic deposits, except near the coast, had hardly been touched. The prin-

[6] Strabo, III, ii, 14.

cipal gold mining was still that of the placers. Says Strabo, III, ii, 8: "At the present day more gold is procured by washing, than by digging it from the (quartz) mines." Henceforth we shall find its mines, both alluvion and quartz, worked with the energy of madness. The valleys and bottoms of the Guadalquiver, [7] Guadiana, Tagus, Mondego, Duero, and Minho, the great alluvions of Gallicia and the gold alluvions and quartz mines of Turdetania, where Cæsar afterwards "filled his coffers," the mines of Bastetania, Cotinæ (Cotillas, or perhaps Constantina, near Almaden), and of Cordova, all were worked by slaves, whom the Romans sacrificed by the myriad. Strabo, who supplies some of this information in his Third Book, adds that "The gold mines nearly all belong to the State," whilst "the silver mines are no longer the property of the State, but of private individuals." Gold was coined exclusively by the State: silver was still permitted to be coined by the Gentes.

But although the gold mines were owned by the State they were in point of fact, during the decadence of the republic, worked by publicans or contractors, to whom the State farmed the privilege of working so-and-so many slaves upon the mines. [8] The profits made by these middle-men, most of whom were idle absentee "nobles," living in Rome, was in direct proportion to the cruelty of the agents employed by them to work the slaves. The sacrifice of human life was frightful: the profit enormous: the infamy eternal. The Spanish mines are graves in which were buried entire races. Their produce was those vast fortunes which were spent in gluttony, drunkenness and debauchery. It is hardly too much to say that conquest of Spain wrought the ruin of Rome in much the same way that the conquest of America afterward ruined Spain. It begat a necessity for further acquisitions of gold and silver, which in turn occasioned the neglect of agriculture and promoted a passion for empire. It was the demand for plunder that sent the armies of Rome to Asia, Greece, and Carthage, and the capture of plunder that built up those wealthy adventurers who, inflated with riches and power, demanded, like Flamininus, Scipio, Sylla, and Cæsar, to be worshipped as gods.

"Polybius, speaking of the silver mines of New Carthage (in Spain)

[7] The Tartessus (Strabo) is the Guadalquiver. In 1888 a great flood in this river destroyed many lives and a million pesetas worth of property in the province of Malaga. The storm was very severe in Seville. Was the derangement of the river due to ancient mining?

[8] Possibly the origin of the encomienda system, to which further reference will be made in the text. Under the Emperors, all of the gold mines and some of the silver mines were worked by the imperial fisc. Jacob, "Hist. Prec. Met."

tells us that they were distant from the city about 20 stadia (2½ miles) that they were very extensive, occupying a circuit of 400 stadia, that there were 40,000 men continually employed in them and that they yielded a daily revenue of 25,000 drachmæ."[9] The process was smelting and cupellation. If the produce had all been divided among the workmen, which is of course out of the question, it would not have amounted to more than five-eighths of a drachma per diem each, scarcely sufficient to keep them alive; to say nothing of the expense of machinery, tools, timber, lights, and other supplies, nor of transportation, management, taxes and mint-charges. Hence, it is quite safe to conclude, with Mr. Jacob, that "the silver produced must have cost more than its current worth." The fact that some people grew rich from these mines does not controvert this general statement.

Another neighbourhood rich in silver was Mons Argentorius or Silver Mountain, the Sierra Cazorla, in which the Guadalquiver, "the silver-bedded Tartessus," takes its rise. Montesquieu, in reading Strabo's account of this district, thought that it must have been to Spain what Potosí afterwards was to America; but there is scarcely warrant for this conjecture. With regard to the Pyrenean mines, between Spain and Gaul, Montesquieu states that in the war for the succession of Spain the Marquis of Rhodes, who had lost a fortune in gold mines and recovered it in the management of certain hospitals, applied to the court of France for leave to search the mountains for gold and silver; and that there he sank his second fortune; for he found nothing.[10]

Very extensive auriferous workings, presumably Roman, have been found in the quartz district in the Sierra Jadina, about 30 miles from the Talavera de la Reina highway and 20 miles from the Madrid-Lisbon railway. The auriferous seams are in ferruginous quartzite. They are usually thin, but in places swell to six feet in thickness. These mines were re-opened in 1855 and 1860, but both times at a loss.[11]

An abandoned Roman quartz mine, 75 feet deep to water, now called Monte del Oro, near Brandomil, about 18 miles from Coruña and the same distance from Corcubion, showed 2 to 2½ feet of very low grade ore, practically worthless; yet this hole in the ground was in 1885 rigged up by London promoters and sold to the public for £75,000; and when the enterprise failed, as fail it was bound to do, these Pharisees blamed the wicked Spaniards.[12]

[9] Strabo, III, ii, 10. [10] Michelet, Rom. Rep. 144.

[11] Private information from J. T. Browne, M. E.

[12] London Financial News, Feb. 21, 1887.

The Roman poets spoke of the Astures as a people of Spain who spent all their lives in digging mines. One might suppose from this that they took a pleasure in this occupation, but in fact, they were slaves, who were dragged to the Astures from many distant lands, from Britain, from Judea, from Mauretania; and forced to end their lives in the unrequited toil of the Spanish mines. [13]

Silver mines were worked by the Romans in Turdetania and at Ilipas, and Sisapo, in the Sierra Morena, also in Leon and the Pyrenees. Says Pliny: "Some have stated that Asturias, Gallicia and Lusitania furnish two thousand pounds of gold annually: but Asturias supplies the most; nor in any other part of the world during the past has so great a quantity been obtained. In every species of gold there is a proportion of silver; a tenth, a ninth, or an eighth. In the gold called albicavense, there is only one thirty-sixth part of silver, on which account it is more valuable than any other:" [14]—a statement which indicates that the art of separating gold from silver was either imperfect or expensive. Yet with all this wealth about them, the natives, like those of Sonth Africa to-day, were poor and miserable. In Bastetania (province of Malaga) they did not even enjoy the convenience of money, but exchanged their commodities for bits of silver, cut rudely from pieces of plate. [15]

It is much the same to-day. In the Sierra Nevada, less than a hundred miles from Malaga, there stands a mountain of auriferous material, which, notwithstanding that it has been worked successively by several ancient races, still contains millions of tons of auriferous gravel. The water supply is abundant, the dump capacious, the gradient ample, the climate charming and labour very cheap; indeed all the conditions of placer-mining are said to be favourable. Yet the mines have been long abandoned and the peasantry of the vicinity are miserably poor. [16] The reason is that, at the level of prices created by the plunder of America and the subsequent working of the American mines by Indian slaves, the mines of Old Spain have not paid to work. A similar condition of affairs occurred in Rome fifteen or sixteen centuries previously. There, the level of prices created by the plunder of Greece, Spain, Gaul, etc., had to be sustained by slave-mining. When that broke down, there was nothing left but retrogression, decay, and the Dark Ages. The modern world is more fortunate. It has been enabled at intervals to arrest the fall of prices with private paper-money and credit. It can never arrest the fall of

[13] Lucan, IV, 298; Silius Italicus, I, 231.
[14] Pliny, Nat. Hist. XXXIII, 23.
[15] Strabo, III, iii, 7.
[16] Skertchley's Report, 1894.

prices permanently without resort to a monetary system resting upon government paper-money and credit.

Shortly after systematic mining was begun by the Spaniards in America, the king of Spain issued an edict forbidding mining for the precious metals in the mother country. Sismondi in his "Underground" says that this edict was issued by Ferdinand; Roscher (I, 410) dates the edict in 1535, which would throw it into the reign of Charles I.; while Kilpatrick says it was issued by Philip III., in 1604, and that it remained in force until May 25, 1823.[17] It is probable that all of these authorites are correct, and that the edict was issued, or repeated, by all of these monarchs. Its object was to stimulate mining in America, and by this means to increase the king's share, which, originally, was one-fifth, (the Quinto.) Its repeal was due to the loss of the Spanish-American colonies. The repeal was followed by a mining mania in Spain, which Macgregor, writing in 1844, alludes to in vague terms, but which does not appear to have been followed by any practical results, so far as the precious metals are concerned.

Within the past forty years the total number of precious metal mines in Spain has been officially stated at 2274 silver and 6 gold, together 2280 mines: but this leaves out of account gravel mines and mines which have not been at all worked in modern times.[18] The enormous preponderance of silver mines shown in this statement lends great significance to that Roman legislation, which abandoned silver-mining and the privilege of minting silver coins to the Gentes. Of this great number of mines but very few are worked at the present time.

The chief impediment to mining in Spain—the same may be said of all the Spanish and Portuguese-American states—is due to that class of persons who, in the name of religion, levy a tribute upon all industries suspected of being profitable and especially those which are conducted in places remote from the protection of the civil law. Unless the local clergy and the provincial bishop are propitiated in advance, mining simply becomes impracticable, obstacles arise on all sides, the peasants will refuse to work, the gold will disappear and the miner will be fortunate who escapes without an accident. These results are not the consequences of any particular form of religion: they would follow any religion which kept the people in ignorance and reposed a large discretionary power in the hands of local curates.

[17] Engineer Kilpatrick, in a MS. report on the mines of the Rio Sil, about 1885.

[18] London Statistical Journal, vol., XXIII.

The current annual production of Spain is about six million ounces; about one-half of one per cent. of which is gold and the remainder silver. If modern mining methods and processes were introduced into Spain she might yet become once more one of the most productive countries in the precious metals; but if the true welfare of the country is to be consulted it were better that the ordinance of 1604 were re-enacted and made a part of the constitutional law. Active mining for the precious metals would rob Spain of the many advantages she has gained of late years in industrial habits, economy, and wealth.

The following is a partial list of the gold, silver, and quicksilver mines in Spain and Portugal, which are known to have been worked in ancient times:

District.	Province.	Character of Mines.
Sierra Nevada	Malaga	Gold gravel
Banks of the Duerna	Astorga, Leon	Gold gravel
Banks of the Sil	San Miguel, Leon	Gold gravel
Castropodame	San Miguel, Leon	Gold quartz
Banks of the Sil	Ponferrada, Leon	Gold gravel
Las Medulas	Las Medulas, Leon	Gold gravel
Banks of the Sil	Orense, Galicia	Gold gravel
Cantabrian Pyrenees	River Burbia	Gold gravel
Bebulo (Guadalcanal)	Near Cordova	Silver
Carthaga Nova	Carthagena, Murcia	Silver
Basin of Guadalquiver	River washings	Gold
Banks of streams	Near Cordova	Gold gravel
Guadalajara	Near Madrid	Gold gravel
Constantina	Andalusia	Gold & copper
Basin of Tagus	River washings	Gold
Sierra Morena	Mountains of Jaen	Silver
Banks of the Baza	Near Granada, Andalusia	Gold gravel
Sisapo, in Bætica	27 miles N. N. W. Seville	Quicksilver
Pyrenees Mountains	Aragon	Silver
Rio Tinto	Huelva	Gold gravel
Rio Tinto	Huelva	Silver
Rio Tinto	Huelva	Copper
Almaden	La Mancha (Al-Makrisi, p. 266)	Quicksilver
Banks of Burbia	Near Villa Franca	Gold gravel
Banks of Oria	Near Cuevas	Gold gravel
Banks of Eria	Affluent of the Duero	Gold gravel
Mt. Teleno	Near Las Medulas	Gold gravel
Jadeña [19]	30 m. from Talavera de la R.	Gold quartz
St. Domingo	Logroño, Old Castile	Gold quartz
Tamaya	(On the Tamega?)	Lock, cit. Livy
Douro	Near Oporto	Gold quartz

[19] The gold quartz mines of Jadena were worked by the Romans and probably also by the Saracens.

There were probably many others worked; but mining is a secret art and miners give as little information as possible to the public.

The St. Domingo mines of Spain were worked by the Romans in what may be termed a scientific manner. In some of them adits were excavated for draining purposes nearly three miles in length, whilst in others with deep shafts the water was raised by means of a series of gigantic wheels. Eight of these, all of them in a high state of preservation, have been recently found by the miners who are employed in workings contiguous to these old mines. The wheels are made of wood, the axles, where the greatest strain would come, being of oak and the other parts of pine. It is estimated that the wheels are 1,800 years old, although the wood is still in a perfect state of preservation, owing to the water being charged with metallic salts, including those of copper and iron. The wheels must have been worked on the treadmill principle, by men standing upon one side of them, the water being raised by one wheel into a large basin, and then lifted another stage by a second wheel, and so on in the same way by others to the surface. [20]

About the year A.D. 66 Pliny was appointed Procurator in Spain, an office intrusted with the management of the revenue. It is to be presumed that in such capacity he must have been more or less well informed as to the sources of revenue, and particularly that important portion of it which was derived from the gold mines. The guarded manner in which he states the output of gold furnishes ground for the supposition that, like America under Spanish rule, the Spanish mines yielded more than the Roman treasury received, or than was reported as having been yielded. He does not say that "the accounts in my office show the yield to have been so much," but that "according to some authorities" the annual yield from Asturias, Galicia and Lusitania, chiefly from Asturias, was 20,000 Roman pounds. In round figures and assuming the gold to have been impure, this was worth in modern money about $4,000,000. Pliny vacated his office as Procurator a little before A. D. 73 and in A. D. 79 he died from the effects of exposure and the sulphurous vapors emitted during that famous eruption of Vesuvius which buried for seventeen centuries the cities of Herculaneum and Pompeii.

Pliny wrote a great number of works, only one of which remains to us. This is his "Natural History," which is divided into thirty-seven books. The text of this work had become so much corrupted before the invention of printing, and has since become so muddled

[20] London Mining Journal, 1887.

through the ignorance of commentators who knew nothing of mining and mining terms, that nothing can be made of it without visiting the mines which he describes and seeing how the work was done. With this aid it becomes intelligible. Mr. Marsh, the gifted author of "The Earth as Modified by Human Action," supposed the hydraulic system of mining as practiced in California to be "substantially identical with that described in an interesting manner by Pliny the Elder;" but he was mistaken. The essential peculiarity of the hydraulic system is the breaking down of the auriferous gravel by powerful streams of water ejected through iron pipes; the Romans had no iron pipes and they did not practice the hydraulic system; but the system which they did practice was substantially the same as the hyraulic system except in the one important respect mentioned. How they broke down the gravel banks will now be stated on the authority of Pliny, backed by the evidence of the works themselves.

Says Pliny: "The third method of obtaining gold (he has already described river washing and quartz mining) surpasses even the labour of the giants. This is by the aid of galleries driven a long distance into the mountain, the men working by torchlight, the length of the torches regulating the shifts and the men never seeing the light of day for many months together. Not infrequently slides occur, the earth caves in and the men are crushed beneath." The text then diverges into a description of quartz mining and stamp mills. The galleries that Pliny alludes to were not driven into the mountain—*i.e.* gravel banks—but parallel to the faces of the banks. At Las Medulas these banks rise precipitously to the almost incredible height of 750 feet above bed rock. There is nothing like them in all California or South America. Generally speaking the gravel is not cemented, but it is so firm that the banks stand to-day quite perpendicular, although many centuries have passed since they were last worked. The object of driving these galleries was to undermine an up-and-down slice of the bank, precisely as gravel banks are undermined to-day by means of hyraulic jets and cross-jets. In this undermining work the hardest part was where cement was encountered. Pliny describes it as "a kind of potter's clay, mixed with gravel—*gangadia* by name—which it is almost impossible to overcome. This substance has to be attacked with iron wedges and and hammers, and it is generally considered that there is nothing more stubborn in existence, except, indeed, the greed for gold, which is the most stubborn of all things."

Having told us how the face of the cliff is undermined, Pliny goes on to state how the mountain, or rather the undermined slice of it, falls down, but here he goes too fast. He entirely omits to show what causes it to fall and this is precisely what the mountain stands ready to tell us to-day. On top of the bank, parallel to its face and situated at the same distance from it as the undermined gallery below, the workmen drove a trench or canal—in some cases merely a line of pits—and filled these with water. Now we can take up Pliny again: "When these operations are all completed they cut away the wooden pillars which support the roof [of the undermined gallery]. The coming downfall gives warning when the time comes and this is perceived by a sentinel, who is set to watch for it upon a suitable elevation hard by. With voice and gestures he directs the workmen in the open sluice-way far below to stand from under, and at the same time takes care of his own safety. The mountain, rent by art, is cleft asunder, hurling its debris to a distance with a tumult which it is impossible for the imagination to conceive, and from the midst of a cloud of dense dust the victorious miners gaze upon this downfall of nature." After describing how water is brought to the brow of these cliffs—of which more anon—and how the fallen earth is run through ground and wooden sluices, arranged with riffles of *ulex* to catch the gold, he follows the earth to the sea, and affords us a glimpse of the slickens question in the first century of our æra. Says Pliny: "The earth, carried downward by the rivers, arrives at last at the sea, and thus are the shattered mountains carried away. These works have greatly helped to extend the shores of Spain by their encroachments upon the deep. By the same agency the tailings of the quartz mines are carried away," thus adding to the sum of slickens.

There was a time when the Minho, Douro, Tagus, Guadiana and Guadalquiver all were navigable rivers, but owing to the operations of gold miners throughout many centuries, they have long ceased to be such. We shall not readily forget when we first caught sight of these gigantic and striking works of the ancient Romans. We had dined the previous evening at the wretched *posada* of Carrucedo. After dinner we started for the village of Las Medulas, intending to sleep there over night and accomplish the ascent of the gravel banks in the morning. From Carrucedo to Las Medulas is two miles and three-quarters; the ascent is from altitude 1500 to altitude 2225, a rise of 725 feet, or about one foot in twenty. The road winds through the "dump" of the mines, which is interspered with immense stacks of bowlders and cobble-stones, arranged with great precision and covered with the gray moss of ages. Between these stacks, the "dump" is now laid out in cultivated fields, where, under the shade of hoary old chestnut trees, a dwarfed and impoverished growth of wheat struggles toward the sun. The road is steep, but it is bordered with wild roses and lavender and one forgets the fatigue of the journey as he inhales their sweet perfume. Half-way up we stopped and gazed over the spacious fields which once formed the main sewer of Las Medulas and where myriads of Roman captives and slaves must have breathed their last in the unremitting and savage drudgery of the mines. Here were then assembled a vast host of unfortunates from all parts of the world, captives from Carthage and Judea, from Britain and from Egypt, flaxen-haired whites from the Baltic, wooly-haired negroes from Africa, freedmen from Rome (those who had been "ungrateful" to their patrons) and even legionary soldiers. This was the Siberia of the ancient empire, the golden prison beyond the sea, the "undiscovered country from whose bourne no traveler returned." Here all their laments were united; here all their hopes dissolved; here all their sufferings were entombed. The stacks of bowlders which they themselves erected are their monuments, the gray moss is their epitaph. So long as time shall endure will these remain; for so vast is their bulk that it would require another army of slaves to remove them. In something over an hour after leaving Carrucedo we reached Las Medulas. This pueblo consists of say forty or fifty houses, scattered about the upper portion of the ancient sluice-ways, which had many ramifications. The air of the place, when one does not venture too near a house or one of the inhabitants, is pure and refreshing. The women are the prettiest to be met with in Northern Spain. It may be very absurd, but we fancied that their countenances were of a Roman cast. Is it possible that the poor victims of the mines have transmitted their features through eighteen centuries of descendants? Spain is a very conservative country, and gold mines, even after they have "petered out," are hard things to abandon. Witness White Pine and Virginia, Nevada.

At Las Medulas the women, instead of wearing *madrenyos*, or wooden shoes, wore leather ones, as in more civilized countries. The men were almost as handsome as the women. Ramiro, the son of our hostess, a youth of 20 years, was as neat a looking fellow as one might find in the Abruzzi. Roman, too, oh, decidedly Roman. It was no longer a fancy—we were sure of it now. Some of the old legionary soldiers must have remained here with their families after the mines were abandoned by the Government, just as the Hessian soldiers settled in Pennsylvania after the Revolutionary war.

After a walk through the village which was just as dirty as the villages previously encountered, we put up at the only house where the traveler can find entertainment at Las Medulas—that of the widow Carlota Ascasvero Lopez, who prepared the supper. In this house the hearth was composed of a raised platform of slate slabs, framed or bordered by logs. The fire was built in the centre of this platform, and over it was suspended an iron pot. Long wooden benches were arranged on each of the two sides of the hearth, and here the family sat while the pot boiled. The smoke, as in the other houses of this country, found it way out the best way it could. There was no chimney. In the bedroom there was a cot, a chair, a table and a chest, all of heavy wood hewn with adzes. In the chest were kept the bed linen and towels. The washstand was a wooden bench on the porch. The fare at supper was coarse, but we fancied it was much cleaner than elsewhere throughout Leon. To bed at 9 o'clock and up at 4:30 o'clock in the morning, when, with Ramiro for a guide, we started to surmount Las Medulas.

It took an hour to reach the top, but the view well repaid the toil of the ascent. Towering up from the faintly descried village, their summits being the plateau upon which we stood, were the monstrous gravel buttresses of Las Medulas. Far below, on the right, wound the river Sil, on the left the Cabrera, and at their junction the town of Domingo Flores, bearing southwest by west. The intermediate country consists of lofty hills and broken ridges, destitute of trees. Near the edge of the gravel cliffs wound the road from Orellan to Buena Yrenes and as we followed this around and crossed a saddle that led from the summit of the gravel cliffs to the mountain behind them, we came in sight of the Roman ditches. Six of them are plainly discernible; some observers have made out a seventh. They are arranged one above the other, like so many zones carved upon the rocks. The lowest one was no doubt constructed first and the other ones subsequently, as a greater and greater altitude of water was needed to wash the mines.

These ditches are cut in slate, the stratification of which is on edge, as we have seen it in many parts Nevada County, California. That it was not difficult to cut, is proved by the fact that the cart-wheels on the Orellan road have cut deep ruts in the same rocks by merely passing over them. We jumped down into one of the ditches and examined it closely. It measures exactly five feet in width on the bottom. The width at the top is not so clearly defined, the elements having worn away the rock and broken down the outer bank, but we fancy it must have been over seven feet wide, with a rise of about three feet, exclusive of the bank formed by heaping up the excavated contents. The grade appears to be about twelve feet to the mile, although this could not be determined with precision without tracing the ditch for a much longer distance than we deemed it necessary. The grade appears to be uniform, and to make it so the Romans must have used some sort of leveling instrument. In fact, the work appears to have been thoroughly well done, and we take this opportunity to compliment the Roman engineer who executed it. Without the aid of gunpowder it seems wonderful that it should have been accomplished at all.

These ditches are twenty-six miles long, and as there are seven of them, the amount of rock-cutting was no less than 182 miles! The water was brought from the upper portion of the river Cabrera. Of course the ditches have no water in them at present, nor have they had for many ages, they having been broken down long since. But enough of them remains to evince the skill, the resolution and the indomitable energy of those who constructed them. Every inch of these 182 miles must have been chiseled out by hand. Myriads of lives must have been sacrificed in this work. And what remains of it all? A few coins of Octavius, of Tiberius, of Caligula, of Claudius, of Nero. One of these we secured. It is an *aureus* of Nero, and was found in the sluice-way of the mine. In weight and size it is something less than the sovereign or half-eagle. In design it is totally different, the coin being much thicker in the middle than at the edges.

Before descending from the summit of Las Medulas we measured with the eye the extent of ground which the Romans had worked out. This must be upwards of 1000 acres, and the average height of the worked portion ranges from fifty to 250 feet, say a general average of 150 feet. Only a small portion of the very high gravel banks—those ranging from 500 to 750 feet in height—was worked off by them. They left this portion to puzzle the engineering skill of a remote posterity. In some portions of these high banks are seen tunnels through which the Romans conveyed the ditch-water to the summit of the lower banks. These tunnels the Spaniards called *cuevas*.

Half an hour after leaving the summit of Las Medulas we had paid our bill and adieus at the village and were on our way back to Carrucedo. Below this village—between it and the river Sil—more wonders remained to be explored. To appreciate these, it must be premised that the immense mass of debris that flowed down from Las Medulas was arranged to fall into river the Sil, but that on its way thither it accumulated in vast heaps, which the water from the base of the cliffs was inadequate to carry off. In order to wash it away and carry it into the Sil, resort was had to the device known in California as "booming," and it is with the booming arrangements of the Roman miners that we shall close this long, but we hope not uninteresting description. Near Carrucedo and at an altitude of more than 1000 feet below the bed-rock at Las Medulas, there anciently ran a stream which emptied by an acute angle into the Sil. This stream, which we will call the Rio Carrucedo, was dammed up by the Romans at a point above Carrucedo and its waters were allowed to accumulate so as to form a lake, which still exists and is of enormous dimensions, several miles across. Below the dam they sank the river-bed fifty to seventy-five feet below its natural level; so that when the lake was let down, its flood might not damage the adjacent country. Into this sunken river-bed and at right angles to it, flowed the debris of the mines. Thence all the lighter portions were swept into the Sil, but the heavier portions remained and clogged up the sunken river-bed. To remove this deposit the floodgates of the dammed-up lake were periodically removed and the whole of its vast contents let down at once. The effect was not only to sweep away the deposit of bowlders and cobble-stones in the sunken river-bed and to carry them into the Sil, but afterward to urge them, together with the deposits of pebbles and sand in the Sil, down, down, down that swift river, until the entire mass was hurled through the Minho into the ocean.

You can see the marks of these "booms" to-day along the rocky walls of the Sil. The operation must have been indescribably awful. Imagine an impetuous rush of waters, 100 feet or more in height, charged with stones greater than cannon-balls, flying through a rocky defile, which in many places, is not more than an eighth of a mile in width! If the precision of the Roman rock-ditches commands our admiration, the magnitude of this booming apparatus must elicit something akin to reverence for its authors. One can scarcely regard one of our battle ships, the Brooklyn bridge, the Chicago water works, the hydraulic jets of California, without being impressed with some sort of reverence for the men who planned such gigantic works of art. When we say that they all dwindle into insignificance beside the colossal booming works of Las Medulas, it will begin to be understood what sort of miners the Romans were.[21]

The comparative poverty and backwardness of certain portions of Spain has been ascribed herein to mining for the precious metals, a

[21] The Author's letter in the San Francisco Chronicle, Sept. 23, 1883.

species of mining that from various causes, has always proved in the long run, to be unprofitable. It would be well if in view of these circumstances a notice somewhat like the following were printed and posted up in the public places: "This fair land was drenched in blood and its people enslaved by the Carthaginians and Romans in pursuit of the precious metals. It recovered only when that pursuit was forbidden by the Goths and followed but feebly by the Arabians. With the Discovery of America and the conquest and loss of the colonies, the same senseless pursuit was resumed, to the destruction of Spain's timber lands, the diversion of its waters courses, the ruin of its agriculture, and neglect of the arts. It is only since the independence of the Spanish American colonies that Spain has had an opportunity to recuperate. But a vicious system of money still debars her from that more rapid progress to which the genius of her people entitles her. Let them abjure this system of money. Let them improve upon the example of their Gothic ancestors and entirely interdict the use of gold and silver for money. Let Spain fabricate her own money. Let the state furnish such money as commerce needs in order to maintain stable prices and stimulate trade. Let her emancipate herself from the designs of usurers and the dangers of foreign intrigue and she will become once more, what she has been more than once, a foremost power in Europe!"

Gold mining in Gaul, as in all other countries, commenced with the river washings. The Cote d' Or on the Rhône derives its name from this fact. The Isère and other affluents of that river were once rich in gold, but this has long since ceased to be the case. The silver mines of L' Argentiere in the Hautes Alpes are known to have been worked by the Romans and such was also probably the case with the mines of La Gardette and Allemont, which yielded a small amount of gold and silver: but the period of their opening is uncertain. The later history of all these mines is alluded to in the first edition of Del Mar's "History of the Precious Metals." Cape Finistère "possesses more valuable silver and lead mines than any other department in France,"[22] but this is not saying much; for the silver mines of Gaul were not important. It is not known if any of the Finistère mines were worked by the Romans. One called "Huelgoat" is said to have been discovered in the last century and worked until 1866, when it ceased to pay. It was then permitted to fill with water, which reaches to the surface. It is 600 feet deep. The auriferous river Auriegne, in Languedoc, is mentioned by Boisard 1,

[22] Malte Brun, Geography.

267–72. In 1548 a gold mining fever broke out in France.[23] This lasted until 1602 and possibly longer. Gold mines were opened or re-opened in the Pyrenees, the mountains of Foix, in the Lyonnais near St. Martins and in La Brie, and Picardy. Silver mines were opened in the Pyrenees, Carcassan, the Lyonnais, Normandy, La Brie and Picardy.[24]

The Garonne was auriferous in ancient times, but we have no record of its produce. The Cevennes Mountains produced both gold and silver in small quantities. Says Strabo: "The Tecto-Sages, a tribe of Aquitaine, dwell near to the Pyrenees, occupying a small district on the north (northwestern) flank of the Cevennes, a land rich in gold." It was equally entitled to be regarded as rich in silver; for when the Romans under Caepio plundered Toulouse they claimed to have obtained "15,000 talents, a part of which was hidden in the chapels (the natives were worshippers of the Matrem Deorum) and the remainder in the sacred lakes. This was not coined (money) but gold and silver in bullion."[25] . . . "When the Romans obtained possession of the country they put up these lakes for sale at public auction, and many of the purchasers after draining them found therein solid masses of silver. In Toulouse there was a temple held in great reverence by the people and on this account loaded with riches, the offerings of the pious which no one dared to touch," until the Romans pillaged it. "The Tarbelli (another tribe) of Aquitaine possessed the richest gold mines: masses of gold as big as the fist, and requiring but little purifying, being found in shallow diggings, together with pellets and gold dust." . . . . The Ruteni and Gabales, tribes of Aquitaine, inhabiting Rouergue and Gevaudan, on the northern flanks of the Pyrenees, near Narbonne, possessed silver mines. The whole of Aquitaine, from Marseilles and Narbonne to the Loire, seems to have been subject to the Arverni, a nation differing in many essential respects from both the Celts and Belgæ. "Their king, Luerius, the father of Bituitus, who fought against Maximus Æmelianus and Domitius Ænobarbus, exhibited his opulence by scattering a profusion of gold and silver coin from his chariot, as he drove across his native plains." A similar display of opulence was made in the same vicinity so late as the 13th century.[26]

Strabo also informs us that at Lugdunum (Lyons on the Rhône)

[23] Calvert, "Rocks of Great Britain," p. 201.

[24] La Septennaire, I, 109, cited in Sully, III, 56*n.*, ed. 1770.

[25] Posidonius in Strabo, IV, i, 13. The Mexicans also sank their treasure in a lake.

[26] Del Mar's "Ancient Britain," p. 158.

there was a temple dedicated to Augustus by all the Galatæ. We can add to this that there was also a temple dedicated to Augustus at Vienne on the same river and quite near Lyons, which temple is still standing and that its dedication can still be read. The Roman prefects of Lyons, continues Strabo, coin both gold and silver money. To this statement, it may be added, that the gold coins were struck at an imperial mint and under the express authority of the sovereign-pontiff, who never permitted any infraction of this prerogative. The silver coins may have been struck either for account of the hierarchical fisc, or for the proconsul, or for others. Both of the last named practices were in vogue. As elsewhere stated, the ratio of value between silver and gold coins, from the reign of the first Cæsar to the downfall of the Empire in A. D. 1204, was always 12 for 1 in weight.

Upon reviewing the statements in Strabo, Pliny and other ancient writers, and comparing them with what subsequent explorations have shown, it is evident that Gaul was comparatively poor in mines of the precious metals, especially silver; and that what the Romans obtained of these metals in that country was the accumulation of ages rather than the current product from mines. The conquerors tore the golden torques from the necks and arms of the vanquished chieftains, they robbed the women and children of their petty trinkets, they even dug into the graves of the dead and despoiled them of the amulets which love and piety had consigned to an eternal repose. But nowhere do we find any evidences that the mines of Gaul yielded them anything more than a meagre return for the wars which were waged upon its poorly-armed and defenceless inhabitants, nor for the myriads, nay the millions, of lives that were sacrificed in the Roman searches for gold. As for the cost of the metal to the Romans: it was only the cost of cutting so many throats. In money, it cost practically nothing.

Suetonius, in his life of Julius Cæsar, says that: "Having renounced all hope of obtaining for his province, Egypt," (a gold-mining country,) "he stood candidate for the office of chief-pontiff. At the expiration of his prætorship he obtained by lot Further Spain and pacified his creditors, who were for detaining him, by finding sureties for his debts." Further Spain was Hispania Bætica. He got this office B. C. 61. According to Appian, he owed, when he went thither, to use his own words: "Bis millies et quingenties centena millia sibi adesse oportere, ut nihil haberet," *i. e.*, he was nearly 2,020,000 sesterces worse than penniless.

"Being now supported by the influence of his father-in-law, Lucius Piso, and son-in-law, Cneius Pompey, he made choice, among all the provinces of Gaul, of the one most likely to furnish him with means and occasions for triumphs. At first, indeed, he received only Cis-Alpine Gaul" (the gold regions of Italy) "with the addition of Illyricum" (another gold region) "through a decree proposed by Vatinius to the people; but soon afterwards he obtained from the Senate, Gallia Comata," (the most auriferous region of France) "the senators being apprehensive that if they refused it to him, it would be granted to him by the people," (in the Comitia.) It will be observed that Cæsar successively tried to get control of all the great gold regions, or rather what were then esteemed to be great gold regions, namely: Egypt, Spain, Piedmont, and Gaul; and that he actually got three out of four of them. In Gaul he imposed "an annual tribute of forty million sesterces," after plundering the country of all the treasure it contained: an intolerable exaction.

"With the money raised from spoils of war, he began to construct a new Forum, the ground plot of which cost him forty million sesterces." Upon entering Rome after the Civil War: "To every foot soldier in his veteran legions, besides the 2000 sesterces paid him in the beginning of the Civil War, he gave 20,000 more in the shape of prize money. He likewise allotted to his veterans, lands, but not in continuity, that the former owners might not be entirely dispossessed.[27] To the people of Rome, besides 10 modii of corn and as many pounds of oil, he gave 300 sesterces per man, which he had formerly promised them and 100 more to each for the delay in fulfilling his promise. He likewise remitted a year's tax due to the Treasury for such houses in Rome as did not (could not afford to) pay above 2000 sesterces taxes a year; and throughout the rest of Italy, for all such as did not exceed in taxes 500 sesterces each. To all this he added a public entertainment and a distribution of meat and—after his Spanish victory," (B. C. 46) "two public dinners: because, considering the first one as too sparing and unsuited to his profuse liberality, he, five days later, added a second one, which was most plentiful. It is likewise reported that he invaded Britain in the hopes of finding pearls." His favourite mistress was Servilia, the mother of Marcus Brutus, "for whom he purchased, in his first consulship, after the commencement of their intrigue, a pearl which cost him six millions of sesterces. He practised no abstinence in pe-

[27] These appear to be what would now be called feudal tenures. See, on Roman feudalism, Del Mar's "Ancient Britain."

cuniary matters, either in his military commands or civil offices; for we have the testimony of some writers that he extorted money, from the proconsul who was his predecessor in Spain, and from the Roman allies in that district, for the discharge of his debts; and that he plundered at the point of the sword some towns in Lusitania, notwithstanding they attempted no resistance and opened their gates to him upon his arrival before them. In Gaul he rifled the temples and chapels of the gods, which were filled with rich offerings," (as Cortes afterwards did in Mexico and Pizarro in Peru) "and demolished cities oftener for the sake of their spoil than for any ill they had done. By these means, gold" (meaning no doubt jewelry-gold and therefore alloyed or impure gold) "became so plentiful with him that he exchanged it throughout Italy and the provinces, for 3000 sesterces the pound. In his first consulship he purloined from the Capitol 3000 pounds weight of gold and substituted for it the same quantity of gilt brass. For gold he also sold to foreign nations and princes the titles of allies and kings; and in the name of Pompey and himself he sqeezed out of Ptolemy alone nearly 6000 talents. He afterwards supported the expense of the Civil Wars and of his triumphs and public spectacles, by the most flagrant rapine and sacrilege." Suetonius omits to add that besides the 3000 pounds weight of which Cæsar robbed the Treasury during his first consulship, he afterwards plundered it of 15,000 pounds weight of gold, 30,000 pounds weight of silver, and 300,000 sesterces in coin. [28]

There is a class of pretended philosophers who never condescend to read history or to study law, who are ignorant of mining, who know nothing of metallurgy, who could neither mix an alloy, nor fabricate nor assay a coin, yet who know all about the precious metals and their influence upon the economy of the world. These sciolists reply to all questions by means of a formula. The formula is "cost of production." If you ask them what was the value of the gold of which Cæsar robbed the temples of Spain and Gaul, their reply will be "the cost of its production." That cost was in fact some millions of human lives, some rivers of human tears, some oceans of human blood, some immeasurable measure of human anguish. Such was the cost of gold to Rome in B. C. 48.

The precious metals production of Gaul or France in modern days has been inconsiderable, and is limited at the present time to about half a million ounces of silver; no gold at all being found.

[28] Pliny, XXXIII, 17.

CHAPTER VIII.

PLUNDER OF EUROPE BY THE ROMANS.

Lack of a satisfactory history of the Roman Commonwealth—Defects of Belot's work—False principles deduced by Adam Smith—Early history of Rome—Civil Code—Corporations—Monetary System—Its beneficent working interrupted by the Oriental Conquests of Alexander and Seleucus—Rome led by their example—Roman conquest and plunder of Carthage, Spain, and Greece, B. C. 300 to 150—Religious impostures employed in these wars—Rise of prices in Rome—Its important consequences—Halcyon Age—Elevation of religious ideals—Legislative reforms—Celebrated jurisconsults—East Indian trade—Shipping—Military engines—War horses—Organization and strength of the army—Arbitration—Treasury—Trade guilds—Public works—Edifices—Aqueducts—Post-houses—Canals—Canal locks—Sewers—Roads—Theatres—Basilicæ—Mechanical arts—Chariots—Harvesters—Threshers—Steel plows—Steel drills—Silos—Carriages—Embroidery—Linen underwear—Musical organs—Public bakeries—Sulphur matches—Stamp mills—Mine-pumps—Booming apparatus—Mica window-panes—Loadstones—Polarization of iron—Steam engines—Sphericity of the earth—Measurement of a meridian—Armillary sphere—Gnomon—Charts—Sun-dials—Clocks—Plants, fruits, vegetables, and flowers—Rise of the medical profession—Literature—History—Agriculture—Philosophy—Oratory—Rhetoric--Commercial institutes—Banks—Safe deposits—Accounts—Tablets—The drama—Poetry—Sudden arrest of the elements of progress—Fall of prices—End of Halcyon Age—Civil wars—Fall of the Commonwealth and Rise of the Empire.

WHEN Augustus destroyed the ancient literature of Rome he committed to twenty centuries of oblivion the most prosperous age of the Commonwealth. In vain did the court annalists substitute the fabulous adventures of Romulus, borrowed from Fabius Pictor and Diocles; in vain did the court poets invent the myths of Æneas and Anchises; in vain have modern historians, following these misleading guides, sought to distil from fables the sources of Roman progress and of Roman power. The history of the Commonwealth has yet to be written; and to him who shall body it forth from the scant materials that time and superstition have left unmarred, there will probably appear no more brilliant period in the annals of man than the one which is made the subject of the present sketch.

In a recent work, the author, Mr. E. Belot, alludes in its title to what he terms the "Revolution economique et monetaire qui eut lieu a Rome au milieu du IIIe siecle avant l' ere chretienne;" but while he fully accounts for this great social revolution he omits to adequately describe it or paint its consequences. He claims that in consequence of the vast spoils of gold and silver which emptied into Rome from the conquest of the Levant and Orient, prices rose in a comparatively brief period to ten times their former height; so that what was previously bought with an ace, had now to be purchased with a denarius. But he neglects to show to what tremendous social results this rise in prices led. Yet it is in these results, rather than in the mere rise of prices by itself, that posterity is chiefly interested.

It has been claimed by Dr. Adam Smith and it is still claimed by his school (that is to say, the duke of Buccleugh's school) and by the commercial communities who have been unconsciously trained in that school, that when money is made of the precious metals prices cannot be raised by increasing the currency, because redundant coins will be exported, or melted down into the arts. Their view is that it is not the Volume of money that influences commerce: but commerce that influences the Volume of money. This subject has been fully discussed in another place. [1] Dr. Smith's doctrine is not only opposed by the events which followed the Plunder of America and by many other events of like character which have happened since that time: it is contradicted by the very examples which that learned but sophistical Scotchman adduced in its support.

During the early Commonwealth the political affairs of Rome were managed with a degree of skill and prudence that wholly belies the story of her evolution from a band of Latin or Alban adventurers. Rather does it point to an organized colony from some old and well ordered State, such as the Spaniards sent to Florida or the French to Louisiana; a colony carrying from some mother country a long established Religion and Code of Civil Laws. Rome's peculiarity as a colony chiefly consisted in the early establishment of a Code of Civil Law and an ideal Monetary System; not the unwieldly copper ingots which bespeak political senility and which Pliny borrowed from the Etruscans and blunderingly fitted to a newly seated community, but such a system as Plato dreamed in his Model Republic; such a system in short as a model colony would most likely wish to try in a new Elysium.

This system has been fully described elsewhere. In brief, it was an emission of State notes forming a measure of value—(mensura publica)—the notes being printed (stamped) on copper bronze owned by the State, and issued by and for the purposes of the State, which regulated the volume of money and by that means ensured the stability of prices and the equitable diffusion of wealth and advant-

[1] Del Mar's "Science of Money," ed. 1900, chap. XVI.

age. The copper bronze pieces were greatly overvalued, they were executed with the highest resources of art, they were stamped with the emblems of government and religion, and it was made a sacrilege to counterfeit them. Somewhat similar systems had been tried before in Sparta, Byzantium, and Clazomenæ, but from one cause or another they had failed. The Romans, guided by Plato's masterly advice on the subject, were to try again and succeed. [2]

Gibbon, though he may not have grasped all the sources of Rome's prosperity during the Commonwealth, nevertheless, thus tersely sums up the result: "From the establishment of equal freedom to the end of the Punic Wars the City was never disturbed by sedition and rarely polluted with atrocious crimes." [3] It was her Numerical System of Money that enabled Rome not only to establish civil freedom, but to maintain it, and to keep herself, so long as she preferred such a course, free from foreign entanglements. It was this system that preserved the equality of fortune which distinguished her citizens in the earlier ages of the Republic. And it was this system that placed her always in a position to take advantage of the blunders or misfortunes of her neighbours. The history of Rome, down to the period when the whole of Italy was added to her sway, is that of a State, or of a people, consisting of various antagonistic elements, yet acting with a harmony which is inexplicable under any other than a Numerical system of money. The solidarity, the discipline, the unselfishness, the patriotism, the stubborn courage and exalted firmness exhibited by the Romans in the memorable series of conflicts, in which she engaged with her neighbours, are inconceivable under political systems which inequitably or unequally dispense the favours of fortune. Such qualities have never since been exhibited by any other people: not even by the Romans themselves. And it is safe to predict that they will never again be exhibited until a like Institution of Equity is made to form the foundation of the social and political structure.

Unfortunately, the very success which resulted, as we believe, from the working of this Institution, indirectly proved to be the cause of its downfall. Alexander and Seleucus had plundered India and Persia: for a time they had cut off the Oriental trade of Egypt and Carthage. The Greek States, though enriched with the spoils of the East and in possession of the gates to its lucrative commerce,

[2] The experiments of the Greek States and Colonies with Numerical moneys and the Numerical System of Plato appear in the author's "History of Monetary Systems."

[3] Gibbon, "Decline and Fall," ch. XL.

could not be brought to unite together. Here was Rome's opportunity, not merely to conquer the Levant, but the World. It was this attempt that brought into her coffers the accumulated spoils of ages and it was out of these spoils that a Metallic system arose to the prejudice and ruin of the Numerical. Another circumstance which contributed to the downfall of the latter was the presence of Hannibal's army in Italy during the Second Punic War. "Soldiers on furlough," (says Livy, XXIII, 29) "generally carried money in their purses for the purpose of trading." With a considerable portion of Italy in the hands of the enemy and the very existence of the Republic in jeopardy, it is evident that the soldiers could buy no supplies from the peasants with Numerical money. When the State is in danger of falling, every institute must correspondingly fail which depends upon the credit of the State. With Hannibal's army at the gates of Rome, the Numerical System virtually came to an end.

Before we enlarge upon the Plunder of Europe, there is another Roman institution whose regulation exercised a tremendous influence upon her affairs and which deserves some mention in this place. The creation of Corporations both religious and commercial was an ecclesiastical device which can be traced back to the period which invented the mythical Numa. During the fourth century before our æra, the Roman people evidently suffered from the existence of these artificial bodies; for among the measures of this period, B. C. 306, was an act which swept out of existence both religious and commercial corporations, new or old. [4] It must therefore be understood that during the Halcyon Period which we are about to describe, there were neither Corporations nor Trusts to rob the private citizen of the fruits of his industry, or to screen him from that personal responsibility which more than aught else promotes commercial integrity, stimulates commercial enterprise, and furnishes the foundation of commercial credit. From this digression we return to the Plunder of Europe.

"Between the years B. C. 320 and B. C. 250 when they defeated Hasdrubal, the Romans conquered the whole of Magna Græcia and Sicily, together with the islands of Sardinia, Corsica and Malta. The defeat of Phyrrhus and Hasdrubal, both of whom carried valuable military chests, the spoil of the rich city of Tarentum and its 14 subject towns, the plunder of the silver mines of Sardinia [5] and sack of the opulent Carthaginian towns of Sicily, brought a considerable

[4] "History of Monetary Systems."

[5] The author's "History of the Precious Metals," ed. 188c, pp. 7 and 21.

quantity of silver into Rome; and as this metal seems to have been especially coveted by the victorious commanders for their personal share of the spoil, it fell chiefly into their hands, or that of the patrician families to whom such commanders belonged." [6]

To those who brought large spoil into the Roman treasury was awarded the glory of a triumph, or such other distinctions and advantages as the State could afford; to those who brought none, was awarded obscurity or disgrace. Under such stimuli the Roman commanders vied with each other in filling the public coffers. Nothing was permitted to stand in their way. Whole nations were put to the sword: neither age nor sex were spared; death, fire and pillage spread from Rome to the confines of the Levantine world; whilst back to Rome, as to a common centre, flowed the blood-stained metals whose acquisition involved this fearful havoc. There was but one object in the soldier's mind; that object was gold and silver, no matter at what cost of dishonour or crime. The worst passions were let loose. Cupidity reigned paramount. Altars were raised in the cities and little children were taught to kneel and pray to Jove predatori—the god of plunder. [7]

In B. C. 293 Papirius Cursor bore in his triumph over 2,000,000 Roman pounds weight of bronze, and 330,000 pounds weight of silver. [8] It was probably this spoil and that of the following twenty years, when Pyrrhus was defeated, that led to the new coinage of silver in B. C. 269.

In B. C. 241, at the close of the First Punic War, Lutatius Catulus obtained, under treaty with Carthage, a bond covenanting to pay 2200 Attic talents talents of silver annually for 20 years. In addition to this, the Roman Comitia demanded 1000 talents which Carthage consented to pay down. [9]

At that period, when the coining press was unknown, the work of coinage was done altogether by the hammer, shears and file. A workman could scarcely finish more than twenty coins per day. Owing to this circumstance, though bullion was plentiful, coined money at times was scarce; and we read in Livy, of an occasion, B. C. 218, when temporary loans were resorted to by the government to meet the immediate demands of the troops. This difficulty had to be surmounted by adding as rapidly as possible to the active force of moneyers.

In B. C. 212 M. Manlius Marcellus captured Syracuse, with im-

[6] The author's "History of Money in Ancient States," p. 246. [7] Blanqui, 48.
[8] Livy, x, 46. [9] Lenglet de Fresnoy, I, 426; Polybius, I, 62, 69, 88; III, 27.

mense spoil of gold, silver, statues, paintings, and other works of art. [10]

In B. C. 209 Fabius Maximus retook Tarentum, with 87,000 pounds weight of gold, a vast quantity of silver bullion and coins, thousands of statues and pictures and 30,000 slaves. [11]

In B. C. 208 Publius Cornelius Scipio captured in Spain and carried to the treasury in Rome 14,342 pounds weight of silver and of coined silver, "a great sum." [12]

In B. C. 202, at the close of the Second Punic War, Scipio contributed to the Roman Treasury the bond of Carthage to pay 10,000 talents of silver within 50 years, besides 25,000 pounds weight of silver down and a vast number of ships, elephants and implements of war. At his triumph he covered into the treasury 123,000 pounds weight of silver. [13]

But these were the least of Scipio's spoils. Before he defeated Hannibal at Zama he had overrun the whole of Spain, which of all the countries of the then known world was richest in the precious metals and in slaves to work them. Bebulo alone had furnished Hannibal with 300 pounds weight of silver daily. What it was made to yield under the stimulus of the encomienda system and the Roman lash we are not informed. [14] In his trial, which took place in B. C. 190, Scipio expressed his indignation in being accused of embezzling four million sesterces, after having paid, as he said, 200 millions into the treasury. Such was the public gratitude to Scipio for bringing this long war to a conclusion, such the elation created by his brilliant success and the triumph that marked it, and such the effect of the example of Alexander the Great, who had set himself up for a god, that Scipio, too, claimed the honours of divinity. This impious pretension was made in B. C. 201, during the consulate of Cn. Cornelius Lentulus, one of his own gens and P. Ælius Pætus, his intimate friend. Happily for Rome, it still had a Cato to expose imposture and punish impiety. [15]

In B. C. 199 Lucius Manlius Acidinus returned from his proconsulship of Spain and covered into the Roman treasury 1200 pounds weight of silver and 200 pounds weight of gold. [16]

[10] Livy, XXV, 40; XXVI, 21; Dunlop, II, 157. [11] Livy, XXVII, 16.

[12] Livy, XXVIII, 38. [13] Livy, XXX, 37.

[14] The Spanish-American encomienda system of the 16th century consisted of grants of vassals (or slaves) to the conquerors and there is reason to believe that its origin was Roman. It is fully explained further on in the text.

[15] The account of Scipio's imposture is given by Aulus Gellius, out of Oppius and Hyginus. Alexander was deified in B. C. 332. [16] Livy, XXXII, 7.

In B. C. 196 Titus Quintius Flamininus returned from the conquest of Thessaly and Epirus with an enormous quantity of "armour, weapons, and bronze and marble statues, of which he had taken greater numbers from Philip, than from the States of Greece. On the second day (of his triumph) appeared the gold and silver, wrought, unwrought and coined. Of unwrought silver there were 18,000 pounds weight and of wrought, 270,000 pounds, consisting of vessels, (mostly engraved,) several of exquisite workmanship; also many others of bronze, and ten shields of silver. The coined silver amounted to 84,000 Attic tetradrachmas, containing each four denarii. Of gold there were 3714 pounds (weight) and one shield of solid gold, and of gold Philipusses (coins) 14,514. On the third day were carried 114 gold crowns presented by the various States." [17]

In B. C. 192, two years before the trial above mentioned, Scipio concluded a treaty with King Antiochus, which provided that the latter should pay 15,000 Eubœan talents of silver; 500 down, 2500 upon the ratification of the treaty by the Senate, and 1000 annually for twelve years. It was especially in relation to this transaction that Scipio was afterwards charged with embezzlement. [18]

In B. C. 192 Manius Acilius returned from his proconsulship of Ætolia and covered into the Roman treasury "3000 pounds weight of silver bullion, 113,000 Attic tetradrachmas, 248,000 cistophoruses and, of chased silver vessels, a great number. He bore also the King's plate, furniture and splendid wardrobe, also golden crowns and presents from 45 allied States, together with spoils of all kinds." [19]

In B. C. 189 Marcus Fulvius returned from the conquest of Ætolia and Cephalenia with 243,112 pounds weight of gold, 83,000 pounds weight of silver, 118,000 Attic tetradrachmas, 12,422 gold Philipusses, besides other spoil, among which were 285 bronze statues, 230 marble statues, together with arms, weapons, catapults, ballistas and engines of various kinds. [20]

In B. C. 186 Cneius Manlius captured from the Gauls of Asia (the Galileans) and brought to Rome, 4603 pounds weight of gold, 220,000 pounds weight of silver, 127,000 Attic tetradrachmas, 250,000 cistophoruses, 16,320 gold Philipusses, together with abundance of Gallic arms and spoils in chariots. [21]

"This army, returning from Asia, was the origin of foreign luxury imported into the City. These men first brought to Rome gilded couches, rich tapestry, with hangings and other works of the loom,

[17] Livy, XXXIV, 52. [18] Livy, XXXVIII, 38, 55. [19] Livy, XXXVII, 41.
[20] Livy, XXXIX, 5. [21] Livy, XXXIX, 7.

and, what were then deemed, magnificent furniture, buffets and single-footed tables. At entertainments likewise were introduced players on the harp and cimbrel, with buffoons for the diversion of the guests. Their meats began also to be prepared with greater care and cost; while the cook, whom the ancients considered as the meanest of their slaves, both as to value and use, rose into great estimation. Nevertheless these instances of extravagance, as they were then deemed, were no more than the seeds of that greater luxury which was afterwards to spring up."[22]

In B. C. 182 Terentius entered Rome in ovation for his success in Hither Spain, carrying 9320 pounds weight of silver, eighty pounds weight of gold and two golden crowns weighing 67 pounds.[23]

In B. C. 181 Lucius Æmilius Paulus triumphed over the Ligurians and carried 25 golden crowns in the procession. He distributed 300 aces to each soldier.[24] This must have agregated an immense sum.

In B. C. 179 "Sempronius Gracchus triumphed over the Celt-Iberians and their allies; next day Lucius Posthumius Albinus triumphed over the Lusitanians and the other Spaniards in that vicinity. Tiberius Gracchus carried in the procession 20,000 pounds weight of silver, and Albinus 40,000. They distributed to each of their soldiers 25 denarii, double to a centurion, triple to a horseman; and the same sums to the allies as to the Roman."[25] These payments are colossal.

In B. C. 167 Lucius Æmilius Paulus having completed the conquest of Macedonia, brought to Rome in gold and silver no less than 120 million sesterces, besides immense spoil of furniture and works of art, the gold and silver alone filling 150 wagons.[26] Besides this, he levied tribute on the inhabitants of Macedonia and appropriated all the gold and silver mines of the kingdom.[27] In a single hour, says Plutarch, and by a preconcerted signal, this plunderer sacked 70 cities and reduced 150,000 Greeks to slavery. Such was the magnitude of the spoil that, after rewarding the entire army, enough remained to absolve the citizens of Rome from the payment of taxes until the period of Julius Cæsar.

In B. C. 169 Lucius Anicius conquered Illyria and brought to

[22] Livy, XXXIX, 6.

[23] Livy, XL, 16. So says the text; but the meaning is probably that the two crowns contained as much gold as there was in 67 "libras" of gold. This would make the crowns weigh about five Roman pounds each. Cf. "Hist. Monetary Systems," p. 57.

[24] Livy, XL, 34. [25] Livy, XLI, 7. [26] Livy, XLV, 40. Blanqui, 49.

[27] Arnold's, Hist. Rom. I, 17: Hist. Money Anc. 282 n. These were the spoils of Asia which had been brought into Macedon by Alexander the Great,

Rome 72 pounds weight of gold, 19 pounds weight of silver, 3000 Roman denarii and 120,000 Illyrian denarii, besides other valuable spoil. [28]

These instances might be greatly multiplied; but it is unnecessary. All who have written on the subject, both Romans, medievals and moderns, have united in testifying to the vast flood of wealth, especially of the precious metals, which flowed into Rome after the campaigns of Alexander and Seleucus and especially between the first and third Punic wars. "So abundant was money at that period (B. C. 186) that the people assessed themselves for a contribution to L. Scipio to defray the expenses of the games which he celebrated." [29] The following passages appear in Livy; either upon his own authority or that of Cato the consul, or Lucius Valerius, the tribune: "The public prosperity is daily increasing." . . . . "The Commonwealth daily grows more prosperous and happy." . . . . "We have begun to handle even royal treasures." . . . . "The Commonwealth is flourishing and happy." . . . . "Every class of persons, every individual, feels the improvement in the State." . . . . "The Oppian law (restricting the use of gold in the arts, etc.,) was repealed within twenty years after it was enacted." [30] Both Tiberius, Livy and Tacitus trace to this period the birth of luxury; Pliny curses both the gold and silver that in the end brought in their train so many evils to Rome; Bodin, Davenant, Barthelemy, Garnier, Letronne, Guérard and Lacour-Gayet descant upon the enormous spoil and the increase of money which marked the period; whilst Dupré de St. Maur, Rome d' Lisle, Dureau de la Malle and Belot attempt to measure the rise of prices and trace some of its effects upon the body politic. However wrongly gained, this much must be said of the Roman spoil: it was widely and rapidly diffused. Neither crown nor church possessed at that period, as they did at later ones, the power to take the first and best fruits of the spolia opima. The public requirements were first satisfied, the gods were appeased with sacrifices and games, the temples were propitiated with generous donations, the taxes were remitted and the remainder of the spoil was distributed to the soldiery. Ten, twenty, forty, sixty, and sometimes an hundred aces, sesterces, or even denarii per man, were successively paid to the entire army.

In B. C. 269–8, according to Pliny, a new coinage of silver was authorized; in B. C. 207–6 a new coinage of gold was begun. In these coinages the ratio of silver to gold was 10 for 1. A Roman

[28] Livy, XLV, 43. [29] Pliny, N. H. XXXIII, 48. [30] Livy, XXXIV; year B. C. 193.

pound weight, 5250 English grains, of pure gold, contained about the same quantity of metal as is now contained in $225 of the United States, or (roughly) £45 sterling. A pound weight of silver was worth by law one-tenth as much. The purchasing power, or value of gold and silver coins (as against commodities) was far different from what it is now; but this is a matter with which we have no present concern.

To measure the rise of prices which took place during the century that intervened between the fall of Tarentum and the destruction of Carthage is no easy matter. The chief difficulty lies in the fact that in this interval Rome changed her money from bronze aces, owned ab initio by the State, to silver sesterces, which originally were private property: whilst the relation between them, not as to the separate pieces, but between the total sum or volume of the denominations of one money, as against the other, is unknown. The next difficulty is that the ancient authors quote but few prices. Finally, the prices of wheat and other breadstuffs, which these authors have preserved, exhibit so little variation, as to excite the suspicion that they are not natural but artificial, that is to say, that they were fixed either by maximum law or monetary regulation.

In B. C. 456 Manius Marcius, ædile of the people, was the first to supply Rome with wheat-corn at one ace the modius or peck.[31] In that year and in B. C. 427, 408 and 252, the price of wheat was always one ace the modius. During the intervening periods we have no quotations. So unvarying a stability of prices, if such it was, might have resulted either from a maximum law, or the sale by the government to the public, of wheat at a fixed price, irrespective of cost. This policy belongs to the Imperial and Dark and Middle Ages, when the peasant was a slave and agriculture unprogressive, rather than to the Commonwealth of Rome, when the peasant was the peer of any man and agriculture furnished sur-names for the families whose subsequent nobility grew out of its abundant returns.[32] It is related that at the Council of Frankfort, A. D. 794, Charlemagne proposed to permanently fix the price of grain at four deniers the boisseau, so that it would remain the same in times both of abundance and scarcity.[33] On the other hand, stability of price can be secured by regulating the volume of the currency, the Nume-

[31] Pliny, XVIII, 4. [32] Pliny, XVIII, 3.

[33] The boisseau of Paris was equal to about ⅓ of a bushel; that of Bordeaux, about 2⅙ bushels, indeed it varied in every province of France. What it was as given in the capitularies of Charlemagne is uncertain. Garnier, p. 47.

rato; and there are many proofs that such was, in fact, the device employed by the Romans. In the earlier history of the Republic the agricultural class was too powerful to permit of corn being sold by the State to the public at less than cost; and in view of this circumstance, it may be safely concluded, if corn fetched no more than an ace the modius, that such was its full value. When the Numerical System of Money was overthrown, when silver money came into common use (B. C. 269) the price of wheat rose to double. In B. C. 205, it was 2 aces the modius; in 202, it was 4 aces; and in 201, it was 2 to 2½ aces the modius. There are no futher quotations of grain within the period embraced by this sketch. [34]

The Rise in Prices during the third and second centuries has been traced, by Dureau de la Malle, Garnier and Letronne, in the price of a day's labour, in the pay of soldiers, the cost of slaves and other data to be found in ancient authors; but the quotations are too few and far apart to afford any safe general conclusions on the subject. Belot's comparison of the census classifications of private fortunes, of the fines imposed by the State for public offenses, and of the rewards paid for public services has led to more satisfactory results, although we cannot altogether follow his view of the rise of prices. He is of the opinion that between the first two Punic wars prices in Rome rose ten times. Our own conclusion are that the influx and coinage of gold and silver during the century and a half under review resulted in an augmentation of prices amounting to not less than 2½ fold, (a new sesterce for an old ace) and possibly to as much as five fold, or two sesterces for an ace. The proofs of such

[34] In B.C. 154, spelt was one ace the modius (no quotations of wheat); B.C. 126 (Lex Sempronia) wheat was supplied to the populace at a fraction under the ancient price, viz., at five-sixth of an ace the modius, evidently under cost. This was one of the celebrated agrarian laws of the Gracchi. It was repealed five years later by the Lex Octavia. In B. C. 74 (Lex Cassia) wheat was purchased by the State at 4 sesterces the modius and a second quality at 3 sesterces. At the same period 4 sesterces was the price in Sicily. Cicero, contra Verres, III, 10. In B. C. 65, (Maximum law), wheat was fixed at 2 sesterces the modius; after the great fire in Rome, A. D. 64, Nero "reduced" the price of wheat to tres nummos (three aces) the modius (Tac. Ann. XV, 39); in the time of Martial, say A. D. 73, wheat was 4 aces the modius (Ep. XII, 76); in the reign of Constantine wheat was 10 sesterces the modius. Many works have been written on the subject of prices in Rome by people who knew little or nothing of its Numerical system of Money; and who therefore were unable to reconcile the apparent contradictions of the classical texts. From one monetary theory Mommsen deduces 6⅓ aces, whilst from another theory Belot deduces 16 aces, for the modius of wheat, in the time of the Gracchi. Garnier admits a "monnaie de compte" during the Republic, but fails to comprehend its relation to the Metallic System which supplanted it. Cf. Pliny, XVIII, 4; Gibbon II, 417*n*.

rise of prices do not rest entirely upon the quotations of commodities, though these alone are perhaps sufficient to establish our conclusion; they rest also upon that notorious increase of luxury which furnished a theme for every Roman moralist from Cato down to Tactitus; whilst they derive further corroboration from the continually increasing rewards of valour, of industry, and, it may be added also, of imposture.

A miracle occurred, such a miracle as without the aid of superstition had never happened before and has never happened since. A petty state of Italy, unknown to Herodotus or Thucydides, and scarcely mentioned by any Greek writer previous to the æra of Alexander, [35] a state which, if its own annals are to believed, had fought for centuries to maintain an obscure but sturdy independence in the valley of the Tiber, suddenly became a continental power, and still more suddenly turned from the mastership of the Italian peninsular to assume that of world. Many ingenious theories have been put forward to account for this miracle, but none of them have proved quite satisfactory. What enabled a few thousand Spaniards to to conquer the whole of civilized America? The belief of the Americans in a Messiah that was to come, their hierarchical government, their feudal condition, their internal dissensions, their institutions of domestic slavery. Both Columbus, Cortes, Cabeça de Vaca and Drake lent themselves to the grave imposture of pretending to be that expected Son of God to whom the natives of America had been taught to yield every obedience and concede all the rights of property. [36] The Tlascalans were employed by the Spaniards to subjugate the Mexicans; the slaves who should aid the Spaniards were promised freedom; the lure of the precious metals heated every Spaniard's mind to phrensy while it chilled his heart to stone. Such were the forces which won America for the Spaniards; and such, also, it can scarcely be doubted, were the forces that won Carthage, Spain, Greece, and Asia for the Romans. Scipio was not the only military commander who claimed the honours of messianic divinity. There were others, both before and after him. Alexander borrowed the

[35] Herodotus, who travelled in Italy about B. C. 460, makes no mention of Rome; neither does Thucydides, who described the great war between Athens and Sicily, B.C. 414. It is first mentioned by Theophrastus, B. C. 370–287. Thompson's Hist. Rom. Literature p. 330.

[36] For pretensions and worship of Columbus, consult Irving's life of the great captain; for Cortes, see Help's "Spanish Conquest in America," ed. 1854, II, 321, 359; for Cabeça de Vaca, see his voyages "Naufragios" c. 31; and for Drake consult "Mineral Report of the United States" ("blue book") for 1867, p. 271. A similar pretense appears in the legend of Hiawatha.

imposture from India and carried it back to its source; Titus Quintius, Sylla, Sertorius, and others repeated it and spread it over the Empire from Armenia to Gaul. With regard to slave emancipation, the Romans in foreign wars always carried the insignia of Liber Pater; whilst in peace and in Rome they hid them away. [37] They conquered Spain with Gauls, Carthage with Spaniards and Numidians, and Greece with the aid of the helots. In the mad hunt for gold and silver they stripped the living, they even tore from the bodies of the dead their heritage of these ominous metals. With such alluring rewards for valour their armies never lacked recruits. Rome's Halcyon period began substantially with the spoil of Tarentum; it ended substantially with the spoil of Asia Minor and Macedon. After that period and until Cæsar overran Gaul the influx of stolen metal fell away; prices ceased to rise; and neither valour, industry, nor imposture earnt more than a scant reward.

The conquest of Italy, Carthage and Asia Minor brought to Rome not only the most precious products of learning and art, it brought also the scholars and artists. Moreover it drew to a common focus the philosophy and religious speculations of the world. The characteristic features of Christianity, its broad philosophy, it recognition of equality before the law and the right to liberty, its consideration for women and children, these were not born under the Empire. Their germs can betraced in Rome back to the third century before our æra and in Greece and the Orient to still earlier periods. They distinctively belong to the Halcyon Ages of Greece and Rome. The elevation of religious ideals is the work of freedom and prosperity; never of slavery and adversity. Abundant supplies of money no sooner made themselves felt in the daily traffic and exchanges of the Romans than the whole body politic awoke to new life, both civil and religious. This is examplified in the course of legislation which during the prevalence of the Numerical system was chiefly in the direction of securing for the people (plebs) increased consideration in the State; such as the right of appeal to the Comitia, the appointment of Tribunes to check the authority of the Senate, the requirement that the Senate should approve the acts of the Comitia and that the latter should bind all classes of the people. The authorization of marriages between the classes and masses; the admission of the plebs to the sacred rites; their privilege to elect one of the consuls and certain proportions of the other magistrates; the repres-

[37] During the Second Punic War they liberated and enlisted their own slaves. Livy, xxxiv, 6.

sion of simony; the publication of the laws, and of the forms of procedure and the calendar; the forbiddance of excessive usury; the repression of bribery; the interdict of attachment upon the person for debt; of imprisonment except by virtue of legal process and prescription; of tampering with the public records; of mortmain; and of binding, scourging, or killing a Roman citizen; all these are reforms which belong to this Halcyon Age. The summit of progressive legislation and series of reforms was reached, when, in B. C. 254, the learned and pious Titus Coruncanius, a plebian, was elected Pontifex Maximus. The Jus Ælianum by Sextus Ælius Catus, securing to the people a knowledge of the Code of Procedure, which notwithstanding the previous laws on the subject had hitherto been mystified by the Church and the Senate, was published in B. C. 201.[38] The adoption of the Rhodian maritime code, distinct traces of which are to be found in the laws of the Roman republic, completes a magnificent edifice of Civil Rights and jurisprudence, whose erection began during the æra of the Numerical System and was continued to the midst of the Halcyon Age which sprang from the plunder of Europe.

The great jurisconsults of the day were Titus Coruncanius, B. C. 254, M. Portius Cato, 234–149, Sextus Ælius Catus and Sextus Pætus, who was consul in 188, the son of that always faithful follower of the Cornelian family, Publius Ælius Pætus, consul B. C. 201.[39]

That knowledge of the Orient which the Romans acquired by the conquest of Tarentum and through their alliance with Hiero II., of Syracuse, B. C. 263, was greatly increased after the capture of that great city by Marcellus in B. C. 212. The Indian trade was at that time largely conducted by the way of Syracuse and Alexandria, but after Syria became a Roman province, it was again carried across the Desert. Some evidence of the Indian religious myths which reached Rome anew at this period will be found in the proceedings of Scipio already mentioned.

In a few years after their conquest of Sicily and Greece the merchant ships of the Romans were to be seen in all the ports of the Levant. The first Roman fighting ships were probably built in B. C. 337, but the first organized fleet was built under L. Valerius Flaccus, B. C. 261, the second in 256, the third in 254, the fourth in 250, and the fifth in 242.

These five fleets comprised nearly 1000 ships. They were all built, equipped and manned within the period under review.[40] Great im-

[38] Dunlop, II, 136. [39] Cicero, Brut. [40] Lenglet de Fresnoy.

provements were also made in battering rams, balistæ, catapults, and other engines and weapons of war. The number and breed of war horses was also much improved. In B. C. 225 (Æmil. Paulus and Att. Regulus, coss.) and upon the report of a rising in Cisalpine Gaul, Italy alone was able to enlist and arm 80,000 cavalry, besides 700,000 foot. [41]

With rising prices every department of human activity was set in motion: all sorts of reforms came to the front and were no sooner discussed and found to be meritorious than they were adopted. Governmental institutions kept pace with conquest and trade. International arbitration was initiated by the Treaty between Scipio and Antiochus.[42] In B. C. 182 the Romans acted as arbiters between Numidia and Carthage. [43] The public funds, hitherto stored in the temples, were now deposited in the Treasury (ærarium); and the first account of them appears in B. C. 157, when they consisted of 17,410 pounds weight of gold, 22,070 pounds weight of silver, and 6,135,000 sesterces in coins. [44] These sums contain the same weight of metal (reckoning silver as 10 for 1 of gold) as $5,463,605, or about £1,100,000. Additional trade guilds seem to have been formed at this period. Numa is credited by Plutarch with having established nine, viz., the Braziers, Potters, etc. In B. C. 493 the guild of Merchants was incorporated. [45] During the period under review were added the Fullers and several others. [46]

The Public Works erected in Rome during the Rise of Prices were both numerous, splendid, and durable. In after times Augustus boasted that he had found a city of bricks and had left one of marble. Fabius Maximus could have boasted that he found one of boards and left one of bricks. Before the conquest of Tarentum most of the private houses of Rome were of wood, roofed with shingles. They were soon afterward built of bricks and stone and roofed with tiles. [47] The Appian duct, the first acqueduct which brought water to Rome, was begun by Appius Claudius Cæcus, B. C. 313; the second, called the Anio (Vetus), which brought the water 43 miles, was commenced by M. Curius Dentatus, B. C. 273, and the third, called the Aqua Tepula, was constructed B. C. 127. The notion has been often advanced by modern writers that the Romans were ignorant of the phenomenon of hydrostatic pressure, the tendency of water to seek its level, and the qualities of the syphon. All these

[41] Pliny, III, 24. (20.) [42] Livy, XXXVIII, 38. [43] Livy, XL, 17.
[44] Pliny, XXXIII, 17. [45] Livy, II, 27. [46] Pliny, XXXV, 57.
[47] For shingles, see Pliny, XVI, 15, (10) quoting C. Nepos; for tiles, see Vitruvius and Dr. Adam.

phantasms have been dispelled by the discovery by Father Secchi of an aqueduct built at Altari, in Italy, about B. C. 200. Says the discoverer, in a letter to the French Academy of Sciences: "It is an inverted syphon, the lowest depression being 101 metres (338 feet) below the point from which the water flowed into the town. At the lowest point it sustains a pressure equal to eleven atmospheres. The pipes are of earthen-ware imbedded in a heavy work of concrete. The length of the pipe line is eleven and a half miles. The whole is still in a good state of preservation." [48] A service of posts followed the construction of the great highways. Ambassadors from the government of Rome were entitled to demand "a horse at each of the towns through which their journey lay." At a later date, B. C. 173, the consul Lucius Postumius Albinus, being ordered by the Senate to fix the boundaries between the public and private lands in Campania, required the authorities of Præneste to provide him with both relays and lodgings. [49]

A machine for measuring and marking long distance levels was certainly in use when the auriferous gravels of Leon and Galicia were first worked by the Romans, and this is not likely to have been much later than the period their acquisition, B. C. 206—for one of these deposits was supplied with water fetched from a distance, over hill and valley, of 28 miles. The means of conveyance was an open ditch line, cut for miles through laminated slate and complemented with high tressle-work flumes, or wooden viaducts, which bridged over the depressions in the line. This work (the ruins of which were personally inspected by the writer) was done with a precision that is impossible without the aid levelling instruments. The use of such instruments is more than implied in Pliny the Younger's letter to Trajan, where he suggests the employment of locks for a canal near the city of Nicomedia, and writes to Calpurnius Macer for a water-engineer (Librator) to decide upon the proposed improvement. [50] The land engineer was called Agrimensor. [51] Pliny's letters only carry the Librator and his instruments of precision back to the imperial age: the works still standing on Mount Teleno carry them back to the æra of Rising Prices. The great Sewers of Rome, the first of which is credited by tradition to the Tarquins, were increased during the Rise of Prices to the number of seven, all built of solid masonry and together completely draining the site of the metropolis. [52]

The magnificent paved Roads, which diverged from Rome in

[48] Trans. Fr. Acad. of Sciences, before 1887. [49] Livy, XLII, 1.
[50] Pliny, Ep. X, 50, 69. [51] Pliny, Ep. V, 15. [52] Pliny, XXXIV, 24; Livy, XXXIX, 44

every direction, some of which, after 2000 years of service, are still in use, were begun by Appius Claudius in B. C. 313. The first road, the Via Appia, originally went only a little beyond Terracina; in B. C. 185, during the Rise of Prices, it was extended to Formiæ and Capua [53] and eventually to Brundusium, a distance of 350 miles. The Via Numicia went also to Brundusium; the Aurelia ran along the coast of Etruria: the Flaminia was built by Caius Flaminius in B. C. 220 and extended to Ariminum, now Rimini, a distance of about 220 miles, where it was afterwards joined by the Via Æmilia; the Cassia went between the Aurelia and Flaminia and ran through Etruria to Mutina; the Æmilia was built by the Consul Marcus Æmilius Lepidus, B. C. 187 and ran from Ariminum to Placentia. Besides these there were the Præenestina, Triburtina, Laurentina, Salaria, Latina, and others of lesser note.

The Decennovium canal, 19 miles long, from Terracina to Appia Forum (through the Pontine marshes) was begun in B. C. 312, but it is evident that its completion must have awaited the extension of the Appian way. In fact, the canal was substantially the work of Cornelius Cethegus, B. C. 160. To use the pithy words of Gibbon, it was afterwards ruined, restored and obliterated. This canal is mentioned by Strabo. Lucan, Dion Cassius, and Cassiodorus.

The first regular theatre at Rome was constructed for Livius Andronicus, on the Aventine, about B. C. 230: the theatre of M. Æmilius Lepidus was built in B. C. 179: the theatre demolished in deference to the pious scruples of P. C. Scipio Nasica, was constructed in B. C. 152. [54] In B. C. 185 Cato the Elder erected a spacious Court of Justice: in B. C. 179 the Basilica Fulvia was erected; and numerous other temples first reared their heads to the sky in this Halcyon Age of Rome.

The arts and sciences exhibited the same sudden and noticeable activity. Pliny's remark that "the Roman people has never shown itself slow to adopt all the useful arts," must have been intended to apply especially to this period. [55] The war chariot armed with scythes first appeared B. C. 281. [56] After the close of the Second Punic War, when estates in Italy had become too large for petty culture, a modification of this machine was used as a harvester. [57] A threshing machine (tribula) was also invented at this period. The steel plough was in use when Cato the Elder wrote. The seed drill was in use about B. C. 175; silos for preserving grain were employed about the same

[53] Livy, XXXIX, 44. [54] Livy, Ep. 48; Dunlop, I, 338. [55] Pliny, XXIX, 7.
[56] Putnam, Ency. Chron. [57] Pliny, XVIII, 72.

time; the rheda, afterwards called the carruca, was a four-wheeled carriage whose lightness enabled it to be driven over the great highways at top speed. This was also a product of the period. The carrus, or curricle, is mentioned by Plautus in Trinummo, III, 4, a comedy which was performed at Rome shortly after the close of the Second Punic war. Sheffers' work is full of allusions to the improved agricultural and vehicular machines of this period.[58] The Oppian law of B.C. 214 referred to "carriages drawn by horses" and in the debate upon its repeal, B.C. 193, the wives of the Latine confederates were depicted by Lucius Valerius, the plebian tribune, as "riding through the city in their carriages, decorated with gold and purple."[59] Embroidery was first made at Rome during this period. The use of linen underclothing was introduced by the Dionysian priests, or priestesses, who accompanied the Matrem Deorum from Pessinus to Rome, B.C. 204;[60] the cortina, or hydraulica, a musical organ, employed in the temples and theatres, belongs also to this period and is mentioned by Plutarch in Phocion and described in the "Organon" of Pub. Optatianus.[61] Bread was first made by public bakers in B.C. 174.[62] Sulphur matches appear to belong to the same period. Pliny describes the sort of sulphur employed in their fabrication, whilst Martial depicts the poor Hebrew children who in his day were employed to peddle them in the streets of the metropolis.[63]

The apothecaries' labels engraved upon stone and printed on bottles and jugs (Wright, in "Celt, Roman, and Saxon,") and the Roman stencil plates and branding irons collected by the British Museum prove that the Romans, had they possessed the secret of making suitable paper, might have anticipated the invention of printing by many ages. The specimens of iron, steel and brass work still extant evince great proficiency in the arts. Their surgical instruments, joiners tools, locks, keys and other implements and mechanisms fell but little short of modern perfection.

The wonderful progress exhibited in these various arts and inventions did not fail to extend itself to the machinery and other devices employed to economize labour in the mines. A passage in Martial has

[58] Pliny, IV, 106; Varro, I, 52. For tribula, Adam, Rom. Ant. 469, 529; for silos, ibid, 530; for vehicles and farming machines, De re Vehiculari Veterum by Joannis Scheffers, Frankfort, 4 to. 1671, p. 3, et passim. [59] Livy, XXXIV, 7.

[60] Adam, 428; Cory's Fragments ed. 1876, p. 162. Strabo. lib. XV.

[61] Plutarch in Phocion. Athenæus attributes the invention to Ctesibius, B. C. 250. Busby's Hist. Music.

[62] Adam, 436. [63] Martial, Ep. XII, 57; Pliny, XXXV 51.

been interpreted as an allusion to the stamp-mill.[64] Possibly the wheel pump employed for the deep levels of the St. Domingo gold mine in Castile and certainly the "booming" apparatus employed to sweep away the tailings of Las Medulas, are of the same marvellous age.[65] Mica was imported from Spain and used for house and carriage windows about B. C. 200.[66] The qualities of the loadstone and various sorts of magnets are described by Pliny. It is suspected by Ajasson that the polarization of iron was also known to the Romans; and as the knowledge seems to have resulted from observing the action of loadstone upon iron and the former was obtained in Spain and Greece, the discovery of polarization is most likely to have taken place soon after the Roman conquest of those countries. That the Greeks before this time were well aware of magnetic attraction is proved by the project of Timochares, who in the reign of Ptolemy Philadelphus, B.C. 285-47, "began to erect a vaulted roof of loadstone in the temple of Arsinoe at Alexandria, in order that the iron statue of that princess might have the appearance of hanging suspended in the air."[67] Another Greek invention, the steam-engine of Hero, B.C. 284-41, also belongs to this period. Draper says that twenty centuries of retrogression separated this invention from that of Watts, and he rather indicates the reason by remarking that while one revolved in a temple (the Serapion) the other drove a woollen mill.[68] The conquest of Greece by the Romans must have communicated these inventions, together with the entire body of Greek scientific knowledge, to Rome.

A degree of the earth's meridian was determined by Erastothenes of Alexandria, who died B.C. 194, the base of the measurement being the distance between Alexandria and Rhodes. Assuming with Aristotle that the shape of the earth was necessarily spherical, this measurement enabled Erastothenes to calculate the circumference of the earth at 252,000 stadia, which Pliny converted into 31,500 Roman miles, equal, as we may say, to 28,832 English miles. The

[64] Such, at least, is the opinion of Depping, II, 77. The passage is as follows: Martial Ep. XII, 57: Hinc otiosus sordidam quatit mensam Neroniana nummularius massa; Illinc paludis malleator Hispanæ Tritum nitenti fuste verberat saxum.

[65] See Chapter on Spain and Gaul.

[66] Pliny mentions the use of mica, but omits the date of its introduction. This is believed to be most probably that of the conquest of Spain.

[67] Pliny, XXXIV, 42; XXXVI, 25 and Note by Ajasson.

[68] In his "Pneumatica," Hero describes various methods of employing steam as a power. To him is ascribed the Æolopile, a toy steam engine. A translation of his treatise appeared in Bologna, 1547. Best edition, Paris, 1693. It is Draper's "Intellectual Developement of Europe," I, 387, which says that the machine was exhibited in the Serapion.

commercial intercourse, which continually went on between Alexandria, Rhodes, and Rome, soon made this determination the common property of the three cities. Indeed, it was not long afterwards publicly discussed by Hipparchus of Athens, Dionysodorus of Melos, Artemidorus of Ephesus and Posidonius of Apamea, all of whom made similar measurements of their own.[69] The armillary sphere, the gnomon and a chart of the world were among the other inventions of Erastosthenes. They were all products of this period. What gave rise to them, what gave rise to all the inventions of this age, was that encouragement, that reward for effort, which is only extended because it can only be extended, to all persons during a period of rising, or at least Stable Prices. The knowledge that a ready and appreciative market was open, in Rome, to every meritorious device, stimulated invention, not only within the Commonwealth, but also beyond it. The stimulus was felt in Rhodes, in Athens, in Alexandria, in short, in every mart and every school of science with which Rome was in communication.

The first sun-dial was erected at Rome in the temple of Quirinus, by Papirius Cursor, the consul, in B. C. 292, from which time the days began to be divided by hours. The first sun-dial (a faulty one) was exposed to the public gaze in B. C. 262. The first correct sun-dial was erected in Rome B. C. 163. The first clepsydra, or water-clock, which gave the hours irrespective of the weather, was erected by P. C. Scipio Nasica, B.C. 158.[70] Even the division of the day into 24 hours is a product of this period.[71]

Many new plants, trees, vegetables, fruits, herbs and flowers were introduced into Rome from the various countries which fell under her sway at this period and are enumerated by Varro and Pliny among the ancients and by Dunlop and Carpenter among the moderns.[72]

Down to this age cures were worked by charms and incantations, or by vows to images or rich offerings to the gods. The introduction of herbs and medicinal plants gave rise to the medical profession, whose Nestor in Rome was Archagatus, B.C. 220. The first Roman work on medicine was published by Cato the Elder, about B.C. 190.[73]

We still use the signs which he copied from the Greek physicians. Our chemists' shops are denoted by ☿ the Bacchic cross; our prescriptions are prefixed with the Roman symbol for, "Receive this in

[69] Pliny, II, 112.

[70] Pliny, VII, 60.

[71] Brady, Clavis Calendaria, I, 34; App. Cyc. 1st ed. IX, 300.

[72] Pliny, XIX, XX; Dunlop, II, 14, 21, 23, 30.

[73] Dunlop, II, 20, 21.

the name of Jove;" while the prescriptions themselves are in both Greek and Roman.

If we turn from the arts to literature the evidence of a sudden and prodigious growth at this period meet us in every department of thought and fancy. The earliest histories of Rome were published about B.C. 260 by Timæus, about B.C. 220 by Q. Fabius Pictor and about B.C. 200 by Lucius Cincius Alimentus.[74]

The first Roman work on agriculture was De Re Rustica by Cato the Elder, B. C. 200; the second was that of Mago, the Carthaginian, in 32 books, translated into Latin by D. Silanus, under an order of the Senate; the third was a work brought from Macedonia among the spoils of King Perseus; all within the space of less than half a century.[75] Cato's work begins with the following characteristic passage: "It would be advantageous to seek profit from commerce if it were not hazardous, or by usury, if that were honest; but our ancestors ordained that whilst a thief should forfeit double the sum he had stolen, the usurer should forfeit quadruple what he had taken, whence it may be concluded that they thought the usurer the worst man of the two. When they wished to greatly praise a man they called him a good farmer. A merchant may be zealous in pushing his fortune, but commerce is perilous and liable to reverses. Farmers, too, make the bravest men and the stoutest soldiers. Their gain is the most honest, the most stable, and the least exposed to envy; whilst those who practice the art of agriculture are of all persons the least addicted to seditious thoughts."

The first philosophers in Italy, Metrodorus of Macedon and Panætius of Rhodes, arrived in B.C. 168, during the Halcyon Age.[76] Many others followed shortly after them; but they were all, together with the Greek physicians, who appeared in Rome at the same period, banished from the city by a decree of the Senate in B. C. 162, the only exceptions being Panætius the philosopher and Polybius the historian. The rest took up their residence in the municipal towns of Italy, where they remained until B.C. 156, when they all returned to the metropolis, this time not as adventurers, but as ambassadors from their native countries. Among them were Diogenes the Stoic, Critolaus the Peripatetic and Carneades of Cyrene, who now held the place of Arceilaus in the New Academy. During the period of their embassy at Rome they lectured to crowded audiences in the most public parts of the city, with such effect as to excite in the Roman youth

[74] Dunlop, II, 68; Middle Ages Revisited, App. "S," on Ludi Sæculares.

[75] Pliny, XVIII, 3; Columella; Dunlop, II, II, 68.

[76] Dunlop, II, 209–10.

an ardent thirst for knowledge, which now became a rival to military glory.[77] Many principles and divisions of the Civil Law are founded on their maxims.

The first Roman work on Oratory was by M. Cornelius Cethegus, who was consul in B.C. 204; the second, by the indefatigable Cato, about B.C. 200 and the third by Ser. Galba, about B.C. 190.[78]

The first rhetorician who taught at Rome was Crates Mallotes, of Pergamus (son of Timocrates), B.C. 169. He was soon followed by C. Octavius Lampadius, Q. Vargunteius and Q. Philocomus.[79]

The prudent use of credit in any form is an indication of security and progress. The employment of bankers implies an extension of credit. In B.C. 218, upon the motion of Marcus Minucius, a plebian tribune, Lucius Æmilius Papus, ex-consul and censor, Marcus Atilius Regulus, ex consul, and Lucius Scribonius Libo, plebian tribune, were appointed "public bankers," (mensarii.) Their functions probably included the collection, disbursement and repayment of the popular loan mentioned by Livy. In such case they constituted a sort of Monetary Commission. In B.C. 212 we read in the same author of "the banking houses which are now called the New Banks." These were no doubt the buildings or offices in which the popular loan was being refunded.[80] Lanciana, the antiquarian, informs us of the remains of an ancient safe-deposit recently discovered in Rome, whose rules and regulations, engraved upon bronze, strikingly resemble those employed by modern institutions of the same class. This safe-deposit was of the time of Hadrian; but there is scarcely reason to doubt that similar ones were constructed at a much earlier date, in short, so early as the time when the Roman grandees drew their spoils from the conquest of Carthage, Greece and Spain.

Various improvements were also made at the same period in the mode of writing and of keeping accounts. The first books written on vellum appeared in Pergamus, B.C. 198 and shortly afterwards in Rome. Wooden tablets, covered with thin sheets of wax, designed to be written upon by the stylus, belong to the same period. Two such tablets, small 8vo. in size, the wax being inscribed with accounts relating to the mines, were, about half a century ago, taken out of the Abrudbanya gold mines of Hungary, where they had lain for centuries. About the same time that these wooden-backed wax writing-tablets were invented in Rome, Ennius, B. C. 200, invented shorthand writing.[81]

[77] Plutarch, in Cato. [78] Cicero, Brut.; Dunlop, II, 220. [79] Dunlop, II, 36.
[80] Livy, XXIII, 21; XXVI, 27; XXXIV, 6; Pliny, XXI, 6. [81] Dunlop, App. p. 4.

The lighter literature kept pace with the teachings of philosophy and history, the extension of credit and the improvement of scriptures and books. The first Roman dramatic authors were Livius Andronicus, whose plays were acted in B. C. 240; Nævius, B. C. 235; Plautus, B.C. 224; Ennius, B.C. 209; P. Licinius Tegula, B.C. 198; and Terence, B.C. 164; at which dates these authors were each thirty years of age. Terence's Andria was first produced in B.C. 167; the Hecyra, 165; the Heautontimorumenos, 163; and the Eunuchus and Phormio, 161. These authors passed away with the Halcyon Age and Rome was left without a single poet.[82] At a subsequent period—about B.C. 123—when Lucilius, the satirist, was 25 years of age and his patrons, Scipio Africanus Minor and Cæius Lælius were each about 63 years of age, there was what may be called a feeble literary revival, a fact that is evidenced by the works of Lucilius himself; but, as from his time to B. C. 80, no other poet appeared in Rome, it is evident that such revival, if indeed it may be so termed, was sporadic and ephemeral.[83] The influx of spoil had ceased; the mines were unable to sustain the customary value of commodities; the Plunder of Europe was accomplished; the Rise of Prices was stopped; and Rome's Halcyon Age was over.

The period that followed was that of falling prices, agrarian disturbances, civil wars, and the overthrow of popular liberty and republican government. The apotheosis of Cæsar and the degradation to which it consigned the people of Rome were supported, but not caused, by superstition. Their cause was the Fall of Prices and the policy of Plunder and Imperialism into which Rome was forced in order to mitigate the consequences of that catastrophe.

[82] Thompson's Rom. Lit. 447.

[83] Dunlop, I, 249.

CHAPTER IX.

BRITAIN.

Preëminence of Britain in mineral resources—Policy of the Romans—Monetary Systems of Britain—Her gold and silver product shipped to Rome—Provincial coins of bronze—Base silver issues—Rising prices and prosperity—Breaking down of this system—Establishment of Roman gold and silver coins—Contraction of the currency—Unprogressive period—Abandonment of Industry—History of mining—Aulus Plautius conquers and Claudius visits, Britain—The search for gold—Placers and veins—Streaming for tin—Mining debris—Silver-lead mines—Immense works at Cheddar—Roman mining laws of Derbyshire—Iron mines in the Valley of the Severn—Jet—Chalk—Coal mines near Newcastle—Construction of highways to convey the mineral produce to the coast and to Rome.

WHAT Spain was to the Roman Commonwealth, Britain was to the Empire: the richest and most productive of the mineral provinces. The splendid gifts of nature, though separated by more than two centuries of time, were no less abused in the one case than the other. Spain was exploited to enrich the patricians; Britain was exploited to enrich the emperors; after which it was abandoned to six centuries of anarchy and decay. To bring these features of Roman avidity clearly into view it is necessary to describe the monetary policy which Rome pursued in respect of the provinces.

It is worthy of observation that the three most distinctive periods of Roman conquest and territorial policy correspond with equally distinctive phases of her monetary systems. These are the bronze-numerical, the silver-coin, and the gold-coin æras. In the bronze-numerical system the money of Rome was issued by the Senate, that is to say practically by the people; in the silver-coin system it was issued by the Gentes, or the patrician families and the corporations subject to their control; in the gold-coin system it was issued by the Sovereign-pontiff, the silver and bronze coins being now subsidiary, or else of geograpically limited circulation. The bronze-numerical æra extended from about the time of the Gaulish Invasion of Rome to the Second Punic War. During this period the conquests of Rome were confined to Italy. Her political policy towards the conquered countries was federation and assimilation; her commercial policy was the extension of agriculture and cultivation of the mechanical arts. This was the period of her most healthy growth; when all that was progressive in her developement and beneficent in her influence made its appearance.

The æra of silver-money embraced that of the civil wars and extended down to the advent of Cæsar. It corresponded with the conquest of Carthage, Hither-Spain, Greece, and Asia Minor, all of which were silver-producing, or silver-using countries, chiefly the former. The political policy of Rome towards these countries was that of sovereign dominion and control; her commercial policy was the exaction of heavy tributes, chiefly of the coveted metal which now formed the basis of her monetary system. Before this period drew to a close, Rome had reached the maturity of her industrial developement.

The distinctive æra of gold-money began with the Imperial period and embraced the conquest or exploitation of Further-Spain, Gaul, Pannonia, Dacia-Ulterior, Britain, and Egypt; all auriferous regions. These countries, like the silver producing states, were at first subjected to heavy metallic tributes; but as natural deposits of gold are more quickly reached and exhausted below the level of profit than those of silver, this policy was afterwards changed, and Rome planted and repopulated some of these provinces, seeking to derive its revenues from them through the more lasting resources of agriculture, commerce and the arts.

A policy similar to that of Rome afterwards governed the Spanish Empire in America. The silver producing states, Mexico, and Peru, were governed from first to last with the sole purpose of exacting from them as much metal as possible. It was for this purpose that their inhabitants were subjected to the Encomienda system and the Mita. Centuries of loss sustained by the mother country and of cruelty and misrule endured by the natives had operated to weaken the control of the former, before the Napoleonic wars in Europe inspired the latter to throw off their yoke. On the other hand, the gold placer countries, Hispaniola, Cuba and Panama, were soon exhausted of their auriferous deposits and began to be planted with sugar and maize as early as the first quarter of the 16th century. It was a necessity of the situation either to plant or abandon these provinces, for there was plainly no more gold to be got; whilst as to Mexico and Peru, whose silver deposits seemed to be inexhaustible, agriculture was never pursued beyond the barest requirements for food.

The policy of Rome toward Britain was influenced by considerations of a mixed character. Britain was opened as an auriferous province: but before Agricola was appointed governor, A.D. 77, the gold mines had become unprofitable. By this time, however, the silver-lead mines were very productive and the work of exhausting them was pushed on with all the ardour of cupidity. As, afterwards in America, the whole of the product was obliged to be sent to the mother country. Thus Britain, metalliferously rich by nature, was kept metalliferously poor by man.

Under these circumstances we should naturally look for a feeble developement of provincial resources, a wretched growth of temporary habitations and a neglected land. On the contrary, we find evidences of great cities, splendid temples, numerous factories and immense agricultural and pastural wealth.

It must not be supposed that such great industrial progress lacked the incentive of rising, or, at least, steady prices; for with falling prices there could have been no adequate reward to either planter, artisan, or trader; no inducement for the continuance of his labours. One by one the Roman colonists and proprietors would have abandoned the province, and left it to the soldiery and the natives. The former, occupied with military works, such as constructing the coast castles, the fortifications of the cities, and the great walls which defended the northern frontier; the latter degraded to mere slaves of the soil, in order to secure food and subsistence for their masters. That such was not the character of the Roman colonization of Britain, is sufficiently attested by the progress and developement of the province.

What sort of a monetary system then did the British province enjoy, that, notwithstanding its deprivation of the precious metals, rendered it so prosperous during the first two centuries of the Roman occupation? The answer is: overvalued bronze numeraries, such as Mæcenas advised Augustus to adopt. That this advice, so far as Britain is concerned, was faithfully carried into practice, is evident from the immense quantities of these coins which were exported from Rome and the numerous and large finds of them in the provinces which have rewarded the researches of modern antiquaries. Dr. Henry is of the opinion that during the Roman occupation of Britain there was more money in circulation than at any time afterwards, until the discovery of America. In view of the absorption and monopoly of the precious metals by Rome, there seems little room to dispute that this money consisted of those symbols which had served the mother country so well in times past and which it was now hoped would equally well serve the infancy and promote the growth of her provinces. In later times, when the power of Rome declined and her provincial system, together with many other of her institutions, was overthrown, the numerical money of Britain was supplanted by gold and silver coins; but from the conquest of Claudius until near the beginning of the third century of our æra, there can be little doubt that the monetary system of that province consisted in the main of a series of overvalued bronze numeraries issued by the Senate of Rome. When first introduced into Britain these numeraries ("sesterces") were subject to limitation of numbers: but if Dr. Henry is right, this regulation had but little practical result in Britain; and the volume of bronze and debased silver money probably continued to increase until after the middle of the second century, when the bronze pieces, having fallen to their metallic value and proving too heavy at this valuation to be commodious, prices were expressed once more in imperial gold and silver coins, which had meanwhile become more and more common in the provincial circulation. The overvalued or debased silver coins, with the effigies of Claudius, Nero, Vespasian, Domitian, Trajan, Hadrian, Antoninus and other emperors, were probably struck in the province. A similar practice obtained in Gaul. The quinarii (half denarii) of Calletes are stamped with the numerical value "X," meaning one denarius.[1]

It is during this æra that we must suppose that all the progress made by the Romans in Britain took place, because it is difficult to account for these results without admitting a rise of prices.

During the contraction which afterwards existed there was no permanent incentive to industry; so progress came to an end. Says Mr. Wright: "We have no traces of a Roman mint in Britain until the reign of Diocletian and Maximian."[2] Yet at Polden Hill near Edington, in Somersetshire, several hundred moulds were recently found for casting debased denarii, with the effigies of Septimius Severus and his wife Julia, of Caracalla, Geta, Macrinus, Heliogabalus, Julia Paula, Alex. Severus, Maximinus, Maximus, Plautilla and Julia Mammæa. Others have been found at Lingwellgate near Wakefield, Yorkshire; at Ruyton and Wroxeter in Shropshire and at Castor in Northamptonshire, the last three places being the sites of the Roman towns Rutunium, Uriconium and Durobrivæ. There can be little doubt that the Romano-British colonists of these remote times did precisely what the British-American colonists did sixteen centuries later: they transgressed the impolitic laws of a distant suzerain and set up their own mints.

[1] Noel Humphreys' "Coin Collectors' Manual," 172.

[2] The coins of Carausius, A.D. 287-93, are the earliest that are known to have been struck in Britain by an authority, which, though at first usurped, was afterwards recognized as imperial.

Mark that we are not here advocating a policy, but formulating an historical inference. Britain was not only occupied by Roman troops and by the turbulent inhabitants whom the former were employed to keep in check; it was occupied also by a large body of Italian and Gaulish colonists, attracted to the new country by the favourable conditions attached to its occupancy, and by the accounts, true or magnified, of its natural resources, which reached Rome through the medium of the provincial authorities. These colonists were composed at first chiefly of the adventurous classes, miners, frontier-traders, camp-followers and pioneers, like the miner, Curtius Rufus, who had nothing to lose and everything to gain.[3] As time wore on it was perceived that Britain was not a mere gold and silver country, like Spain, to be plundered, devastated and abandoned to adventurers; it was also a pastoral, agricultural and commercial province, like Gaul, which already had become a prosperous colony. From this time onward the class of immigrants greatly improved and instead of the bludgeon, chain and whip, symbols of mine-slavery, they brought with them the plough, the mattock, the pruning-knife and the economy and thrift of agricultural life. This Italian occupancy of Britain is proved by the numerous and extensive mechanical works, the remains of which are to be found in nearly every county of England and Wales, as well as in the lowlands of Scotland. The Walls of Agricola and Severus and the coast-castles, were probably built by the army, assisted by native labourers; but the cities, temples, villas, baths, statues, altars, ornamental tombs, tasselated pavements, mosaics, etc., are of a character that bespeak the presence of an artistic class, who would hardly have left Rome to live in Britain, without the incentive of high wages and permanent employment. Under the metallic system, which Rome had devised for her own advantage—a system which withdrew from the provinces their product of the precious metals, to heap it up at the world's moneyed centre, such conditions of prosperity in Britain would have been impossible without a provincial money substantially independent of Rome. This means money not composed of gold or silver, or, if so, only in small part; so that the exportation of the metallic product of the province to Rome, which if Rome had had her own way would have kept it forever poor, was by these means divested of its injurious influence.

With this understanding, we are better prepared to follow the details of the metallic production. The expedition under Aulus Plautius evidently followed the Valley of the Thames to the source of that

[3] Tacitus, Annals, XI, 20-21.

river at Cirencester, probably panning for gold all the way along and finding little or none. After establishing a camp at Cirencester, which eventually developed into an important manufacturing city, the Romans retraced their steps down the valley of the river to Wallingford, forty-six miles west of the future Metropolis, where the ocean tides made their appearance. They established camps both here and at the confluence of the Thames and Lea; near London.[4] Soon after this, the sovereign-pontiff, Claudius, arrived to "complete" a conquest which was in fact already effected. After a stay of but sixteen days in Britain, that divinity returned to Rome.

The nature of the instructions left with Aulus Plautius is evidenced by the expeditions which were soon after sent into the Cotswold Hills and subsequently into North Wales, in which last named locality the natives had evidently reported the existence of gold-placers. We have but scant means of measuring the commercial success of these expeditions; and can only conclude from the present appearance of the ground, that the deposits were small and soon exhausted. Down to this time tin had been produced in Britain chiefly by "streaming," a process similar to washing placer-gold. As yet, no subterranean tin mines had been developed; but streaming had been and was yet to be, conducted on so large a scale, as to foul the harbours of Plymouth, Dartmouth, Tynemouth, Falmouth and Fowey, with mining debris, and render it necessary, at later periods, to repair the injury thus occasioned. On this subject Anderson cites the statute of 23 Henry VIII. (1532) which provides for the mending of these havens, choked up by the gravel from stream works.[5] As these works had been going on since the earliest period of British history [6] and most actively during the Roman period, it is not unreasonable to suppose that the fouling of the havens by mining debris, had begun ages before the enactment of the statute; most probably during the Roman æra.

[4] There are various monkish legends to account for this name, the popular one, from Geoffrey of Monmouth, who connects it with the fabulous King Lud, has the disadvantage of being complicated with the Trojan Æneas, Brutus, etc. We know for certain that London had its present name, that is to say, Londinium, the Latin form of it, so early as the period of Tacitus; for it is so written in his Annals, XIV, 33. Among several remains of Norse council rings, Mallet found one at Lunden in Scania. There are several other Lundens in Sweden and Denmark. The Trinobantes, who occupied the country north of London, when the Romans first took possession of it, were a Gothic tribe; and it may be reasonably conjectured that following an almost universal custom, the name of the place was brought by their forefathers from their ancient home in Gotland.

[5] History of Commerce, II, 56.

[6] Herodotus, Thalia, CXV.

Roman placer mining for gold, and streaming for tin, was followed by subterranean mining for these and other metals and minerals, namely, gold, silver, tin, lead, copper, iron, jet, chalk, building-stone, and mineral coals. The gold quartz mines opened by the Romans are too numerous to mention. Remains of their prospecting and mining works are to be found at Gogofan, indeed all over Cærmarthenshire, in the valley of the Mawdatch and other parts of Wales. The gold mines were all worked for, or on account of, the Imperial fisc. All gold found in the earth and all "treasure trove" belonged by right to the sovereign-pontiff.[7] The coinage of gold was also a sacerdotal prerogative, strictly monopolized by the sovereign-pontiff, and always conducted near the place of his residence.

Silver-lead ores were excavated in Derbyshire and Monmouthshire and calamine and silver-zinc ores in Shropshire, where the deposits were followed far into the hills. Silver, or silver-lead mines were worked in Northumbria and especially at Allendale and Alston-Moor; in Derbyshire, at Wirksworth and near the Peak (opened tempo Hadriano) at Shelve in Shropshire (opened tempo Trajano), in the Mendip Hills of Somersetshire (opened A. D. 49) where pigs have been found cast with the words "Britanicus" and "Vespasiano;" and at Kangie, near Cheddar, (opened tempo Vespasiano) where the reduction works were of unusual size. Here, the refuse, which the imperfect appliances of the miners forced them to cast aside as worthless, has been smelted and re-smelted, in modern times, with profitable results. It is only a few years since some of these re-smelting works were in operation.

A Roman ingot of silver weighing 320 metric grams (4931¾ English grains) was found in 1777 within the Tower of London and is described in "Archæologia," vol. V. If this was a talent, then divide its weight by 12 and 5=60 and we have the aureus of 82⅓ English grains. This weight would assign the ingot to the period of Aurelian's second coinage, A. D. 274. Should this conjecture prove to be well founded, this is the oldest "dollar" in existence. Other Roman silver ingots have been found in Coleraine, Ireland, with a large hoard of Roman silver coins.[8]

The frequent disturbance of the ground has left but few memorials of the Roman occupation on or near the surface, but enough remains

[7] Such was the law down to the reign of Nerva and probably down to that of Hadrian who divided treasure-trove equally between the discoverer and the Crown. Del Mar's Middle Ages Revisited, pp. 105, 255, 359.

[8] "Mining operations and metallurgy of the Romans in England and Wales," by Rev. J. Chas. Cox, of Derbyshire: Read before the Archæological Institute; July 27' 1894: Archæological Journal, 1895, p. 25.

to show that Kangie was once an important mining camp. Foundations of buildings have been traced, and fine Samian wares, rude native pottery, and Roman coins, ornaments, tools and weapons have been exhumed in the "Townfield." A broken tablet records that the Armenian Legion was once garrisoned here.

Huge pigs of lead still exist bearing the names of Vespasiano and Antoninus. Little else remains near the ancient camp; but in all the country around—

> Some trace of Imperial tenure, now
> Clashes at times on the peasant's plough.
> Fragments of graceful vases
> With gods and heroes traced;
> Records of Roman triumph
> In letters half effaced;
> A tarnished ring whose fiery gems,
> Still on its circle set,
> From the far sands of Indus brought,
> Gleam through their setting rudely wrought
> As if the skies their hues had caught,
> Flamed in their glory yet.

Copper mines were worked by the Romans in Montgomeryshire, Denbighshire, Shropshire, and the Island of Anglesea. Like those of silver-lead, most of the copper mines of Britain yielded a small proportion of silver. The lead mines were free to be worked by any Roman citizen. The silver-lead mines were farmed or leased by the Roman procurators to favoured individuals, or companies, who worked them on their own account, paying certain taxes and dues to the fisc, based upon production.

At first, the deposits of lead ore were near the surface; these proved so profitable that, like the old hand-placers of California, the holdings were divided into small properties, or, as Pliny puts it, the quantity of ore permitted to be worked by any individual was limited by miner's law.[9] Mr. Wright thinks the lead ores thus alluded to were those of Shelve in Shropshire. There is a reason, which possibly escaped the observation of that distinguished antiquarian, for believing that the law alluded to by Pliny, related to the lead mines in the Peak of Derby. The reason is that the law is still in force; and so far as is known, it has continued to be in force for upwards of eighteen centuries.

The law of the Commonwealth permitted any Roman citizen to search for and locate or "denounce" mines. He was authorized to sink and dig mines or veins of ores, in or under all, except consecrated, lands, no matter to whom they belonged; and to follow his

[9] "Natural History," XXXIV, 49.

vein whithersoever it led. The condition attached to this privilege was that of continuous working. If the citizen abandoned his mining claim for a given time, it was liable to be taken up and worked by any other citizen. These principles of the Roman mining law will be found established to-day in all those countries in which such law has not been disturbed by the imperial and medieval system; for instance, in all of North and South America, and in some portions of Africa, countries which derived them direct from the Civil Code. After the wars of Marius and Sylla, this mining law, so far as it covered gold and silver mines, was modified. During the Empire, the gold mines were regarded as belonging exclusively to the sovereign-pontiff; the silver mines were resigned to the nobles; while those of the baser metals, including lead, continued, though not entirely as before, to be open to location, "denouncement," and working by citizens. When the Roman laws were codified by Justinian, the mining rule underwent some further modification.[10]

The mining rule, as modified by Justinian, became the law of the province of Britain, so fast as that country was rescued from Gothic (pagan) to christian control. In those portions which longest remained subject to paganism, notably the kingdom of Mercia (formerly Flavia Cæsariensis) the ancient mining law of the Roman Commonwealth continued to prevail; and under that law, any citizen, or subject of Mercia, was at liberty to locate and work lead mines on, or under, all except consecrated lands, no matter to whom they belonged; provided that he did not abandon his work for more than a certain interval.[11] It was in this manner that the Peak of Derby was worked. The limit of mining ground permitted to be taken up by a miner is believed to have been 300 Roman feet, on the ledge or vein, without any limit as to his right to follow the course. Mercia was converted to christianity in the ninth century. By this time, however, the Roman Civil law had become partly submerged beneath the Canon law, and the former was kept out of view until after the Fall of Constantinople. Thus it failed to supplant the ancient Roman mining law of Mercia, as it had supplanted the law in the other kingdoms of the Heptarchy. Hence the ancient law, in the guise of "immemorial custom," continued to rule the miners of the Peak.

[10] It appears in the Code, II, 7, (6) under the title: De Metallariis et metalis et procuratoribus metallorum.

[11] This interval in the Peak of Derby is now three weeks, not counting any time during which the mine may be under water, etc. The interval of abandonment, necessary to vacate a claim, in any Spanish or American State, is one year; in Portugal and its colonies, it is one year and a day. This last, was probably the law of the Roman commonwealth.

In 1287 (16 Edw. I) a Mining Commission, the first of its kind since the Roman period, consisting of twelve persons, was authorized by the King to examine into the laws and conditions under which mining was prosecuted throughout the realm. Among other matters, this Commission reported, as to the rights and customs of the Peak of Derby, that "the miners claimed by no charter, but immemorial custom." These rights seem to have been now for the first time curtailed. Under the Roman law the miner was entitled to work any unlocated claim to the extent of say, three meers, or 300 Roman feet. Under the modification established by Edward I., he was only entitled to work two meers, the third meer pertaining to his location, going to the King. In case the King either directly or through lessees, failed to work his portion, (which portion he was at liberty to choose from either side of the located claim,) then the locator had the right to purchase the third meer, at a price to be fixed by arbitration. It is probably from this period that the district over which this regulation prevailed was called the King's Field, or Fee. Mr. Tapling, from whose works some of these details are derived, follows the Crown lawyers by insinuating that under a rescript of Valentinian and Valens, all mines were held by the Emperor, to whom was due a "royalty," consisting of a portion of the ores extracted; but the use of this term finds no support either in the Roman institutes, or the customs of those countries or districts where the Roman institutes prevailed, or still prevail. The basis of the Roman law was the right of the citizen to pursue mining on any, except consecrated, ground; the basis of a royalty is the assumed proprietorship or paramountship of the land on which the mining is done.

From 1287 to 1851 no further alteration was made in the customs of the Peak, but by this time, their extreme antiquity and lack of congruity with the mining laws of the remainder of the Kingdom, had led to conflict and litigation. To remedy this, the Act XIV and XV Victoria, c. 94 was passed, which pretends, without essentially altering the custom, to give it more precision and uniformity. The preamble to this act contains the following rule: "It is lawful for all the subjects of this realm to search for, sink, or dig mines or veins of lead ore, upon, in, or under, all manner of lands, of whose inheritance soever they may be; churches, churchyards, places for public worship, burial grounds, dwelling houses, orchards, gardens, pleasure grounds, and highways excepted." Here is the old Roman law again, this time with several more limitations attached to it, but essentially the law of the Commonwealth, which, without any interruption whatever, continued to be, and still remains the law of the Peak of Derby, which was in force when the leaden pigs with the Imperial stamps, that now adorn our museums, were cast in the province of Flavia Cæsariensis.

There was probably another feature in the Roman mining law, as it existed when Britain was conquered, which, with equal probability, was greatly modified about the time of Vespasian, during whose reign Britain was pacified. This was the right of the conquerors, like that one exercised at a later date by the Spanish conquerors of America, to claim and select a certain number of the natives and compel them to work in the mines, upon condition of affording them protection, shelter and subsistence. This system, called in Latin *commendatio*, and in Spanish *encomienda*, was probably the origin of the Medieval-English custom of commendation. It gave rise to great cruelties on the part of the colonists. The sordid exactions of an alien church occasioned other cruelties. Together, these causes are responsible for many of the revolts of natives, which distinguish the Roman occupation of Britain.

The silver extracted from the British mines, when separated from all impurities and cast into bars, and after the payment of taxes and of other charges for refining and stamping, was purchased by the Roman mints, at its lawful value, which was always one-twelfth that of gold, weight for weight.

Remains of Roman iron mines have been found in many parts of Britain, in Maresfield, Sedlescombe and Westfield, Sussex; in Luxborough, in the Brendon Hills, in Luccombe on the confines of Exmoor Forest and in other parts of West Somersetshire; in Lanchester, near Hadrian's Wall, Durham; and at several places in Monmouth, Hereford, Worcester and Gloucester, such as Berry Hills, near Ross and in the Forest of Dean. These last named mines appear to have been the most extensive. The Romans used chiefly the hematites, nodules and bog-iron, not touchiug the more difficult clay iron-stone of the carboniferous series. According to Yarranton, who employed himself in reworking the ancient cinder-heaps, the Romans used the foot-blast. Vast numbers of coins, chiefly bronze, were found in the refuse.[12]

The jet obtained by the Romans in Britain was what is now known as cannel-coal, and of this substance, numerous articles of use and ornament were made, either by carving or turning. Great numbers of these objects are preserved in our museums. The Roman quarries

[12] Yarranton, "England's Improvement," II, 162.

of chalk and building-stone can no longer be distinguished, on account of modern extensions of their works; but we have abundant remains to prove how industriously they were developed.

The most interesting of the mining remains are those of coals. There can be no doubt that the Romans opened coal mines at South Benwell, Chester, (North Tyne;) at a place near Wirksworth in Derbyshire; at another near Grindon Lake, by Sewing Shields, in Shropshire (Uriconium); and elsewhere. Some of these ancient workings remain untouched. Cinders and soot from mineral coals have also been found with Roman remains, (in Britain) other than those of the mines.[13]

Yet for nearly a thousand years after they had abandoned these mines, the use of stone-coals appears to have been unknown; and had to be rediscovered or reintroduced. This may be due to the fact, already noticed, that subterranean mining, whether of gold, silver, iron, or coal, involves large capitals, permanent works, systematic procedure and good government; and that all these advantages were lost to Britain when its Roman government began to decay.

Before this came to pass, that is to say, during the early Empire, and in order to convey the produce of the mines to the mills and smelting furnaces, roads were opened between them, which had the further use to bring the agriculturists and townsmen into communication; and thus promote commercial intercourse and the growth of new industries. So rapidly did these various industries attract and so liberally did they maintain, an industrious population in Britain, that within eighty years from the period of its conquest, it was sufficiently important to merit a ceremonious visit from the divine Hadrian, and opulent enough to sustain a vast expenditure for new fortifications and fleets. But within a century of this period the decline of Britain began and in this decline, mining fell, to rise no more, until it shared in the general Renaissance of all European industries.

[13] Address of Sir J. L. Bell to the North of England Institute of Mining and Mechanical Engineers, August, 1887. A license to dig stone coals near Newcastle is said to have been granted by Henry III., in 1239. Sea coals, (the same as stone-coals,) were forbidden to be used for fuel, for the reason, alleged by Stow, that they were "prejudicial to human health." Haydn. There may have been a far different, an ecclesiastical, reason for the inhibition. However this may be, a man was executed in London during the 14th century for infracting this law. Proc. Soc. Scotish Antiquaries, 1886-7, p. 95. The Scotch-Norsemen, who were less amenable to ecclesiastical restraint than the English, used stone-coals for fuel, so early as 1291. Ibid.

CHAPTER X.

EUROPE DURING THE MIDDLE AGES.

The Barbarian irruptions fail to account for the decline of mining—The gold mines were controlled by the Roman hierarchical fisc—Copper mines by the Senate—Silver mines by the proconsuls—These various classes of mines flourished or declined with the powers or policy that controlled them—The gold mines rose and fell with the hierarchy—The copper mines declined with the Senate—The Roman silver mines alone never stopped—Mines of Thrace—Mœsia—Illyria—Pannonia—Hispania—Gaul—Britain—In the sixth century the silver mines of Central Europe passed from the control of the Romans to that of the pagan Avars, Czechs and Saxons—In the eighth century those of Hispania and Lusitania passed to the Saracens—Sudden revival of mining after the Fall of Constantinople—The legal value of silver raised by the independent princes—Carlovingian Conquest of the Saxons and Avars—Medieval plunder of the Moslems—Sporadic plunder of the Lombards and Jews—Gold washings of the Ebro, Darro, Douro, Sil, Minho, Elbe, Rhine, Danube, Rhône, Guadalquiver, Tagus and Garonne—Mines opened by the Moslems in Spain, Africa and Asia—Despoilment of Moslem Spain.

IN his work on the Precious Metals, Mr. William Jacob attributes the relinquishment of mining by the Romans to the Barbarian Irruptions of the fourth and fifth centuries of our aera; and he declares that during the entire interval, from the sixth to the fifteenth centuries, "the precious metals were sought, not by exploring the bowels of the earth, but by the more summary process of conquest, tribute and plunder." This hardly states the case correctly. Though there is no positive evidence to sustain the conjecture, it is quite likely that the barbarian revolts and disturbances led to the abandonment by the Romans of many of their mines; but these disturbances must have lost their influence many centuries before the fifteenth: and while it is not doubted that mining as a whole declined during this interval, it is believed that it did not decline at the time, nor for the reasons, nor in the sweeping manner indicated by Mr. Jacob.

Down to the fourth century the Romans worked gold, silver and copper mines in nearly every province of the Empire. Comparatively few of their mines were north of the Danube, and chief among them were those of Britain and Pannonia.[1] The Roman mining industry was at its height in the Augustan period: in the reign of Diocletian it had greatly declined, especially gold mining: in that of Constantine, mining generally was in a moribund condition, although, as we shall presently see, silver mining continued with less abatement than gold. From the Augustan æra, the gold mines were under the exclusive control of the Hierarchical Fisc, the copper mines were subjected to the Senate, the silver mines were resigned to the proconsuls or viceroys, and each of these classes of mines seems, in fact, to have followed the fate of the power which encouraged, patronized or protected them. The Senate expired in the reign of Claudius; the Hierarchy, though it held the monopoly of gold, was unable to work the gold mines at a profit after the reign of Theodosius;[2] whilst the proconsuls, who gradually developed into counts, dukes and exarchs, and at a subsequent period into feudal sovereigns, continued the search for silver with scarcely diminished energy. As a matter of fact, we hear of little or no copper mining after the Augustan, and little or no gold mining after the Theodosian æra. On the other hand, silver mining never stopped, although it fell away, as all things fell away, within the Empire during the Medieval Ages, including population, commerce, the arts, and consequently the demand of jewelers and other artizans for the precious metals. Gibbon assures us that the mines of Thrace, Mœsia (Servia), and Illyria were still worked by the Romans during the fourth century. So, too, were the mines of Orsova in the Banat; Guadalcanal, Constantina and Novo Cartago in Spain; the district of Isère (L'Argentiere, La Gardette and Allemont mines) and others in Gaul, as well as those of Britain. Dr. Adolt Gurlt in the "Metallur-

[1] Tacitus, Annals XI, 20, says that in the reign of Tiberius, Curtius Rufus, a Roman of mean extraction, but afterwards proconsul in Africa, discovered a silver vein in the country of the Mattiacci. This is believed to be the same with Marpourg or Marburg in Hesse Nassau. For early Roman mining in Britain, see the author's "Ancient Britain," Calvert's "Rocks of Britain," and the address of the Rev. J. Chas. Cox, to the Archæological Institute, in their Journal for 1895, p. 25.

[2] The reason why the Roman Fisc was unable to work the gold mines at a profit was due to its own policy. It fixed the ratio between silver and gold at 12 for 1, and exacted its tributes at this ratio. It was, therefore, cheaper for it to procure gold for silver in India, where the ratio was but 6 or 6½ for 1, than to delve for gold in Europe. The adoption of the Indian ratio by the Saracens rendered the commercial aspect of this policy profitable so long as the Arabian Empire survived in Egypt and Spain, both of which were gold-producing countries, and both willing to exchange their gold for the silver of the Romans. When the Hierarchy fell, many of the ancient gold mines of Europe were re-opened.

gical Review," also includes the silver mines at Wiesloch, near Hiedelberg, at Blissenbuch, near Engelskirchen, and others near Halzappe and Ems, but without giving dates; so that it is quite possible that he is mistaken as to who worked them, and that they were opened by the Saxons and not the Romans. In the fifth century we only know of silver mines being worked by the Romans in the Felsobanya district of Pannonia, or Hungary. These contained some gold. In the sixth century mining, within the Roman Empire and under the Roman Government, almost came to an end. The Roman Government itself gradually shrank to the limits of Greece and Asia Minor.

On the other hand, A. D. 550, the pagan Avars opened the electrum (gold and silver) mines of Kremnitz, and the silver mines of Chemnitz and Transylvania (Vorrspatak)[3] In the seventh century these mines were still working, and to them was now added the silver district of Rothausberg in Bohemia.[4] In the eighth century the heretical Arabs opened the gold and silver mines of Spain, while the pagan Czechs and Avars worked those of Bohemia (Rothausberg), Transylvania, Croatia, the Banat (Orsova) and the Carpathians; and the pagan Saxons, those of the Hartz and Tyrol—for example, Andreasberg and Zell.[5] In the ninth century there are few or no records of European mining except in Arabian Spain. In the tenth century (A. D. 941), the Rammelsberg and Clausthal silver mines of the Hartz were opened by the pagan Saxons, and Kuarzim, in Bohemia, A. D. 998, by the Czechs. In the eleventh century, in addition to numerous gold and silver mines in Arabian Spain, the sands of the Tagus, in Portugal, were washed for gold by the Arabs. In the twelfth century the Freiberg silver mines of the Erzebirge were opened by the Saxons. L'Argentiere, in the French Alps, was also re-opened. In all these cases, except the unimportant one last men-

[3] At Vorrspatak, gold, silver and gypsum are worked in "veins traversing a tertiary sandstone, being about the only known instance of such a mode of occurrence," Encyc. Brit. ed., 1886. The writer has inspected similar mines in Southern Utah ("Dixie") and Mexico. From the Utah sandstone silver mines, live frogs have been taken, proving that the formation is very recent.

[4] Rothausberg or Rathausburg is near Gastein, Austrian Alps, 9,000 feet above sea level.

[5] Zell is believed to have originally yielded more silver than gold, and afterwards more gold than silver, which is uncommon. The silver mines of Nrakonya, between Dubova and Ogradena (Orsova district), were worked by the Romans, as also were the gold mines above Bogshan (Carpathians); others at Wiesskirchen, near the Danube; and still others in Transylvania. The writer inspected those of Bogshan some fifteen years ago, and found in them undoubted marks of Roman workmanship.

tioned, these mines were opened or re-opened by heretics or pagans, none of them by the Romans or Christians. This circumstance has already been noticed, commented upon and explained in our "History of Monetary Systems."

In the thirteenth century there occurred a sudden and extraordinary revival of mining within the boundaries of Christendom. For example: Marienberg; Annaberg; Schneeberg (from 1320 to 1350 this district produced 100,000 lbs. weight of silver per annum);[6] Johann-Georgenstadt; Joachimsthal; Wiestock in Baden; the silver mines of the Lahn (Nassau);[7] Blissenbruck and other mines in the valleys of the Sieg and Agga; Kitzbuchel and Rohrerbuchel (Tyrol); the Kaurzim district of Bohemia (A. D. 1284), the Adelbert mine of the Prizibram district of Bohemia; Neusohl, Schmolinitz, Nagybanya, Abrudbanya, and other districts in Hungary; Farehajer; Salzburg (silver); Altenberg (electrum); Shellgadin (called from its richness "The Throne of Plutus"); the silver mines of Sardinia, re-opened by the Pisans, 1283; the mines of Servia and Illyria re-opened by the Venetians; the silver mines of Guadalcanal, and others in the Sierra Morena of Spain; the Almaden quicksilver mines of the same country. In France, the silver mines L'Argentiere and others re-opened in the Isère district, Bouxeda and Pugalduc in Languedoc, Paleyrac

[6] Between the time of Charlemagne and the end of the twelfth century the mines of Bohemia, the Hartz and other parts of Germany, were discovered" (or re-discovered) . . . "Others were found before the discovery of America. . . . Masses of silver were often met with just beneath the surface. . . . One such, long used by the Margrave Albrecht of Meissen as a table, yielded 450 pounds of pure silver. . . . Bohemia was described by the king of the country, in 1295, as being sufficiently productive of the precious metals to supply the wants of the world, The Saxon mines yielded an annual land tax of 100,000 Bohemian groschen, while those of Schneeberg, on the Erzebirge, produced, during the first thirty years of their working, an annual average of 10,000 cwt. of silver. Other mines as rich were afterwards opened in the Tyrol and at Salzburg," Yeat's Hist. Com., 188. All these were old Roman mines re-opened. The product of Schneeberg was not ten thousand, but one thousand quintals per annum, not $18,000,000 but $1,800,000 per annum for thirty years. A far better summary of mining during the Middle Ages, than that afforded by Mr. Yeats, will be found in the "Museum Metallicum" of Ulysses Aldrovandi, Bononiæ, 1648, folio, pp. 41–9.

[7] Malte Brun, v, 102, says that the silver mines in the duchy of Nassau, near Branbach and Holzappel, yielded during the 18th century about £8,000 ($40,000) per annum; and that from the evidence afforded by numerous relics of antiquity exhumed near the baths of Wiesbaden, it is to be presumed that the latter were known to the Romans. The writer may add that he has panned gold from the river gravel near Wiesbaden, and that from all these evidences and others he is of the opinion that the Romans had several mining camps in the vicinity.

and Termanez in the Pyrenees, Viviers in the Vivarais, the mines of Dauphine at Brandes, near Grenoble, the silver mine d'Ouzals near Toulouse, the cupreous silver mines of the Seigneur Rosselin de Foix, in Provence; and (in the fourteenth century) the silver-lead mine of Lacroix, in Lorraine. In Britain, numerous silver and silver-lead mines were opened in Cærmarthenshire and other provinces. Most of these mines were of silver, a few of electrum, still fewer of gold, and none of copper, unless it formed the by-product of a silver mine.[8]

What had occurred to cause this revival? The Fall of the Sacred Empire in 1204. This event loosed the venerable but feeble grasp of the Basileus upon the prerogatives with which Cæsar had invested his office, including those of mining, coinage, and the ratio of value between silver and gold. With the fall of the sacred ratio of 12 for 1 the independent kings raised the gold value of their silver coins and thus encouraged numerous silver mines to be opened or re-opened, which previously, and at the sacred ratio of 12 for 1, did not pay to work.

The conquest of the pagan Saxons and extermination and despoilment of the pagan Avars by Charlemagne in the eighth century, the crusades against the Saracens of Spain and Portugal in the 12th to the 15th centuries, and the repeated despoilment of the Lombards and Jews, who were the intermediaries everywhere between the Romans and the pagans, and had grown rich not so much upon usury, which was the charge against them, as upon the profits of legitimate commerce; these were also among the sources of the precious metals which found their way into the mints of the medieval princes.[9]

Besides the more important and permanent sources of supply already mentioned, the auriferous streams of Europe probably furnished in all ages a small but steady supply of gold, as, indeed, some of them do still, for we have ourselves seen gold-washing conducted on the Tiber in Italy, the Darro, Douro, Sil, Minho, and other rivers in Spain, and the Elbe and the Rhine in Germany. Besides these rivers, there were the Danube, Rhône, Guadalquiver, Tagus and Garonne, all of which were worked during the Middle Ages.

The banks of the Tagus in Portugal contain three classes of auriferous sand: 1, cascalhao, or gravel beneath the water; 2, medao, or sand bank above the water; and 3, malhada, or furrow sand washed on top of the valley soil, beyond the medao. The Tagus

[8] For other mines of the Middle Ages, see Aldrovandi cited in a previous note.

[9] For plunder of the Lombards and Jews by Edward III., in 1336, see Jacob.

sands were washed by the Romans down to the Augustan period, and probably until the conquest of the peninsular by the Goths, who, being in some measure governed by Buddhic or Bacchic tradition, seem to have refrained from searching for the precious metals. After the Moslem Conquest, the Tagus was worked by the Saracens down to A. D. 1147, and after that by the Portuguese down to A. D. 1550. It is during the period last mentioned that we have any quantitative record of the product. In these four centuries the Tagus' sands yielded about £378,000, or less than a thousand pounds sterling ($5,000) per annum, of which the King of Portugal received one-half as royalty, besides charging the miners for licences and compelling them to sell their gold bullion to the Royal Mint at less than its legal or mint valuation in coins.[10]

The influx of the precious metals from America and the Orient, which occurred during the 16th century, the rise of prices and wages which followed, and the demand for miners in the newly-found countries, put a stop to the humble efforts of the Tagus artilleseros; nevertheless, these sands were again worked from 1814 to 1833. During this period (less than sixteen years of actual work), the aggregate product was 46,270, expenses 43,142, and profit 3,128 milreis. The number of men employed was about twenty; the product of each man was about 2s. 6d., or 62 cents; the expenses, royalty, licenses, etc., 2s. 4d., or 58 cents, and the profit 2d., or four cents, per diem.[11]

The sands of the Rhine, chiefly between Strasbourg and Phillipsburg, were worked for gold during the Middle Ages. At one time the magistrates of Strasbourg farmed out the right to work them. By the year 1718 the washings had become so meagre that the share of these magistrates was only four or five ounces per annum. In 1846 the annual yield was estimated at 36 pounds troy. The washers usually made from 1s. 3d. to 1s. 8d. (30 to 40 cents) a day. Occasionally the profits rose to several times these sums.

The Moslem mines of the Middle Ages have been so fully described in the author's previous works[12] that nothing needs to be added in this place except the details which have rewarded his researches

[10] For plunder of the Byzantine temples in France, by Charles Martel, about 730, see Raynal, II, 292; for plunder of the Lombards and Jews, by Edward III., in 1336, see Jacob, p. 179; and for numerous despoilments of the Jews during the Middle Ages, consult Madox's "History of the Exchequer."

[11] Report of the United States Monetary Commission, of 1876, I, 457.

[12] Del Mar's History of the Precious Metals, ed. 1880, p. 41; History of Money in Ancient States, ed. 1885, p. 133; Money and Civilization, ed. 1886, p. 18; and History of Monetary Systems, ed. 1901, p. 125.

since those works were written. Lady Callcott gives a date to the working of the mines of Jaen, Bulache and Oroche, etc., by the Moslems. This is A. D. 976; but she omits to state whether this was the earliest working of those mines or not. The gifted authoress also asserts that the Moslem gold mines, in Spain, were worked for the account of private individuals and not for that of the State or the Crown [13]

Among the mines of Spain opened by the Moslems, Al-Makkari mentions the gold washings of the Lerida, an affluent of the Ebro; a very rich silver mine near Santiago, the capital of Gallicia; one at Tadmir, another in the mountains of Al-hamah (La Seca); besides gold mines in Cordova, Andalusia, and other provinces; and quicksilver (cinnabar) and tin mines both in Spain and Portugal.[14] We are also informed that the Moslems opened or re-opened numerous mines in Mauritania and interior Africa, in Sardinia, in Egypt and down the East Coast of Africa as far as Sofala, as well as in the Hedjaz, Syria and Armenia.

In A. D. 933 the Arabians, from the Hedjaz and from Muscat, which latter place Anderson includes under the general name of "Persia," re-opened the mines of south-eastern Africa down to the Mozambique, or else renewed the trade for gold upon such terms or under such conditions that to this period is ascribed the building of Brava, Mombaza, Quiloa, Mozambique, Magadoxa, Sofala, and other Arabian towns on the African main and in Madagascar, as emporia for the prosecution of the gold trade. The introduction of mercury for amalgamation, or of some other improved process for obtaining or treating gold ores, may have furnished the immediate stimulus for this revival; but it is mediately due to the Moslem Conquest of India during the tenth century. This encouraged the erection not only of the emporia mentioned, but also of others not immediately connected with the gold trade: those which stretched from the Red Sea to Cape Comorin, the southernmost point of India. When the Portuguese doubled the Cape of Good Hope in the 15th century, they found the Arabs in full possession of a systematic and prosperous commerce, which extended in an unbroken line from the Mozambique Channel by way of Zanzibar, Muscat, Laristan, and Beluchistan to the Indies.[15]

If we examine the dates when the oldest existing cathedrals and churches of Europe were erected it will be found that but few of

[13] Callcott, History of Spain, I, 249. [14] Al-Makkari, I, 89-90.

[15] Anderson's Hist. of Commerce, I, 92.

them are earlier than the 13th or later than the 15th century. St. Denis, St. Germain L'Auxerrois, St. Geneviéve, St. Martin des Champs, St. Julien le Pauvre, St. Pierre-aux-Bœufs, St. Antoine des Quinze-Vingts, La Sainte Chapelle and the foundation of La Sorbonne, of Paris, are all of the 13th century; Westminster Abbey and Salisbury Cathedral were founded in 1220; the Cathedral of Cologne 1248; of Worcester, 1281; of York, 1291; of Canterbury, 1378; of Milan, 1386; of St. Peters at Rome, 1450. Whence came the vast sums of Money represented by these structures of a period when money was so scarce that two pence purchased a day's labor, and three pence a bushel of wheat? There can be but one answer to this question: they were built from the metallic spoil of the Eastern Empire, which fell in 1204, and of the Arabian Empire, in Spain, which succumbed to a series of conquests that began in 1095 and ended with its downfall in 1492. In the author's "Middle Ages Revisited," chapter XII, he has described the plunder of the Eastern Empire; in the present work he will endeavor to trace that of Arabian Spain.

The Moslem Empire in Spain was to the medieval world what Rome had been to the ancient: the seat of learning, the home of science and the arts, the emporium of riches. In a previous work on Money, the populousness, brilliancy and opulence of Moorish Spain was described at some length, and that such an Empire should have fallen beneath the arms of the petty state of Castile is only explicable when we discover: Firstly, that among the Moslems there existed internal sources of decay, and Secondly, that Castile was a mere euphemism for all Christendom.

One of the sources of the decay of Moslem Spain is adverted to in the following passage:—"When the mines ceased to pay—an event that came about much sooner with the Saracens than with the Romans, because the former were forbidden by their religion to enslave any but infidels—the supplies of the precious metals diminished, the level of prices began to fall, commerce became depressed, industry gradually ceased, numbers of people were thrown out of employment and reduced to beggary, the rich became relatively richer, the poor relatively poorer, the central government lost authority, and hundreds of wealthy proprietors, rendered arrogant by their wealth, claimed political privileges which formerly it would have cost them their lives to assert. The powers of the Moslem government fell into the hands of jealous nobles and hostile sheiks, and, in this condition, it became an easy prey to the enemy." [16]

[16] Money and Civilization, 83.

The generic source of decay was the Mahometan religion. Unlike Christianity, which adapts itself to all countries and conditions of men, and is not for a day but for all time, Islamism is suitable only to certain countries and social conditions; thus, while it may continue for ages to remain the religion of a race, it must sooner or later cease to be that of a nation. The Koran permits no new laws to be passed, and where it prevails, there can be no such thing as representative government, or a legislature. Therefore, when the growth of a Mahometan nation attains that phase of maturity which demands the substitution of constitutional or representative, for despotic government, either the Koran must be thrown overboard or the Moslem ship of State must take in sail and drop astern, motionless and inactive; while the rest of the world, with eager wings, races on to discovery, riches and power. Such was the case with Moslem Spain.

If to these internal sources of decay, of themselves sufficient in time to ruin any Empire, there is added the combined hostility of the entire Christian world, nothing loth at any time to taste the spoil of the heathen, but rendered especially eager for spoil during the Middle Ages by its own internecine struggles and the general backwardness of industry, there is little necessity to look further for the causes that led to the downfall of Moslem Spain.

The first attack upon the Moslems in Spain was made by Charlemagne. His acquisition of the Spanish March, though it severed but little territory from the Moslems and added but little riches to Christendom, yet it afforded a basis for the military operations of subsequent invaders. His son, Louis le Debonnaire (814--17), lent great assistance to the Christians of Spain by granting them lands in France.[17] About 875, Alfonso III., King of Asturias and Leon, married Amelina, cousin to the King of France, and niece of Sancho, King of Navarre, thus forming a French alliance.

The footing obtained through the conquest of Charlemagne, the assistance of Louis, and the avidity of those bands of Free Lances, or mercenaries, who at this period wandered through Western Europe in quest of employment and plunder, had much to do with the foundation of the petty monarchies of Northern Spain.[18] Through inter-marriages, they became gradually merged into one or two principal kingdoms; and, being continually assisted by the Christian monarchs of France and other countries, they at length secured

[17] Robertson's Charles V., I. 215.

[18] "Spain was aided by volunteers from all western christendom." Freeman, Hist. Saracens, 142.

important conquests from the Moslems. Such was the nature of the acquisitions made by Ferdinand I. of Castile and his Christian allies from France and Italy; and from this æra (eleventh century) is to be dated the first serious step in the decline of the Moslem power in Spain. The operations of Ferdinand and his allies had been greatly assisted by the Normans, who, in the early part of this century sacked many of the maritime cities of the Peninsular, the island of Sicily and other Moslem possessions, and who, with the spoil thus obtained, enriched those nobles, their masters, who were soon also to become the masters of anarchical England.

In 1085, the Christian allies had pushed their conquests to the province of Toledo, the capital of which yielded in that year to the strategy of Alonzo. It at once became the capital of Christian Spain. The general policy of the victors was never to keep faith with Moors or Jews; to extort confessions from them of hidden treasures by means of torture, and to exile them from the conquered lands and fill their places with Christains from France and other states. The burning alive and hewing in pieces of Moorish and Jewish prisoners was begun by the heroic Cid Campeador, while the wholesale deportation of these people was carried out by the valiant Alonzo I. of Aragon. The Cid was brought up by his godfather, a priest of Burgos, in the belief that with Jews and infidels there was no faith to be kept. His conduct to them was both cruel and perfidious; and while he is said never to have failed in his word to a Christian, he raised money from the Jews, pretending to have, in pawn, chests full of plate, to be opened a year after they were delivered; while the boxes, in fact, contained nothing but stones and sand. For his mercy and tenderness to Christians, he made up by the burnings alive and hewings in pieces of the Moors, to extort confessions of hidden treasures.[19]

This was in strange contrast with the conduct of the Moors toward the Gothic princes from whom they conquered Spain nearly four hundred years previously. The following is the text of one of the treaties made on that occasion:

"Treaty of peace between Abdul Asis ben Muza ben Noseir and Theodomir,[20] Gothic King of the Land of Tadmir:[21] In the name of

[19] Callcott, I, 292.

[20] Theo=god and Mir=the Sea, or God-of-the-Sea, suggesting the sacerdotal Buddhic title of a Prince of Mingrelia and the custom of Marrying the Sea, mentioned in the author's "Worship of Augustus Cæsar," 93, 219. Tad, or Tat, is also a Buddhic or Bacchic sacerdotal name. An ancient image of Bacchus as a sea god, bearing a trident, has recently been exhumed in France, and is depicted in the Chicago "Open Court" for December, 1900.

[21] The Land of Tadmir included Murcia and Carthagena.

the most merciful God: Abdul Asis and Tadmir (Theodomir) make this covenant of peace, which may God confirm and protect; that Theodomir and no other shall have the command of his principality; that there shall not be war between them, nor shall they take captive each other's wives or children; that they shall not be molested in their religion, nor shall their temples be destroyed, nor shall other services or obligations be imposed, beyond those herein contained. That this covenant shall likewise extend to the seven towns of Orihuela, Valentola, Alicant, Mala, Bosara, Ota and Lorca; that he shall not receive our enemies, nor fail in his fidelity, nor conceal any hostile design which he may learn; that he and each of his nobles shall pay every year one piece of gold (dinar), besides four measures each of wheat, barley, wine, vinegar, honey and oil; and that the vassals shall pay one-half of the like tax. Written on the 4th of Regib, in the year of the Hegira ninety-four (equivalent to our A. D. 712). Witnesses: Othman ben Abi Abda; Nabib ben Abi Obeida; Edris ben Maicera; Abulcasim el Mezeli."

In 1095, the Pope of Rome authorized the first general crusade, and in 1136 a special crusade against the Moors and Jews of Spain, and he incited or authorized the Genoese to begin the attack. This crusade resulted in the conquest of Almeria with great booty, part of which was reserved and awarded to the Holy Father. In 1173, the Genoese assisted in the siege of Tortosa and shared its spoil with the King of Castile and the Count of Barcelona. In 1139, the Emperor Conrad III. sent a fleet of German (Bremen), French and English ships, apparently to Palestine, but really to Portugal, to assist Alfonso in the capture of Lisbon from the Moors. When they had effected this object the allies defeated the Moslems on the plain of Ourique, and thus established the Christian kingdom of Portugal. After extirpating or banishing all the Moors, the country was populated by planting numerous colonies of Christians. The execution of these measures gained for Sancho I. (1185–1212), the surname of Poplador, or populator.[22]

"Their various connections with foreign nations and the gradual increase of their dominions, encouraged the christain kings of Spain to form more decided plans than they had hitherto done for the final reduction of the Moors. Accordingly, at a conference held for that purpose in 1179, they laid out the whole of the Moorish territory into partitions (spheres of influence), and to each of the christian kings one of these partitions was to belong; and although another should

[22] Callcott, I, 339, 372.

make war and overcome the Moors within its limits, still the conquest should belong to the king to whom the portion was originally assigned."[23]

About 1185, the kings of Navarre and Castile agreed to refer a dispute, concerning their conquests from the Moors, to Henry II. of England, the father-in-law of the latter. The ambassadors and their retinues found Henry at Windsor, and there the dispute was settled to the satisfaction of both parties.[24] At another time Henry convened a plenary court at Beaucaire, on the Rhône, to setttle a dispute between the King of Aragon, as Count of Provence, and the Count of Toulouse. On this occasion, and in order to exhibit their opulence, 100,000 silver pennies, paid by the Count of Toulouse, were distributed among 20,000 Free Lances; the Baron Bertram Raimbant sowed the land around the castle of Beaucaire with 120,000 silver farthings; William de Martel, who had 300 knights in his train, caused their meat to be cooked over wax tapers; the Countess of Magel presented a crown of immense value to the assembly; and Raymond de Rons proved his Gothic lineage by sacrificing to the flames, and before the whole court, thirty valuable horses.[25]

Between 1184 and 1212 the Pope of Rome instigated several crusades against the Spanish Moors. Bands of adventurers and Free Lances from Italy, France, and even England, poured into Spain and Portugal, and together they defeated Mohamed Abu Abdallah at Tolosa in 1212, killed 100,000 Moors, took 60,000 prisoners, whom they reduced to slavery, and thus virtually crushed and terminated the Almohade dynasty.

In 1217 the christians gained Cordova. In 1229 James I., of Aragon, assisted by forces from Genoa, Marseilles and Provence, took Majorca from the Moors. Marseilles (then a republic) had for its share of the spoil 300 houses in the capital of Majorca, besides other houses and lands in other parts of that then rich island. In 1231 the Genoese took Ceuta from the Moors. In 1237 the military forces of James I. of Aragon were greatly strengthened by a band of chosen men-at-arms from France, under the Bishop of Narbonne, and by numerous English knights and men-at-arms. Together, these warriors and mercenaries carried the crusade into Valencia. This campaign began with an act of treachery to Seid, the Moorish chieftain, and ended with the sack of Valencia and the banishment of the entire population. In 1240, when Murcia fell into the hands of the crusaders, a similar policy was carried out: despoilment and exile.

[23] Callcott, I, 362. [24] Callcott, I, 365. [25] Callcott I, 388.

In 1275 Edward I., King of England, assisted Alfonso of Castile (his brother-in-law) in his military operations against the Moors. Alfonso X., of Castile, 1252–84, was assisted by Louis IX. and Philip III., of France, to whom the former was related. Peter the Cruel of Castile, 1350-69, was assisted by both of these princes in his dynastic and Moorish wars, and by those companies of adventure, who, after the pacification between France and England, had lost the occupation of war and retained only that of plunder. Finally, Henry III., of Castile, 1391-1406, was assisted by Charles III., of Navarra, to defend christianized Murcia against the Moors of Grenada.

Thus, province by province, nearly the whole of that magnificent empire which the Moslems had conquered from the Goths, and enriched with seven centuries of agriculture and mechanical industry, fell into the hands of the christians. The spoil was immense; and it was this spoil that, upon being distributed among the various christian states of Northern and Western Europe, laid the foundations of their prosperity. Following the spoil came the lands, improvements, manufactures, commerce and scientific acquirements of the Saracens, all of which fell to the conquerers. The Moors were banished or reduced to slavery and compelled to till the lands, or practice the trades, or prosecute the commerce, which was once their own, but now went to enrich the hated Romans.[26] A Jew, whom Alfonso XI., of Castile, 1312–50, had appointed to collect tribute from the inhabitants of a province of Spain which he had wrested from the Moors, told Ibn-Kaldun that he had collected no less than 65 komt (cuentos), each komt being five kintars (or hundred weights) of gold: that is to say, nearly 200 (325?) kintars of gold from his unhappy compatriots.[27]

Some of the various agencies through which a share of the plunder of Spain reached England have already been indicated. These were chiefly the Norman maritime raids of the eleventh and the crusades of the twelfth century. Both of these were against Moorish Spain. The Jews, banished from Spain, also flocked to England under the Normans. A number of Jewish names, such as Ben (?), José, Sancto, Santon, Vives, etc., will be found in the Norman exchequer rolls, cited by Madox. Never had England been so rich as when she was flooded with the golden mancusses brought in by these refugees during the reign of Henry III. Some instances of the comparatively

[26] To this day the Moors regard all christians as Romans. A passport recently granted by the sheriff of a town in Morocco, to an American traveller, described the latter as a "Roman."

[27] Al-Makkari, 402*n*.

large sums of gold and silver found at this period in the castles of English prelates and barons are given in the author's "History of Monetary Systems." Other and larger sums were sent from England to Rome, either as Peter's Pence or to pay for special privileges or indulgences; and still others to defray the expenses of further crusades.

A few examples of Moslem opulence will serve to convey an estimate of the wealth gained by christendom from the despoilment of the Moors. During the reign of Abdurahman, Caliph of Cordova, 912–61, Abdul Melik ben Said was appointed privy councillor, upon which he sent the following presents to the Caliph: 400 pounds weight of pure gold of Tibar, silver in bars to the value of 420,000 sequins (the sequin or dinar contained about the same quantity of gold as 12s. or $3 of the present day), 400 pounds of lign-aloes, 500 ounces of amber, 300 ounces of precious camphor, 30 pieces of silk and gold interwoven; 110 fine ermine cloaks from Khorrassan, 48 horse caparisons of silk and gold woven at Bagdad, 400 pounds of spun silk, 30 Persian carpets, 800 horse champrons of burnished steel, 1,000 shields, 100,000 arrows, 15 thoroughbred Arab horses with gold embroidered housings, 100 Spanish and African horses, well clothed; 20 sumpter mules with pack saddles and curtains, and 40 male and 20 female slaves, well matched.[28] Towards the end of his reign, Alkahem II., 961–76, ordered a cadastre and census of his caliphate. The returns showed that it contained six large cities, the capitals of provinces; 80 cities exceedingly populous, 300 of the third class, and villages, hamlets, villas and farms innumerable. The valley of the Guadalquiver alone contained 12,000 farms. Cordova had 200,000 houses, 600 mosques, 50 hospitals, 80 public schools, and 90 public baths. The yearly revenues of the state were 12 million mithcals of gold (weighing about 10s. each), besides the taxes paid in kind. The gold mines belonging to the king, or to private persons, produced great sums. Those of Jaen, Bulache, Aroche and of the mountains near the Tagus, in the west of Spain, were very productive. Rubies were found at Beja and Malaga, coral was fished on the coasts of Andalusia, pearls were found on those of Tarragona, reservoirs and irrigating canals were built in Granada, Murcia, Valencia and Aragon, and the most illustrious knights devoted their leisure to tillage and horticulture.[29] In the twelfth century the pulpit and pew of Abdelmumen were constructed of aromatic wood, sculptured in scrolls and flowers; the clasps and hinges were of solid

[28] Callcott, I, 241. [29] Callcott, I, 249.

gold; the coffins of the distinguished dead were wrought of cyprus wood and adorned with gold; while in the fourteenth century (battle of Guadelcito) the quantity of gold and silver captured by the christians from Muley Hassan was so vast that its coinage and distribution caused a general rise of prices in christain Spain.[30]

The Abbé Raynal (III, 246) says that the kingdom of Cordova, previous to its conquest in January, 1492, was for its size the richest in Europe. It contained no less than three million industrious inhabitants. Nowhere else throughout the civilized world were the lands so well cultivated; manufactures so varied, numerous and improved; nor navigation and commerce so extensive, regular and profitable. The public revenues in money alone amounted to a sum equal to three million livres of the writer's time, which his translator converts into the equivalent of £292,000 sterling; "a prodigious sum at a time when gold and silver, compared with now, were exceedingly scarce."

Besides the revenues of the Caliphate, there were local revenues which probably amounted to even a larger sum. In every house there was some token of opulence, some service of plate, some article of jewelry, some precious memento. The bodies of the dead were adorned with gold and jewels. All this wealth became the spoil of the christian conquest. Even the graves were violated in search of gold.

The manner in which the Empire of China is being invaded, parcelled out into "spheres of influence," and plundered by christian troops at the present day, is not so very unlike the invasion and despoilment of Moslem Spain during the Middle Ages as to save from a blush of shame the professed worshippers of a benignant and merciful God. The future boast of the christian nations will not be which one conquered the most of China, but which one robbed her the least.

[30] Callcott, I, 323–5; II, 31.

THE AMALGAMATION PROCESS.

It is a common belief, one which will be found in most works of reference, that the employment of mercury for the purpose of catching, saving, or recovering the precious metals, called the Amalgamation Process, was invented by one Medina, in 1557. This belief is erroneous. The fact that mercury, when it touches gold or silver, instantly combines with it, and in case the latter is in the form of powder or grains, forms a compound which we call amalgam, was known and utilized in the most ancient times. These circumstances are attested by the evidence adduced on page 8 of the present work; but we shall presently bring forward other evidences. Meanwhile, it is necessary to observe that there is another, a complicated metallurgical process, called by English writers "amalgamation," but known to Spaniards as the "patio process," patio referring to the court or yard in which the process is carried on. In the patio process the pulverized ores of silver, after being spread over the court, receive a charge of mercury, common salt, copper pyrites and water; the combination being called "magistral." The mixture results in converting the argentiferous portion of the ore into metallic silver and rejecting the worthless remainder.

While there is nothing to show that the ancients were aware of this process, there is abundant reason to believe that simple amalgamation was quite familiar to them. At the present day the natives employ this process in Assam (276), Bhandara (322), Dharwar (318), Jhilam (347), Malabar (337), Moradabad (344), the Punjab (344), Rawalpindi (348), and other mining districts of India, the numbers affixed to each district indicating the pages of Lock's work on Gold, in which the practice is mentioned. It is confidently believed that this practice has continued since the Moslem conquest of the tenth century. It may even be of the greatest antiquity.

Sir Gardiner Wilkinson, A. E., ed. 1878, II, 213n, says that mercury in flasks has been discovered in ancient Egyptian tombs. This metal was sacred to Hes-Iris, or Osiris, who was the same as Buddha, Bacchus, or Mercury, whose name it bears. It was probably used by the Egyptians for recovering the precious metals. In the Greek silver mining districts of Laurium there were deposits both of cinnabar, the ore of mercury and copper, the former of which were worked to advantage. Boeckh, Polit Econ. Ath. ed. 1857, p. 416. These deposits, found in such close conjunction, could scarcely have failed to suggest the "magistral" of the Spaniards. At all events, it is quite likely that the Greeks employed simple amalgamation, for it is treated familiarly by the Romans. It is distinctly mentioned by Vitruvius about B. C. 27, and described by Pliny the Elder about A. D. 73. Says the latter, N. H., XXXIII, 32: "Gold is the only substance that quicksilver attracts to itself; hence it is that it is such an excellent refiner of gold, for on being shaken in an earthen vessel with gold it rejects all dross, clinging only to the gold. The dross being expelled, it simply remains to separate the quicksilver from the gold. This is done by enclosing the mixture—amalgam—in a well-prepared skin, which, being squeezed, exudes the quicksilver, like a sort of perspiration, leaving the pure gold behind." This is precisely the process employed by common miners to-day; the skin used being a chamois bag. Edrisi—12th century—informs us that during the eleventh century, and "long previously" quicksilver amalgamation was employed by the Arabian and negro miners of Western Africa, Abyssinia, Bodja, and Nubia. Humboldt, "Fluctuations of Gold," ed. 1900, p. 27. Amalgamation was employed by the gold miners of Portugal during the reign of King Diniz, 1279-1325. Rep. U. S. Monetary Com. 1876, I, 458. It had probably been learnt from the Arabs. In 1525, "a miner, named Paolo Belvio, was sent with a provision of quicksilver to Hayti, in order to expedite the gold washings by means of amalgamation." Humboldt, op. cit. 26. Herrera, in his history of Spanish-America, says that, although previous to the opening of Potosi the Spaniards knew the art of amalgamating mercury with gold, they were unaware that mercury would amalgamate with silver and that then they discovered it for the first time. Anderson, Hist. Com. II, 76. Herrera must have borrowed his metallurgy from Pliny. It is practically impossible to know that mercury will amalgamate with gold and yet not know that it will also amalgamate with silver, because in subterranean mines both metals are commonly found in the same matrix. Even in alluvial or placer gold there is often a small proportion of silver. That mercury amalgamates with both metals must therefore be a fact familiar to every one who has used it. What Herrera should have said was, that previous to the opening of the Mexican silver mines by the Spaniards they were familiar with simple amalgamation, but not with the complicated patio process, which may have been the discovery of Medina in 1557, and that in 1571 Pedro Fernandes de Velasco carried this process to Potosi, in Peru. And. Hist. Com. sub an. 1572, and the authorities therein cited; Beckmann, Hist. Inv. I, 17. I am, however, inclined to the belief, that Medina's invention was merely a modification of, or improvement upon, a patio amalgamation process employed by the Arabian and afterwards by the Spanish metallurgists in Europe. Agricola, who was well acquainted with simple amalgamation, also with the reduction of silver ores by roasting, indicates a third method which could scarcely have failed to suggest the patio process.

CHAPTER XI.

THE PLUNDER OF AMERICA.

Brief review of the financial condition of Europe at the period of the Discovery—Dearth of metallic money—Motive of Columbus' expedition: to discover and obtain gold—Expeditions of Cortes and Pizarro—Expedition of De Soto—This was essentially a charter to murder, torture and enslave the natives of America, in order to obtain gold for the Crown of Spain—Opinions of Baron von Humboldt and Sir Arthur Helps.

THE History of the Precious Metals in America can best be told after clearing the ground with a brief review of the monetary condition and circumstances of the states of Europe at the time when America was discovered. Strictly speaking, these circumstances would carry us back quite to the beginning of metallic money in Greece; but of this event a full account will be found in the author's previous works. Suffice it to say in this place that after many experiments with coins of gold and silver in the Greek and Roman republics, these metals had been so far abandoned as money that the measure of value in those states was eventually made to depend less upon the quantity of metal contained in the coins, than upon the number of coins emitted and kept in circulation by the state. The integer of these systems was called in Greek, *nomisma*, in Latin, *numisma*, both of which terms relate to that prescription of law which conserved and emphasized the numerical feature of the system in each state. Such integer consisted of the whole sum of money; not upon any fraction of it. [1]

Upon the relinquishment of these systems, both the Greek states and the republic of Rome again committed themselves to metallic systems, this time with open mints or private coinage; the consequence of which was the gradual concentration of wealth in the hands of a few persons, a circumstance which powerfully assisted the downfall of the state. Upon the assumption of imperial power over the European world by Julius Cæsar and especially by Augustus, private coinage, or the issuance of "gentes" coins, was at once forbidden

[1] Paulus, in the Digest.

and the state once more assumed the control of money, which, although the pieces were still made of the precious metals, was so regulated as to constitute a more or less equitable measure of value: the principal means employed in this regulation being the imposition of mine-royalties and a seigniorage or "retinue" upon coinage, coupled with a localization of the bronze and sometimes also of the silver issues. This highly artificial system, though it lasted several centuries, gave way when the subject kingdoms and provinces of Rome revolted from her control and established themselves as independent or partly independent states; a movement that began with the so-called Barbarian uprisings of the fifth century and was completely consummated when Constantinople fell in 1204.

At this period the quantity of money in circulation, outside of the Moslem states, was extremely small; according to Mr. William Jacob, the vast acquisitions of the Empire had disappeared chiefly through wear and tear, coupled with the lack of fresh supplies of the precious metals from the mines. Much of these metals had been taken as spoil by the Moslems and transported to their various empires in the Orient; much had been absorbed and sequestered by the temples and religious houses of the West; and much had also been hidden and lost in secret receptacles. It was estimated by Gregory King in 1685 that the whole stock of the precious metals, in coin and plate, in Europe, at the period of the Discovery of America in 1492 did not exceed £35,-000,000 in value; and to this estimate Mr. Jacob, after the most careful researches, lent his full support. The population of Europe at that period could hardly have exceeded thirty millions; so that the quantity of coin and plate did not much exceed in value £1, or say $5 per capita. Of this amount it could hardly be supposed that more than one-half consisted of coins. The low level of prices at this period fully corroborates this view. Moreover, there was nothing to alleviate the scarcity of money; no means of accelerating its movement from hand to hand, and so of increasing its velocity or efficiency; no substitutes for coins; no negotiable instruments; no banks except those of the Italian republics; few or no good roads; no rapid means of communication; little peace or security; and no credit. Since the fall of the Roman empire every device by means of which this inadequate and always sinking Measure of Value could be enlarged had been tried, but in vain. The ratio of value between gold and silver in the coins had been altered by the kings of the western states with a frequency that almost defies belief. The coins had been repeatedly degraded and debased; clipping and counterfeiting were offences so common that notwithstanding the severest penalties, they were often committed by persons of the highest respectability, by prelates, by feudal noblemen and even by sovereign kings. The emission of leather moneys had been repeatedly attempted, but the general insecurity was too great and the condition of credit too low to admit of any extensive issues of this kind of money. Bills of exchange known to the East Indians as *hoondees* and familiar to the Greeks and Romans of the republican periods, had from the same cause almost entirely fallen into disuse. The cause was the low state of credit. The social state itself, so far as it depended upon that exchange of labour and its products which is impossible without the use of money, was upon the point of dissolution, when Columbus offered to the Crown of Castile his project for approaching the rich countries of the Orient by sailing westward.

What was the object of thus seeking Cathay and Japan? To discover them? They had long been discovered and were well known both to the Moslems, who had established subject states in the Orient, and to the Norsemen, who traded eastward with Tartary and India, and had even voyaged westward to the coasts of Labrador and Massachusetts. The Italians had long traded with the Orient through Alexandria and had even sent Marco Polo into China. No. The voyage of Columbus was not to discover Cathay, but to plunder it; to plunder it of those precious metals, to the use of which the Roman empire had committed all Europe and from the absence of which its various states were now suffering the throes of social decay and dissolution.

The terms which Columbus demanded and the Crown conceded in its contract with him, is a proof of this position. He demanded one-eighth of all the profits of the voyage. To this the Crown consented, after making a better provision for itself, by requiring that in the first place one-fifth of all the treasure found or captured in the lands approached should be reserved for the king. The terms of this contract are given more fully in the author's "History of the Precious Metals," and therefore they need not be repeated here. From beginning to end it was essentially a business bargain; its object was not geographical discovery, but gold and silver; its aim was not the dissemination of the Christian religion, but the acquisition of plunder and especially that kind of plunder of which the Spanish states at that period stood in the sorest need.

Said the illustrious Von Humboldt: "America was discovered, not as has been so long falsely pretended, because Columbus predicted another Continent, but because he sought by the west a nearer way to the gold mines of Japan and the spice countries in the southeast of Asia."[2] The expeditions of Cortes and Pizarro had precisely the same objects: to discover and acquire the precious metals, without permitting any considerations of religion or humanity to stand in the way of these objects.

Forty-five years after the Discovery of America the Crown of Spain made a contract with De Soto similar to that with Columbus. It will be instructive to examine its details. This document is dated Valladolid, April 20, 1537. It provides that De Soto shall be paid a salary of 1500 ducats (each, of the weight of about a half-sovereign or quarter-eagle of the present day,) and 100,000 maravedis for each one of three fortresses which he is to erect in the "Indies." To the alcade of the expedition it awards a salary of 200 gold pesos. De Soto may take with him free of duty (almojarifazgo) negro slaves to work the mines. All salaries except that of the alcade are to be paid from the proceeds of the enterprise, so that in case of its failure, there will be nothing to pay. Of gold obtained from mines, the king is to receive during the first year one-tenth, during the second year one-ninth, and so on until the proportion is increased to one-fith; but of gold obtained by traffic or plunder, he is always to receive a fifth. De Soto shall not be required to pay any taxes. He shall have the entire disposal of the Indians. There shall be reserved 100,000 maravedis a year for a hospital for the Spaniards, which shall be free from taxes. No priests or attorneys shall accompany the expedition, except the alcade and such priests as may be appointed by the Crown. After the king's fifth is laid aside from the spoils of war, and the ransom of caciques, etc., then one-sixth shall go to De Soto and the remainder divided among the men. In case of the death of a cacique, whether by murder, public execution, or disease, one-fifth of his property shall go to the king, then one-half of the remainder also to the king, leaving four-tenths to the expedition. Of treasure taken in battle or by traffic, one-fifth shall go to the king; of treasure plundered from native temples, graves, houses or grounds, one-half to the king without discount, the remainder to the discoverer. Signed, CHARLES, The King.[3]

Here is a charter to murder, torture and enslave human beings, to despoil temples and to desecrate graves. It is signed by the King

[2] "Fluctuations of Gold," Berlin, 1838. American edition, 1899, p. 10.

[3] This document appears in full in the New York Historical Magazine for February, 1861. (Br. Mu. Press mark, P. P. 6323.)

of Spain who was also the Emperor of Germany; it is committed to a swash-buckler who by the most infamous means had made his fortune with Pizarro in Peru; it is as sordid a document as ever was penned; a disgrace to Spain, to Christianity, to civilization. It plainly and unequivocally lays bare the motive of this expedition. This was not to discover or explore North America, but to plunder it of gold and silver, to replenish the coffers of the king, to provide those blood-stained metals out of which man, in retrogressive periods, is obliged, through his own degeneracy and distrust of his fellow-men, to fabricate his Measure of Value. Said Sir Arthur Helps, the accomplished historian of the Spanish Conquest of America: "The blood-cemented walls of the Alcazar of Madrid might boast of being raised upon a complication of human suffering hitherto unparallelled in the annals of mankind. . . . Each ducat spent upon these palaces, was, at a moderate computation, freighted with ten human lives."[4] Let us be still more moderate and say one human life to the ducat: even this was sufficiently atrocious.

[4] "The Spanish Conquest in America," London, 1857, III, 215.

CHAPTER XII.

HISPANIOLA.

Gold the first inquiry of Columbus—Its fatal significance to the natives—Columbus' second voyage—The mines of Cibao—Columbus proposes to ship the natives as slaves to Spain—Sufferings of the colonists—Their search for gold—Their disappointment and cruelty—Columbus ships four cargoes of natives to Spain as slaves—He hunts the natives with bloodhounds—Despair of the natives—Columbus reduces them all to vassalage—Their rapid exhaustion and extinction—Story of the cacique Hatuey—The golden calf—Cruelty of Ovando—Death of Queen Isabella—Her terrible legacy to Ferdinand—Columbus dies in poverty and debt—Forty thousand natives dragged from the Bahamas and condemned to the mines—Character of the gold-seekers.

NO sooner had Columbus taken formal possession of the island of Hispaniola than he asked the wondering natives for gold. This fatal word, so fraught with misfortune to the aborigines that it might fittingly furnish an epitaph for their race, and so tainted with dishonour to their conquerors that four centuries of time have not sufficed to remove its stigma, seems to have been literally the first verbal commuuication from the Old World to the New.

Some of the islanders had a few gold ornaments about them. "Poor wretches" (says Navarette) "if they had possessed the slightest gift of prophecy, they would have thrown these baubles into the deepest sea!" They pointed south and answered, "Cubanacan," meaning the middle of Cuba.

Shortly after the discovery, Columbus was wrecked on the coast of Cuba, and he sent to the neighbouring cacique, Guacanagari, to inform him of his misfortune. The good chief was moved to tears by the sad accident, and with the labour of his people lightened the wrecked vessel, removed the effects to a place of safety, stationed guards around them for their better security, and then offered Columbus all of his own property to make good any loss which the latter had sustained.

Touched by this unparallelled kindness, Columbus thus expressed himself of these Indios: "They are a loving uncovetous people, so docile in all things that, I assure your Highnesses, I believe in all the world there is not a better people or a better country; they love their neighbours as themselves, and they have the sweetest way in the world of talking, and always with a smile."

In return for their hospitality and loving kindness, the Spanish captain resolved to establish a colony among them, having found such goodwill and such signs of gold. He built a fort, called it La Navidad, left forty adventurers in it, among them an Irishman and an Englishman, and sailed to Spain.

The first thing done, after his return home—the recital of his wondrous story, his reception at the Court of Spain, and the Te Deum—was to obtain a grant of the newly-found domain and all its contents, animate and inanimate, from the Pope of Rome. These objects were effected by a Bull, dated May, 1493.

In September, 1493, Columbus set forth again, this time with seventeen vessels and 1500 men.

He found La Navidad destroyed, and his forty colonists missing. According to the cacique, Guacanagari, the Spaniards had made a raid, probably for gold, upon a tribe of the interior, and notwithstanding the advantages of their arms, had been defeated and killed to a man. Columbus built another fort in another part of the island, called it Isabella, and at once gave his attention to the subject of gold.

"Hearing of the mines of Cibao, he sent to reconnoitre them; and the Indios, little foreseeing what was to come of it, gave gold to the Spanish messengers. Columbus accordingly resolved to found a colony at Cibao."

In January, 1494, Columbus sent to the joint sovereigns of Spain, by the hands of Antonio de Torres, the Receiver of the colony, an account of his second voyage, with recommendations for the consideration and approval of Los Reyes.

After the complimentary address, it begins with the reasons why the admiral had not been able to send home more gold. His people have been ill; it was necessary to keep guard, etc. "*He has done well,*" is written in the margin by order of Los Reyes.

He suggests the building of a fortress near the place where gold can be got. Their Highnesses approve: "*This is well, and so it must be done.*"

He then suggests to make slaves of the Indios, and to ship some of them to Spain, to help pay for the expenses of the expedition. The answer to this atrocious project is evasive, as though Los Reyes did not wish to wound so valued a servant by a point blank refusal. It is: "*Suspended for the present.*"

Money was very welcome at the Spanish Court, where there was more show than maravedis; but Los Reyes were not yet prepared to obtain it by sanctioning the enslavement of an innocent and friendly people. On the other hand, Columbus was eager for the measure.

While de Torres was at the Court with these recommendations, Columbus' colony fared badly on the island. The provisions which they had brought with them failed, and white men were threatened with starvation, where the Indios lived without effort. To their great disgust the Spaniards had to go to work, and till the earth for bread, instead of scouring it, as they had expected, for gold.

"The rage and vexation of these men, many of whom had come out with the notion of finding gold ready for them on the sea shore, may be imagined. . . . The colonists, however, were somewhat cheered, after a time, by hearing of gold mines, and seeing specimens of 'ore' brought from thence; and the admiral went himself, and founded the fort of St. Thomas, in the mining district of Cibao."

It is needless to say that, without the establishment of any permanent sources of supplies, the gold hunters failed in their enterprise, and most of them lost their lives. "They went straggling over the country; they consumed the provisions of the poor Indians, astonishing them by their voracious appetites; waste, rapine, injury and insult followed in their steps."

Worn out with their sufferings, the miserable Indios "passed from terror to despair," and threatened the Spanish settlement. Columbus sallies forth, routs the Indios of Macorix, and captures the majority, four shiploads of whom he sends to Spain, February 24, 1495, as slaves. These were the very ships that brought out the evasive reply of Los Reyes to Columbus' request for leave to enslave the natives.

After this, Columbus starts upon another expedition, at the head of 400 cavalry, clad in steel, armed with arquebuses, and attended by bloodhounds. He is opposed by 100,000 Indios. Their soft and naked bodies not being proof against horses, fire-arms, or ferocious dogs, a horrible carnage ensues, and another bloody installment is paid towards the cost of gold. Columbus captures the cacique, Caonabó, through the vilest treachery, and imposes a tribute of gold upon the entire population of Hispaniola.

The tribute is as follows: Every Indio above fourteen years old, who was in the provinces of the mines, or near to these provinces, was to pay every three months a little bellfull of gold; and all other Indios an arroba of cotton.

When this unreasonable tribute was imposed, Guarionéx, cacique of the Vega Real, said that his people did not know where to find the gold, and offered in its place to cultivate a huge farm, fifty-five leagues long, covering the whole island, and to produce therefrom enough corn to feed the whole of Castile. Poor Indio! This was, indeed, a suggestion of despair. Hispaniola, at the utmost, did not contain more than 1,200,000 Indios, man, woman, and child. Castile contained a population of 3,000,000 or 4,000,000. An attempt to feed a population so large by one so small, and at a distance of 4,000 miles, could only have ended in failure. But Guarionéx might as well have made this as any other proposal. What their Catholic Majesties wanted was not bread but gold; and this is what, in their names, Columbus was bent upon obtaining. Yet however much he desired it, the gold could not be collected, simply because there were no gold mines of any consequence, only some poor washings, in Hispaniola, from whence it might be got. Columbus was, therefore, obliged to change the nature of his oppressions. This was done by reducing the whole native population to vassalage; and thus, in the year of our Lord 1496, was begun the system of *repartimientos* in America. [1] Such was the reward for the unparallelled kindness of good Guacanagari, and for his loving, uncovetous people, "who always spoke with a smile."

Reduced to a condition of vassalage, infinitely worse than slavery, the Indios fell into the profoundest sadness, and bethought themselves of the desperate remedy of attempting to starve out their masters by refusing to sow or plant anything. The wild scheme reacted upon themselves. The Spaniards did, indeed, suffer from famine; but power, exercised in the cruelest manner, enabled them to elude the fate which had been intended for them; whilst the Indios died in great numbers of hunger, sickness and misery.

In the early part of 1496, Columbus discovered a gold mine in the south-eastern part of Hispaniola. On his return to Spain in the same year he sent out orders to his brother Bartholomew to build a fort there. This was done and the place called San Domingo. From this port Bartholomew sailed out to Xaragua, east of the modern

[1] The repartimiento, afterwards the encomienda, was derived from the feudal tenures of Spain. It was a grant of Indios (not including land) to render fixed tribute, or personal services, or both, during the life of the encomiendero or suzerain. This was afterwards extended to two, three, four, five and six lives, and was greatly abused. Consult Irving's "Conquest of Granada," vol. i, pp. 145, 164, 173, 197, 198, and iv, p. 353 *et seq.*

Port-au-Prince, the only unconquered portion of the island. He reduced it to vassalage and demanded tribute in gold. The cacique Bohechio pleaded that there was no gold in his dominions; so the tribute had to be commuted in cotton and cassaba-bread. Returning to Fort Isabella, Bartholomew found that 300 of his followers had died from hunger and disease, the first considerable installment of the myriads of Spaniards who subsequently perished in the same criminal search for the precious metals.

In 1498 Columbus again set forth from Spain—this time with eight ships and about 900 men. Upon his arrival at San Domingo he sent five of these ships to Spain laden with 600 slaves.

The Court of Spain—at first conditionally, as though it hesitated to thwart its favourite commanders, afterwards absolutely, when it found that none of them were above the practice, and that all evaded the conditions—disapproved of enslaving the Indios. Its objection to this transaction of Columbus was that the captives were not taken in war, and it marked the severity of its displeasure by superseding Columbus in his command and ordering him home.

The officer choosen to replace him was Ovando. In the instructions given to this knight A.D. 1501, he was ordered to treat the Indios justly, and pay them one golden peso a year for their labour in getting gold. Between subjecting themselves to these conditions and living in a state of slavery, there could have been to the Indios but little choice, even if it had been accorded to them. It is due to the Spanish Crown to say that deceived by the reports of the over-sanguine gold-hunters, it supposed that gold was easy of acquisition in the West Indies, and that a moderate amount of involuntary labour on the part of the natives would suffice to produce what was demanded of them.

Ovando left Spain in 1502 with a score or more vessels, and 2,500 persons. As these vessels neared the shore of San Domingo, the colonists ran down to hear the news from home, and, in return, to narrate that a lump of gold of extraordinary size had recently been obtained on the island. It had been picked up by a native woman and was estimated to have been worth 1,350,000 maravedis. Nothing more clearly reveals the character of these expeditions and the persons who composed them, than a brief relation of the fatal consequences of this announcement. Ovando's people no sooner landed than they ran off to the placers, where, in a short time, more than 1,000 of the 2,500 perished miserably from hunger and disease.

"Here it may be noticed that, in general, those colonists who devoted themselves to mining, remained poor; while the farmers grew rich. When melting time came, which was at stated intervals of eight months, it often happened that after the king's dues were paid, and those who had claims upon the produce for advances already made to the miners, were satisfied, nothing remained for the miner himself. *And so all this blood and toil were not paid for, even in money;* and many still continued to cat their meals from the same wooden platters they had been accustomed to in the old country; only with discontented minds and souls beginning to be imbruted with cruelty." (Helps.)

At this juncture, Columbus, authorized to make further explorations in the New World, suddenly appeared at San Domingo. The orders of the Crown forbade him to disembark at the island, for fear that the course of administration for which he had been rebuked would be persisted in; but a violent hurricane was apprehended, and the safety of his fleet afforded him sufficient excuse to seek a harbour. In this storm, which took place as the admiral had forseen, the greater part of a large fleet of vessels which had recently set sail for Spain were lost, with all on board—another sacrifice to the thirst for gold.

Shortly after this, a force of 400 men was sent to reduce the Indios of the province of Higuey. These unfortunates were hunted with firearms and bloodhounds. Of the captives taken, those not wanted as slaves had both their hands cut off, many were thrown to the dogs, and several thousand put to the sword.

Ovando, finding that, under the merciful instructions of Los Reyes about dealing with the Indios, he could get no gold—for they shunned the Spaniards "as the sparrow the hawk" and fled to the woods, there to avoid them and die—transmitted to the Court a report to this effect. In a reply dated December, 1503, Ovando was directed "to compel" the Indios to have dealings with the Spaniards; and thus the slave system begun by Columbus, was re-established by the Court.

It may not be uninteresting in this place to hear what the Indios themselves thought about the conquest of America and the motives which impelled the Spaniards in its prosecution. Something of this is embodied in the story of Hatuey, cacique of a province of Cuba.

Apprehensive that the Spaniards would come, as they afterwards did come, to his territory, Hatuey called his people together and recounting the cruelties of the white men, said they did all these things for a great God whom they loved much. This God he would show them. Accordingly he produced a small casket filled with gold.

"Here is the God whom they serve and after whom they go; and, as you have heard, already they are longing to pass over to this place, *not pretending more than to seek this God;* wherefore let us make to him here a festival and dances, so that when they come, He may tell them to do us no harm." (Herrera.)

The Indios approved this council, and to propitiate the God whom they thought their enemies worshipped, they danced around it until they were exhausted; when the cacique turned to them and said that they should not keep the God of the Christians anywhere, for were it even in their entrails it would be torn out; but that they should throw it in the river that the Christians might not know where it was; "and there," says the account "they threw it."

In 1503, Ovando set out with 70 horsemen and 300 foot-soldiers to visit the friendly Queen Anacaona of Xaragua, who hospitably received him with feasting and rejoicing. In return, Ovando, whose object was to terrify the unhappy natives into submission and slavery, invited the chiefs to a mock tournament, where, at a signal from himself, the queen and her caciques were all treacherously captured, the former was put to death by hanging and the latter were burnt alive.

Shortly afterwards, in an expedition against the Indios of the province of Higuey, the Spaniards cut off the hands of their captives, hanged thirteen of them "in honour and reverence of Christ our Lord and his twelve Apostles," and used the hanging bodies of their miserable victims as dumb figures to try their swords upon. At another time, the Indios were burnt alive in a sort of wooden cradle. "Todo esto yo lo vide con mis ojos corporales mortales." All this I saw with my own corporeal mortal eyes. (Las Casas.)

Queen Isabella of Spain died in November, 1504. Could she with her dying eyes have seen into the Far West, she would have "beheld the Indian labouring at the mine under the most cruel buffetings, his family neglected, perishing, or enslaved; she would have marked him on his return, after eight months of dire toil, enter a place which knew him not, or a household that could only sorrow over the gaunt creature who had returned to them, and mingle their sorrows with his; or, still more sad, she would have seen Indians who had been brought from far distant homes, linger at the mines, too hopeless or too careless to return."

Isabella's will contained a bequest which unfortunately removed all restraint from the oppressions visited upon the Indios. She left to her widower, the Regent Ferdinand, one-half of the revenues of the

Indies as a life estate. In the methods which were resorted to for the collection of these revenues, this meant one-half of the gold which could be extorted by the sweat and blood of the Indios; and Ferdinand, needy and thus endowed, withheld no licence to the adventurers in America, which they alleged was needful in order to swell the Fifths due to the Crown, and the importance of the Queen's legacy.

Upon the death of Isabella, Ferdinand, not being the immediate heir to the crown of Spain, retired to his kingdom of Naples, and was succeeded in the government of Spain by King Philip. This monarch died in 1506, and Ferdinand then became King of Spain. A few months before this, Columbus had died, and, as we shall see of all the Conquistadores, in poverty and debt.

At this period the Indios had become "a sort of money" which was granted in repartimiento to favourites at the Spanish court. "The mania for gold-finding was now probably at its height, and the sacrifice of Indian life proportionately great." So few of the Indios remained alive that negro slaves began to be imported from Africa to fill their places at the mines.

The king was told that the Bahama Islands were full of Indios who might be transported to Hispaniola in order that "they might assist in getting gold, and the king be much served." Ferdinand, who was fully as mindful of his interests as the adventurers upon the islands, gave the required licence, and the evil work commenced. In five years time, forty thousand of the Bahamians, captured under every circumstance of treachery and cruelty, were transported across the sea, all of them to die lingering deaths at the gold mines.

This was among the last acts of the Ovando administration, which closed with the appointment of Diego Columbus in 1509. Only seventeen years had elapsed since the discovery of the island. According to Humboldt's "Fluctuations of Gold," the amount of gold thus far obtained was scarcely more than five million dollars. The cost of its production was several expensive expeditions with their outfits, some thousands of Spanish lives, and at least a million and a half of Indios!

Such was the cruelty of the gold-hunters, and the terror they inspired in the natives, that according to the Abbé Raynal, when Drake captured San Domingo in 1586, he learned from the few survivors of what had once been a populous country that, rather than become the fathers of children who might be subjected to the treatment which they had endured, they had unanimously refrained from conjugal intercourse.

It must not be supposed that these atrocities were peculiar to the Spaniards: rather was it peculiar to the class of adventurers to be found in all countries who, in the hope of rapidly and easily acquired fortunes, coupled with the fascination of a career of adventure, licence, and rapine, are the first to brave the dangers and seek the profits of a miner's life. Similar cruelties have been related of the ancients, who were not Spaniards. Similar ones can also be told of the Portugese, the English, the French, as well as the Americans. They are narrated here of the Spaniards simply because these instances are connected with the greatest supply of the precious metals known to history.

"I swear that numbers of men have gone to the Indies who did not deserve water from God or man," wrote Columbus to the home government, and it was the same with those who went from other countries than Spain.

The vilest scoundrels in Europe were let loose upon the unoffending aborigines of America, and the darkest and most detestable crimes were committed in the sacred name of Jesus Christ.

To these cruelties the necessities of the Crown opened the door. A letter of King Ferdinand to the colonists of Hispaniola is thus fairly paraphrased: "Get gold: humanely if you can; but at all hazards get gold; and here are facilities for you."

CHAPTER XIII.

EL DORADO.

The legend of Dorado—Religious ceremony of the Muiska Indians of New Granada—Version of Martinez—Simon—Orellana—Sir Walter Raleigh—Vain searches for the Golden Country of the legend—The real gold country of South America found by the Portuguese in Brazil—Terra Firma mistaken for Dorado—The former an earthly paradise—Gold money and trinkets of the natives---Desolation caused by the European gold hunters---The pearl fishery---Slave hunting---Las Casas---His despair and retirement---Cruelties of the Spaniards---The actual gold region of Terra Firma, or Venezuela---Its present condition.

EL DORADO means "The Golden," or "The Gilded." It was applied by the Spaniards to that country of limitless gold which their avid imagination had located in South America, some of them fixing it in the Valley of the Essequibo, others in that of the Orinoco, and others again among the Muiska Indians of Bogotá, whose high-priest, it was said, clothed himself during a religious ceremony, with the dust of the metal so much coveted by the Spaniards. Martinez saw El Dorado in Manoa, a city of Guiana, whose buildings were roofed with gold; while in 1540, Orellana recognized it in the valley of the Amazon, whither Raleigh afterwards went to seek it, but found it not. In truth, it never existed at all. It was a myth created by cupidity and nourished by credulity, the search for which cost the lives of myriads of natives and not a few adventurers, both Spanish, English and others.

One of the legends of El Dorado is given by a Spanish monk named Simon, who says that a Spanish captain named Sebastiano Belalcázar having invaded the district of Lake Guatavita near Bogotá, questioned the natives about gold, asking the spokesman if there was any such metal in his country. "He answered that there was abundance of it, together with many emeralds, which he called greenstones," and added that "there was a lake in the land of his overlord which the latter entered several times a year, upon a raft, advancing to its centre, he being naked, except that his entire body was covered from head to foot with an adhesive gum, upon which was sprinkled a great quantity of gold dust. This dust sticking to the gum became a coating of gold, which upon a clear day shone resplendently in the rising sun; such being the hour selected for the ceremony. He then made sacrifices and offerings, throwing into the water some pieces of gold and emeralds. Then he caused himself to be washed with saponaceous herbs, when all the gold upon his person fell into the lake and was lost to view. The ceremony being concluded, he came ashore and resumed his ordinary vestments. This news was so welcome to Belalcázar and his followers that they determined to penetrate this golden country, which they called La Provincia del Dorado—that is to say, the Golden province, where the cacique gilds his body before offering sacrifices. Such is the root and branch of the story that has gone out into the world under so many different forms by the name of El Dorado."

Another version of the Dorado appears to have originated with Francisco Orellana, a companion of Pizarro in Peru. When in 1540 Gonzalo Pizarro started to hunt for gold and slaves east of the Andes, Orellana was second in command of the expedition, which comprised 350 Spaniards, 4,000 Indian porters, and 1,000 blood-hounds for hunting down the natives. After crossing the mountains, the Spaniards discovered the Napo, one of the upper affluents of the Amazon. Despairing, for lack of provisions, of being able to return by the route they had taken, the adventurers constructed a "brigantine" large enough to hold a portion of their numbers and the baggage. The command of this vessel was given by Pizarro to Orellana, with instructions to keep in touch with those who intended to follow the course of the stream afoot. Their provisions becoming at length entirely exhausted, Orellana was instructed to drop down the stream with 50 soldiers, to a village reputed to be some leagues below and return with such provisions as he could secure. In three days Orellana reached the Amazon, which here flowed through a wilderness destitute of human food. To return was difficult; to abandon his commander and continue down the stream, was a course that promised many advantages. This course he adopted. Starting from the confluence of the Napo and Amazon, in February, 1541, Orellana reached the ocean in the following August, thence he sailed to Cubagua and afterwards to Spain. Broken in fortune, health and reputation, Orellana had still a card left to play; this he found in his fertile imagination. He reported that he had voyaged through a country only inhabited by women, and where gold was so plentiful that houses were roofed with it. In Manoa, the capital, the temples were built with the same costly material. Nothing was too extravagant to be believed by the greedy ears of cupidity. His tale spread so fast and received such wide belief that several expeditions were organized, some within the same year, to subdue the fair huntresses of South America and carry home to Spain the sheathing of their golden temples and dwellings. One of these expeditions was headed by Philip de Hutten, a German knight, who started late in 1541 from Caro, on the Pearl Coast, with a small band of Spaniards. After a brief absence he returned to the coast with the story that he had penetrated to the capital of the Omegas, that the roofs of the houses shone like gold, but that he had been driven away by the natives and therefore required a larger force and "more capital" to prosecute the adventure. There was no tribe of Omegas in South America. There was a tribe of Amaguas on the banks of the Amazon, but it does not appear that Hutten had traversed so great a distance. Should it be admitted that this gold hunting "noble" was capable of drawing the long-bow, his story might be explained without discussing this objection. However, he succeeded in obtaining what he wanted—more men and more capital; but while preparing his second expedition, he perished by the hand of one of his associates. (Malouet's "Guyane.")

In 1545 Orellana, having procured sufficient capital for the purpose, set sail from Spain with a large and well equipped force, to conquer the haughty Amazons and pillage their imaginary Dorado. He was fortunate in dying peaceably on the voyage, for his companions would assuredly have murdered him when they came face to face with the dismal truth. It is needless to say that the expedition miserably failed; but though Orellana died and his expedition perished, his lie lived a long life; and it is possibly not quite dead yet.

The most famous of the numerous expeditions to discover and pillage this figment of Orellana's brain was organized by Sir Walter Raleigh. After massacreing the Spaniards who aided Desmond in Ireland, seizing for himself twenty thousand acres of Desmond's lands and debauching one of Queen Elizabeth's maids of honor, and thus rendering himself quite eligible for an enterprise of this character, he prepared for a voyage to the land of gold. In 1595 he set sail with five ships. After spending several months in roaming the country between the Amazon and Orinoco, he returned to England with Orellana's tale embellished. In his "Discovery of the Large, Rich and Beautiful Empire of Guiana," he describes the gilded King of this favored country (el rey Dorado) whose chamberlains, *every morning*, after rubbing his naked body with aromatic oils, blew powdered gold over it, through long sarbacans! It has been shown in a previous work that Raleigh never had the least intention to prospect or mine for gold; and that he was not even equipped with the picks and shovels which would form the most elementary tools needed for such purposes; in short, that the expedition was designed to plunder and enslave the natives and not to prosecute any legitimate industry. But indeed Raleigh was not the only adventurer of this type. In the Spanish records and in Rodway's volume we find whole catalogues of ruffians who had no word but "gold" upon their lips, no thought but of greed and murder in their heads. Juan Corteso, Gaspar de Sylva, Jeronimo Ortal, Father Iala, Alonzo de Herrera, above all, that prince of monsters Lope de Aquirre, colour the pages with the darkest hues of bloody emprise. As for Aquirre, there is no more terrible story in all the history of the Spanish Main. A companion of the dashing Pedro de Ursua, he set out in the year 1560 to search the Amazon for treasure cities, and within a month he had murdered his captain and all those who stood by him. Two days later he cut the throat of the beautiful Donna Iñez de Altienza, who had followed Pedro from Spain to share the dangers and hardships of his undertaking.

> Aquirre was about fifty years of age, short of stature and sparsely built, ill-featured, his face small and lean, his beard black, and his eyes as piercing as those of a hawk. When he looked at any one he fixed his gaze sternly, particularly when annoyed; he was a noisy talker and boaster, and when well supported, very bold and determined, but otherwise a coward. . . . He was never without one or two coats of mail or a steel breastplate, and always carried a sword, dagger, arquebuse, or lance. His sleep was mostly taken in the day, as he was afraid to rest at night, although he never took off his armour altogether, nor put away his weapons.

It is a curious fact that those who searched for El Dorado never found it; whilst those who never searched for it, found not indeed El Dorado, but the only great gold bearing districts of America south of California. These were neither in New Granada, Terra Firma, the West India islands, nor New Spain, nor indeed in any part of America invaded by the Spaniards; but in Brazil. After the Spaniards had plundered the Indians of their trinkets, after they had worn them out in the petty gold placers of Hispaniola, Mexico, the Isthmus, New Granada and Peru, after they had dug into the ancient graves of the Indians and robbed the dead of the ornaments which had been devoted by the hands of piety, they came to a halt.[1] There

[1] The Peruvian and Central American graves were classified by those who dishonoured them into "Huacas de Pilares" and "Huacas Tapadas," or graves without tombstones, the former being richer than the latter and commanding a higher price in the Spanish markets!

was evidently no more gold to be got; and but for the adventitious discovery of the silver of Potosí, they would probably have abandoned the countries they had ruined and permitted the remains of the native races to recover from their devastating presence. But Potosí, together with the subsequent discoveries of rich silver mines in Mexico, sealed the fate of the Indians.

Meanwhile, something resembling a Dorado had been discovered by the Portuguese in Brazil. This was in 1573, when the placers of Minhas Geraes were discovered by Sebastiaõ Fernandes Tourinhõ. A quarter of a century later, 1595-1605, occurred the great discoveries at Ouro Preto. From first to last these mines produced no less a sum in gold than £180,000,000. Dr. Southey's estimate is upwards of £250,000,000; but this appears to be excessive. It was estimated in 1880 that, weight for weight, Brazil had produced only a fourth less gold than either California or Australia. "When it is considered how much less gold there was in the world's stock of the precious metals at the period when Brazil threw her auriferous product into Europe, than there was when California and Australia began to be productive, the importance of the Brazilian mines is seen to have been even greater than that of the great placers of the present century."[2]

Among the smaller placers of the world whose importance was great enough to exercise some influence upon the history of money in America, were those of the Apallachian Range of North America which yielded from first to last—1824 to 1849—about ten million dollars in gold. It was these mines and the Russian placers of the same period which called forth that remarkable but little known treatise of the illustrious Von Humboldt, "The Fluctuations of Gold," than which no more fascinating monograph on the subject has ever been written.[3] Here it was, in North Carolina, that the writer enjoyed his early experience as a mining engineer. A number of Spanish relics, such as spear-heads, horse-shoes, etc., of ancient types, picked up near the gold placers of Salisbury, testify to the presence of the early gold hunters much farther North than they are commonly supposed to have ventured.

But let us return to the terrible and pathetic history of El Dorado; terrible in respect of the desolation and ruin which, in the pursuit of gold, the Spaniards wrought upon this beautiful land and its innocent

[2] Del Mar's History of the Precious Metals, 1880, p. 124.

[3] Originally published in Berlin, 1838; republished in New York by the Cambridge Encyclopedia Company, 1899.

inhabitants; pathetic with respect to the hopeless efforts which one good man among them made to avert this ruin and lead the natives by pious methods into the fold of Christianity. Those who would peruse this story in detail should consult Sir Arthur Helps' admirable work. The scope of the present history compels us never to lose sight of the precious metals and their immediate surroundings.

Yet there is one more reflection which this history enforces upon us and for which we must beg the reader's indulgence. From the moment when America was discovered by the Spaniards down to the present day, it does not seem to have possessed any further interest for them or for the rest of Europe beyond that of seeing it exploited for the precious metals. The first rude conquerors who visited it from the older world, ravaged it for golden spoil; the race of men who followed afterwards, dug into its mines and gutted them of their precious contents, only to transport these to Europe; the alien financiers who to-day are permitted to influence so largely the polity of America, exercise their power largely for the sake of the gold it produces. The means employed by the Spaniards was pillage; by the Creoles, slavery; and by the aliens, who are permitted to mould its laws at the present day, a chicanery misnamed "finance." The object has been the same with all of them, gold; the destination of the gold has always been the same, the mints of Europe; there to enrich classes who are already rich and keep the remote regions which produced this wealth, in comparative indigence. It is not, as has been falsely claimed, the Catholic religion, which keeps South America poor, nor a republican form of government which subjects the vast resources and energies of North America to the designs of the arch-intriguants who govern the banks and exchanges of Europe. It is that European System of Money, which, whether the coins were made of one metal or two metals, has never failed, so long as those metals were gratuitously coined and free to be melted down on both sides of the Atlantic, to withdraw the bulk of them to Europe and place the American states at the mercy of the European mints and melting houses and the classes who control them.

The native name for that portion of the South American continent which stretches from the Orinoco to Cumaná was Paria; whilst from Cumaná to the Gulf of Venezuela it was called Cumaná. Together, these districts were called by the Spaniards, the Pearl Coast, whilst the interior portion was at a later period called El Dorado. That larger tract of coast which stretches from the Amazon to the Magdalena, was called Terra Firma, and within it are comprised the present states of the three Guianas and Venezuela. It is described by the early voyagers as an earthly heaven; indeed Columbus told his men, when his ship was in the Gulf of Paria, that he thought it must be "a Continent which he had discovered, the same Continent of the East of which he had always been in search; and that the waters, (which we now know to be a branch of the river Orinoco,) formed one of the four great rivers which descended from the garden of Paradise." He added that "they were in the richest country of the world," a remark, which, it seems, however, was not applied to the fertility of its fields, but to his expectations of gold. (Oviedo, Hist. Gen. Ind., XIX, i.)

The Chimay Indians who inhabited the coasts, were not savages, but agriculturalists, fishermen and hunters. They lived in permanent dwellings, sat upon chairs, dined at tables and, alas, for their own happiness,[4] they wore ornaments of gold and necklaces of pearls. It was these trinkets that attracted the cupidity of the Spaniards and doomed the native races to destruction.

Columbus described these people as "tall, well built, and of very graceful bearing, with long smooth hair, which they covered with a beautiful head dress of worked and coloured handkerchiefs, that appeared at a distance to be made of silk." Everywhere he met with the kindest reception and hospitality. "He found the men, the country and the products, equally admirable. It is somewhat curious that he does not mention his discovery of pearls to the Catholic Monarchs and he afterwards makes a poor excuse for this. The reason I conjecture to have been a wish to preserve this knowledge to himself, that the fruits of his enterprise might not be prematurely snatched from him. His shipmates, however, were sure to disperse the intelligence; and the gains to be made on the Pearl Coast were probably the most tempting bait for future navigators to follow in the tract of Columbus and complete the discovery of the earthly Paradise."

The natives cultivated maize, cassaba and cotton, weaving the latter into clothing, hammocks, and other articles of utility. They even manufactured a sort of wine, or beer, from the maize. "The trees descended to the sea. There were houses and people and very beautiful lands which reminded him (Columbus) from their beauty and their verdure, of the gardens, or huertas, of Valencia in the month of May." Not only this, but the lands were well cultivated, muy labrada. "Farms and populous places were visible above the water

[4] F. G. Squier, cited in Century Magazine, 1890, p. 890.

as he coasted onwards; and still the trees descended to the sea—a sure sign of the general mildness of the climate, wherever it occurs." . . . "The expedition proceeded onwards, anchoring in the various ports and bays which there are on that coast, until it came to a very beautiful spot, near a river, where there were not only houses, but places of fortification. There were also gardens of such beauty that one of the voyagers, afterwards giving evidence in a lawsuit connected with the proceedings on that coast, declared that he had never seen a more delicious spot." (Helps, II, 113.)

Upon this happy shore, at Paria, Columbus landed in 1498, setting upon it that great cross which was the symbol of the sovereignty claimed by him for Ferdinand and Isabella, and should have also been that of hope and salvation for the natives. But from the moment of its erection everything changed for the worse. The first enquiries of the admiral were for gold; the next for pearls. The proceeds of his voyage in these coveted objects did not in the end amount to much, but they served to stimulate other adventurers. In December of the same year the news of his discovery reached Spain; in the following May, Alonso de Ojeda, started with an expedition from Spain with the object to exploit this beautiful land. A few days later another expedltion started with the same object, led by Per Alonso Niño and Christóbal Guerra. This last one came to the island of Margarita (Pearls) where they procured some pearls in exchange for glass-beads, pins and needles. At Mochima they obtained in an hour 15 ounces of pearls for trumperies that cost in Spain but 200 maravedis. At Curianá, on the Main, they met with "the most gracious reception, as if it were a meeting of parents and children." The houses were built of wood and thatched with palm leaves. Every kind of food was abundant—fish, flesh, fowls, and bread made of Indian corn. Markets and fairs were held, in which were displayed all the bravery of jars, pitchers, dishes, and porringers of native manufacture. But the Spaniards cared nothing for these, only for those fatal ornaments of "gold made in the form of little birds, frogs and other figures, very well wrought." When the strangers, with affected carelessness, asked where "that yellow dirt" came from, they were told Cauchieto, some forty leagues off. Securing what gold they could obtain at Curianá, the adventurers voyaged to Cauchieto, where they found that pearls were dear and gold was cheap. At Chichiribichi, a place near the present port of La Guayra, Alonso de Ojeda had anticipated them, by attacking and plundering the natives, who did not receive Niño's expedition with the usual amiability. Returning to Curianá they found such a supply of pearls ready for them, some as big as filberts, that they purchased as much as 150 marks weight, at a cost of not more than ten or 12 ducats worth of trinkets. In February, 1500, this expedition returned to Bayona in Galicia, "the mariners being laden with pearls as if they were carrying bundles of straw." In a few months time the news spread all over Spain and flowed back to Hispaniola. An expedition at once started from that island, which occupied the sterile islet of Cubagua, between Margarita and the Main. There was conducted that pearl fishery which afterwards gave its name to the Coast of Terra Firma.

Thus far there had been comparatively little friction between the Spaniards and Indians, but Hispaniola, now nearly depopulated of the natives by the rigours of the gold mines, was too much in need of new victims and too near to Terra Firma, to induce the Spaniards to forego the advantage of kidnapping the inhabitants of the Main. In 1512 they carried off a cacique, with 17 of his men, to the mines of Hispaniola. This cruel and treacherous act was avenged by the Indians, who, after affording the Spaniards an ample opportunity to return the captives, put to death the unhappy Dominican monks who had erected a pioneer mission on the Main. In 1518 the Franciscans and Dominicans of Hispaniola, nothing daunted by the fate of their brethren, erected two new monasteries on the Pearl Coast, the Indians receiving them kindly. Scarcely had these amicable arrangements been made when a Spaniard named Alonso de Ojeda—not the one previously mentioned—started from Cabagua to kidnap natives on the Main. Four leagues beyond the monasteries, at a place on the coast named Maracapána, Ojeda treacherously attacked a band of 50 Indians, whom he had employed to carry maize; and after slaughtering a number of them, carried the remainder away in slavery. This act roused the natives of the coast to fury. They attacked the monasteries, dispersed its inmates, tore the emblems of their religion into shreds and killed 80 of their companions. Not content with this, they started for Cubagua, where there were 300 Spaniards getting rich with the pearl fishery, put the latter to flight and plundered their mushroom city of New Cadiz. When this news reached St. Domingo a punitive expedition, under Ocampo, was organized to chastise the Indians. Having discharged this mission with cruel fidelity, Ocampo made use of the occasion to secure a large number of slaves, "carrying his incursions into that mountainous country, the abode of the Tegares," a place south of the present city of Caraccas.

It was in the midst of these scenes that the benevolent Las Casas made that memorable but vain attempt to establish peace and the Christian religion upon Terra Firma. The Indians were docile and willing enough, but the Spaniards wanted gold and slaves, objects which were irreconcilable with either peace or religion. Even the subordinates of the clerigo could not forego these temptations; and taking advantage of his absence in St. Domingo, his lieutenant, one Francisco de Soto, "sent away the only two boats the colony had, to traffic for pearls, gold, and even for slaves." The result was another rising of the Indians, the destruction of the mission and the dispersion of the Dominicans. When intelligence of these occurrences reached Las Casas, he lost heart and retiring to a convent, renounced the Christian world forever.

Freed from the restraint which this worthy man and reformer had imposed upon their cupidity, the Spaniards now commenced in earnest that dread work of devastation which eventually rendered this once smiling land a desert. In 1522 Jacomé Castellon "fought the Indians, recovered the country, restored the pearl fisheries and filled Cubagua and even St. Domingo with slaves." (Gomara, Hist. Ind., c. 78.)

By the year 1541 the pearl fishery had ceased entirely, or else had ceased to be productive, and we now again hear of El Dorado, which was the name mentioned by the governor of Cubagua as that of an interior province of Terra Firma, where gold and slaves were to be had in plenty. In the same year an expedition with these objects in view was started from Cubagua under the leadership of Ortal, which moved eastward along the coast and there "commenced a hunt, that led the Spaniards through the wildest tract of country which Belzoni, (who was present and writes the story,) thinks that foxes would have hesitated to enter. The cruel hunters, like wild beasts, made their forays more by night than by day, and in the course of a march of a hundred miles they succeeded in capturing 240 Indians, males and females, children and adults." Returning to the coast, the Spaniards adopted another mode of planting religion and civilization in El Dorado. "When the Indians came down to fish, the Spaniards rushed out of their hiding places and generally contrived to capture the fishers, who appear to have been mostly women and children." (Belzoni, Hist. Novi Orbis, I, ii.)

One of these expeditions after travelling 700 miles returned to Maracapána, bringing no fewer than 4,000 slaves. These represented but a portion of the natives who were torn from their homes; for many of them, who were found to be unequal to the journey, were put to death on the road. "That miserable band of slaves," wrote Belzoni, "was indeed a foul and melancholy spectacle to those who beheld it; men and women debilitated by hunger and misery, their bodies naked, lacerated and mutilated. You might behold the wretched mothers lost in grief and tears, dragging two or three children after them, or carrying them upon their necks and shoulders, and the whole band connected together by ropes or iron chains around their necks or arms and hands." These unhappy victims were carried to Cubagua, where a fifth of their number was taken for the king of Spain and branded with the initial of his name, King Charles the First of Spain and the Fifth of Germany, both of glorious memory. "The great bulk of the captives were then exchanged for wine, corn and other necessaries; nor did these accursed marauders hesitate to make a saleable commodity of that for which a man should be ready to lay down his own life in defence—namely, the child that is about to be born to him." (Helps, quoting the words of Belzoni.)

Such were the crimes committed in El Dorado to obtain the gold of Hispaniola and St. Domingo. When Columbus first visited the Coast of Terra Firma, namely, in 1498, it was a scene of fertility and happiness. "When I came there," says Belzoni, in 1541, "it was nearly reduced to a solitary desert." Yet less than 300 miles from the scene of this wickedness lay one of the richest gold mines that the world ever saw, the "Callao." But the Spaniards did not visit El Dorado to prospect or dig for gold; they came to plunder gold and to extort it from slavery.

The only region of Terra Firma which, down to the present time, has proved to contain gold in any considerable quantity and accessible to the natives, before the introduction of European arts, that is to say, placer gold, is in Venezuela (or Guiana) in the valley of the Cuyuni, an affluent of the Essequibo. This is the territory in dispute between Venezuela and Great Britain, the origin of the so much vaunted arbitration treaty of 1897. The air is humid, the climate is fatal to whites, and for their labour the Indians demand sixty cents to one dollar, or 2s. 6d. to 4s. per day in gold, beside certain allowances of food and raiment. The total product at the present time is about one million dollars a year, at a cost of about one and a quarter millions.

Whilst exploring the countries of the Upper Orinoco in the early part of the present century, Baron von Humboldt was informed that the placers of that region were "the classical soil of the Dorado of Parima." This is quite possible.

The "Callao" mine is in the Caratal district, department of Roscio, State of Guiana, Republic of Venezuela. The district is about 160 miles E. S. E. of Ciudad Bolivar, or Angostura, on the Orinoco, and it contains, besides the "Callao," numerous other quartz mines, most of which, although productive, have failed to be profitable. The mines, whose surface had long been worked as placers, were opened for quartz about the year 1866. Commencing in that year with a product of 15,000 ounces, this gradually increased in 1880 to 130,000 ounces, about one-half of the whole product (900,000 ounces) having been obtained from the "Callao" alone. According to the Report of the British Consul at Ciudad Bolivar, for 1880, gold is the chief and, it may be said, almost the only industry of the State of Guiana, on which both public and private incomes more or less depend. "Absorbing, as it does, almost all the labour of the state, by offering superior inducements to labourers, it renders every other enterprise hopeless. Gold-mining is the sole pre-occupation of all minds. In this vice-consular district, as an industry, it only dates, it may be said, from 1866, when companies were formed for working this hitherto undeveloped source of wealth. But whether from the enormous expenses which have been incurred in importing and setting up suitable machinery, the transporting of it to Caratal, a distance of about 150 miles from Port Las Tablas, by bullock-wagons, or the exceptional dearness of labour, provisions, and fuel, which latter has to be procured from the adjacent forests at great outlay, for the working of steam machinery, the fact is that until now, only one, the Callao Company, has returned dividends to its shareholders." Since the year above mentioned, the produce of the district has greatly declined.

CHAPTER XIV.

DARIEN.

Ojeda and Nicuesa summon the Indios to supply gold—Unable to do so, they are tortured, robbed and enslaved—Miserable end of the Spaniards—Cruelties of Vasco Nunez de Balbao—Discovery of the Pacific Ocean—Religion and plunder—The Pearl islands—Gold fishery—Indios thrown to the dogs—Frightful mortality of the natives—Cruelties of Ayora—The bloodhounds' share of spoil.

LET us now transfer the scene to the Isthmus of Darien. This country had been discovered by Columbus in 1502. In 1509 Ojeda was appointed governor. In that year this adventurer sailed from San Domingo with two ships, two brigantines, 300 men and twelve horses; his object being to found a colony at Darien, and prosecute the search for gold. He failed in the enterprise, and was supplanted by one Enciso, who, with another expedition, arrived at Darien in 1510. The Indios, as usual, received the white men kindly. Being asked for gold—always the first demand of the "heaven-descended" strangers—the Indios gave up all they had, which of course was not much, seeing that they had none in use as money, no diggings of any account wherefrom to obtain more, and no knowledge of mining. The white men then asked them for more gold. Being unable to comply, their cacique was tortured and their town captured and pillaged. Some golden trinkets found among their simple effects furnished a presumption that they knew whence to obtain more of the coveted metal. This cruel suspicion sealed their fate; many of them were tortured and foully put to death; but with little avail to their masters, for in fact no more gold was obtained. Enciso's expedition proving as unsuccessful as Ojeda's (probably for the same reason: his failure to get gold), he was supplanted by Nicuesa, who also failed, and the latter was followed by Vasco Nuñez de Balbao. Meanwhile Ojeda died in poverty, and Nicuesa perished in a desert. Of several hundred gold-seekers only seventy odd remained.

The first move of Vasco Nuñez, after his arrival at Darien, was to send seven men to the province of Cueva to search for gold. A wretch named Juan Alonso, who, a year and a half before, had found refuge and relief from starvation among the compassionate and forgiving Indios, now delivered his benefactors over to Vasco Nuñez, who, with 130 armed men, had entered their territory. Vasco Nunez pillaged their town, devastated their country, and dragged their cacique to Darien, there to be used, poor simpleton, as an instrument to point the way to other native settlements where gold might be captured. Some of these settlements were in the province of Comegra. Entering this province and treating it with the utmost cruelty, the Spaniards obtained in all 4,000 pesos of gold, which seemed so great a prize—they thought not of the thousands of lives which they had cruelly sacrificed to obtain it—that forthwith they quarrelled amongst themselves over its division. Observing this, the son of the cacique Comogre dashed the gold disdainfully to the ground, and told the Spaniards that if that was the object of their expeditions, and their cruel treatment of the Indios, he could show them where "they could get their bellies full of it." The land he spoke of was six suns' journey to the southward. He meant Peru.

Either supposing that Comogre's son wished to save his people by leading the Spaniards so far away from their fort and supplies as to endanger their safety, or being unwilling to hazard so long a journey, the latter failed to act upon this suggestion. Upon one of them, however, the statement of Comogre's son made a deep impression, and led to the most important and extraordinary results. This man's name was Francisco Pizarro, who at that time was one of the adventurers in Vasco Nunez's band.

Vasco Nunez returned to Darien, whence he sallied forth at intervals to pillage the country. His plan of operations was to put the Indios to the torture, make them reveal the villages where there was any gold, and at night to attack these villages, in order to secure the coveted prize. He hanged thirty caciques, destroyed a vast number of lives and devastated the valleys of the Isthmus in every direction. Everywhere he sought for gold, asked for gold, tortured for gold and murdered for gold. Down to 1512 he had secured but 75,000 pesos; for in that year the king's Fifth for the whole period of the occupation of Darien was remitted to Spain, and this amounted to only 15,000 pesos.

Not only did the Spaniards maltreat the Indios, and quarrel among themselves about the spoil, they even mutinied against their leaders. In 1513 "they accused their commander of unfairness in this division, and as there was a sum of 10,000 castellanos just about to be divided, this was the cause, or they made it the pretext, of their intention to seize upon him." Vasco Nunez escaped this danger only by relinquishing his share of the booty.

In this same year reinforcements were received from Spain of two more ships and 250 men, whereupon Vasco Nunez started with 190 of the latter to cross the mountains, hunt for more gold, and perhaps reach the South Sea, of whose existence the Indios had apprised him. Rambling through the humid defiles of the Isthmus, he comes upon many new settlements, destroys a great many lives, on one occasion no less than 600, and captures a gratifying amount of golden trinkets. The scenes upon this journey remind eye-witnesses of the shambles. On September 25th he beholds the Pacific Ocean from the summit of the mountains, lifts up his hands, steeped in innocent blood, to return thanks for this famous discovery, reminds his hearers of the gold which Comogre's son had advised them was to be found beyond this sea, and promises to lead them to this treasure.

Vasco then descends the Sierras, kills a few hundred Indios and gets 400 pesos more of gold. Hearing of a temple full of gold in the caciquedom of Dabaybe, he proceeds thither and pillages it. He conquers Coquera and demands gold; he declares the object of his expedition to be gold, to enable the kings of Castile to propagate the true faith. A *veedor* attends every expedition to secure the king's Fifth of the gold. After robbing the cacique Tumaco, who yields to him not only gold, but also pearls, Vasco writes to the king of Spain concerning the riches of Peru (of which he now hears again from Tumaco), and he prepares to return to Darien with many gold-hunting projects in his cruel mind.

The simple caciques shed tears at his departure. On his way he captures the cacique Pacra, who, because he fails to produce gold, he throws to the dogs to be torn to pieces. He captures the cacique Tubanamá, whom he threatens with death if he does not procure gold, and whom he releases on the payment of 6,000 pesos; all that the poor wretch could find in his petty dominions. Vasco himself then questions him as to the origin of this gold. Trying the gravel of the streams he finds it to be auriferous, and orders Tumanamá to collect more gold on pain of death. He then departs for Darien (this is in 1514) and reaches the port, where he finds two more ships from Spain, awaiting his orders. In his letter to the king, accompanied by rich presents, Vasco states that he has not lost a man in this expedition. He asks for more men in order to penetrate a country of the Indios close to the South Sea, where gold can be got by fishing for it with nets; and the king responds to this exciting intelligence by sending out an imposing expedition to exploit this newly-found country of Panama.

This expedition consists of a new governor (Pedrarias), a new *veedor* (Oviedo, the subsequent historian), twelve or fifteen vessels, and 1,500 adventurers, amongst them "not a small number of avaricious old men," who were anxious to take part in the gold-fishery; besides several nobles and priests. The latter were furnished with a Royal Proclamation, or Requerimiento, addressed to the Indios, claiming their lands, gold and services, as vassals, and as the property of the Pope and the king. This proclamation was to be read to the Indios on all occasions before giving them battle. The Spaniards used to read it to themselves and the trees, as they marched in ambush upon the devoted natives. "Entre si leian el Requerimiento á los arboles." The following is the text of the "Requerimiento," as furnished by Dr. Palacios Rubios, jurist and member of the Royal Council for the Indies:

"On the part of the King, Don Fernando, and of Doña Juana, his daughter, Queen of Castile and Leon, subduers of barbarous nations, we, their humble servitors, hereby notify and make known to you, as best we can, that the Lord our God, Living and Eternal, created the heavens and the earth, and also one man and one woman, of whom you and we, and all mankind, were and are the descendants, as well as all those who come after us. But on account of the multitude which has sprung from this man and woman in the five thousand years since the world was created, it was necessary that some men should go one way and some another, and that they should be divided into many kingdoms and provinces; for in one alone they could not be sustained.

Of all these nations God our Lord gave the charge to one man, named St. Peter, that he should be Lord and Superior of all mankind, who should obey him, and that he should be the head of the whole human race, wherever men should live, and under whatever law, sect, or belief they should be; and the Lord gave the world to St. Peter for his kingdom and jurisdiction.

And the Lord commanded him to place his seat in Rome, as the spot most fitting from which to rule the world; he also permitted him to have his seat in any other part of the world, and to judge and govern both Christians, Moors, Jews, Gentiles, and all other sects. This office of Peter was called Pontifex Maximus, or the Pope, as if to say Great and Admirable Father and Governor of men. Men who lived in that time obeyed St. Peter and took him for Lord, King, and Superior of the Universe; so also have they regarded the others, who after him have been elected to the Pontificate, and so has it been continued even till now, and so will it continue till the end of the world. One of these Pontiffs who succeeded St. Peter as Lord of the world in the dignity and office before mentioned, made Donation of these Isles and Terra Firma and all contained therein to the aforesaid King Fernando and Queen Juana and to their representatives, our Lords, as is shown in certain writings upon the subject, which writings you may examine if you wish.

Thus their Highnesses are the rightful Kings and Lords of these Isles and Terra Firma by virtue of this Donation. Some Isles, indeed, almost all those to whom these presents have been notified, have acknowledged and done homage to their Highnesses, as Lords and Kings, in the way that subjects ought to do—namely, with alacrity and good will and without remonstrance or delay, as soon as they were informed of the aforesaid circumstances.

And also they received and obeyed the priests whom their Highnesses sent to preach to them and to teach them our Holy Faith; and these of their own free will, without any reward or condition, have become Christians, and remained so; and their Highnesses have joyfully and benignantly received them, and also have commanded them to be treated as their subjects and vassals; and you, too, are held and obliged to do the same. Wherefore, as best we can, we ask and require you that you do consider what we have said to you, and that you take the time that shall be necessary to understand and deliberate upon it, and that you do acknowledge the Church as the Mistress and Superior of the whole world (por Señora y Superiora del Universal Mundo), and the high priest called the Pope, and in his name and stead the King Don Fernando and Queen Doña Juana, as superiors and lords and kings of these Isles and Terra Firma, by virtue of the said Donation, and that you consent and agree that these religious fathers should declare and preach to you the aforesaid. If you do so, you will do well, and that which you are required to do to their Highnesses, and we, in their name, will receive you in all love and charity, and shall leave you your wives, and your children, and your lands, free, without servitude, that you may do with them and with yourselves freely that which you like and think best, and you shall not be compelled to turn Christians, unless you yourselves, when informed of the truth, should wish to be converted to our Holy Catholic Faith, as almost all the inhabitants of the rest of the Isles have done. And besides this, their Highnesses award you many privileges and exemptions and will grant you many benefits. But, if you do not do this, and maliciously make delay in it, I certify to you that, with the help of God, we shall forcibly enter into your country, and shall make war against you in all ways and manners that we can, and shall subject you to the yoke and obedience of the Roman Church and of their Highnesses; and shall take you and your wives and your children, and shall make slaves of them, and as such, shall sell and dispose of them as their Highnesses may command; and we shall take your property, and shall do you all the injury and damage that we can, as vassals who disobey and refuse to acknowledge their lord and resist and thwart him; and we protest that the deaths and losses which shall accrue from this are your fault, and not that of their Highnesses, nor ours, nor of these noble cavaliers who accompany us. And to prove that we have proclaimed this to you and duly made this Requisition, the Imperial Notary[1] here present will affix hereunto his certificate, in writing, and the rest who are present will be witnesses subscribing to this Requisition."

The new gold fishing expedition arrived at Darien in 1514, whereupon Vasco Nuñez turned over the government to Pedrarias, giving him at the same time an account of the land and his own administration. The native population numbered two millions; the mountains and streams contained gold. Pearls were to be found at the Rich Isle, a rock in the Bay of Panama. Vasco Nunez also reported that his force consisted of 450 men. Marauding parties were at once organized by Pedrarias to pillage the adjacent countries; but before they could set forth, the seething and humid climate of the Isthmus, coupled with a lack of provisions adequate for so great a number of persons as were under his command, combined to very nearly destroy the whole party. In less than a month there perished 700 Spaniards, who thus contributed another quota of lives towards the disastrous search for gold.

"Men clad in silks and brocades absolutely perished of hunger, and might be seen feeding like cattle upon herbage. One of the principal hidalgos went through the streets saying that he was perishing of hunger, and in sight of the whole town, dropt down dead." The condition of despair and ferocity to which these gold-hunters were now reduced may be easily imagined. They were ready for any cruelty.

An expedition was sent along the coast, under "Juan de Ayora with 400 men in a ship and three caravels, to get gold," and provisions. The friendly caciques Comogre, Poncha, and Pocorosa, "came with their gold to this new Spanish chief; but their people were harassed and made slaves, and their wives were carried off." The same cruel and piratical acts were visited upon the hapless Tubanamá.

The licentiate Zuazo thus describes Ayora's method of dealing with one of the caciques, whom Vasco Nunez had previously terrified into the condition called "friendly." The Indios received Ayora with hospitality, providing roast beef, game, bread and wine, no small evidences of civilization. After dinner, Ayora sent for his host, the cacique, and ordered him to bring gold, on pain of being burnt or thrown to the bloodhounds. The cacique sent for the little gold which could be obtained by massing together the paltry trinkets of his tribe, and presented it to Ayora. The latter being dissatisfied, demands more, and seizing the cacique ties him up and compels him

[1] A notary of the Holy Roman Empire.

to order his people to make a further search. This being done, a few more trinkets are added to the fatal store. Ayora, still insatiate, thereupon orders the unhappy cacique to be burnt alive before his eyes, and this was actually done in sight of the miserable natives, his followers. By this and similar means Ayora amassed together a large amount of gold, though he never got back with it alive; for he and his whole force were surprised and cut off by the outraged and indignant Indios. In these transactions the lives of 400 Spaniards, and it is impossible to say how many natives, were sacrificed.

Before Ayora's defeat was known at Darien, Hurtado set forth to inquire about him, and on his way kidnapped 100 peaceable and inoffensive Indios, whom he reduced to slavery and carried off to Darien, where they were divided among the Spaniards, six each to the governor and bishop, and four to the treasurer, etc.

Oviedo, an eye-witness, informs the king of the extortion and dishonesty of the bishops and priests in these words, "Quanto estorbo el obispo, é sus clerigos, quán exentos, é deshonestos." Bernal Diaz also informs us that for the frightful atrocities committed in Mexico under Cortes, the Pope of Rome offered for sale indulgences sent by the hand of a certain friar named Pedro de Aria, who so managed his business that in a few months he amassed great riches, which he remitted to Spain. "Traxo unas Bulas de Señor S. Pedro, y con ellas nos componian, si algo eramos en cargo en las guerras en que andavamos; por manera que en pocos meses el fraile fué rico y compuesto á Castilla." In Hurtado's expedition, the king's Fifth (twenty slaves) was not forgotten. These slaves were sold at auction and branded for exportation, to work in the gold mines of Hispaniola. Even the dogs got their share of the spoil; for be it known that to the owners of certain ferocious dogs which accompanied these expeditions was accorded a share of the spoil equal to that given to a foot soldier. Oviedo says that Vasco Nuñez owned a dog, named Leonçico, who earned for him in this way upwards of a thousand crowns.

CHAPTER XV.

PANAMA.

Murderous expeditions—Morales captures the native women and stabs them to death on the march—Pizarro—Espinosa stabs or throws to the hounds 40,000 victims and brands 2,000 others—Vasco Nunez de Balbao makes a partnership with Pedrarias to search for gold on the Pacific—He builds four vessels at Darien, transports them in pieces across the mountains to Panama, and ravages the Pacific coast—The partners fall out and Pedrarias orders Vasco Nunez to be executed.

MANY similar expeditions are sent out by Pedrarias; among them, one under Becerra. This captain comes back laden with gold, and is accompanied by captives taken by force from friendly caciques, and branded as slaves. From one cacique he takes all his daughters, three or four in number, whom he uses as concubines; another cacique he burns alive for bringing an unsatisfactory amount of gold, and so on, and so on. Morales, another captain, goes with eighty men to the Isle of Pearls. He steals all the desirable females from a native town; kills a vast number of the men, and throws twenty caciques to his dogs, who tear them to pieces, and eat their quivering bodies. The injured Indios pursue him on his return; when Morales, to divert them, and quicken his retreat, commits an abominable act. At intervals on the march he stabs the women whom he has ravished, and thus puts to death another ninety or one hundred persons. Even Vasco Nuñez himself, one of the cruelest of men, speaks of this as the vilest deed ever heard of. Oviedo stigmatizes it as Herodian. Pizarro was in this expedition, and may have derived from it some of those sinister views of policy which he afterwards carried out in Peru. One of these views was that it was permitted by the Church not to keep faith with heretics.

Badajoz now goes out with a gold foraging expedition. He obtains 80,000 castellanos of gold, and loses the whole of it through an Indian surprise. Espinosa next tries his fortune at gold-hunting. On this occasion, for the first time in the history of the world, we learn something definite and immediate concerning the cost of gold obtained by conquest. In Espinosa's expedition there was a Franciscan monk, named Francisco de San Roman. After this priest returned to Spain, and while in the Dominican College of San Tómás of Seville, whither he had retired in disgust at the world, he stated that he had seen with his own eyes, killed by the sword, or thrown to savage dogs, in this "murderous" expedition of Espinosa's, above 40,000 souls. "Que habia visto por sus ojos matar á espada y echar á perros bravos en este viaje de Espinosa, sobre cuarenta mil ánimas." In addition to this, Espinosa brought into Darien from the same expedition 2,000 Indios, whom he branded for shipment as slaves to Hispaniola, all of whom perished in a short time, some at Darien, some on the voyage, and the rest in the mines of Hispaniola. The net proceeds of this foray were 80,000 pesos of gold, so that the immediate cost of every two pesos was more than one human life. It would be curious to learn how the fashionable politico-economical maxim that "value is determined by cost of production" can be reconciled with such an instance of the cost of gold by conquest.

Notwithstanding these experiences, the Crown of Spain still believed in the advantage of searching for gold in the Indies. Yet the failure of the Darien colony was so complete that the gold smelting house at Darien had to be closed for want of supplies; and it was so obvious, that the failure was acknowledged even by statesmen in Spain, who were too remote from the scene of operations to learn much about the cost or the nature of gold forays, and were always the last to abandon, because they reaped the most from, these expeditions. They said that the colony had led to nothing—meaning no gold; and had founded nothing—meaning the abandoned Casa de la Fundicion or smelting house.

Vasco Nunez now comes to the front again. He conducts a godl-hunting expedition to the country of Dabaybe, but without success, except that of perceiving good signs of gold. Shortly after this, his appointment as Adelantado comes from Spain, and he is granted the government of Panama. From this place he may be able to reach the wonderful land of gold mentioned by Comogre's son, and reported by Vasco in his earlier letters to the king. This is the express and only real object of the appointment; it is the express and only object of the expedition.

Vasco first effects an understanding with Pedrarias, whose daughter, in Spain, he espouses by words in Darien, and agrees to allow his newly-made father-in-law a share of the expected booty. In return, Pedrarias forwards Vasco's enterprise. The latter prepares to roam the South Sea, and reach Peru, by building four brigantines at Acla, a port about 100 miles south of the modern Aspinwall, on the Atlantic, assisted by the forced labour of the natives. His plan is, after completing them, to take these vessels to pieces, and transport them on the bare backs of the Indios over the mountains to the Pacific Ocean, and there to put them together again.

Any one who has been on the Isthmus, and felt its hot and humid climate, will be able to understand the terrible difficulties of such an undertaking. How many of the gentle and affectionate Indios were mercilessly used up in hewing the trees we know not; but we do know that five hundred of them perished in the first portage of the timber, a distance of twelve leagues. At this stage the wood turned out to be worm-eaten, and the whole work had to be done over again. It may not be too much to say that, in the end, the undertaking cost several thousand native lives.

No sooner is it completed, the heavy timbers cut and transported over the mountains, and two of the four vessels put together and launched at Panama, than Vasco starts upon his foray for gold. He sails down the coast of the hitherto peaceful South Sea, and lands and despoils the natives everywhere. The two other vessels are subsequently completed, and now he has four of them, with 400 armed and desperate men aboard.

But as though an avenging Nemesis followed behind this enterprise, the robbers fall out among themselves, and justice gets her due. Pedrarias, hearing that Vasco Nunez intends, when once fairly away, to cruise on his own account, and without dividing with him, sends for him from Darien, accuses him of treason to the Crown (their common stalking horse), and puts him, his son-in-law, to a disgraceful death. This occurred in 1517, and for the present it deferred the projected pillage of Peru. We shall see the attempt to carry this enterprise revived seven years later, upon a regular gold-hunter's basis, the conductors being Pizarro and Almagro, one of whom could not read, nor the other one write; one a ruffian and an outcast, the other an assassin and a fugitive from justice in Spain. Pedrarias, as before, was one of the partners, his share being one-fourth, after deducting the king's Fifth; the balance going equally to Pizarro, Almagro, and De Luque; the latter a ranchero, on the river Chagre, who had saved up some money, which he now advanced for the purposes of the proposed expedition.

Meanwhile, the order of events renders it necessary to turn to the conquest of Mexico.

CHAPTER XVI.

MEXICO.

Expeditions of Bernal Diaz and Juan de Grijalva to Yucatan—These lead to Cortes' expedition to Mexico—Character of Cortes—His expedition departs without authority and commits piracy—His invasion of Mexico—Demands for gold—Compliance of Montezuma—Cortes destroys his own fleet, not from heroism, but fear of punishment at home—Massacre of the Tlascalans and alliance with the survivors—The city of Mexico—Hospitality and credulity of Montezuma—Cortes regarded as the Messiah—Treacherous seizure of Montezuma—His forced profession of vassalage to Spain—Cortes demands all the gold in the Empire—His search for mines—Montezuma, undeceived, asks him to depart—Arrival of reinforcements for Cortes—Death of Montezuma—Cortes besieges and captures Mexico, and puts it to the sword—Unexpected smallness of the booty—The Mexican mines worked by the Conqueror—Frightful mortality of the condemned natives—Terrible picture by an eye witness—Prehistoric mining—Mining under the Spaniards—Recent Mining—Production since the Conquest—The future of the Empire.

RATHER more than six years had passed since the first exploration of the Isthmus for gold before it was perceived that its glittering title of Castilla de Oro was undeserved. In 1516 the Casa de Fundicion at Darien was closed, and several of the men in Pedrarias' command asked leave to go to Cuba. Among these was Bernal Diaz, who has written an account of the conquest. Diaz was allowed to go. Upon his arrival in Cuba he asked the Governor, (Velasquez) for an encomienda of Indios.[1] As owing to the almost complete extermination of the Indios, encomiendas were now scarce, and Diaz was impatient to make his fortune, he effected a partnership with some other adventurers, among them Francisco Hernandez and the Governor Velasquez,[2] to seek new lands, and capture gold. Leaving Santiago de Cuba, in 1517, with three vessels and 110 men,

[1] The remark of Sir Arthur Helps (I, 222), which we have quoted on a previous page—namely, that the Indios had become a sort of money, was as true of Mexico as of Hispaniola. Its apology will be found in Exodus XXI, 21, where a man's slaves are regarded as "his money."

[2] "I must remark here upon the deplorable manner in which all these expeditions were managed; the Governor descending to the condition of a merchant adventurer, and being concerned in the profits of each enterprise." Helps, vol. II, p. 252.

they soon sighted the coast of Yucatan, where they landed and commenced their usual operations—robbery, torture, murder, and spoliation. After obtaining a few gold ornaments, the natives, discovering their visitors' character and objects, refused to have any further dealings with them, and compelled them to go away. Upon their return to Havana they whetted the appetite of Velasquez with the sight of the gold they had secured; and he assisted them to fit out another expedition. This was commanded by Juan de Grijalva; Pedro de Alvarado, being in command of one of the vessels. They made the mainland (1518) as before, raided upon the natives, got some gold, which they sent to Velasquez by the hands of Alvarado, and waited for reinforcements; for in this part of the country the natives were highly civilized, and lived in stone dwellings, impervious to bullets and bloodhounds. Velasquez, more eager than ever, fitted out a considerable armanda, the command of which he entrusted to an adventurer and gold miner, named Hernando Cortes.

Cortes' occupation in Cuba was getting gold by means of an encomienda of Indios. "How many of whom died in extracting this gold for him, God will have kept a better account than I," says Las Casas. "Los que por sacarle el oro murieron, Dios habrá tenido mejor cuenta que yo." Cortes was much addicted to gambling, in which occupation his composure and coolness were remarkable. He was neat in his person, and wore a beautiful gold chain and a diamond ring.

The armada consisted of ten vessels (one of them a brigantine), 550 Spaniards, 200 or 300 Indios, 12 or 15 horses, 10 brass breech-loading cannons, a number of falconets, and a large quantity of small arms and ammunition. One of the breech-loaders is still preserved. The outfit cost several thousand castellanos, without reckoning the vessels or stores: Cortes contributing 5,000 castellanós, seven of the vessels, and certain stores, obtained upon a pledge of the future profits of his encomienda.

On the eve of departure, Velasquez, suspecting from certain preparations of Cortes that the latter intended, when once away, to conduct the expedition for his own profit (and to violate the partnership), revoked his official authority for its completion and departure; but before this could be prevented, Cortes, who was apprised of Velasquez' intentions, weighed anchor and sailed out of the harbour.

The expedition departed from Santiago November 18, 1518. On the voyage Cortes pillaged the King of Spain's stores at Macaca, and a Spanish vessel at sea. After his arrival at Trinidad, orders came from Velasquez to supersede him—but he refused to give up his command. At Havana, where he also stopped, similar orders reached him, but these he also disobeyed. In fact, to use his own words, he had become a pirate, and Bernal Diaz says he carried the Black Flag. "Su estandarte era de tafetan negro, con cruz colorada, etc." The voluntary destruction of his "fleet" at Vera Cruz is an ancient yarn which we read of Tiriadates, Julian and many other heroes of antiquity. If true, in the case of Cortes, it must not be mistaken for an heroic action; for there was nothing more heroic in attacking a post of naked Indios with breech-loaders than in the recent murder of Lobenguela's Zulus with repeating rifles and Maxim guns. Assuming the narrative to be correct, the motive of Cortes was probably the despair of his followers. In brief, there was nothing left for them but ruin, or such ample success in gold-hunting as should efface their piratical actions in the eyes of the king.

Their commander had no authority from Velasquez to make a colonial settlement, but only to seek gold, and this authority Velasquez had twice revoked, to Cortes' knowledge. In order to give his acts a semblance of legality he caused his followers, after they had landed in Mexico, to request him, in writing, to form a colony and appoint officers; this was a cunning move, but it could not alter the illegal and piratical character of the expedition.

Cortes lands in Mexico at Tobasco, whence he carries off an Indian princess (Dona Marina) to fill the double part of concubine and interpreter. Upon this incident the ungallant Helps remarks: "It is clear that throughout the conquest of America the Indian women several times betrayed their country, under circumstances which do not seem to me to indicate so much a love of truth (as Herrera says of women generally) as a love of what is personal and near, and an indifferance to what is abstract and remote, a disposition which has been noted equally of all women in all countries. In a word, they loved their lovers and did not care much about their country."

Cortes next lands at San Juan de Uloa, where he sees the messengers of Montezuma, of whom he at once asks if their king has any gold. Being answered in the affirmative, he said, "Let him send it to me, for I and my companions have a complaint, a disease of the heart, which is cured by gold." Awed by his cannon, his steel armour, and above all, his horses, the frightened messengers conveyed this insolent speech and grim jest to Montezuma, and brought back, alas! for the peace of their country, a sum of gold and an abundance of civil words.

Cortes first builds a fort at Vera Cruz, makes an alliance with the dissatisfied cacique of Cempoala, and sends what gold he had obtained to Spain, in the hope of getting authority for his expedition; but the Court treats his acts as piratical. Then he destroys his ships. This last act was not his own. His followers demanded it, because each one knew he dared not return to Spain or any of its colonies, and feared treachery and desertion by the others, who might get away before him, and, turning king's evidence, inform the authorities where he might be found.

After this, Cortes marches towards the city of Mexico. On his way he rides down the Tlascalans with his iron-shod horses, encountering 149,000 of them in one field, brings them to terms and pillages them,[3] the gold and silver of the city of Tlascala especially exciting his cupidity. He then marches to Cholulu, where he strikes terror into the townspeople by slaying and burning them right and left. After pillaging the town he continues his march, and comes in sight of the great valley of Mexico. Looking down upon the wondrous cities of that magnificent plain, the adventurers thought of the booty it contained, and recalled a proverb well known in Spain: "The more Moors the more spoil." "Mas Moros, mas ganancia."

Approaching the city, Cortes is met on the way by the king's ambassadors, who furnish at two places lodgment and banquets for his followers. Then he beholds Montezuma in a litter covered with a pall of green feathers adorned with gold, silver, and pearls, and precious stones: his mantle being similarly adorned; on his head a mitred diadem of gold, and on his feet, golden sandals. They exchange presents, Montezuma giving to Cortes two collars ornamented with golden craw-fish, and Cortes to Montezuma (somewhat significantly) a collar of false pearls and diamonds. The procession moved toward the city, the people admiring the glistening armour of the Spaniards and the wondrous animals they bestrode.

They entered the city November 8, 1519. It contained from 300,000 to 800,000 inhabitants; Cortes had about 450 men. He is given a palace, and every provision is made for the comfort of his men. He opens proceedings with Montezuma by averring that he and his band are messengers from God and His deputies, the Pope and the King of Spain; and are come to redeem Montezuma and his people from sin. This was a crafty move. Upon being reported in Spain, and in case he triumphed in Mexico, it would serve to make the peace of

[3] See reference to Spanish demand for gold in the Tlascalan Council. Helps, II, 290. Tlascala contained 500,000 heads of families, or, say, 2,500,000 people.

Cortes with both Crown and Mitre, for it would prove that he had proceeded in the name of authority. It had a wonderful effect upon Montezuma, who, as Cortes well knew (from the ambassadors with whom he had previously conferred), believed, in common with his people, in the coming of a Messiah.[4]

Said the king, "We hold it for certain that you are the personages of whom our ancestors spoke, who would come from where the Sun rises; and to your King I am greatly beholden and will give him of all that which I may possess." Cortes perceiving the effect of his talk, followed it up with other of the same sort, and in effect made out an abstract of title to Montezuma and his whole empire. The freebooters about Cortes must have grown tired of this palaver and felt anxious to come to business, for Bernal Diaz writes (c. 90) that Cortes turned to his men and said to them all: "We will soon finish with him. This is only the first touch, you know." "E dixonos Cortes a todos nosotros, que con el fuimos; con estoc cumplimos, por ser el primer toque." Afterwards they got a lot of gold and trinkets, and thus ended the first day.

Next day Cortes asked Montezuma to see the temple, which request was granted. Here he saw the God of War, "covered with gold, pearls, and precious stones," and girt about with golden serpents. A golden shield and the faces of men wrought in gold and their hearts in silver surrounded the shrine. This sight intoxicated the Spaniards, who were impatient to be let loose upon the devoted city. Cortes resolves to begin the work by seizing upon Montezuma.

On the third day the king gave his treacherous guest some golden ornaments and one of his daughters. In return, Cortes asked the king to go and live with his band in their quarters, in short, to become his prisoner. To this ungrateful and audacious demand, the king replied with dignity: "I am not one of those persons who are put in prison. Even if I were to consent, my subjects would never permit

[4] The incarnations of Quetzalcoatl, whose other names were Votan, Cukulcan, Iesona, Bacob, and Papachtic, or "Him of the Flowing Locks," are variously fixed in B. C. 955, B. C. 297, A. D. 722, and A. D. 895. According to one system of astrology his re-appearance was due in the year corresponding to A. D. 1553: according to another system, it was due in the year corresponding to A. D. 1527, which was only seven or eight years after the landing of Cortes. The details of the myth, as given by the various authorities quoted in "The Worship of Augustus," p. 205, and elsewhere, are very surprising. The god was to make his advent upon a White Horse, the horse being an animal unknown to the Mexicans; yet clearly depicted in their picture-chronicles. Cortes had but little difficulty in persuading the credulous Mexicans that he and his horse constituted the fulfilment of the expectation. Another detail of the myth was that upon the advent of the Messiah "the Mexican Empire was to cease." (Helps, II, 360.) It was owing to this delusion that the rights of sovereignty and lordship (señoría destas tierras) fell so readily before the demands of the crafty Spanish adventurer.

it." Cortes endeavours to urge him, and Montezuma shows him how absurd his demand is; when the colloquy is interrupted by one of the Spaniards who says, "What is the use of all this talk? Let him yield himself our prisoner, or we will this instant dispatch him." The upshot of it was that Montezuma was carried a prisoner to the Spaniards' quarters, and there immediately put in irons. The credulity and infatuation of Montezuma were so great, that after being himself made a prisoner he assisted to make one of his nephew, Cacamatzin, the only one of the royal family who saw through the pretended Messiah, and real pirate, Cortes. After this (oh, the infatuation of a false belief!), at Cortes' request, Montezuma publicly recommended his nobles and people to declare themselves "jointly and severally" vassals to the king of Spain.

With this act fell the Mexican Empire. The time for plunder and massacre had now come. "As might be expected, one of the first things demanded of Montezuma after this act of vassalage, was gold, of which a great quantity—no less than to the value of one hundred thousand ducats—was handed over to Cortes by the king." This was indeed a great quantity, for Mexico produces very little gold, her main yield of the precious metals being of silver. Even if the Mexican's knew how to obtain this metal from the ore, which is doubtful, silver was too heavy for Cortes' views. What he wanted was very portable property. "He first took care to ascertain where the Mexican gold mines were to be found, and forthwith sent Spaniards, accompanied by Montezuma's officers, into the several provinces designated as gold-producing." He then obtained from Montezuma a complete map of the coast, and ascertained where the best harbours were situated.

By this time the simple monarch began to understand something of the character of the Gods and Messiahs on horseback who had dropped down upon him, and begged Cortes to depart, offering him a load of gold for each of his men and two for himself. On pretense of assenting, Cortes induced Montezuma to order his people to assist him (Cortes) in hewing the timber for four ships (brigantines). These he really wanted for service on the lake in which the city of Mexico was situated, and they were actually built and subsequently used in the attack on the place.

At this juncture eighteen vessels and 800 men arrived at Vera Cruz under Narvaez, to arrest Cortes. Cortes first tries to bribe, then he marches against the king's forces, whom he cajoles and defeats, then he enlists the whole command under his flag. After this he returns to Mexico. During his absence the city had revolted against the forces, which, under command of Alvarado, had been left to watch it. In this strait the latter induced Montezuma to exhort his people to forbearance, and the Emperor complied. This act so exasperated the Mexicans that they stoned the feeble monarch to death. Cortes, on his return, besieges the city, is defeated, loses all his gold and silver, retires to the country of the Tlascalans, obtains 150,000 Indian allies, again besieges Mexico, destroys it bit by bit, slaughters 300,000 of its inhabitants, and after a series of carnages which lasted seventy-five days subdues the city, August 13, 1521.[5]

An orgie with the women who were captured alive, celebrated the victory, and then came the division of the spoil; but "the conquerors were entirely disappointed with the smallness of the booty." They put both the captured Emperor of Mexico, Montezuma's brother, and the King of Tlacuba to the torture, to reveal the whereabouts of the coveted gold. The only reply they got was the disappointing one that during the siege the Emperor had caused "whatever gold, silver, precious stones and jewels" remained, to be thrown into the lake.

The scope of the present work does not render it necessary to pursue the history of this conquest any further. How the city of Mexico was compelled to be rebuilt by the Indios at their own expense of materials, labour, and food; how the land was parcelled out among the conquerors, and the inhabitants given in encomiendas; and how the search for gold, which was commenced with murder by fire and sword, was continued with murder by the lash and the mine —these are proceedings which, if related in detail, would of themselves suffice to fill many volumes. That the reader may be able to form some judgment with regard to the sacrifice of life in the mines, an account of them, twenty years after the country was conquered, will now be given from the relation of an eye-witness.

In the library of Sir Thomas Phillipps, Bart., of Middle Hill, is an original manuscript letter from Fray Toribio Motolinia de Paredes, to Don Antonio Pimentel, Conde de Benavente, dated February 24, 1541. It is from this letter that the following quotation is made. First, however, it is necessary to state that Father Motolinia de Paredes was a monk who had joined the Spanish colony in Mexico, and was "greatly honoured by his contemporaries and trusted by Cortes."

[5] Torquemada says that the city and suburbs contained 120,000 houses. This would imply a population of say 750,000.

In the letter above mentioned, "this excellent monk gives an account of what he considers to have been the Ten Plagues of New Spain. 1. The small-pox.[6] 2. The slaughter during the Conquest. 3. A great famine which took place immediately after the capture of the city. 4. The Indian and negro overseers. 5. The excessive tributes and services demanded from the Indians. 6. The gold mines. 7. The rebuilding of Mexico. 8. The making of slaves in order to work them in the mines. 9. The transport service for the mines. 10. The dissensions among the Spaniards themselves."

In the description which he gives of the Ninth plague, Father Motolinia dwells upon the loss of life amongst the Indians employed in the transport service of the mines. "They came from seventy leagues and upwards," he says, "bringing provisions and whatever was needful, and when they had arrived, the Spanish mine masters would detain them for several days to do some specific work, such as blasting a rock or completing a building. The provisions they had brought for themselves were soon exhausted, and then the poor wretches had to starve, for no one would give them food, and they had no money wherewith to buy it. The result of all this atrocity and mismanagement was that some died on their way to the mines; some at the mines; some on their way back; some (and these were most to be pitied) just after they had reached home." "Volvian tales que luego se morian."

The number of deaths was so great that the corpses bred pestilence; and, mentioning one particular mine (or mining district), Motolinia affirms that for half a league round it, and for a great part of the road to it, you could scarcely make a step except upon dead bodies or the bones of dead men. The birds of prey coming to feed upon these corpses darkened the sun. "Y destos y de los esclavos que murieron en las minas fué tanto el hedor que causó pestilencia, en especial en las minas de Guaxacan, (Oajaca) en las quales media legua á la redonda y mucha parte del camino apenas se podia pisar sino sobre hombres ó sobré huesos. Y eran tantas las aves y cuervos que venian á comer sobre los cuerpos muertos que hazian gran sombra á el sol."

The history of the precious metals in Mexico has been made the subject of so many excellent treatises familiar to the European world, prominent among them being the works of the Abbé Raynal, Baron Alexander Von Humboldt, and Sir Henry George Ward, that only a

[6] There is reason to suspect that a far graver disease is meant, and that the Spaniards were responsible for its introduction.

brief sketch of the subject needs to be included in the present chapter. This sketch will be arranged under the heads of Prehistoric Mining, Mining under the Spaniards, and Recent Mining.

With regard to Prehistoric mining it must be premised that the Aztecs had no knowledge of iron, and therefore, that subterranean mining, below such rare surface deposits of visible native metal as may have been accidently discovered and worked by the people, is practically out of question. It is true that modern prospectors have found in Mexico (here this includes Arizona, New Mexico, Utah and California, etc.,) old shafts and remains of mining works which appeared to them to have been the scene of prehistoric mining, and that this opinion has even found its way into works of reference—for example, in Appleton's Encyclopedia, ed. 1862, v, 679. and an official publication entitled "Mineral Resources of the United States." But the appearances were probably misleading. What was found was ancient workings coupled with volcanic upheavals, or else with deposits of lava or trap, both of which last were regarded as evidences of vast antiquity. This inference is hardly warranted. There are volcanic formations in Utah from which plants of existing species and even live frogs have been taken, and which, therefore, must be of very recent formation. Most of the territories named are so desert and tenantless that in many parts a volcanic disturbance and a plutonic alteration of the rocks may even now occur (overnight) without its being known and recorded. In the Abo country in New Mexico there are evidences of volcanic eruptions which overwhelmed dwellings and buried the inhabitants in ashes and lava. Old mining shafts, the ruins of rude smelters, heaps of slag and blackened float have been found in the Manzana Mountains. (Eureka Leader, 1881.) The prospectors regarded these remains as of a geological antiquity; whereas, they may have been and probably were the ruins of mining prosecuted long since the Spaniards plundered Mexico.

Mining under the Spaniards began shortly after the Conquest. Beside what gold the ill-starred Aztecs wore upon their persons, or had accumulated in their temples, there was comparatively little of that metal to reward the followers of Cortes. The river beds and such few other gold deposits as were known to the natives or discovered by their conquerors, were soon exhausted, and but for the opening of the silver mines Mexico might have been spared by the marauders as Florida and Louisiana were spared, because they contained but little to gratify their ferocious cupidity.

Among the districts first opened were Tasco, Zultepec, Tlalpujahua, and Pachuca. These were in the vicinity of the capital and had probably been superficially worked by the natives. Zacatecas was opened in 1532, and the Veta Grande of the same district in 1548. In the same year San Luis Potosí (not to be confounded with the Potosí of Peru) was opened; Sombrerete in 1555; and Guanajuato in 1558. Two centuries later this last-named district was again in bonanza. Bartolomeo de Medina discovered or improved the patio process while working at Pachuca in 1557. During the 17th century but few new districts of importance were opened, but in the following century Colima, Biscaina, Jacal, Real del Monte, Valenciana, Catorce, Guarisamey near Durango, Ramas and other rich districts swelled with their product the ample yield of silver in Mexico. In round figures, it may be stated, that from the Conquest to the close of the 19th century, the total coinage of Mexico was in silver about 3,270 million dollars of 8½ to the mark, and in gold about 130 million dollars of 16 to the ounce; together about 3,500 millions. The present annual coinage of silver is about 21 millions, and of gold about half a million dollars, both of the same weights as the foregoing. The "product" of late years has been "estimated" by the Director of the United States Mints at very much larger sums, especially of silver; but as the estimate is put forward evidently to sustain a monetary theory, it is regarded as much safer to adhere to the stastistics of the Mexican coinage. If the product for the entire period since the Conquest is desired to be deduced from the coinage, something will have to be added for bullion used in the arts or exported uncoined. On the other hand, something needs to be added for plate coined and for coins melted and recoined in Mexico. If we add to the 3,500 millions coined in Mexico, an equal sum produced in Peru and the other countries of Spanish and Portuguese America, we have the enormous total of 7,000 million dollars of the precious metals from these sources alone, to say nothing of the rest of the world.

An eminent writer has said that "the only mode of procuring the services of others, on any large scale, in the absence of money, is by force, which is slavery. Money, by constituting a medium in which the smallest services can be paid for, substitutes wages for the lash, and renders the liberty of the individual consistent with the maintenance and support of society." Such was the effect produced by the billions exhumed from the American mines. They cost the lives of 30 millions of natives; they have emancipated ten times 30 millions of Europeans, in whose customs and literature were stored the accumulated knowledge of countless ages. The expeditions of the conquerors of America were sordid and ferocious; the unlooked for result has been to confer upon the entire human race the blessings of freedom and the elements of future progress.

Until Mexico achieved its independence in 1821, the principal, almost the sole industry of its masters was the pursuit of gold and silver. Agriculture was followed but little beyond the point of securing sustenance for the men and animals employed in mining. The arts were almost entirely neglected and manufactured goods were nearly all imported. Wherever the mines gave out, everything in the vicinity went to ruin. For example, the city of Chihuahua had neither pastural, agricultural, manufacturing nor commercial resources; almost its sole reliance was the silver mines of the vicinity, of which the chief ones were Santa Eulalia, Batopilas, El Parral, Mordas, and Jesus Maria. The mines built the city and afterwards ruined it. Chihauhua once contained 70,000 inhabitants. Until recently this number had fallen to 10,000. National independence and the introduction of railways have since increased it to 20,000. San Luis Potosí once had a population of 75,000 to 100,000. In 1642 the mines caved in, burying 400 miners, after which the town went to ruin. The population fell to less than 10,000. National freedom and the promotion of trade has since increased it to 30,000.

Though the Revolution put an end for a time to mining on a large scale, this was afterwards resumed, and Mexico is to-day as great a producer of silver as it ever was. It is also a great agricultural and pastural State, and within recent years the construction of railway lines has afforded tremendous impetus to manufactures. Under the present admirable administration encouragement has been afforded to every department of industry, and increasing prosperity has marked the development of the country in every direction. The blood of its aboriginal inhabitants was not shed in vain. It has built up an empire which may yet be destined to play an important part in the drama of the world's development.

PRODUCTION OF GOLD AND SILVER IN MEXICO.

According to a paper read before the Scientific Society of Mexico, by Antonio Alzate, the product of gold in Mexico during the Eighties was about three-quarters of a million dollars, of 4.86 to the £ sterling, while during the Nineties, until 1894-5, it was about one million dollars. In the last-named fiscal year a law, introduced June 4, 1894, "reduced the annual mining tax up to 10 per cent. and abolished all other federal taxes on gold, except the Stamp tax." Under this encouragement, the annual product of gold rose in 1894-5 to about $5,000,000, in 1895-6 to $6,500,000, and in 1896-7 to $7,250,000. On the same basis the product in 1897-8 was $8,000,000, and in 1898-9 $9,500,000.

According to the United States Consular Report, No. 222, the product of gold in Mexico in 1892-3 was about $1,270,000; 1893-4, $1,250,000; 1894-5, $4,750,000; 1895-6, $6,000,000; 1896-7, $5,860,000; 1897-8, $7,500,000; and in 1898-9, $9,500,000.

Mr. Valentine, of Wells, Fargo & Co., estimates the product from 1877 to 1894 at about a million dollars a year; 1894-5, $4,750,000; 1895-6, $5,500,000; 1896-7, $8,500,000; 1897-8, $9,225,000; and 1898-9, at $10,000,000.

As it is well known that during the prevalence of the taxes, much of the gold produced in Mexico was smuggled out of the country, either as gold or as gold mingled with silver, it is deemed fairer to estimate the gold product during the Seventies at a million a year; during the Eighties at two millions, and during the Nineties three millions, until 1895, when it rose to six millions, and has since increased at the rate of about a million a year, until at the present time, 1901, it is about twelve millions.

The best statistics which we have thus far seen concerning the silver product of Mexico are those published by Mr. Valentine, who, however, has cast them into American dollars of 371 1-4 grains fine each, so that the product in ounces can only be deduced by calculation. If the dollars in Mr. Valentine's tables are multiplied by the decimal fraction, 0.773,437, the result will be as follows, the last two years being estimated from data furnished by the government of Mexico:

Production of Silver in Mexico, fine ounces, Troy.

Year.	Ounces.	Year.	Ounces.	Year.	Ounces.	Year.	Ounces.
1877-8	19,209,855	1883-4	24,514,686	1889-90	32,097,636	1895-6	42,113,645
1878-9	19,432,604	1884-5	25,698,218	1890-1	33,257,791	1896-7	46,934,477
1879-80	20,728,112	1885-6	26,383,483	1891-2	35,384,743	1897-8	52,226,333
1880-1	22,610,657	1886-7	26,760,920	1892-3	37,511,695	1898-9	50,513,170
1881-2	22,684,134	1887-8	27,002,233	1893-4	36,544,898	1899-00	50,000,000
1882-3	22,869,759	1888-9	31,483,527	1894-5	41,939,621	1900-1	50,000,000

It will thus be seen that while during the last quarter of a century the production of gold in Mexico has increased from one to twelve million dollars a year; that of silver, after increasing from 19 to 52 millions, has fallen to about 50 million ounces. When the 21 million dollars silver annually coined, as mentioned on p. 180, are reduced to weights, namely, about 16 1-4 million ounces fine, and these are deducted from the estimated total product, it follows that Mexico now annually exports about 33 to 35 million ounces fine silver, while the export of gold amounts in weight to over half a million ounces fine, with a tendancy to increase.

CHAPTER XVII.

YUCATAN AND HONDURAS.

The spoil of Mexico—Cortes sends the King's share and the regalia of Montezuma to the Court of Spain, and obtains legitimacy for his acts—His depredations renewed—Aversion to lawyers—Search after gold mines—Enslavement of the Indios—Cortes sends Olid to raid Honduras—Olid turns buccaneer—Is pursued by Cortes—Montezuma's brother and another royal captive put to death on the march—Cortes ravages Yucatan—Olid is overtaken and killed—Cortes returns to Mexico to find himself superceded in command—The new governor of Mexico, Ponce de Leon—He impeaches Cortes—Is assassinated—Succeeded by de Aguilar, who also dies suddenly—Estrada, the new governor—Cortes picks a quarrel with him and is sent to Spain—He obtains favour at the Court and is permitted to return to Mexico as a subordinate—His riches and subsequent poverty—He again returns to Spain, where he dies in obscurity and indigence—Mortality of the Mexicans occasioned by the gold-seeking expeditions of the conquerors—The huacas of Chiriqui—The mines of Honduras.

IN the early part of 1521 the Spanish Crown sent out a governor for New Spain (the modern Mexico) named Cristobal de Tapia; but Cortes and his armed associates drove him away. In 1522 Cortes, having gathered what he deemed sufficient spoil for the purpose in view, sent 88,000 pesos in gold bars [1] and Montezuma's regalia and wardrobe to the Court of Spain, with the request that in consideration of his having conquered a new country for the Crown his acts might be legitimated. Although these treasures never reached Spain, having been captured at sea by a French corsair, the "Jean Florin," Cortes was recognized by the Crown, in a despatch dated Valladolid, October 15, 1522, as governor and captain-general.

Notwithstanding this act of legitimation, the predatory character of Cortes and his band remained unchanged. In 1523 Cortes asks the Crown that no lawyers shall be sent to New Spain, or, if any should get there, that they may not have authority to advocate causes. Similar requests were made by Vasco Nunez de Balbao, from Terra Firma, in 1513; by the Commissioners of Cuba in 1516; by Pizarro, from Peru, in 1529; and by Cabeça de Vaca, from La Plata, in 1541. Consult Helps, III, 17–20.

[1] If this was the king's quinto it shows that 440,000 pesos was all the spoil that had been obtained in Mexico down to that date.

In 1524 Cortes left the now rebuilt city and capital of Mexico in a defenceless state for the purpose of despatching a party of Spaniards to a reported gold mine in Mechoacan. The report came from Alvarado, who at the same time was ravaging Mechoacan, by torturing its inhabitants and plundering its graves. In the same year Cortes granted encomiendas of Indios to his followers, in defiance of an express prohibition from the Crown. Also in the same year, he sent Cristobal de Olid to make a raid on the Indios in Honduras. Olid, following his chief's example, turned buccaneer on his own account. Cortes tired of the comparative inactivity which followed the capture of the capital, and anxious to add to those spoils which, as yet, had fallen far short of his expectations, determined to follow Olid, and he set forth with an expedition from Mexico, to sail and march 1,500 miles along an unknown coast. Fearing to trust the men behind him, he carries along with him the large quantities of gold and silver that remained in his hands. He also takes with him Dona Marina, as interpreter, and his royal captives, the Emperor of Mexico (Montezuma's brother) and the King of Tlacuba; and on one of the Carnival days in February, 1525, and in the name of Jesus Christ, he puts the two captives to death in the obscurity of an Honduran forest. When led to execution the Emperor of Mexico exclaimed, "Oh, Malinché (Cortes), it is long that I have known the falseness of your words, and have foreseen that you would give me that death, which, alas! I did not give myself when I surrendered to you in my city of Mexico. Wherefore do you put me to death without justice? May God demand the like of you!"

Cortes then marches through and ravages Yucatan, a country whose fertility, government, and civilization compared favourably with those of Spain itself. A detachment of his band, under command of his cousin, Francisco de las Casas, overtakes Olid, and assassinates him; when Cortes, securing all the spoil, resolves to return to Mexico by the sea. A storm drives him to Havana, and he only reaches Mexico in 1526, there to learn that his piratical adventures had scandalized the Crown of Spain, and induced it to supercede him in his lately granted authority. In November, 1525, Ponce de Leon had been created governor of New Spain, with orders to proceed thither immediately, and take a residencia of Cortes.[2]

[2] Helps, III, 61. The residencia (a relic of the Roman government in Spain, the form of which survived the dark ages) was an inquiry by any official into the actions of his predecessor. It amounted to an impeachment before the sternest of judges, although in practice it was often circumvented. The Roman residencia in Venice is mentioned as of the year 1229, in Hazlitt, II, 242, ed. 1858.

The history of Cortes does not terminate here; for ten years later we find him, with the incurable habit of gold-seekers, expending his gains in new expeditions, this time to California, the lower portion of which he was actually the first to discover. But here we can conveniently agree to epitomise the close of his career in Mexico.

On the seventeenth day of the residencia of Cortes, the governor, Ponce de Leon, dies suddenly of poison, and the residencia is broken off. Marcos de Aguilar, his successor, dies two months after his appointment—Bernal Diaz says of sickness—but Diaz was one of Cortes' original companions, and might himself have assassinated Aguilar, for he was none too good for such a business. To Aguilar succeeds Alonzo de Estrada. With him Cortes picks a violent quarrel, when Estrada, probably fearing the same fate as his predecessors, sent the truculent conquisador to Spain. There Cortes contrives to make favour with the Court. His "specimens of the riches and the curiosities of his new country dispelled at once the vapours of doubt which had lately obscured his name and deeds."

He is made Marquis del Valle, appointed captain-general, or military commander, subject to the civil governor of New Spain, and permitted to return thither, which he does in 1530. In a recent publication called "The Silver Country," Cortes is said to have obtained £1,200,000 of spoil in Mexico; but we have not been able to find satisfactory authority for this statement.

Being charged in his residencia with possessing 200 cuentos of rent—that is, with encomiendas of Indios yielding 200 cuentos gold tribute per annum—he offers to commute his encomiendas for 20 cuentos of cash rent in New Spain, or 10 cuentos in the mother country. In September, 1538, he complains that he has not means enough to live in the city of Mexico, and must reside in the country. His Indios had probably been worked to death by this time, and the "rental" of his encomiendas had fallen away. In 1547, having meanwhile again returned to Spain, there to find himself shunned at the Court, he died in obscurity and poverty in the sixty-third year of his age, leaving behind him more orphan witnesses of his cruelty and rapacity than probably man ever did before.

Under date of August 14, 1531, the Spanish auditor, Quiroga, sent to the Council for the Indies in Spain an account of the patience, diligence and docility of the natives, which, in view of the atrocious cruelty of the Spaniards, is most affecting. Alluding to the orphaned condition of a vast proportion of the people, he says: "They are numerous as the stars of heaven and the sands of the sea; an immense number of orphans whose fathers and mothers have perished in the mines through the rigour of our Spaniards."

Sir A. G. Ward (II, 697) says there are no gold or silver mines in Yucatan; a circumstance which may account for its ancient wealth and splendour, in which respects it greatly exceeded all the native States of America. This is not said in disparagement of mining as an art, but of mining when relied upon too exclusively as a source of national prosperity. Many of the Spanish economical writers of the 17th century warned their countrymen that the pursuit of mining was a certain road to national poverty. To say nothing of their own country, the many recent examples before them of Hispaniola, Mexico, Darien and Peru, rendered this conclusion irresistible. Their opinions were afterward confirmed by the observations of Antonio de Uloa, Adam Smith (I, xi, 2), J. R. McCulloch, Dr. Joseph Townsend (*Travels in Spain*, III, 345), and others, some of whom, like Uloa and Townsend, were practical engineers and miners, as well as philosophers and men of affairs.

But although there were no mines in Yutacan, the conquerers of the country, the defacers and destroyers of its ancient monuments, managed to extract from it a considerable quantity of gold. This they got by plundering the graves of the natives. In recent years a number of native graves, which had been overlooked by the Spaniards, were opened at a place called Chiriqui, in the district of Veragua, on the Chiriqui river in Costa Rica, lat. 8 10 N. This was formerly included in Darien, but as the discovery led to others farther north, within the limits of Yucatan, it has been deemed appropriate to record it under the latter head.

In Veraguas were found a great number of sepulchres of the aborigines, many of them rich in relics, and which soon obtained great celebrity as the *huacas* of Chiriqui. Considerable quantities of gold ornaments were found in them, and for a time the region in which they occur was thronged with adventurers in eager search for these hidden treasures. The amount was soon discovered to be exaggerated, and the Chiriqui fever abated as rapidly as it rose.

The so-called *huacas* (a name borrowed evidently from Peru) occur in the plains as well as on the slopes of the Cordilleras of Veraguas, Chiriqui and Azuero, and on the islands of the coasts, and are divided into two classes, *Huacas de Pilares*, or pillar graves, marked by rows of upright stones, sometimes carved in imitation of men or animals, and of varying dimensions; and *Huacas Tapadas*, covered graves, consisting of mounds overlaid by water-worn stones. The deposits, whether of human remains, pottery or objects of gold, are always to be found at the bottom of the grave—which varies in depth from six to fifteen feet, being invariably sunk to a hard substratum —and contained in a rough coffin or box of flat stones.

The pottery and the metal ornaments deposited in these graves are now about all that remain, the bodies of the dead to whom they belonged, having, in most instances, wholly disappeared. Some of the vases are of good design and material. They are often accompanied, in the graves of females, by *metlatls*, or grinding-stones, coinciding precisely with those now in use for crushing maize. The golden articles are various in shape—in all cases cast, but with certain portions afterwards wrought into shape. All have projections or are pierced, for suspension, and many present evidences of having been worn for long periods. In shape they are generally representations of mythological and natural objects, peculiar to the region in which they are found. Similar graves have been found in Yucatan, at Chichen, Izamal, Macoba, and Uxmal. These places, however, have since become more interesting and valuable for their architectural remains. Seventy years ago the ruins of Uxmal were so completely unknown that the re-discovery of the *casas grandes* by Dr. Mitchell, of Sisal, and Baron de Waldeck, was as complete a surprise to the citizens of the neighboring town of Merida as the exhumation of Pompeii to the burghers of Nola and Castellamare. But since 1830 Uxmal has been the Mecca of American antiquaries, as well as the hunting-ground of countless seekers after buried treasures. Yet, though the general character of the ruins is now tolerably well known, it is clear, from the recent researches of Plongeon, that any reliable interpretation of the inscriptions is still a desideratum. The traditions of Mexico are as silent about a vast and wealthy city in Western Yucatan as are the chronicles of the Conquistadores. Uxmal is a corruption of Huasacmal, "the Main City." A quarry not far off is known as the "Man Killery," and hard by, in the Sierra de Macoba, is a plateau named "The Field of Defeat," where, if not closely watched by the priests, the natives still celebrate a festival enigmatically known as the Week of Deliverance.

In Honduras the history of the precious metals has been the same as in all the Spanish-American countries: first, plunder of the natives; second, robbery of the graves; third, placer mining with slaves (encomiendas); fourth, vein mining with slaves; fifth, revolution, independence, exhaustion and civil commotion, at the end of which, foreign capital with hired labourers renews the search for gold. The oldest vein mines in Honduras are those of the Yuscaran, and among these the Malacate mine is the most celebrated. This is practically a silver mine, although one-twelfth of the value of the ores was derived from gold, both metals being found in the same matrix. After continuing productive for centuries, this mine was abandoned in 1845, on account of a slide. About 1875 a new company "expended great sums" in clearing away the stone and rubbish with which the shafts and galleries were filled; but without recovering the vein. They then sank a new shaft 450 feet deep, and tunnelled 1,800 in a horizontal direction, with no greater success. After continuing the search for seven years it was finally given up in despair. Although vein mines are still being worked in various places throughout Honduras, the chief mineral product is gold derived from placer washings by the natives, who sell their dust to local traders for $12 an ounce, its value at the mint being $20. At best, this product is too small to merit further notice in the present connection. The State of Honduras is poor and owes a large debt, upon which it fails to pay the interest; while the currency consists of base silver and copper coins.

Although, after Yucatan and Honduras had been plundered, they were too poor in mines to yield any further contributions of the precious metals to Spain, yet it was along their coast lines that the plate ships, laden with the metallic produce of Chile, Peru and Darien, were sometimes conveyed to Vera Cruz, there to be augmented by the tribute of Mexico. From Vera Cruz the fleet sailed to Havana, where it was joined by vessels direct from Honduras, Carthagena, and other places. "The ships then sail through the Straits of Bahama; they continue their course to the height of New England, and after sailing for a long time in this latitude of forty degrees, they at length veer to the south-east, to come in view of Cape St. Vincent and to proceed to Cadiz . . . From 1748 to 1753, one year with another (annnal average), New Spain sent to the mother-country, by the way of Vera Cruz and Honduras, $114,910 in gold, $8,724,300 in silver, and $3,693,084 in merchandise, at the price in Europe"—these sums being deduced, at five to the dollar, from the French livres, in which they are given by the Abbé Raynal, III, 423.

CHAPTER XVIII.

GUATEMALA.

Its ancient populousness, wealth and civilization—All destroyed by a few gold-seekers—Origin of its invasion—Native feuds—Alvarado dispatched to Guatemala—His demands for gold—Wholesale slaughter of the natives—Enslavement of the survivors—Plentifulness of spoil—Prices in the Spanish camp—The native children taxed for gold—The adults forced by cruel labour in the mines to consume one another for food—Alvarado's death-bed repentance.

THE populousness, wealth, and civilization of Guatemala were attested by many evidences, yet all these were destroyed in a few months by a band of 280 gold-hunters under Pedro de Alvarado.

After the capture of Mexico by Cortes in 1521, a number of chieftains sent ambassadors to him to treat for peace. Among these was the "King" of Mechoacan, a province about seventy leagues to the south-west of Mexico. From these ambassadors the eager ears of Cortes learnt of the South Sea and its islands abounding in gold, pearls, precious stones and spices. These stories formed the basis of Cortes' expeditions some years afterwards, in one of which California was discovered. At present he sends a party of soldiers to view the sea, explore the intervening country, and take possession of both. This is done in 1522, nine years after the discovery of the same sea further south by Vasco Nunez. To get at the sea the Spaniards passed south of Mechoacan through Tehuantepec, where the terror of their deeds induced other native chiefs to offer their submission to Cortes. In order to propitiate his favour the envoys from Tehuantepec mention a country with which they are at enmity, and whose riches might well reward the researches of their dread visitors. It was through these native quarrels, due to the prevalence of feudalism, that Cortes, like his ancient prototype Cæsar in Gaul, was enabled the more readily to plunder the country to which he had conducted his soldiers.

Cortes accordingly sends Alvarado to Guatemala in 1522. His proceedings are of the usual character. The innocent natives receive him with kindness and hospitality. He then demands gold. This is given him, when he demands more. The cacique exclaims that there is no more—that he has given all. Whereupon Alvarado threatens him with death if he does not bring more. The alarmed chief scours the whole country for gold, and collects together all there is to be found, about 30,000 pesos. This is given to the Spanish captain, who, in return, treacherously puts the cacique in prison, where he dies. Even Bernal Diaz exclaims against the treachery of this proceeding on the part of Alvarado; but the conquistadores had yet something to learn of treachery from Pizarro.

Alvarado then ravages the country, terrifying the inhabitants of the towns, from one of which (Guatemala) he obtains "magnificent presents of gold, jewels, and provisions, which, it is said, required no fewer than 5,000 men (natives) to carry."

He then puts to the sword as many of 30,000 natives arrayed between him and the town of Quezaltenango as fail to secure their safety in flight; and in a second "battle" nearer to the town, commits such carnage that he himself says, although he had seen some of the fiercest "battles" in the Indies, he caused on this occasion the greatest destruction of life that had ever before been known in the world. "Nuestros amigos, i peones hacin una destruicion la maior del mundo." Alvarado's "Relacion." He then enters Utatlan, where he burns the chiefs, razes the city to the ground, and brands the inhabitants as slaves; in his own words, "the lords to death, and the rest as slaves."

Here the freebooters, flushed with spoil, wine and women, resolve to rest and enjoy their ease. They build a fortified town, whose site, near the Volcan de Fuego, can be seen from the deck of every steamer that plies between Panama and San Francisco, and founded the place with a grand parade, in which they turned out and marched, adorned with plumes of feathers, and gold and jewels. The plentifulness of spoil is evinced by the prices that obtained among them even in this land of bountiful crops; for a pig brought seventeen to twenty pesos of gold; the tailor of the band demanded such prices for his handiwork that each movement of the needle was worth a real; and the shoemaker from his earnings might go shod in the precious metals. Even a year later eggs were worth a real apiece.

Alvarado, now appointed by Cortes lieutenant-governor and captain-general of Guatemala, remains with his crew in their new settlement for several years, compelling the natives to supply them with everything they required, and much more than they needed.

"The unvaried tradition of the Indians relates that the lieutenant-governor imposed upon the inhabitants of Patinamit, or Tecpan-Guatemala, a burden that could not be borne. It was, that a number of children, boys and girls (one account says 800), should each of them bring him daily a reed full of golden grains. The children played about like children, and failed to bring in the required tribute. The extortionate governor punished or threatened to punish the adult population." These, already overburdened with similar exactions, could endure their sufferings no longer, and revolted, with, of course, the usual consequences—death on all sides, death by thousands and tens of thousands.

"They are most poor," writes the Bishop of Guatemala to the Emperor Charles V. in 1539, "having only a little maize, a grinding stone, a pot to boil in, a hammock and a little hut of straw with four posts, which every day is burnt down."

When Alvarado had with merciless cruelty drained Guatemala of its pitiful trinkets of gold, he abandoned the country, and, contrary to the orders of the Crown, joined the standard of Pizarro in Peru.

It is related of this monster that his exactions of gold and the labour that he compelled the Indios to perform in other ways gave them no rest, not even enough to enable them to cultivate food for themselves, and that he thus forced them to eat one another.

After receiving his death-blow[1] Alvarado lived long enough to make a will, which, being drawn by a priest, contains many confessions of his misdeeds, all of which are sought to be atoned for by offerings to the Church. He had unjustly enslaved and branded the natives, therefore he leaves them in an encomienda, whose revenues shall go to the Church. This is reparation, indeed! As for the slaves in the mines, also unjustly captured and branded, he piously declares that they shall continue at the work until his debts are paid; for it seems that, like all the gold-hunters, he died poor. The murders he has committed he atones for by a present of 500 golden pesos to be sent to Castile, and there used for the redemption of christian captives from the Moors.

[1] Alvarado went to Spain in 1527, was appointed governor of Guatemala and ennobled. He returned to Guatemala in 1530. In 1534 or 1535 he declared his intention of joining Pizarro, the fame of whose booty, including the Ransom of the Inca, was noised all over the world. Although Alvarado was forbidden by the Crown, he nevertheless proceeded in spite of the interdict. He joined Pizarro and was well paid for his assistance. He returned to Guatemala in 1535, whence he set out upon other expeditions (on the Californian Coast), and was killed in 1541. After his death, his wife, Doña Beatrice de la Cueva, whom he had brought out from Spain on his last voyage, was chosen by the Provincial Council as Governadora—the first instance of a woman having obtained that office, or, perhaps, any other, amongst Europeans in America.

CHAPTER XIX.

PERU.

Pizarro's momentous gold-seeking partnership—Andagoya reaches Peru—Garcia's expedition—The conquest is abandoned for a time—Pizarro undertakes it—Sufferings of his command—Forays—The partners fall out—The memorable chalk line that was drawn by Tafur—Privations of Pizarro—He reaches Tumbez—Spoils of gold and silver—The richness of Peru—Pizarro returns to Panama to prepare for further conquests in Peru—Pizarro goes to Spain for troops, supplies and authority—He obtains all these, and in addition a title of nobility—Returns to Panama and sails for Peru—He lands upon and sweeps through the devoted country—Spoils—Civil feuds in Peru—Death of Huayna-Capac and dissensions between his sons—Pizarro destroys the peaceful city of Tumbez—Messengers from Atahualpa—Pizarro artfully espouses his cause and betrays him—Atahualpa's naked followers put to the sword and himself taken prisoner—Pillage of the Peruvian camp—Atahualpa's ransom—Enormous booty—The Inca treacherously murdered—The empire desolated by the Spaniards—Amazing destruction of the natives—The sword, the bloodhounds, and, worse than all, the mines—Encomiendas—The mines of Potosí—Cost of gold obtained by conquest and slavery.

IT has already been shown that the search for Peru, instigated by the chance words of the cacique Comogre's son, was purely a commercial speculation, with no other object in view than the acquisition of gold and silver. The fate of the partnership of Pedrarias and Vasco Nunez has also been related, and so have been the details of the subsequent partnership arrangements between Pedrarias, De Luque, and the illiterate gold-hunters Pizarro and Almagro. We have now to relate what came of this bargain, the most momentous the world has ever known; for it involved the lives of ten to fifteen millions of people.

Between the death of Vasco Nunez (1517) and the last-named partnership arrangement (1524), Pedrarias, in 1522, had sent one Pascual de Andagoya along the coast. This man succeeded in reaching Peru, and there gave the natives an indication of the fierce and reckless character of the strangers with whom it was fated that before many years they should have to contend in greater numbers. Pascual becoming disabled, and Pedrarias being at that time intent upon a gold foray into Nicaragua, the Peruvian project was laid aside. In 1523, Alexis Garcia, a Portuguese commander, traversed Brazil and the Gran Chaco of La Plata, scaled the lofty Andes and succeeded in pillaging Alto Peru. ("Hist. Monetary Systems," p. 411.) In 1524, possibly through having received some knowledge of Garcia's achievement, the project of invading Peru from the sea was again entertained by Pizarro, and the famous quadripartite partnership was formed. Pizarro set sail with one vessel (one of those built by Vasco Nunez), two canoes, eighty men, and four horses. These forces were procured at great expense—the men employed to get the ships ready for sea receiving two golden pesos a day and rations, and those who were enlisted, each receiving advances of fifty to a hundred pesos or more.

Off the coast of (the subsequent) New Granada, where the shores were nothing but desert, the expedition ran short of provisions. After eating their horses, the men were reduced to chewing a dry cowhide. Here twenty-seven of them perished from starvation. Succour arriving from Panama, they resumed their voyage, and landing at Puerto de la Candaleria, scoured the country, and captured a village, where they made their first haul of gold, in the shape of some paltry ornaments. How many native lives were sacrificed to obtain these baubles is not related.

Frequent repetition of these incursions, in which it may reasonably be conjectured that every ounce of gold cost at least a hundred native lives, afforded the Spaniards in the end enough gold to warrant their sending the treasurer, Nicolas de Rivera, back to Panama with the spoil. This was done in order to procure reinforcements from Pedrarias. The latter, it seems, was very angry when he heard that so many Spaniards had perished for so small a return of treasure. As for the Indios who had been massacred, these do not appear to have been worth a thought. Before this, however, Almagro had set sail with the second vessel belonging to the partnership, and finding Pizarro at one of the little ports on the coast, the two worthies joined forces, and made an attack upon the first native town they found large enough to promise any gold. This was on the San Juan river (lat. 4 N.). They slaughtered a number of Indios, captured others, and got altogether the material of 15,000 pesos in gold of an inferior description.

This spoil being sent to Panama by the hands of Almagro, procured them in return a further reinforcement of forty men and more provisions, contributed, for a consideration, by the new governor of that province, Pedro de los Rios.

This was in 1526. Down to this time the Peruvian exploration company had been nothing except loss. Upon Almagro's return from the San Juan river raid he had evidently concealed some of the spoil from Pedrarias, for instead of paying him part of the 15,000 pesos, he came to ask for aid. The fact that Pedrarias was going out of office also made him insolent, for he told the governor that his only contribution to the funds of the partnership had been a single she-calf. This made Pedrarias angry. He reminded Almagro of the many soldiers whose lives had been sacrificed in the enterprise, and whose aid, but for his (Pedrarias') official countenance, the expedition would not have secured, and wound up by demanding 4,000 pesos to go out. After some hot altercation, Almagro signed a bond to pay Pedrarias 1,000 pesos "to renounce all rights and claims to the enterprise of Almagro and Company." (Oviedo, "Hist. Ind.," lib. 29, cap. 23.)

Meanwhile, Bartolomé Ruiz, Pizarro's pilot, pushes down the coast as far as Zalongo (near Puerto Viejo?), captures some castaway natives on a raft, from whom he gets some gold trinkets, hears of the great king Huayna-Capac and the famous city of Cusco, where there is much gold; and returns to San Juan where Pizarro has remained. The commander has meantime lost fourteen men in foraging for gold; and the rest are much reduced from sickness and privation. Reinforced by Almagro's recruits, they all re-embark, and push down the coast to Tacamez, where they again raid on the natives, procuring a little gold, and as many women as they desire.

At this point Almagro and Pizarro quarreled—probably over the division of the gold—Pizarro claiming that all the privations had been his, and declaring that he would return to Panama. High words passed between them, and swords were drawn; but after a stormy scene, and much crimination, they became reconciled. They returned to the island of Gallo, a little to the north of Tacamez, whence Almagro sailed to Panama for additional succour. Whilst the commanders were sustained by sanguine hopes of "discovering that which would enrich them all," the men, worn out by privation, desired to return to Panama. One of them contrived to send a secret message to the Governor, in which, alluding to the exposure of himself and his comrades (not an allusion to their continual slaughter of Indios), he calls Almagro the salesman and Pizarro the butcher.

This message resulted in the despatch of a lawyer, named Tafur, to the island of Gallo, there to enable the men freely to choose whether they would return or remain with the commanders. Tafur reaches Gallo, goes on board Pizarro's vessel, draws a chalk line on the deck, and tells the men to choose. Fourteen men, amongst whom was a native of Crete, and a mulatto slave, stood by the side of Pizarro; the rest returned to Panama.

With his reduced company Pizarro went to a small island called Gorgona; and Almagro returned to Panama to endeavour to obtain other recruits and supplies. Three months passed away before he succeeded. On his return to Gallo he found Pizarro and his company in a pitiful condition, having during many weeks subsisted on shellfish found upon the beach, and other precarious supplies of food.

The marauders now pushed further down the coast, passing in view of the towering summits of Chimborazo and Cotopaxi, though nothing of this was said in their narratives; their minds been fixed less upon such sublime marvels of nature as upon "gold, rich stuffs and precious stones." In the course of twenty days, having meanwhile frequently landed and scoured the country to little purpose, they arrived at the town of Tumbez, in the bay of Guayaquil; and here for the first time they came upon undoubted signs of the objects of their search.

At a small island which they had passed the day before, the Spaniards had found and pillaged a native temple, containing a stone image and rich offerings of gold and some silver pieces, wrought into the shape of "hands, women's breasts, and heads, a large silver jug, which held an arroba (four gallons) of water," etc. At Tumbez, where they were received with wonderment and hospitality (the place was too strong to attack), they beheld a fortress with six or seven walls, aqueducts, many houses of stone, and vessels of silver and gold. Being invited into the temple and palace (for Tumbez was a watering-place where the Inca, Huayna-Capac, occasionally dwelt), they perceived that the former was lined with plates of gold, the latter filled with gold and silver vessels, furniture, etc., and the gardens ornamented with golden statues.

The Spaniards resolved to start for Panama, and return with forces enough to destroy this peaceful and hospitable town, slaughter its inhabitants, and capture the gold it contained. Before putting this design into execution they sailed down the coast a little farther, went ashore, where they were received with the usual hospitality, and obtained a couple of native boys, who were taught on the voyage and at Panama to speak Spanish, with the view of employing them as interpreters in the projected expedition against Peru. This being done, they sped back to Panama, where they arrived at the end of the year 1527, freighted with great news.

At Panama the partners Almagro, Pizarro, and De Luque, discussing their prospects, agreed that before all things it was necessary that Pizarro should go to the Court of Spain. What was wanted was a clear title to their discovery, free from any claims on the part of the governor of Panama; and more aid in the shape of men and provisions than either the governor could afford them or their now nearly exhausted means could procure. Pizarro accordingly went to Spain, where he was so successful that in January, 1530, he sailed from Seville with two ships and 125 men, with the latter of whom he duly reached Panama. The story he had to tell his partners was not over-pleasing to them, particularly Almagro, for it seems that in his representations to the Court, Pizarro had omitted their part in the enterprise, and had so enlarged upon his own that he had been created a Knight of Santiago, and had obtained the sole Governorship (Adelantado) of Peru[1] for himself. For De Luque he had obtained the bishopric of Tumbez (the Inca's little sea-side resort, as yet unconscious of the fatal distinction conferred upon it); for Almagro, who had lost an eye in one of their gold forays, and who had therefore fared worse than either of his partners, nothing. More than this, Pizarro had brought out with him four of his own brothers, a circumstance that did not help to allay the discontent and jealousy of Amagro.

However, a peace was again patched up between the confederates, and their new expedition was prepared for sea. Consisting of three vessels, 183 men, and thirty-seven horses, it set sail December 28, 1530 (Feast of the Innocents), and in three days reached San Mateo, where Pizarro landed his forces and commenced that march through the country which was to end in the subjugation of an empire and the destruction of millions of human lives.

The story can henceforth be told more rapidly. Pizarro now had force enough, especially in the thirty-seven horses, to sweep all before him. He was resolved to grasp at every grain of gold and silver which the country contained; no matter at what cost of life or suffering. He had toiled and hungered in this enterprise for six years, and now that his reward was within sight he resolved to take it.

From San Mateo he marches upon the town of Coaque, which he puts to the sword, capturing trinkets amounting to 15,000 pesos in gold, 1,500 marks in silver, and many emeralds. These he sends to Panama and Nicaragua, to procure further reinforcements of men and horses. It is seven months before the latter arrive. Meanwhile a civil war

[1] "Two hundred leagues down the coast from Tenumpuela (island of Puña) to Chincha." The wording was important, for it gave rise to the feuds between Almagro and Pizarro, in which much blood was shed.

in Peru, unknown to Pizarro, is sapping the basis of the empire and preparing it for his hands; Huayna-Capac is dead and his two sons are quarrelling for the throne. Upon the arrival of his reinforcements Pizarro ravages many villages on his way to the bay of Guayaquil. Arriving there, he puts off to the island of Puña on rafts, and is received with hospitality and presents of gold and silver by the chief curaca, a Peruvian title corresponding to cacique. In return, he seizes upon the curaca and his sons under pretence that an attack upon him was being planned, and after duly terrifying them, sets the chief free, in order to avoid alarming the people of Tumbez. He then sets off for Tumbez, establishes himself in a couple of adjacent forts, and after some preparation attacks and puts the town and the surrounding country to the sword, plundering right and left, and distributing the surviving Indios into repartimientos. This is in May, 1532. More reinforcements arrive from Panama, and Pizarro sends by the return vessels the king's Fifth of the spoil, and with his own share pays for his supplies.

In September, 1532, Pizarro marched upon Cassamarca, plundered the Indios on the way, and applied torture to compel them to reveal where he might find more gold. Midway between San Miguel and Cassamarca messengers come from Atahualpa, that son of Huayna-Capac who had succeeded in possessing himself of this portion of Peru. They bring presents of gold and precious stones and tenders of welcome. Pizarro in return offers to espouse the cause of Atahuallpa against his usurping brother. It was the artifice of the ages: *divide et impera.*

Pizarro then pushes on, and entering Cassamarca, there occupies the palace, fortifies himself and sends an insolent message to the Inca, who, with his army, is in the field a league distant, bidding him to come to him. His envoys are Fernando de Soto and Fernando Pizarro, the latter of whom tells the Inca that a single one of their enchanted horses was sufficient to subdue the whole country. The Inca smiles at the threat implied in these words, but promises to come. This he does, and the Peruvian and Spanish chiefs meet at Cassamarca, Atahualpa attended by 5,000 naked and unarmed men, led by captains wearing (alas!) golden crowns and armour, themselves carrying arrows barbed with the same metal. It may be stated in this place that the most highly esteemed metal of the Peruvianswas copper, and the Inca wondered, after seeing that the Spaniards possessed glass, which he considered far more desirable than gold, why they had come so far and behaved so ill, for comparatively useless materials like gold and silver. Truly did Aristotle observe that the value of gold is a mere matter of convention.

Having induced the Inca and his principal men to venture within his lines, Pizarro proceeded to execute an abominable act of impiety, treachery and murder. The proceeding is thus related by the Abbé Raynal:

It was arranged that Pizarro was to command the troops, Almagro superintend the commissariat, and De Luque supply the means. This partnership of ambition, avarice and ferocity was completed by fanaticism. De Luque publicly consecrated a Host, part of which he ate and divided the remainder among his two associates; all three swearing by the Blood of God that to enrich themselves they would spare neither age nor sex. . . . At Cassamarca, Atahualpa sent the Spaniards fruits, corn, emeralds and several vases of gold or silver, begging them to depart and that on the next day he would come in person to arrange for their peaceful withdrawal. Pizarro's reply was to place his forces in ambush. He planted his cavalry in the gardens of the palace where they could not be seen; his infantry was stationed in the court; his artillery was pointed toward the gate by which the Emperor was to enter. Atahualpa came without suspicion. He was carried on a throne of gold. Turning to his principal officers, he said: "These strangers claim to be the messengers of God; be careful not to offend them."

Then Father Vicente de Valverdo, crucifix in hand, advanced toward the Emperor and read to him the Requerimiento which was duly translated by the interpreter, Philipillo. The monk wound up by demanding submission and tribute, under penalty of fire and sword. What the Emperor thought of this mingled tissue of falsehood, impudence and avidity we know not, but listening to it with some show of patience and politeness he replied: "I am very willing to become a friend to the King of Spain, but not his vassal. Your Pope must surely be a most extraordinary man to give so liberally what does not belong to him. I shall not change my religion at your bidding; and if the christians adore a God who died upon a cross, I worship one who never dies—the Sun." He then asked Vicente where he got his title to the command of the earth. "In this book," replied the monk, presenting his breviary to the Emperor. Atahualpa took the book, examined it on all sides, fell a-laughing and throwing it away, added, "Neither this nor any other writing conveys a title to the earth." He then reminded the Spaniards that they had already pillaged the country, and he demanded back the spoils they had taken.

Upon this, Father Vicente turned to Pizarro and bade him make no delay. Pizarro then lets loose his bloodthirsty followers (there were about 350 of them), slaughters 2,000 of the Inca's attendants, including women and children, and takes the Inca himself prisoner with 3,000 others; the Spanish injuries amounting merely to one horse wounded; from which circumstance it is evident that the natives, however numerous they might be, had no means of defending themselves against their treacherous invaders. "When the Spaniards returned from this infamous massacre, they passed the night in drunkenness, dancing and debauchery." Next day Pizarra pillages the Inca's camp and gets 80,000 pesos of gold, a lot of silver utensils, emeralds, women and provisions.

Atahualpa, anxious for his liberty, and observing that the chief concern of the christians was to obtain gold, offers for his Ransom to fill the room of his imprisonment as high as he could reach, eight or nine feet, with gold, if they gave him two months' time. This stupendous offer being very readily accepted, arrangements were made to receive the Ransom. Pizarro now obtains from Panama reinforcements of six vessels, 160 men, and 84 horses. Fernando Pizarro goes to Pachacamác for the Ransom and gets 27 loads of gold and 2,000 marks of silver. Three soldiers are sent to Cusco for another installment of the Ransom, and there behave with the greatest insolence, avarice and incontinence. So far the gold alone amounted to 1,326,539 pesos of pure metal. The gold and silver eventually amounted in value to about $4,500,000. When it was all paid, the Inca, instead of being liberated, was loaded with chains, condemned by an almost unanimous vote of the Spaniards to be done to death, tied to a stake and murdered with a cross-bowstring.

It is hardly necessary for the purposes of the present work to pursue the terrible story any farther. The fate of Peru may be summed up in a few words. At the period of the Spanish Conquest it contained perhaps fifteen, certainly from ten to eleven millions of inhabitants. By the year 1550 several millions of them had been destroyed. Antiguedades Peruanas, c. iii., pp. 65, 146*n*. When these people were counted under the general census ordered by Felipe II., their numbers had fallen to 8,280,000. In the course of two centuries they had diminished to fewer than 1,079,122. Father Domingo, who is quoted further on, says one-half to two-thirds were exploited; over such an enormity charity almost prefers to throw a veil.

According to Malte-Brun, the first Spanish census was taken in 1551, and including Santa Fé and Bogotá, the population of Peru at that period was 8,255,000. By Santa Fé and Bogotá is understood all of New Granada, now divided into New Granada and Equador. This included Spaniards and other Europeans, as well as negroes, mulattos and mestizos. The same author gives for the population of New Granada (now New Granada and Equador), in 1808, 1,800,000 souls, including Europeans. If these countries be included in the limits of ancient Peru, there existed in all of them towards the close of the 18th century not more than 2,500,000 Indios and mixed races of partly native blood. Deduct this number from the original fifteen millions and the remainder marks the number destroyed by the sword, the mines and the dogs.

Father Las Casas, who, though "fervid in condemnation, is not noted for inaccuracy or carelessness in his statements of fact," declared ("Destruycion de los Indios," p. 5) that in the first forty years after the discovery of America, twelve or fifteen millions of the natives had been destroyed by the infernal work of the christians, "infernales obras de los cristianos." Sir Arthur Helps prefers to extend his view to the first sixty years and to count the extermination at twelve millions. Of this frightful waste of life about one-half each may fairly be apportioned to Mexico and Peru. When the principal cause of the amazing depopulation is demanded, there is but one answer: the mines. From the moment that the Inca was murdered, the cruel work began. Thousands and hundreds of thousands of the peaceful Indios were murdered in cold blood; the land was ransacked for gold; and in this search neither age nor sex was spared. Before that search ended, by which every ounce of metal which the unhappy nation had ever possessed was secured by the Spaniards, the former were reduced to the most wretched slavery and compelled by millions to work in the mines, while every grave in Peru was violated in the search for gold.

The repartimiento, a grant of the services of the conquered to the conqueror, was probably of Roman imperial origin. In Candia, during the 13th century, the Venetians apportioned the conquered natives among their soldiers, and, parcelled out the land in cavallerias, and fanterias (Hazlitt). A similar practice was pursued by the Spaniards towards the conquered Moors. It is mentioned by Al Makari in his History of Spain. Columbus introduced it into Hispaniola in 1496. Repartimientos, afterwards called encomiendas, of Indios were at first granted by the Spanish authorities to the conquistadores for one life; in 1536 for two lives, *i.e.*, during the lives of the first grantee and of his successor; then they were abolished, then granted again for two lives, then confiscated by the Crown; in 1559 they were granted for three lives, in 1607 for four lives, and in 1629 for five lives. The encomienda did not convey any land: only the personal services of the conquered Indios. Land, however, and until about a hundred years after the Conquest, was given by the Crown of Spain for nothing; the horse-soldier receiving a cavalleria and the foot-soldier a peonia, etc.

At first the slavery of the Indios was absolute. Under this system they died so fast that towards the beginning of the 17th century the Spanish government established the Mita. By this law only the Indios of the ages from eighteen to fifty were compelled to work in the mines, but of these only one-seventh were to work for a period of six months, when they were to be succeeded by another detachment of one-seventh; so that the same individual would only work once in three-and-a-half years. He must, moreover, be paid half a dollar a day. But these regulations were evaded by the Spaniards, and in effect the labour was unremitting. The value of a man was reckoned at three sheep; of a woman and child, a sheep and a lamb. (Charlevoix, II, 103.)

In the Munoz collection of manuscripts relating to Peru, there is a letter written to the King of Spain by Father Domingo de Santo Tomas, dated 1550. In this letter the writer alludes to himself as a poor monk whose duty did not require him to look into such matters, but whom pity and Christian charity did not permit to witness them in silence. He lived in Peru, and he says that in ten years one-half or even two-thirds "of men, cattle, and the works of men" had been destroyed. Hoy ha diez anos que ha que yo entré in ella, hasta ahora no hai al presente la mitad i de muchas cosas dellas ni aun de tres partes la una, sino que todo se ha acabado. In the course of this letter the monk gives an elaborate account of the horrible sufferings and privations of the Indios in the mines of Potosí, and his conclusion is: "Se mueren los pobres como animales sin dueno . . . los que de esto se escapan jamas buelben a sus tierras." The poor creatures died like cattle, and even the few who escaped alive never reached their homes.

Those who may wish to view for themselves the fiendish cruelties which were practised by the Spaniards in Peru will find plenty of material in the Fourth volume of Sir Arthur Helps' "Spanish Conquest in America" My uncle, Don Manuel, who was Sir Arthur's collaborator in the preparation of this great work, told me that his eyes were often filled with tears when he recorded them. But let us hear Giovanni Botero, who lived two centuries and a-half nearer than Sir Arthur did to the period of these cruelties, and who probably conversed with numerous returned friars and other eye-witnesses of the transactions in America.

"The natives (of the Indies) have small regard for gold, silver or minerals," yet "authors who have described this continent inform you of nothing but gold-yielding rivers, seas abounding with pearls and lands with gems. . . . But, alas! avarice and depravity, under the mask of religion and vainglory, had no sooner set foot in this paradise than all things were turned topsy-turvy. Since then happiness hath taken flight and now nothing is recorded of it save the undermining of mountains, disembowelling the earth, transporting the natives, and depopulating the towns, and that by tyranny and slavery." . . . The Spaniards "are the robbers and ravishers of the world. . . . The treasures of the conquered, breeds in them covetousness; their weakness, breeds ambition. Neither the West nor the East can satiate them. They covet the wealth and the power of all nations with equal greed and hypocrisy. On robbery, murder and villiany they impose the false title of Empire. Solitude they term peace, and desolation, tranquility. Had not Charles V. cast strict reins upon their avidity the Indies had been quite depopulated and Spain filled with slaves. Of 400,000 natives living in Hispaniola upon the arrival of the Spaniards, they now can scarce show you 8,000. About the like number you shall find in the Honduras remaining of 410,000 when the Spaniards therein first set footing. If you even read their own books you shall meet with no better accounts concerning Guatemala, Nicaragua, etc., the greater part of whose inhabitants are either slain, enslaved or consumed in the mines. But the progeny of the natives is not quite extirpated, and some day they will arise and throw off the hated yoke of their oppressors." A prophecy which was remarkably verified two centuries later!

The precious metals obtained by the Peruvians previous to the Spanish conquest consisted nearly altogether of gold secured by washing the river gravels. No native shafts were found. A few excavations had been made into the sides of hills with outcrops of native gold or silver. Of this character were the superficial workings at Porcos (district of Potosí), from which Prescott rather lightly says they obtained "considerable returns," implying that such returns were of silver. It is probable, that like many silver leads in the Sierras and Cordilleras (for example, the Comstock Lode of Nevada), the top of the Porcos vein only yielded gold; and that little or no silver was obtained from it until the shaft, begun by Hernando Pizarro, in 1539, carried the workings into silver ore. Between this date and 1544, Gonzalo Pizarro was at Charcas "busily employed in exploring the rich veins of Potosí." (Prescott). A Spanish law relating to the mining district of Potosí, dated 1535, is still extant. The district was registered in 1545. It is this last date which in works of reference is erroneously given for that of the discovery. The story of the Indian who discovered the mines of Potosí by the accidental displacement of a shrub to which he had clung in ascending the mountain, may be dismissed to the realms of fiction. It is possible that the brothers Pizarro worked these mines between 1534 and 1545, on their own account, and without paying the Quinto; because when Francisco Pizarro was slain, at Lima, in 1541, by the partisans of the murdered Almagro, they found a surprising amount of treasure secreted in the house of the conquistador.

Next in celebrity to the silver mines of Potosí were those of the Cerro de Pasco, though these are of a much later æra. They were first worked in 1705. In 1784 the annual product, then about 70,000 marcs (weight), gradually increased to 400,000 marcs (1841), and then fell to 160,000 marcs (1879). From 1811 to 1825, and since 1879, the mines produced nothing. Du Chatenet, who wrote in 1880, says that about eight or nine thousand natives, all gamblers and drunkards, used to be employed in the vicinity at half a dollar per diem, and that the bars of bullion which were worth $200 at the casa de fundicion, cost, on the average, $270. Von Tschudi (1838) says that there are several thousand shafts at Pasco, all in a ruinous condition, and that many caves had occurred and numerous lives lost, in one case 300, a tragedy which has given to the place the name of Matagente or Deadman's shaft. The people are all gamblers, and are so poor that in the province of Janja they use eggs for money, at the rate of two cents each. For bonanza the local term is boya, but there are no longer any boyas at Pasco.

The silver mines of San José were once so productive that on the occasion when the owner's first child was christened, for whom the Vireyna (viceroyess) stood sponsor, the father laid a triple row of silver bricks for the latter to walk upon, which stretched from the palace to the church. Then he presented the entire pavement to the sponsor. How many natives were sacrificed to secure this object of ostentation is not mentioned (Von Tschudi). In the same district are the quicksilver mines of Huancavelica, opened about the year 1570, and worked for two centuries, though after 1684 they declined.

In 1790 the richest of these mines caved in, killing 122 men, since which time the natives have refused to work in them. (Jacob, 266.)

In the year following the murder of Francisco Pizarro, the partisans of the younger Almagro discovered the rich gold placers of Caravaya, from which they took one nugget weighing four arrobas, 129½ lbs., and another of nearly equal weight, besides vast quantities of pepitas and dust. Eventually, the wild Chuncha Indians of the Sirineyri tribe fell upon the gold washers and drove them away. In the following century a band of Spanish mulattos reoccupied these diggings, from which they extracted so much treasure that in return for his ample Fifths the King of Spain granted them the singular privilege (their own asking) of being called Senores, and of riding on white mules with bells and red trappings. In 1767 the town of San Gavan, with 4,000 families, mostly negros, and a large treasure, was surprised and entirely destroyed by the Carangas and Suchimanis Chunchas. (Lock.) In 1849, Caravaga was again invaded by the gold diggers; but the news from California soon drew them away to more profitable and less dangerous regions. The natives have a horror of the precious metals, and will not touch them. Attempts have been made by the whites to explore the ruins of San Gavan, beneath which there is believed to remain an immense sum of gold; but the Indians will not permit it. Vein mines have been opened in this district, and these, together with the scrapings of the placers, constitute nearly the entire auriferous product of Peru, which is now about a quarter of a million dollars per annum. The entire gold product, including plunder, robbery of graves and mining, has been about 190 tons of 2,000 pounds. The present annual silver product is about 4½ million ounces, worth in gold 2¾ million dollars. A comprehensive estimate of the entire product of the precious metals in Peru, from the Conquest to the present time, is given elsewhere in the present work.

Did the Peruvians before the Conquest ever use gold or silver money? We answer this question by asking another one. Did the Chinese ever use gold or silver money? The reply to this question is, Yes; and reasoning by analogy, so must be the reply to the other. The Chinese do not now use money of the precious metals; only bronze t'sien, chuen, sapeques, or "cash"; but they certainly did, on more than one occasion, use gold and silver money, because some of the pieces are still extant, and specimens of them may be seen in several of the great numismatic collections of Europe. Mr. Prescott says that the Peruvians "knew nothing of money," and implies that they never had used money. From certain considerations set forth at length in the "History of Money in America," this opinion appears to be unfounded. In addition to those considerations, which of themselves appear to be sufficient, some further evidences have rewarded the writer's researches, which seem to add strength to his previous conclusions.

"Near Truxillo, Humboldt (in 1802) visited the ruins of the ancient city of Chimu, and descended into the tomb of a Peruvian prince in which Garci Gutierez de Toledo, while digging a gallery in 1576, discovered a mass of gold amounting in value to more than a million of dollars." Bayard Taylor's Life of Humboldt, p. 262. Sarmiento says that 100,000 castellanos, equal in weight to a quarter of a million gold dollars, were sometimes (this implies more than once) got from a single tomb. "The burying of treasures (meaning gold), was an old and very generally prevailing Peruvian custom. Subterranean chambers were often found below many of the private dwellings of Caxamarca." Taylor's Humboldt, 256.

The object in burying the treasures was obviously to preserve them; but why preserve these peculiar treasures unless confidence was felt in the ability of the owners to exchange them at pleasure for such other things as they might desire; and what could have been the basis of this confidence other than the employment of gold as a common medium of exchange, either at the time of the hoarding or at some previous time, the value of the gold at which previous time was connected with its value at the then present time, by means of the intermediate measure of value established by the government? Except a few mementos and these buried treasures of gold, no other kind of personal property was preserved by the Peruvians. All else was used, consumed, devoted to present wants, the existing requirements of the owner. Why preserve gold unless its previous use and previous value as money inspired confidence in its future value?

In the "History of Money in America" it is proved from Prescott himself that the Incas monopolized the native mines of copper, silver and gold; and in the History of the Precious Metals in Chile it is shown that the Incas exacted the tribute of that country in gold. These circumstances point to gold as having been then, or else previously, a common measure of value in Peru; and a measure of value is only another term for money. It may be that at the period of the Conquest the use of gold for money had been temporarily abandoned on account of the fitful supplies of gold from the placers or mines, and the distance and isolation of Peru from other populous countries; an isolation which left it without a vent for its surplus treasures.

Such a previous and familiar employment of gold as is here suggested could alone have afforded the basis of that confidence in its future and readily recognized value which induced the lords of Peru to bury immense sums of that metal beneath their dwellings or in their tombs.

Had the Peruvians been a new and progressive nation we should regard the metal they hoarded as nascent money; but as they were an ancient and moribund nation, we regard it as decadent money; that which had once been stamped or shaped and valued by the State, but was now demonetised; just as we know is at present the case with the gold slugs and silver "shoos" of China. Cortes reported that tin money, chisel-shaped, was used in Mexico; and the writer has seen specimens of this money in the Paris collection. The natives called the pieces xiquipili or siccapili, and a full-sized engraving of one of them appears in the American Journal of Numismatics, Vol. V, No. 2. Taylor, in his Life of Humboldt, p. 276, explicitly informs us that before the Conquest, the Mexicans "usually paid their tributes in two ways, either by collecting, in leathern sacks or small baskets of slender rushes, the grains of native gold, or by founding the metal into bars. These bars, like those now used in trade, are represented in the ancient Mexican paintings."

If the natives of Mexico could guide Pizarro to Peru, the natives of Peru could scarcely have been ignorant of Mexico and its employment of money. It is incredible that so powerful an instrumentality of societary life was employed by the Mexicans and not by a so similiarly constituted and so numerous a people as the Peruvians. The native scale of value employed in Mexico was 20 cocoa beans=1 olatl; 20 olatl=1 zontle; 20 zontle=1 xiquipili or siccapili, the latter being either a tin coin as above described or else a gold slug, weighing about one-fourth of an ounce Troy. As the Peruvians used capsicum pods, which are similar to cocao beans, for small sums of money, and as they exacted their tributes in gold and hoarded gold, it is difficult to believe that they did not also employ gold to represent large sums of money, in the same manner as the Mexicans.

In 1886, Senor Gaston, the Chief of the Bureau of Mercantile Statistics in Lima, made the following report on the social condition of the capital of Peru :—

"There are 100,000 population in Lima; 34 per cent. of these are minors, 47 per cent. single and widowed, 19 per cent. married. This is is not a matter of surprise, owing to the economical features mentioned hereafter, but it must be noted with pain that from the great number of minors it clearly appears that unmarried persons take almost as active a part in propagating their species as those in wedlock. Statistics show that 46 per cent. of the births in 1884 were the fruits of wedlock, and 54 per cent. illegitimate. It is, therefore, more than evident that over two-thirds of the population of Lima live in a state of concubinage, or something of that character, rather than in legal and family ties. According to census reports there is an excess of 2,534 unmarried or widowed women over the same classes of males, and as women are less adapted or educated to gain their living by industry than the male portion, the only means of subsistence left to them is that of depending upon the stronger sex, without regard to law or morality. In a society of only 6,000 families this excess of 2,500 single women, urged by necessity, perverted through ignorance and even influenced by the climate, is an element of danger to public morals and general good. In a word, of the 100,000 people in Lima, but little more than one-third live in family ties.

There are 33,914 persons under age, 34,159 females and only 30,083 active producers, so that every able-bodied man has to support more than three persons. The cost of food, clothing, etc., may be reckoned at $15 per head per month. Each man able to work should earn $45 per month. This amount is only obtained by the most limited number. A condition of the most abject poverty prevails, which shortens the life of the people. The consuming class is constantly increasing, and every man forced to shoulder the musket is but a new burden laid on the feeble shoulders of the producers. It is therefore shown that the population of Lima does not rest, as it should, on a family basis, and that from a lack of labour-power and excess of unmarried females there is wanting that great producing element which is a principal factor in well-established communities. The result of all this is a never-ending struggle between an educated minority and the powerful resistance of ignorant masses, who, almost irresponsible for their acts, entertain a deep-seated hatred for those they hold to be their oppressors. They are a suspicious, selfish mass, refusing all efforts towards their own self-regeneration, without habits of order, unable to respect laws of which they are ignorant, living in intellectual darkness, and generally in moral depravity; their existence passes by without a thought being given to their advancement, their perservation as a race, or to their hereafter.

When it comes to pursuits, 64,956 are without any. Even among those married *not one-fifth of the offspring survive the first year's existence*, and the reason assigned is that the parents cannot provide them with the proper nourishment.

Even more lamentable is the fate of those numerous unmarried women whose children never know their fathers. For them there is no employment, no protection, no support. When their children live, they grow up weaklings, owing to the privations they undergo, and they generally end their days in a hospital. Premiums are given for useless and unsubstantial objects, whilst talent and honourable labour go unrewarded. Immense sums are subscribed to build *sumptuous temples* side by side with the hovels where the poor die in want and misery. Hundreds of people exist in damp, narrow and unhealthy dwellings, whilst in the main street of the city are extensive *convents and monasteries*, covering acres of ground and affording shelter to a handful of useless and idle monks and nuns. On this account there are more churches and convents than municipal schools, more children and women than men, more soldiers and Chinese than active citizens, more beggars than labourers, more priests than men of science, more illegitimate than legitimate children, more concubines than women legally married, more ignorance than enlightenment, more corruption than morality, more gunpowder than bread."

From this terrible picture of Lima, it follows that the city must be largely supported by the country, and that the resources of the latter must be heavily taxed for its support. As these resources are comparatively slender, the condition of the Peruvian republic must be so deplorable that little hope can be entertained of a general amelioration of its affairs without the adoption of reforms which at the present time appear hopeless. These should obviously begin with the assumption of civil over ecclesiastical rule, the taxation of ecclesiastical property, and the absolute abandonment and interdiction of mining for the precious metals. It is much to be feared that before such drastic remedies can be grasped, Peru will have to pass through more than one more revolution.

CHAPTER XX.

CHILE.

Almagro captures the Inca's tribute and invades Chile—Defeated by the Purumanians—Valdivia's expedition—Defeated by the Arucanians—His second expedition proves more successful—Opening of Marga Marga—Details of the mining—Valdivia pushes farther south—Building of the Seven Cities—He is attacked, defeated and put to death by the Arucanians—Whose independence is respected by the Spaniards—Produce of the gold washings since 1548—Mode of working the mines—Attempts of the Dutch and English to wrest Chile from the Spaniards—Revolution of 1817—Independence of Chile—General results of mining in Chile—Current product of gold and silver—Erroneous statistics published by the American government.

CHILE is one of the few countries in America where the Spaniards met with a foe who did not fear them, and who, after having more than once put them to flight, finally succeeded in securing respect for their own independence by treaty. These were the Arucanians. Chile consists of a narrow strip of land, less than 100 miles wide, but nearly 2,000 miles long, lying between the Andes and the Pacific Ocean, and extending from latitude 23 to 56 degrees south. Much of this is auriferous; some of it is argentiferous. Nearly all of it is cultivable. In short, it is the California of the Southern hemisphere, and naught but bad government has made it poor.

In 1450, the Inca of Peru extended his dominion southward through Chile, to the river Rapel, where the resistance of the natives compelled him to halt. Here he built a fort and imposed a tribute upon the natives within his lines, amounting to about 1,400 pounds weight of gold annually. This tribute was cast into small bars, marked with the Inca's stamp and escorted each year to Peru by 400 bowmen. At the principal towns along the road it was welcomed by the Peruvians with rejoicings and festivities, not because there was any expectation that it was going to be of benefit to them, but because it was destined for the decoration of the great Temple at Cuzco, where the Deity was incarnated and worshipped in the person of the Inca. Meanwhile, the Purumanian Indians, who paid this tribute, were obliged to gather it painfully, with many tears and sighs, from the "manta" of gravel which surmounted their coasts and streams, wondering, no doubt, why so much cruelty should be exercised for the sake of an acquisition for which the Peruvians themselves had no immediate nor economical employment The Purumanians (whose name suggests that they were the aborigines of Peru), frequently revolted and were as often suppressed; though not without great losses to the armies of the Incas.

When the Spaniards conquered Peru they intercepted one of the Inca convoys of gold from Chile. Nothing more was needed to doom the unfortunate country to destruction. It was resolved to ravage Chile from end to end. The leader of the expedition was Diego de Almagro, who, with 570 European troops, set forth from Cuzco in 1535. The Spaniards met with no opposition till after passing Copiapó, when, like their predecessors the Peruvians, they were defeated by the brave Purumanians and compelled to retreat to Peru. Five years later an expedition, under Pedro de Valdivia, met with better success, and advanced as far south as the present town of Santiago, where it was stopped by the Purumanians. After obtaining re-enforcements from Peru, Valdivia advanced to the river Maule, lat. 35 S., where he encountered the Arucanians, a foe still braver and more determined, or better equipped, than the Purumanians. Here he was defeated and compelled to retreat to Peru. In 1548 Valdivia headed another expedition to Chile. This time he advanced beyond the Maule, to the place afterwards known as Penco or Concepcion, where he erected a fort. Here he was repeatedly assailed by the natives, but until 1553 he managed to hold his ground and ransack the surrounding country for gold.

The principal field of operations was Marga Marga, a "manta" of auriferous gravel situated in a valley between Valparaiso and Santiago; the means employed was the enslavement of the natives, who, by the ecclesiastical artifice of commendatio, were both religiously and lawfully bound to expend their lives in washing gold for their suzerain the King of Spain, and his suzerain the Cæsar of Germany, or the Pope of Rome, whichever happened to be uppermost at the time.

Here it is to be observed that that oldest of all ruses de guerre, the Divine Origin of the invaders, was again successfully employed. "When Almagro crossed the Cordilleras, the natives, regarding the Spaniards as allied to the Divinity, collected for them gold and silver amounting to 290,000 ducats," (Haydn,) equal in contents to as many half-sovereigns of the present day. So far as Chile is concerned, belief in the incarnation of the Creator appears to have been limited to the Purumanians; the Arucanians repudiating the mythos altogether. Whether the Purumanians learnt this doctrine from the Peruvians, or communicated it to them, is not known. It is most likely to have come from Peru, where it was supported by scriptures, calendars and other ecclesiastical evidences, of which the Purumanians possessed only a derived or secondary knowledge. The fact that it was only entertained by such of the Purumanians as dwelt north of Copiapó lends additional support to the conjecture. In this connection it is rather curious that the Arucanians, "whose religion was a kind of Manichaeism" (Helps), venerated a Genius of War, who bore the Egyptian name of Epon-Ammon. Their caciques were also called Ul-Menes, another Egyptian name. They valued reputation, tolerated philosophers, despised superstitions, and had neither temples, idols, nor priests. (Helps, II, 152–3.)

> Y estos que guardan órden algo estrecha
> No tienen ley, ni Dios, ni que hay pecados;
> Mas solo aquel virir les aprovecha
> De ser por sabios hombres reputados.
>
> —*La Arucana de* ALONSO DE ERCILLA Y ZUNIGA.

The operations of the Spaniards at Marga Marga began immediately after Almagro's Conquest of Chile. The captive Indians, after having been disarmed, were distributed among the conquerors in encomienda and driven like sheep into the gravel mines. At first the average number of encomiendieros to each Spaniard was about 30; the leader getting 600 and the rest according to their military rank. As will be seen below, these numbers were afterwards greatly increased. Within a year after Valdivia began his fell work at Marga Marga the free Indians descended from the mountains upon the province of Coquimbo, which lay between Santiago and Peru, put the Spaniards stationed there to death and cut off Valdivia's land communication. In their alarm at this event the Spaniards at Santiago begged Valdivia to abandon the country and return to Peru, whilst yet the sea route, by way of Valparaiso, was open to them. His reply was that should he do so, the King of Spain would lose his Fifths; and that that alone was sufficient reason for him to remain. Sending to Peru for re-enforcements, Valdivia drove the natives out of Coquimbo, and as Marga Marga was not enough to fill the maws of the newcomers, Valdivia pushed his conquests farther south, subduing and enslaving the Indians wherever he went, and erecting forts as a means of keeping them in subjection. These forts, from their number, were called Septem Civitates, or the "Seven Cities," namely, Angol, Caneta,

Imperial, Aranco, Villarica, Valdivia, and Osorno; the last-named, in the year 1551 and in lat. 41 south. The principal gold field worked within these newly-acquired domains were Quilacoya, where 10,000 to 15,000 naked Indians were kept at forced labour; the total annual output was about $300,000; the King's Fifth was about $60,000. The daily produce of each slave was equivalent to about six cents, say 3d. sterling; the cost of supporting him, or rather of killing him, on a scant supply of maize and beans, was about half as much as his produce. In short, the net yield of a slave, so long as he could be kept at work, was about 3 cents, or 1½d. per day. It was for this precarious reward that the Spaniards deserted their own country, one of the most beautiful on earth; it was for this that they abandoned its mines, which were far richer—indeed, they are yet—than those of Chile; it was for this that they traversed the ocean, to attack, enslave and doom to destruction an innocent and harmless race of Indians; and it was for this that they committed enormities which rendered the name of Spaniard odious to the rest of the world. We forbear to moralize upon the King, whose quinto, or the Pope, whose annats, were derived from such a source. Father Las Casas, the historians of Spain, Father Raynal, and the good Dominican friars, have said all that is necessary on this subject; and we can add nothing to the bitterness and force of their condemnations.

In 1553, the Spanish empire in southern Chile was suddenly extinguished. Valdivia's forces were defeated by the brave Arucanians, the Spanish leader was captured and put to death by pouring molten gold down his throat, the entire colony of gold-hunters was driven out of the country; and carrying their victorious arms to the north, the Indians assaulted and captured Concepcion, which, though afterwards won by the Spaniards, formed, for a century, the boundary line (the river Bio-Bio) between the gold-seekers and their quarry. At length, in 1665, the former were fain to make a treaty with the Arucanians, acknowledging their independence and establishing the limits of their territory at the Bay of Concepcion. This peace lasted until 1723, when it was broken by the Spaniards, who drove the Arucanians into the mountains, where they resided until the Revolution, maintaining their independence and allowing no white man to enter their territory, and no one to "prospect" for gold.

The history of the precious metals in Chile is in its general aspects the counterpart of its history in Peru. First, the plunder of the natives, and the robbing of their temples and graves; second, the gold placers; third, the quartz mines. In Chile, it used to be a saying that he who worked a copper mine might gain; he who worked a silver mine might gain or lose; but he who worked a gold mine was sure to be ruined. The gold is embedded in gravel beaches lying high above the level of the petty streams that drain the Sierras. These beaches were toilsome to get at and expensive to work. The silver mines are in the Sierras, which are difficult of access and intensely cold. Yet, by reducing the defenceless inhabitants to slavery, which was substantially the case down to the Revolution of 1817, the Spaniards managed to extract from these forbidding sources, during the period from the Conquest to the Revolution, about £25,000,000 in gold alone; chiefly from plunder and the Marga Marga and Aconcagua gold washings, El Bronce de Petorca, Illapel and Andacollo, and those of La Ligua on the coast north of Quillota and Catemo, 12 leagues up the river Aconcagua.

Mr Darwin, who visited the gold quartz mines of Yaquil (Jajuel) about the year 1846, thus describes the manner of working them: The labourers or miners are paid about £1 a month, together with food. This consists of 16 figs and two small rolls of bread for breakfast, boiled beans for dinner, and broken roasted wheat grains for supper. They scarcely ever taste meat. Out of their £1 a month they have to clothe themselves and support their families. The "apires" are those who bring up the ore to the surface. Their pay is from 24 to 28 shillings a month. They live entirely on boiled beans and bread; they would prefer the bread alone, but the masters, finding that they cannot work so hard on this, insist upon their eating the beans. The mines are 450 feet deep, and each apire brings up nearly 200 pounds weight of ore each time he ascends. With this load he has to climb up the alternate notches cut in the trunks of trees placed in a zig-zag line up the shaft. The men, who wear only a pair of drawers, ascend with this heavy load from the bottom of the shaft to the top. Even young men, eighteen or twenty years of age, whose muscular development is far from completed, perform this amazing task. They leave the mine only once in three weeks, when they stay with their families for two days. As a means of preventing the men from abstracting any of the gold, the owners establish a very summary and stringent tribunal. Whenever the superintendent finds a lump of ore secreted for theft, its full value is stopped out of the wages of all the men, so that they watch over each other, each having a direct interest in the honesty of all the rest.

The amount of labour they undergo is greater than that of slaves; being to a certain extent masters of their own actions, they bear up against what would wear down most men. Living for weeks together in the most desolate spots, when they descend to the villages on holidays there is no excess of extravagance into which they do not run. The miners dig the ore from the bowels of the earth, while the apires are simply labourers, much like bricklayers' labourers, but the latter carry less heavy loads, and up a much less height. According to the general regulation, the apire is not allowed to halt for breath, until the mine is over six hundred feet deep. The average load is considered as rather more than 200 pounds, and Mr. Darwin was assured that one of 300 pounds (twenty-two stones and a-half), by way of a trial, has been brought up from the deepest mine! At that time, the apires were bringing up the usual load twelve times in the day, that is, 2,400 pounds from eighty yards deep; and they were employed in the intervals in breaking and picking ore. They rarely eat meat, once a week, and never oftener, and then only the hard, dry charqui (dried beef). Although with a knowledge that the labour is voluntary, it was, nevertheless, quite revolting to see the state in which they reached the mouth of the mine; their bodies bent forward, leaning with their arms on the steps, their legs bowed, their muscles quivering, the perspiration streaming from their faces over their breasts, their nostrils distended, the corners of their mouths forcibly drawn back, and the expulsion of their breath the most laborious; each time, from habit, they utter an articulate cry of "ay-ay," which ends in a sound rising from deep in the chest, but shrill, like the note of a fife. After staggering to the pile of ores, they emptied the carpacho; in two or three seconds recovering their breath, they wiped the sweat from their brows, and, apparently quite fresh, descended the mine again at a quick pace.

In 1642, the Dutch West India Company, observing that the natives of Chile were inveterately incensed against their conquerors the Spaniards, flattered themselves that they would make an easy conquest of that country. For this end, they fitted out a squadron of ships, under Admiral Brouwer, hoping to possess themselves of the Chilean gold mines. At first they defeated the Spaniards, and gained over some of the caciques, with whom they entered into an alliance, after furnishing them with arms. The Dutch even erected a fort at or near Valdivia, but found the Indians less tractable than they had imagined. In the end, the Dutch were obliged to retire from Chile without having achieved the object of all this preparation. (Anderson, Hist. Com., II, 398.)

About the year 1655, on apostate Roman priest, named Gage, returned to England from the Spanish West Indies, where he had resided many years, and communicated to Oliver Cromwell so particular an account of the feeble condition of the Spanish garrisons in America as induced the Protector to attempt their reduction. In his relation, Father Gage was amply supported by one Simon de Cafferes, a renegade Spaniard. Accordingly, Cromwell, in 1655, sent Vice-Admiral Penn with thirty ships of war, and four thousand soldiers, under Generals Venables and Holmes. Their first point of attack was near St. Domingo, in Hispaniola; but here they met with so warm a reception, losing 600 men at the outset, that the attempt was abandoned. The expedition next sailed to Jamaica, where it was more successful. The project submitted by Simon de Cafferes was couched in the following language, which, it will be observed, contemplated the conquest of Chile:

With four men-of-war only, and four other ships with provisions and ammunition, and one thousand soldiers, to round Cape Horn, sail into the South Sea (Pacific Ocean), and passing Valdivia, from which port the Spaniards have already been driven by the natives, seize upon the kingdom of Chile; our people to rendezvous at the Isle of Mocha (half-way between Valdivia and Aranco), where they might victual and water, as there are none but Indians there. As Chile abounds with more gold than any other part of America, as well as with provisions and a wholesome climate; as, moreover, the Chileans are the most warlike of any American people, and being mortal foes to the Spaniards by reason of their former cruelties, therefore they would probably gladly side with any people inclinable to drive the Spaniards quite out of their country. If this project should succeed it would distress Spain in the most sensible and and least guarded part. The ships of war above mentioned would serve to seize upon the Spanish treasures going annually from Chile to Spain by the coast of Africa, as well as those which go by Lima and Guayaquil to Panama, and the two-year rich Acapulco ships to the Philippines. (Anderson, II, 432.)

To support this project, Cafferes offered to enlist, in Holland, some of the men who had been with Brouwer's expedition against Valdivia, and whose knowledge of the ground might prove useful. But after considering the difficulties of the enterprise, distance, storms, absence of friendly ports, dubious attitude of the natives, dangers of the climate, etc., Cromwell decided that the expedition should confine its operations to the Atlantic coasts of America and not venture to Chile. Ten years later, the brave Arucanians drove the Spaniards north of the river Bio-Bio, and extorted from them the treaty above mentioned, thus removing all encouragement to Dutch or English marauding expeditions.

Nevertheless, in 1669, Charles II. of England being advised to "attempt a settlement in a country so greatly abounding in gold" as Chile, sent out Admiral Sir John Narborough with a thirty-six gun ship and a pink (tender), with orders to ingratiate himself with the natives and open a trade with them; but on no account to molest the Spanish settlement which still lingered at Valdivia; England being now (temporarily) at peace with Spain. Sir John's attempt at "trade" with the natives was, however, so vigorously resented by the Spaniards at Valdivia, who seized upon his lieutenant and three of his sailors, that he "judged it prudent to return home." (Anderson, II, 501.)

The first plunder of Chile and the graves yielded about $18,000,000. From 1545 to 1560 the gold production of Chile averaged about $1,-000,000 per annum; from 1561 to 1740 it averaged about $250,000 per annum; during the 20 years, 1741–60, it averaged $350,000 per annum; 1761–80, $500,000 per annum; 1781–1800, $600,000 per annum; 1801–17, $1,000,000 per annum. From this period, that of the Revolution, the production greatly declined. During the period 1818–20 the average annual production was only about $140,000; then it rose in 1821–40 to $800,000, and fell in 1841–50 to $700,000; and in 1851–95 to $250,000. During the five years ending with and including the year 1900 the annual average production of gold gradually increased. This was probably due in Chile, as in the United States, Mexico and other countries, to the demonetisation and relative fall of silver and the transferance of the labours of prospectors and miners from silver to gold deposites. The subsequent production of gold is estimated as follows: 1896, $500,000; 1897, $600,000; 1898, $700,000; 1899, $800,000; 1900, $900,000; total for the quinquennial period, $3,500,000.

The production of silver in Chile, which, during the latter part of the 18th century, is indicated by an annual coinage of about $150,-000, fell during the Revolution to almost nothing, and only assumed noticeable proportions after the decline of the Californian and Australian placers, and the return of the Chilean miners to their native country. During the last decade of the 19th century the average annual exports of silver from Chile amounted to about five million Chilean silver dollars, each of 257.7144 Troy grains fine; hence to about 2,684,567 ounces. These exports frequently include a considerable proportion of silver produced in Bolivia—for example, from the mines of Huancha, whose product has usually been exported from Autofagasta in Chile. As owing to the adoption of a paper currency, the coinage of the precious metals in Chile has fallen almost to nothing, these exports exceed the production. The Director of the United States Mints, more than doubles these figures. But as repeated instances prove, objections to the methods of this official are of no avail. Neither he nor the superior officers who employ his statistics in their recommendations to Congress, nor the Congress which accepts them as the basis of its legislation, have ever given them the least scrutiny. His tumid figures agree with the theories of the ignorant majority, and that appears to be sufficient. Vulgus vult decipi—decipiatur.

CHAPTER XXI.

LA PLATA, OR BUENOS AYRES.

Slave-hunting and pillage—Encomiendas—Surrender of the coinage prerogative—Mining—Taxes—Sale of indulgences—The Mita—Disappearance of the natives—African slave trade—The Paraguay missions—The Paulistas attack the missions, and capture the christian converts—Effect of fire-arms—Cost of Gold—Destruction of the missions and enslavement of the converts—Unprofitableness of the mines—Their produce—Spanish moneys—Mining and monetary laws—Quinto, seigniorage and covos—Mints—Ratios—Counterfeit coins—Individual or "free" coinage—Poverty of the Spanish Crown—Excessive over-valuation of billon and copper coins—Increase of counterfeits—Expulsion of the Moors from Spain—Panic of 1714—Revolution of 1732—Mutations of moneys—Revolution of 1810—Value of gold not due to cost of production.

THE Rio de la Plata was discovered in 1516 by Juan de Solis, a Spanish commander; in 1523 Alexis Garcia, a Portuguese commander, traversed Brazil and the Gran Chaco, scaled the Andes, and pillaged Alto Peru three years in advance of Pizarro; in 1527 Sebastian Cabot founded San Espiritu, gave the name of La Plata to the river and country, and petitioned the King of Spain to aid him in opening a route to Peru *via* the Vermejo; in 1535 Mendoza founded Buenos Ayres, his lieutenant, Del Campo, giving it that name; in 1537 Juan d'Ayolas ascended the Paraguay, and struck across the continent to Alto Peru, which he pillaged; in 1542 Cabeça de Vaca, landing in Brazil, sent his lieutenant Yrala to pillage Peru, but he failed to reach it. He returned with 12,000 Indian slaves and a few sheep. Upon being appointed Governor of La Plata, with headquarters at the Fort of Asuncion, Yrala, in turn, sent Nuflo de Chaves to pillage Peru, but was there confronted by a pillaging party from Peru itself under Hurtado de Mendoza. In 1560 these united expeditions founded Santa Cruz de la Sierra.

Meanwhile, in 1535, the mines of Potosí had been opened by the Spaniards of Peru. A line of forts and settlements was soon afterwards established between these mines and the estuary of La Plata. It was now perceived that the natives were too poor to afford further encouragement to pillaging expeditions, and the efforts of the Spanish adventurers were turned to mining for gold and silver. To fill the coffers of the King of Spain, experience had proved that it was necessary to capture and enslave the natives; and as a pretext for this procedure, there was read, nominally to them, but really to the trees (*á los arboles*), a proclamation (*Requerimiento*) from the king, which ended with these words: "But if you will not comply or maliciously hesitate to obey these injunctions, I will enter your country by force; I will carry on war against you with the utmost violence; I will enslave and subject you to the yoke of obedience; I will take your wives and children, make them slaves, and dispose of them at pleasure; I will plunder you and do you all the mischief in my power, treating you as rebellious subjects, unwilling to submit to lawful authority; and I protest that all the bloodshed and calamities which shall follow are to be imputed to you, and not to His Majesty nor to me, nor to the gentle and honourable cavaliers who serve the king under me." This Requisition was the original basis of Argentine civilization, the Magna Charta of South American rights—mine-slavery for the endowment of a foreign church and king; torture, pollution, and death for the natives. Whether the Indians obeyed or disobeyed it, the result of the Requisition was the same; to them it only meant extinction, and but for the Revolution of 1817 would have left South America a solitude.

Clad in mailed shirts, armed with sword and pistol, and mounted upon fleet horses, the "gentle and honourable cavaliers" of Castile now scoured La Plata for the slaves which the king had granted to them in encomienda.[1] "To our esteemed Don Juan, an encomienda of 5,000 Indians; to our beloved Don Enrique, an encomienda of 10,000 Indians; to our cherished Don Manuel, an encomienda of 50,000 Indians"; to have and to hold, to dishonour, to rob, to squeeze, to exploit to death in mines, to torture, to mutilate, to feed to the dogs. Such, practically, was the nature of the encomienda. Its object was to stimulate the production of gold and silver for the Spanish crown and for St. Peter's of Rome.

Under these warrants—the Requisition and the Encomienda—the natives of La Plata were hunted down with bloodhounds, and thrust naked into the frozen mines of the Andes, which at the present time, and with all the aids of science, machinery, and steam, cannot be made

[1] Encomiendas are mentioned by Al-Makkari, in his History of the Mahometan Dynasties, ed. Gayangos, Appendix, p. lxxxiv, in connection with the Arabian Conquest of Spain, as being of Roman or other ancient origin.

to pay the expenses of their maintenance.[2] There the Indians were confined to the work and driven with the lash until they died. The price of a living man was three sheep, of a woman and child, a sheep and a lamb. The number of deaths was so great that the corpses bred pestilence; and in the mining districts of the Andes, one could scarcely make a step except upon the dead bodies or the bones of men. This was the æra when the Coinage Prerogative was stolen from the princes of Europe by that body of intriguants, courtiers and financiers who have since became so opulent and powerful.

During the pillaging æra (1523 to 1550) and the first mining æra (1550 to 1580) the whole of La Plata territory was explored. In the year last named, and together with Bolivia and other territory, it was made part of the viceroyalty of Peru, and Don Juan de Gáray of Lima was appointed lieutenant-governor. In 1620 the territory of La Plata was separated into two governments, both subject to the viceroyalty of Peru; the north-western portion being governed directly from Lima, and the south-western portion from Asuncion. A frontier custom-house was established at Cordova, at which merchandise passing either way paid a duty of 50 per cent. *ad valorem*—a regulation that was not relaxed until 1665. In 1614 the gross annual revenues of the viceroyalty of Peru amounted to five million silver pesos—about a million sterling. Of these revenues about one-third were derived from the taxes on the production and coinage of the precious metals; one-sixth from the Indian tribute; one-tenth from excise on spirits (pulque), playing cards, gunpowder, and cock-fights;[3] and the balance, or four-tenths, from customs duties and ecclesiastical ninths and annats. This did not include thebiennial indulgences, which were sold in the viceroyalty of Peru every other year to the amount of about 1,200,000 silver pesos, or £240,000. What with the alcavala of 4 per cent. on the sale of goods, and the profits of the Crown on the sale of quicksilver, the entire gross revenues of Peru were about seven million silver pesos, or nearly a million and a-half sterling. The expenses of collection were about one-half, and the moiety of these figures may fairly be regarded as the proportion belonging to La Plata. (Robertson's "America," II, 511.)

A century had passed since Garcia crossed the continent, but as yet, beyond the petty maize-gardens of the natives, scarcely a farm

[2] Robertson's "Hist. America," p. 503, and Mackenna's "Libro de Plata."

[3] This strange basis of taxation in La Plata was renewed in 1889, only instead of fighting-cocks it was race-horses. The basis of the tax and the motive for levying it were the same. The basis was a national vice, the legacy of mining days; the motive, an exhausted treasury. (U. S. Commercial Relations, Consular Reports), 1890, p.113.

had been planted, and not a single manufactory erected. The one industry of the country was gold and silver mining, and this was carried on altogether with enforced native labour. Everything came from Spain—horses, gunpowder, weapons, blankets, sombreros, spurs, manacles, whips, playing-cards, dice, bloodhounds, fighting-cocks and papal indulgences. Everything went to Spain, or else—al diablo. Nothing remained in La Plata, not even the Indians. A few escaped to the woods; the remainder, constituting communities, civilized and other, which had once numbered several millions of persons, were being coldly pressed to death in the mines.

Upon the "gentle and honourable cavaliers" who had brought about this dismal tragedy of tragedies, neither the tears of the Indians nor the appeals of pitying priests produced any effect. In deference to the representations of the latter, the Crown in 1548 had enacted the Mita (*mitad* means a half), which sought to limit the proportion of the Indians to be employed in the mines, and ameliorate the conditions of their service; but the law was not obeyed, and it did but little good. In 1550 Father Domingo, writing from Peru, said that from one-half to two-thirds of the native population had been destroyed by the Spaniards. Alluding especially to the Potosí mines, he said, "The poor creatures died like cattle; and even the few who escaped alive never lived to reach their miserable homes."

To ensure the continuance of the mines it had become necessary to follow the example which had been set in New Spain. This was to carry on a systematic slave-trade across the ocean, and to substitute for the exterminated Americans, new races from Africa. This trade soon grew to large proportions. The run from the coast of Africa was comparatively short, and the slaves were landed at the town of Buenos Ayres and driven across the desert, to the Andes. In the beginning of this trade negros were landed in the Rio de la Plata at a cost of about £5 per head—a valuation that will serve to measure the little difficulty of obtaining them, and the lack of care bestowed upon them. Once in the mines, they were exploited as rapidly as possible; it being cheaper to fill their places with fresh recruits than to take any trouble with the old ones. It was in the midst of this new and horrible traffic, and before negros became difficult to obtain—that is to say, in 1607—that the Jesuit fathers of La Plata first saw their way to save what was left of the Indians. Of some millions of this race, but a few thousands remained, either near the settlements or hiding from the slave-hunters and execrating the name of Spaniard. The energetic representations of a few benevolent priests at court, backed by the known ease of obtaining mining slaves from Africa, first met with success in 1609, when the king of Spain authorized their Provincial, Father Diego de Torres, to establish missions in La Plata for the care and conversion of the Indians, forbade these missions from being disturbed by any officers of the Crown, and authorized the Provincial to oppose, in the king's name, any such disturbance.

This edict gave rise to a most interesting experiment in government. Whilst the Puritan fathers were governing New England on the lines of the Old Testament, the Jesuit fathers organized the Paraguay missions on the lines of the New. They went among the Indians at risk of their lives. With infinite difficulty they persuaded them to emerge from their hiding-places, to abandon their fugitive and solitary lives, to trust themselves to the guidance of white men, and to live in social communities. The Jesuit priests taught and encouraged them to work and to pray as Christians; they indulged their native rites and customs; they humoured their beliefs and superstitions; they themselves even spoke the native language, and avoided the use of that sonorous but treacherous tongue of Castile, which to the Indians was only a presage of betrayal, violence and death.

In this experiment the Jesuits were opposed by the entire Spanish population of La Plata; even the officials, contrary to the king's express orders, throwing obstacles in their way and encouraging their enemies. Foremost among them were the Paulistas, those bandits——Mamelucos, they were called—of the Portuguese frontier, who had discovered gold placers in Minhas Geraes, and wanted slaves to work them. In 1628 these bandits broke through all the restraints of law, and attacked the Jesuit missions. Some of them, disguised as Jesuit priests, visited the christianised Indians and beguiled them into slavery; others rode boldly into the Reductions, as the missions were called, and tore the Indian converts away to the mines. Everywhere their steps were marked with blood; the Reductions were raised to the ground, the houses ransacked, the churches pillaged, the altars polluted with innocent blood. (Charlevoix: Raynal.)

Between 1628 and 1630, over 60,000 christianized Indians, chiefly captured in these raids, were sold in the slave marts of San Pedro and Rio Janeiro, and sent to the gold mines of Brazil. Many thousands were slain by the Paulistas, or died from fatigue and privation. Mr. Page has estimated that altogether over "100,000 christain natives were either enslaved or butchered."

The unhappy Jesuits gathered together the scattered remains of their flocks—about 12,000 Indians, exclusive of women and children—and retreated to the north-east corner of Corrientes, where they again commenced their pious task. Their account of these transactions, carried to Spain by Fathers de Montania and Tano, resulted in extorting from the king a reluctant permission that the Indians might be allowed to bear firearms. In 1641, on the occasion of another attempt on the Reductions by the Paulistas and Argentines, the Jesuits distributed 300 muskets among the Guaranís; and these were used with such deadly effect that but few of the white bandits who attacked them escaped alive.

But, though foiled in fight, the Buenos Ayreans were fertile in intrigue. In January, 1640, their governor, who kept them in restraint, mysteriously died. Without waiting for the royal appointment of his successor, they at once effected a revolution, took the royal authority into their own hands, and chose Don Bernardin, a known enemy of the Reductions, as governor. Their champion lost no time. In March, 1649, he authorized the pillage of the Jesuit College at Asuncion. When this design was accomplished, and while meditating further mischief to the Reductions, he was summoned to give an account of his conduct to the Viceroy of Peru, and the sedition came to an inglorious end. In 1651 the Paulistas and Buenos Ayreans made a fresh attack upon the Reductions, but the muskets of the Guaranís again so effectually repulsed them that they abandoned all attempts of this character upon the missions of the Paraguay.

In 1691 the Jesuits established similar missions among the Chiquitos of Bolivia, and these also were attempted to be destroyed and the Indians enslaved by the Paulistas and Buenos Ayreans; but after the latter had met with partial success, and depopulated a few villages, firearms again won the day, and the bandits were driven off.

These transactions render it perfectly plain that the terms of submission which the colonists of South America were ordered to offer to the natives before making "war" upon them, however soothing such terms may have been to the king's conscience, were altogether opposed to the policy of his Spanish subjects, and therefore impracticable. This policy was to plunder and enslave the natives; to squeeze out of them all that perfidy, cupidity, and the torture could extort; to exploit them without pause or mercy; and for the colonists to avoid doing any work themselves. Nowhere do we find them engaged in planting, in rearing herds of animals, in gathering the gifts of Nature, or in utilising her forces. These occupations were relinquished to the natives and the Jesuit priests, against whom they waged unceasing and unrelenting war. When the Jesuits first offered to civilise and evangelise the natives, the colonists opposed; when the king issued peremptory orders that the Jesuits and their converts should not be molested, the colonists disobeyed and revolted; when, notwithstanding their enmity, the Christian Commonwealth of the Indians grew into an imposing fabric, they once more revolted against the royal authority.

An eye-witness informs us what they did with the Mission Indians. In "A Relation to Mr. B. M.'s Voyage to Buenos Aires, from hence by Land to Potosí" (London, 1716), the author says that in this frozen region, where even on the surface it never thaws till daylight, he saw in one place (in the year 1713) 2,200 captured Indians driven into a paddock like sheep, and there parcelled out to various mine-owners, by whom they were driven to the workings under ground. It was death to attempt escape; it was death to remain, for they died in platoons. This was the price of gold, the cost of production—Death.

In 1723, Antiquera, governor of Buenos Ayres, took upon himself to order the banishment of the Jesuits from the country, and he organized another raid upon the Paraguay missions. After the first surprise, the Guaraní converts recovered themselves, met his forces in the field, and stopped his further operations. When this news reached Lima a royal battalion was sent against Antiquera, who was arrested, carried to Lima, and there executed, in 1731.

Several governors of Buenos Ayres now followed each other in rapid succession—Zavala, Barna, and Saroeta. The burning question that divided the colonists was still the mine exploitation of the christian natives. In 1732 this question occasioned another rebellion. The pro-slavery and anti-Jesuit party were called the Communeros, or Home Rulers; the king's party, the Contrabandos. Upon the outbreak of this revolution the Communeros deposed the king's officers, appointed a provincial Junta, and elected, as president of the province, Don Luis José de Barreyo. Upon evincing some hesitation to attack the christian Reductions, this officer was deposed in favour of Don Mannel de Ruiloba, who, for the same reason, was also deposed. The Communeros then dissolved their Junta and appointed a dictator to carry out their designs. Before these could be effected they were met and defeated by the king's forces under Zavala, and all these rebellions and revolutions were brought to an end.

By this time the price of heathen negros brought from Africa had increased to £15 each—a price so great as to be regarded as an intolerable burden by the colonists who owned the mines; and a fresh effort was made to capture and obtain as slaves, and for nothing, the christian Indians of the Paraguay missions. From arms the colonists had recourse to intrigue, and they appealed from the plains of Tucuman to the Court of Madrid.

About the year 1743 the Christian missions were in the enjoyment of unprecedented prosperity and power. Those of the Parana and Uruguay numbered about 140,000 souls; the Chiquitos Reductions numbered 24,000; the Abipones and others about 6,000: total, about 170,000 converts; of whom 12,000 to 14,000 were provided with horses, arms and ammunition. In thirty Reductions the converts possessed 769,590 horses, 13,900 mules, and 271,540 sheep, besides large herds of cattle and other animals. It has been suggested that this prosperity was not accompanied by any increase of population. The Abbé Raynal and other writers have even assigned causes for this phenomenon, such as changed habits of life, the prevalence of small-pox and fevers, etc., but they have one and all omitted to supply any evidence in support of the premiss. Taking into account the small numbers of the natives whom the Jesuits managed to collect together after the tragedy of 1628, and the mendacious reports which their enemies circulated at Court, the present writer is compelled to regard the supposed absence of increase among the natives as having little foundation in fact.

However this may be, the prosperity of the Reductions furnished a weapon to their unscrupulous enemies; and this was sharpened by the straitened condition of the royal finances. In vain the Jesuits pleaded the evangelisation of the natives; in vain their docility, their industry, their sobriety; in vain that priceless gift of nature, the Cinchona bark, which they had discovered and gathered from the trees of La Paz, to lay at the feet of suffering mankind. These sentimental pleas did not fill the king's coffers. The mines were declining for lack of cheap labour; the price of negro slaves had become prohibitive; and but one resource seemed open to the Treasury—to reduce the Reductions and thrust the christain converts into the mines. The annual revenues of the Crown had fallen from 401 million reales de vellon in the reign of Philip IV., to 42 millions in that of Philip V. The annual expenses had risen from 183 millions to more than 200 millions. The king's Fifth from the American mines, which formerly was paid without a murmur, was now much evaded, and had seriously dwindled away. The christian Indians must go into the mines. Besides, were not the Jesuits accused of amassing great treasures for themselves? Were their churches not laden with plate?[4] Had it not been alleged that their christian republic, their *imperium in imperio*, threatened the integrity of the Crown? Assuredly the missions ought to be destroyed, and the produce of the mines increased.

In 1759, Pombal, the prime minister of Portugal, had banished the Jesuits from the Portuguese possessions, and, loading a ship with them in Portugal, had despatched it to Civita Vecchia—in a word, he had sent them back to Rome. Here was example added to reason. The result of these considerations was a triumph for the colonists of La Plata. In 1767 the fiat went forth, and the Jesuits were banished by Charles III. from Spain and its colonies. Within three months' time this edict was enforced in La Plata, and the christian "republic" was levelled to the dust.

In his letter to the Pope, apologising for this transaction, the king terms it a measure of "political economy"; and there is no reason to doubt the royal word. His first step was to ship all the Jesuits of Spain to Rome; his next, to hunt down the 222 Jesuits of La Plata and deport them from Buenos Ayres; then he authorised the colonists to plunder the Jesuit colleges and churches; finally, he let them loose upon the devoted Indians.

The result can be told in a few words: the midnight raid, the chain-gang, the mines, torture, and extinction. In 1801 a census of the Indian population was made by Don Joaquin de Soria. There were in the thirty missions 45,639 souls; less by 98,398 than in the year 1767. In this interval of thirty-four years more than two-thirds of the original number, to say nothing of the natural increase, had disappeared; horses, cattle and sheep were gone; the fields were destroyed; the houses pulled to pieces or burned down, and nought remained save a few hovels and a crumbling adobe church, with faded frescoes and a cracked bell.

In August, 1776, the provinces of La Plata were separated from Peru, and together with several other portions of that viceroyalty erected into a separate viceroyalty, the fourth of that rank in Spanish America. It was called La Plata. This government consisted of the following provinces:—Buenos Ayres, Rio de la Plata, Tucuman to the Andes and the Vermejo, all of Paraguay, and all of the present state of Bolivia, including the mining districts of Potosí, Oruro and La Paz.

[4] Says Postlethwayt (Dict., art. "Gold"), "The richest gold lavaderos (washings) of Chile fall into the laps of the Jesuits, who farm or purchase abundance of mines and lavaderos, which are wrought for their benefit by their servants." The belief that the Jesuits possessed rich mines in which they employed their converts to work in secret is not even yet extinguished. See a curious note printed in 1882 by the "Industrial" of Buenos Ayres, and published in Mackenna's "Libro de Plata."

About the year 1825 two Englishmen, Head and Miers, traveled extensively in Buenos Ayres, and published details concerning the ruin of this country by the early Spanish conquerors and miners, its present poverty, and the extravagant and lying accounts sent to England for the purposes of speculation. Mr. Miers said that the population, wealth, and resources of the country were everywhere exaggerated, that phantoms of wealth and power were conjured up to feed the appetite of cupidity, that the mining companies recently organized in England to work these fabulous resources would probably all come to grief, and that he deemed it his duty to tear off the mask of deception which covered the real indigence of this exploited country. But these revelations went unheeded in 1825, and were forgotten in 1875, when precisely the same sort of deception was practised which had succeeded so well fifty years before. With the collapse of the boom of 1875, mining in Buenos Ayres happily came to an end.

Until the invention of the power-drill and the cyanide process, mining for the precious metals was on the average the most unprofitable of all industries. Many states and communities, aware of this fact, entirely forbade its continuance. The richest mining countries were among the poorest in general wealth. The contents of a mine when once exhausted can never be renewed; the value of the newly-added product (when measured by other commodities) must always tend to diminish, because, at least in peaceful times, the quantity of the former always tends to increase—a fact due to the imperishable character of these metals, and the fabrication of a certain proportion of them into coins. After its initial phases, mining requires large capital and elaborate machinery and plant. As mines are usually in remote and inaccessible regions, it seldom paid to remove the plant; so that when a mine "shut down," the plant was commonly a total loss. When a mine becomes barren, the owners do not abandon their costly plant at once, but keep on in the hope that the mine will improve. For these reasons mines are often worked for long periods at a loss, and with only a remote prospect of gain—a prospect sustained by occasional instances of good fortune, but far more often frustrated by bad.

During the first twenty years after its discovery by the Spaniards there was no mining in La Plata, only plunder of the Indians, who were forced into a few wretched alluvions, or placers, and compelled to produce a stipulated quantum of gold, or suffer the torture. Mining began with the opening of Potosí in 1535. This is a mountainous district, about eighteen miles in circuit and three miles above the sea level. Its opening by the Spaniards cost the lives of millions of Indians. During the first ten years the workings were small and desultory.[5] Systematic working commenced in 1545. From 1545 to 1547 it was considered a bad month when the mines failed to yield $1,500,000; from 1548 to 1551 the product fell below $1,000,000 a month. From 1556 to 1578 the average annual product was $2,000,000; from 1579 to 1736 the average annual product rose to about $3,000,000; from 1737 to 1789 it fell to $2,500,000; from 1535 to 1789 the estimated product was $788,000,000; from official records, the Prefect of Potosí calculated the total product, in 1835, at 734,-000,000 dollars or pesos; altogether, from first to last, it was probably (including smuggling) about 750 millions, or more than twice the product of the Comstock mines.

The principal mining districts in the viceroyalty of La Plata proper were La Paz, Carangas, and Oruro. In the list published by Helars from the records of the Spanish Chancery, it appears that there were no less than twenty-two districts worked for gold and silver, and that in these districts there were simultaneously worked 27 principal mines of silver, 30 gold, 7 copper, 7 lead, and 2 tin. Humboldt estimated their united product, at the period of the revolt from Spain, at $4,-200,000 dollars per annum. It is a significant fact that no native will invest his capital in the working of mines. He will search for mines, develop them until they look promising enough to sell, and then scour the earth for "foreign capital" to purchase or work them.

In 1825, after the independence of the Argentine Confederation was acknowledged by Great Britain, a mining boom was organized in La Plata, and several millions sterling of British capital were invested in enterprises, not one of which ever paid a dividend. In 1851 Dalence said, "In Potosí there are 26 silver mines working, more than 1,800 abandoned; Porco, 33 working, 1519 abandoned; Chayanta, 8 working, 130 abandoned; Chicas, 22 working, 650 abandoned; Lipez, 2 working, 760 abandoned; Oruro and vicinty, 11 working, 1,215 abandoned; also 200 gold mines abandoned; Poopó 15 silver mines working, 316 abandoned; Carangas, 4 working, 285 abandoned; Cicasica, 9 working, 320 abandoned; Inquisivi, 5 working, 160 abandoned; Araca, 4 gold mines working, hundreds abandoned; Soratu, 7 working, over 500 abandoned; Berenguela de Pacajes, all abandoned, although many were rich; Arque, 2 working, 100 abandoned; in Ayopapa, many silver mines, and in Choquecama many gold ones, all abandoned." Altogether there were about 8,300 mines,

[5] Garcillaso de la Vega, who visited the mines in person.

of which 150 were being worked. These numbers must have included prospects as well as mines.[6]

At a later period there occurred a slight revival of mining in these districts; but as they now belonged to Bolivia, and no longer to La Plata, they need no further mention in this place. Among the best data on the subject of the mines of La Plata are the reports of the British consuls in the Consular Reports, and the American consuls in the "United States Commercial Relations," 1875, and the year following, also in "El Libro de Plata," por B. Vicuña Mackenna, Santiago de Chile, 1882, in which the author concludes the subject with these significant words: "There is no more misleading name than that of La Plata: that country's true source of wealth is not silver mines, but wool and hides" (p. 607). To this summary are now to be added live and slaughtered animals, wheat and provisions.

At the period of the discovery of La Plata by the Spaniards the principal coins of Spain were the gold castellano, containing about 63½ grains fine, identical with the Arabian dinar and Byzantine solidus; the gold ducat of about 56 grains fine, identical with the Venetian sequin; the real de plata of 51¼ grains fine silver, and the billon maravedi of 1.52 grains of fine silver. These coins were valued in maravedis as follows:—the maravedi, 1; the real, 34; the ducat, 383; and the castellano, 490. On or before 1579 the ducat was raised to 434 and the castellano to 556 maravedis. The castellano was sometimes called "a piece of gold," and sometimes "a gold peso"; the ducat was sometimes called "a gold real." The "dobla" and the "pistole" were double ducats. The "escudo" was a silver piece of 8 reals, or piesa de á ocho," or "peso fuerte," or, as it was afterwards called, a hard dollar.[7] The "peso sencillo," or soft dollar, contained 304½ grains of fine silver (about 6 reals). This was the value of the Paraguay and River Plate peso, as fixed by the royal ordinance of 1618. The following data, from the year 1535 to the year 1620, are taken from the "Recopilacion de Leyes de los Reynos de las Indias," or code of laws relating to America, published by royal authority at Madrid, 1774.

1492. One-half of all gold or silver obtained in America must be paid to the king. During the government of Ovando in Hispaniola this requirement was reduced to one-third; and in 1504 to one-fifth (quinto), at which rate it continued until 1736. This tax is of very ancient origin. In India the king exacted one-half of gold and silver

[6] "Bosquejo estadistico de Bolivia," por José Maria Dalence, Chuquisaca, 1851, pp. 293–4. [7] Vethake, "Cyc. Americana."

spoil or produce.[8] In Japan the emperor exacted two-thirds.[9] The government of Athens exacted one-twenty-fourth from the mines of Laurium.[10] The temple of Delphos exacted a tenth from all gold mines. The government of Rome levied a similar tax[11]; and although the rate is not mentioned, it was probably one-tenth.[12] One-fifth was the proportion demanded by the Koran and exacted by the earliest Moslem caliphs on both spoil and produce.[13] The same proportion was reserved for the caliph by the Moslems in Spain.[14] The Christian kings followed this example, and even exceeded the Moslems in avidity. From 1147 to 1550 the king of Portugal exacted one-half of their produce from the gold washers of the upper Tagus.[15] In 1379 King John of Castile declared the mines "free of lords and church," and subject only to the royal fifth of the gross produce. In 1578 Queen Elizabeth of England stipulated with Sir Humphrey Gilbert that he should pay the Crown one-fifth of all the gold and silver he might obtain in Nova Scotia.[16] At the present time the Vigra copper and gold mine, North Wales, besides rental, pays the Crown of England one-fifteenth of its produce. The quinto tax had much to do with the monetary systems not only of La Plata, but of other countries, and its study is commended to those misled "economists" who imagine that the cost of producing the precious metals has anything to do with the current value of coins.

1519. All gold and silver bullion obtained in America shall be taken to the governor of the province wherein obtained, or to the justice, or royal assayer, or to the mint-master, if there be one, who, after having retained one-fifth for the king, shall stamp the remainder with its value in Spanish coins of the same metal, enhanced to the extent of the value of the dues (derechos) pertaining to the king. These last appear to have been one-and-a-half per cent. *ad valorem*. This law forbids private coinage; it secured two payments to the king so long as the bullion remained in America, and further payments whenever it was sent for coinage to Spain. It was repeated in 1535 with the penalty of death for infraction.

1535. The law of this date establishes the first mints in America, namely, those of Mexico, Santa Fé, San Luis Potosí and Axiquipilco, all these places being in Mexico; also a mint to coin billon pieces for the king at San Domingo, in the island of Hispaniola. Besides the quinto or fifth on production, there were levied three reals

[8] "Code of Manu," viii, p. 39. [9] "Hist. Money in Ancient States," p. 54. [10] Xenophon, "De Vectigal." [11] Livy, xxiv, p. 21. [12] Adam's "Roman Antiq.," voc. "Decumæ." [13] Al Koran, c, viii, of Spoils. [14] Calcott's "Spain," vol. i, p. 95. [15] Calcott's "Spain," vol. i, p. 66. [16] "Hist. Prec. Metals," p. 37.

on every mark weight of silver, namely, two reals to cover the cost of coinage and one real for the king's seigniorage. The derecho of 1½ per cent., though not mentioned, appears to have remained. All bullion presented for coinage must exhibit proofs of its having paid the quinto, or else it is liable to confiscation. The exportation of coins, except to Spain, is forbidden. The date of this law proves that the celebrated mining district of Potosí, in Peru or La Plata, was opened ten years earlier than is commonly supposed.

1535. Counterfeit money in circulation ordered to be traced up and seized. Mint offices to be overhauled.

1537. The American mints are permitted to coin "reales de á ocho" (pesos), halves, quarters, and eighths, "como en estos reynos," the same as in Spain.

1544. The law of 1537 is repeated more explicitly, "the coins to be of same weight, fineness, and value as those of Castile."

1546. Changes the ratio of value between the precious metals, and therefore the weights of coins.

1550. Forbids private dealings in gold dust or bullion.

1551. Besides the quinto, a duty (derecho) of 1½ per cent. *ad valorem* is made payable to the king on all gold and silver. This law (the "covos") is repeated in 1552.

1565. Silver coins may continue to be struck in America, but neither gold nor billon coins. The explanation of this regulation is found in the Sacred Myth of Gold, and the facility it offered to the Spanish emperor-king to raise the value of that metal by proclamation. After plundering America—the plunder consisting chiefly of gold—the Spaniards opened Potosí, and commenced mining, when the produce became chiefly of silver. Believing that the seigniorage upon gold would thenceforth yield but a small revenue, the king determined to enhance the value of the metal by decree, and to enjoy the entire advantage of this enhancement by coining the gold himself, and forbidding the coinage of that metal in America. The valuation of gold to silver in the Flemish and Austrian coins of Charles, previous to 1546, was 1 to 10.755; in that year it was raised to 1 to 13.333.

1565. Counterfeit money reported in circulation, and ordered to be traced and seized. Mint offices to be overhauled.

1567. The seigniorage law of 1535 is modified by remitting the two reals for cost of coinage, and retaining the one real seigniorage due to the king. The derecho or "covos" of 1½ per cent. also remained.

1579. The law recites that not merely of gold and silver, but of all metals and minerals, one-fifth part belongs of right to the king. In taxing gold and silver there shall first be exacted 1½ per cent. of their weight to compensate the royal smelters, weighers and assayers, and immediately thereafter 20 per cent. of their weight for the king (Law XIX of the Royal Fifths).

1579. In retaining the derechos and the quinto of gold the royal officials shall count at the rate of 24 maravedis to the quilate (carat) of gold, or 556 maravedis to the castellano of 22½ carats, "which is its just and true value." The carat is usually regarded as a measure of fineness; here it is evidently used as a weight. If it was a weight, we can only suppose that it weighed 2.96 English grains, because in weighing gold there were 4 Spanish grains (the silver mark was divided into 4,608 grains) to the carat, and 4,800 grains to the mark, which both as to gold and silver contained 3,550½ grains English. In such case the castellano weighed 66⅔ English grains. This must mean gross or standard weight, including alloy; but we are not at all confident that the calculation is correct.

1579. By the same law silver was valued at 2,050 maravedis "per mark of eight ounces of five pesos." We can make nothing definite of this. Here the maravedi is ordered to contain 1.73 English grains of fine silver; whereas, according to its relation (272) to the dollar, or peso, or piece-of-8, it could not have contained over 1.52 grains fine. Perhaps it was an attempt to change the ratio to 14½, thus: 100 grains of gold coined into 834 maravedis, and 100 grains of silver coined into 57.74 maravedis.

1581. The preamble sets forth the vexatious diversity of weights and measures employed in the various viceroyalties or provinces of America, and substitutes for all of them the weights and measures of Toledo (New Castile), and the vara, or yard, of (Old) Castile.

1589–95. Gold, silver, and billon money authorised to be coined in Hispaniola; the billon money to be legal-tender at fixed rates, the refuser to be punished. This money was apparently intended to circulate in all of the American provinces.

1591. The law of 1550 is modified by repealing the prohibition as to dealings in gold dust or bullion, and re-enacting the prohibition to deal in silver bullion. The viceroys are to furnish to individuals, without limit, coined money in exchange for silver bullion upon which the king's Fifth and other duties have been paid. This appears to be a sort of counter-move to the "individual" or "free coinage" legislation of the Netherlands in 1572. It was the first Spanish step toward individual coinage.

1595. The following coins, struck in Hispaniola under the king's warrant, were declared legal-tender in that island under heavy penalties; peso de plata, 450 maravedis, or 225 quartos; escudo de oro, 400 maravedis, or 200 quartos; real de plata, 34 maravedis, or 17 quartos. Bad (false ?) money was stated to be in circulation in Hispaniola.

1596. The colonial official practice of exacting the king's dues in heavy coins and paying the public expenses with light ones, is forbidden.

1596. The viceroy of Peru is ordered to confine the Indians to the work of mining, and not to permit them to leave the mines ("Recop.," II, p. 257).

1603. The value of all billon and copper coins is doubled by decree of Philip III., in Spain. This decree brought about a virtual suspension of coin payments caused by the exportation from the kingdom of full-weighted gold and silver coins. Premium on silver coins in Spain, 40 per cent. in billon coins. Great confusion in the Spanish monetary system. The Spanish-American viceroys, by being obliged to suspend individual coinage, only made matters worse.

1608. The viceroys of Spain are again permitted to coin money for individual account, and without any more specific limit than they may deem necessary. This appears to be a more complete measure of private coinage than the ordinance of 1591, because it says nothing about the king's Fifth or other dues. It was the second Spanish step towards individual coinage.

1611. Final expulsion of the Moors from Granada. In 1380 Spain had a population of 21,700,000; in 1492 nearly all the Moors were expelled, and the population dropped to 13 or 14 millions; in 1609 the population had increased to 20 millions; in 1610–11 the remainder of the Moors, about one million, were expelled. In the course of a century, through this and other causes, the population dropped to eight millions. Only recently has it increased to the numbers of the fourteenth century

1618. The peso of the Indian tribute, and of Paraguay, Rio de la Plata, and Tucuman, may be discharged with six reals.

1620. From all silver bullion brought to the king's officers shall first be deducted the king's Fifth,[17] and his prerogative (derecho) of mintage, 2 reals per mark, and his seigniorage, 1 real per mark; then

[17] The evasion of the quinto at this period was common. Captain Shelvock captured a Spanish vessel in 1721 which was laden with a quantity of preserved fruit packed in boxes. Upon opening these boxes some of them were found to contain cakes of silver bullion (Mavor's Voyages, IV, 118). Numerous other instances of like kind prove that the smuggling of silver out of Spanish America had become an organized trade.

out of the remaining bullion there shall be coined 67 reals to the mark weight. In this year base silver money made its appearance in New Granada, and is noticed in one of the laws.

1632. Royal taxes on gold, and taxes paid in gold, shall be remitted to the king in the same metal, and not paid with silver or any other metal or commodity (Law XX of the Royal Fifths). This implies that the king derived some advantage from gold which he did not from silver. This may have been in the change of ratio from 13⅓ to 14½, suggested under the year 1579. It certainly did not arise out of the seigniorage on gold coinage. On the other hand, it may only have been the elaboration of a similar decree issued in 1557, designed to prevent fraud on the part of the officials.

About 1640, reign of Philip IV. The value of the billon and copper coins of Spain again doubled by decree; no limit assigned to their fabrication, and no adequate safeguards provided against counterfeits. The consequence was that the country was flooded with base coins made in Germany, Flanders, France, England and Italy.

1643. No modification of royal decrees fixing the value of money is to be permitted or countenanced.

1645. From the Basileus, Alexis II., down to 1572 in Holland, or to 1645 in Spanish-America, when Philip IV. permanently threw open the viceregal mints to "free" coinage, the issuer and owner of the monetary measure was the King of each of the states that had emerged from the ruins of the Roman empire. These princes so frequently and grossly abused their prerogative that the demands of commerce caused it to be finally swept away by the hardly less objectionable device of "free" coinage. The welfare of nations will eventually compel the prerogative to be resumed.

In the reign of Philip V. (1700–45) the floating debt of Spain was funded. This debt consisted of Assignments of Anticipations (exchequer bills), temporary debts created by the Bureau of Loans, Tickets of Subsistence (military scrip), and Mint Bills (bullion receipts ?). In October 1710, all these demands upon the State were ordered to be funded, with or without consent of the creditors, into 5 per cent. stock. At this period (the close of Louis XIV's reign), the finances of France were in such a deplorable condition that 500 patents of nobility were sold by the French government for 2,000 écus each, the currency was depleted, prices declined, and a vortex was being formed which was soon to be filled by the paper emissions of John Law.[18] Under these circumstances there was no market in

18 "Money and Civilization," p. 231.

France for the Spanish stock. Hence its emission in place of the floating debt caused a violent outcry from those who had expected earlier payment of their claims. Indeed, the discredit of the Mint Bills had already—that is, in 1710—caused the failure of Samuel Bernard, at that time the richest banker in Europe. He had twenty millions of these demands upon the Spanish government, and was forced to exchange them for twenty millions of unmarketable stock. The politicians of to-day who are trifling with the dangerous subject of money may glean a lesson from what happened on this occasion. The failure of Bernard brought on a general financial panic in 1714, and, like the forcible closure of the Boston colonial mint in 1694, the panic ended with the revolt of the colonies and their loss to the mother country.

The South American revolution of 1732 was not merely a protest against the State protection of the Jesuits and their system of Indian tutelage; it was a general expression of disappointment and disgust with the royal government. The contraction of currency and credit rendered the taxes not only doubly oppressive, it deprived the colonists of the power to pay them. Indeed, the colonists contended that these circumstances and measures compelled them to coerce the Indians and thrust them into the mines; that it induced them to oppose the Jesuits and their benevolent system; that it forced them either to become smugglers under cover of the royal flag, or else to rid themselves of trickery, deceit, favouritism, bribery and connivance, by taking the field as home-rulers (Communeros). The South American revolution of 1732 was the herald of the North American revolution of 1775. It is true that the Communeros were put down by the royal forces, nevertheless they gained something. By a royal decree of the year 1736, the king's share of the precious metals' produce was reduced from one-fifth to one-tenth of the silver, and to one-twentieth of the gold; and this arrangement continued in force until the revolution of 1810 and the extinction of the royal authority.

The Spanish mint laws were altered so frequently, that an absence of alterations for so long a period as that from 1736 to 1772 seems remarkable, yet the writer can find no data for this period except the following changes in the ratio, which are here shown in connection with those which preceded and followed that period, viz.: 1650, 15 for 1; 1675, 16; 1760, 14¼; 1765, 14⅞; 1772, 16; 1775, 15½. By the mint law of 1772, which Dr. Kelly says was applicable to all the provinces in Spanish America, there were ordered to be coined from a mark of gold 0.89583⅓ fine, 8½ doblons, andfrom a mark of silver

0.89583⅓ fine, 8½ pesos, or hard dollars; the halves and quarters to be of proportionate weight. The doblon therefore contained 374.2 grains of fine gold, and the peso 374.2 grains of fine silver. As there were 16 pesos to the doblon, the ratio was 16 for 1. All these coins were full legal-tender, and were open to illimitable private coinage in the mint after paying the tenth on silver, the twentieth on gold, and the mint charges. On silver the royal dues and mint charges were each one real per mark weight. There was also a "small change" currency, of limited tender, consisting of pesetas and half-pesetas, 0.812½ fine, the former containing 83¾ grains of such debased silver, and the latter one-half that quantity. The half-peseta was called the Mexican or provincial real. These two highly over-valued coins were designed to circulate as quarters and eighths of the peso or duro, and this design succeeded so long as no full-weighted quarters and eighths, and so long as no counterfeits, were struck. The appearance of the latter compelled the Crown to issue full-weighted quarters and eighths; whereupon the pesetas and half-pesetas dropped in value to fifths and tenths of the dollar respectively. These over-valued pesetas were first issued in 1721.[19] Here is an instance where bad money did not drive out the good, but where the emissions of the latter compelled the former to take a lower value. In fact, the so-called "Gresham's law" on this subject is not one of money at all, but of commodities, and it only relates to money when it has been prostituted to private coinage and degraded to the rank of a commodity.

In 1775 the mint ratio of Spain was changed from 16 to 15½ for 1, by coining full legal-tender pesetas of 72.1 grains fine.[20] As the French ratio at the time was 14½, the conflict between these ratios resulted in a mean ratio in the Paris bullion market of 15.08 for 1, a fact which induced the French government in 1785 to recoin its gold at the Spanish valuation of 1775,[21] though meanwhile, that is to say in 1779, the Spanish mint had returned to the ratio of 16 for 1. These ratios, 15½ in France and 16 in Spain, continued until 1873 in France and 1864 in Spain.

Such was the money of La Plata when the revolution of 1810–17 occurred. When the smoke of this conflict cleared off, and the royal forces were driven out of the colony, the currency consisted chiefly of the issues of the Bank of Buenos Ayres, which, being kept within prudent limits, circulated not only at "par in gold," but in fact were worth more than gold coins of the same denominations.

19 Kelly, ii, p. 168. 20 Ibid. 21 Calonne's Report.

Such is the metallic history of La Plata, or Argentina, once regarded as the silver country, *par excellence*, of the world. It has added more than a thousand million pesos to the general stock of gold and silver; it has scarcely one of those pesos left. Its currency consists of depreciated paper notes, and its mines yield so trifling a quantity of the precious metals, less than half a million dollars per annum, as not to be worth further notice. The people of Argentina possess a vast domain, much of which is fertile and full of promise; their climate is salubrious and inviting; they belong to a race that has many glorious memories to dwell upon; they are not devoid of energy and ambition: but if they would win for themselves from other people a just appreciation of these advantages, they must begin by casting aside the financial delusions which are made to rest upon the precious metals, and rear their state anew upon the solid foundations of industry, probity, and truth.

An English critic in a late able and impartial review of the first edition of this work questioned the propriety of devoting so much space to the dreadful details of the Spanish Conquest of America. These details were not recounted without a purpose which was openly expressed in the work itself.

It has been a received axiom of political economy which does not lack support even in the practical administration of government [22] that the value of the precious metals is derived from the cost of their production. America is the only country, the story of whose conquest comes to us in such a form that the motives of the conquerors, the means they employed, and the results they accomplished are all to be clearly perceived. This story proves that the value of the precious metals is not at all connected with the cost of their production.

Whether the acquisition of the precious metals formed, as has been supposed, the chiefest of the motives which actuated the Spaniards, or not, is immaterial. It certainly formed the chiefest, perhaps it may be said the only, result; for besides despoiling Aboriginal America of her gold and silver, Spain accomplished nothing in the new world, except extermination and destruction. She swept away half as many human lives as all Europe contained at the period of the Discovery. She destroyed every memorial of the Aztec and Peruvian civilizations. She disfigured the entire face of Central and South America. And she planted nothing in the place of what she destroyed save a race laden with disadvantages and a few mission churches crumbling to decay.

[22] This doctrine is asserted in President Hayes' veto of February 28, 1878.

The spoil which she obtained amounted altogether to some seven thousand millions of dollars. Now what did it cost the conquerers? Practically nothing. All the expeditions which they fitted out, all the lives (of Spaniards) that were lost in prosecuting them; all the maravedis which these expeditions and the subsequent supplies to the colonists ever cost, let the lives be reckoned at never so high a value, and they were for the most part, the lives of a class of men whom Spain was only too glad to be rid of, amount to nothing compared with the stupendous sum expressed by seven thousand millions of dollars.

We repeat, the metallic spoil of America cost Spain, cost Europe, cost the civilized world practically nothing. Neither did the spoil of barbarian Europe cost anything to the Romans, nor that of India to the Macedonians, nor that of Asia Minor to the still more ancient Persians. If the economical axiom that value is due to cost of production be correct, how came the precious metals which these nations acquired by conquest to possess any value? If they cost nothing, why were they worth anything? Simply because their value is *not* due to the cost of production, but to their usefulness and their quantity; to the relation of supply to demand.

This relation is wholly irrespective of cost of production, and if the view which the present work affords of the conquest of America by the Spaniards has served, though ever so inadequately, to prepare the reader for the reception of this important principle, the author will have reason to believe that he has not treated the subject in vain.

[23] While that portion of the precious metals (about one-half of the existing stock), which was obtained through conquest and slavery, practically cost nothing, the portion obtained through free labour has cost more than it is worth: a fact long familiar to mining men and now admitted by numerous publicists. Such is the penalty which nations must pay for indulgence in violence and crime: their fruits lower the value of the products of free labour. Not until the precious metals have entirely ceased to be acquired through conquest and slavery can free mining labour hope to obtain an equitable reward in the value of its product. This is one of the "harmonies of political economy" which Malthus suspected and Bastiat overlooked.

CHAPTER XXII.

BRAZIL.

Gold discovered in 1573—Plunder and slavery of the natives—Negro slaves—Interruption to mining caused by wars—Slave hunting and the importing of negros resumed in 1670—Description of the mines—Dearness of provisions—Bullion money—Tragedy of Lonira—The Paulistas—Chronology of the mines—Exactions of the Crown—Population of Brazil—Individual product of the mining slaves—Total product of Brazil—Compared with California and Australia—The Quinto—Coinage—Alterations of the coins of Portugal and Brazil.

BRAZIL was discovered by Vincente Yanez Pinzon, a Spanish captain and companion of Columbus, in January, 1500. It was rediscovered, and more completely explored in the following April by Pedro Alvarez de Cabral, a Portuguese navigator, and by Americo Vespucci in 1503. The Bull, issued by Pope Alexander VI., dated May 3rd, 1493, reads: "Decretum et Indultum Alexandri Sexti super expeditione in Barbaros noviorbis, quos Indos vocant," etc. It gave to Portugal all countries *west and south* of a meridian drawn from pole to pole 100 leagues west of the Azores; but in 1494 this blunder was rectified, and the line placed 270 leagues farther west of the Azores by an arrangement between the two powers. (Dwinell's Colonial Hist. of California, pp. 8, 9.) Notwithstanding the equivocal language of the Papal bull, Spain neglected to urge her claims to Brazil, and Portugal retained a nominal possession of the country. By the explorers of that period, whose sole object, under whatever specious disguise, was the acquisition of the precious metals, Brazil was regarded as of no value, until 1549, when, it being found that the natives possessed gold ornaments, the presence of gold in the beds of the rivers was suspected, and the country was regarded with more interest at the Portuguese Court. In the year 1555, Villegagnon, a knight of Malta, applied to Admiral Coligny for leave to invite the Huguenots of France to emigrate to Brazil. This permission Coligny obtained from King Henry II. The Huguenots, who were then bitterly persecuted, availed themselves of the opportunity to send a colony to Brazil; the movement amounting in two years to some 10,000 persons. These colonists founded the city of Rio de Janeiro. The hardships of their new life, aggravated by the deceit and tyranny of Villegagnon, whose solicitude for religion was a sham, and whose sole object was to plunder the natives and the colony for his own ends, drove the Huguenots back to France. Four years later the Portuguese took possession of the settlement, and established their flag along the entire coast. It was following this period, and in the year 1573, that gold placers were discovered by Sebastian Fernandes Tourinho, in Minhas Geraes. The discoveries at Ouro Preto occurred 1595–1605. Then followed the troubles with the Paulistas and the Jesuit missions of Paraguay.

During the latter part of the sixteenth and early part of the seventeenth centuries, that is to say, during the period 1580–1640 when Portugal and its possessions fell to Spain, the English, then at war with Spain and Portugal, attacked and plundered the settlements on the Brazilian coast, their prime object being the acquisition of the precious metals. The Netherlands, also at war with Spain, which at that time included Portugal, and animated by a similar motive, attacked and captured San Salvador in 1624, obtaining a large booty. In 1630, and from 1633 to 1636, they sacked nearly all the Portuguese settlements on the coast of Brazil, and established Dutch colonies in their places. During an interval, which lasted nearly a century, but few discoveries of gold mines were made. In 1654, the Portuguese again obtained possession of Brazil, and except when in 1711, DuGuay-Trouin, a French corsair, took Rio, and ransomed it for 70,000 crusadoes, they remained masters of the country until it achieved its independence in 1822.

Upon the first settlement by Europeans, the natives were plundered of the few gold ornaments or trinkets which they possessed, and commanded under pain of death to seek for more. The boundless extent of the country, and the slender resources of the Europeans, effectually prevented the execution of their threats, and preserved the natives from that extermination which befel them in the West India Islands, in the comparatively narrow empires of Mexico and Peru, and upon the Isthmus. The Portuguese rode along the coasts and rivers terrifying the inhabitants with their horses, their bloodhounds, their arquebuses, and their coats of mail. They appropriated the women, and reduced the men to mine slavery. But their depredations soon wore themselves out, for the natives retreated to the interior, and left them to their own resources and evil thoughts.

Among the productions of the colony which were sent to Portugal was a small amount of placer gold, obtained in insignificant quantities from the sands of the streams near the banks of which the settlements had been made.

Some of the worst of the many bad characters who composed the settlements, tired of the restraints of social life, or else compelled by their companions to retire from the colony, sought a refuge in the interior, where they followed a predatory life, occasionally appearing in the neighbourhood of the settlements to obtain supplies in return for the little gold they managed to pick up or extort from the natives. The unusually large quantity of gold from these rovers which found its way into the settlements about the year 1670 attracted the attention of the colonial authorities; but the gold-finders refused to impart any information upon the subject unless they were accorded a pardon from the king for all offences and full protection for the future.

This being obtained they stated that they had discovered placers at Jaragua, in the province of San Paulo, including nuggets of considerable size. This was a blind, because Jaragua had been discovered a century previously. Some authorities place the date of the discovery of gold by the outlaws of Brazil so late as 1694, in which year it is stated that there arrived at Lisbon about a ton and a-half (over $700,000) of gold obtained by "a body of outlaws in a distant part of Brazil, and who, to get a market for it, submitted to the king's Fifth." (Universal Hist., XIX, 20.) But this information is defective.

These discoveries led to a more systematic exploration of the country by parties from the settlements, and several placers were laid open, the most important being in the province of Minhas Geraes (General Mines). This province was formed into a *capitania* of Brazil, and in 1714, it was districted into four *comarcos*, viz.: S. Joao del Rey; Sabara, Villa Rica, and Cerro do Rio. It was not until after 1693 that the product became so important as to make itself felt in Europe. From this time until after the middle of the following century new placers continued to be discovered, and the product to augment. Afterwards the placers fell off, and during the present century they were succeeded by the exploration of a few quartz mines, which still yield a small amount of gold.

No sooner had the placers of Brazil become productive than they attracted a rush of colonists thither from Portugal. The natives were now hunted down with more system and success, and great numbers of them were reduced to slavery in the mines, where, as fast as they perished, they were supplanted by negros obtained from Africa.

"Nearly all the revolutions that have occurred at Para are directly or indirectly traceable to the spirit of revenge with which the bloody expeditions of the early slave-hunters are associated in the minds of the natives and mixed bloods throughout the country." Kidder's Travels in Brazil, 1844.

The physical aspects of the placer country, the nature of the deposits, the mode of working them, the cost of production, the character of the earlier discoverers and miners, the usages that grew up from these circumstances, and the general condition of society in the mining regions will now be briefly alluded to. The country was mountainous, in some parts sterile and dry, in others clothed with impenetrable forests, and abounding with wild animals and reptiles. The placers were remote from the settlements and difficult of access. Food and supplies had generally to be brought from a vast distance, and upon the shoulders of slaves. For example, from Rio de Janeiro to Ouro Preto—200 miles—took fifteen days. Between the mines of Matto Grosso and the market at Para was 1,000 miles in a straight line: 2,500 miles by river routes. There was a dry and rainy season, for which latter the miners had always to wait; so that mining could only be prosecuted during a portion of each year.

The placers consisted of the gravel banks of small rivers flowing from lofty mountains, the richest beds having been found on the flanks of the Sierras dos Vertentes and the Sierra do Salto in lat. 20 to 21 south, and on the Rio Verdi, near Campanha; though many were found in other localities. One kind of earth washed for gold is described by Mr. Mawe and Dr. Von Spix as a ferruginous sandstone conglomerate called "jacotinga." Generally speaking, the soil is red, and ferruginous. The gold lies for the most part in strata of rounded pebbles, gravel and sand, incumbent on solid rock. This the miners called cascalhâo. The primitive mode of working was with gourds, or wooden bowls: subsequently, and where water of a sufficiently high level could be commanded, the ground was cut in steps about twenty or thirty feet long, two or three feet broad, and one foot deep. Upon each step stood six or eight slaves, who, as the water was allowed to flow gently from above, kept the auriferous earth agitated until it was reduced to the consistency of mud and washed below. At the bottom of a series of these steps was cut a trench into which the precipitation flowed, and where after five days' washing it was sufficiently concentrated. It was then removed by hand to an adjoining stream, and there subjected to the bowl process of separation.

When the placers of Goyaz and Matto Grosso were first opened, slaves employed by the gold-seekers sometimes obtained for their masters three and even four ounces of gold per day; and although this fertility of the mines was not continuous, yet, until after the middle of the 18th century, if not profitable, they were very productive. In 1846, the average product of a miner scarcely exceeded 15 cents a day. At the present time these placers can only be worked profitably by the hydraulic process, when they may again become productive enough to engage the attention of investors.

In the early days the only food to be obtained in the mountains was a few birds, deer and mangabas, the latter a wild fruit. There were instances when the price of corn at the mines was more than a pound weight of gold ($240) per bushel. In one instance, the same price was paid for a pound of salt! A drove of cattle which some adventurer had managed to convey to the mines (of Goyaz and Matto Grosso) sold, flesh and bone together, for an ounce and a-half of gold (about $30) per pound. When they failed to capture slaves it required all the gold which the miners could obtain to keep them in food, and even this was insufficient, for great numbers of them died from leprosy and starvation. Macgregor IV, p. 210.

At the outset the miners were the worst of characters: outlaws, renegados and traitors. As the settlements grew up in the mining regions the disposition of the new comers greatly improved that of the mass. But the mining communities were always noted for their cruel treatment of the natives (whom they captured only to work to death), their proneness to violence, their quickness to use the knife on all occasions, their passion for gambling and their licentiousness. So late as 1846 the slaves in the province of Pernambuco were treated so cruelly that, according to Mr. A. de Mornay, they cultivated the fatal vice of eating earth in order to "put an end to their already worn-out existences." Macgregor, IV, p. 181. This, however, was not the reason for eating earth, because Humbolt assures us that some of the Amazonian Indians, in their wild state, eat small portions of a certain kind of clay, from choice. The Pernambuco slaves eat earth because they were insufficiently fed by their masters.

These characteristics of the Portuguese miners had scarcely become softened by time, the repression exercised by the military forces of the colonial government, and the influence of the priests, when the gold placers lost their importance, and Brazil fell into decay. Then followed the revolution and Independence. In 1844 the arrival of three French mechanics, a carpenter, a joiner, and a blacksmith, in the province of Goyaz, was deemed so important an event as to be stated in the message of the President to the provincial assembly. When the writer worked in these placers forty years later the people crowded about him to see the operation of a chisel, a plane, and a saw; the only tool previously used in the vicinity being a hand-adze.

The currency of the country at first was gold-dust, afterwards bars of uncoined gold; later still, when the colonial mints became capable of supplying the country with coin, the use of dust and bullion was prohibited; for the Crown did not omit to extort a seigniorage from the coin. Says Kelly: The gold dust deposited in the beds of the various streams is a common right, but when found, is by law bound to be carried to the royal smelting-houses (Cazas de Fundiçao) established in various districts, where, one-fifth of it being retained (*in natura*) for the Royal Quinto, a bar is made of the remainder. which is weighed, assayed, numbered, stamped, and returned to the owner accompanied by a certificate, signed by the proper officers, showing the value in money of such bar, calculated at 1,500 Reis per octave of eleven-twelfths (.9167) fine. These bars serve as a circulating medium, but it is strictly prohibited to export them. They are ultimately carried to the Royal Mint at Rio de Janeiro, where they are received at 1,500 Reis per octave, and paid for in gold coin valued at 1,600 Reis per octave; the King thus exacting a seignorage of 6⅔ per cent., in addition to the Quinto (or 20 per cent.) previously taken on the gold dust. Dr. Southey states that gold was first stamped by the authorities of Brazil in 1701; but in a work entitled "The Empire of Brazil," Rio de Janeiro, 1877, pp. 402–3, there are evidences that the King's Fifth was paid and coinage commenced so early as 1694. Upon the decline of the placers gold gave place to silver coins smuggled from La Plata. Here private money ceased and government money began, viz.: base silver coins and coppers. These were partly replaced in 1797 by government paper notes, and at a later date by a system consisting altogether of government and bank paper. This last system exists to the present day, and it has efficiently served all the purposes of money for the progressive empire which has grown up from such rude beginnings.

The system of hunting down the natives, enslaving them as encomienderos and working them to death in the mines, did not operate so smoothly in Brazil as it did in Mexico and Peru, because in the first-named country there were plenty of backwoods into which the Indians could retreat and into which many of the captives managed to retreat when they found that captivity meant death. The escape of their victims and the difficulty of hunting for others compelled the Portuguese to send to Africa for negro slaves, who, originally, cost about $5 each in Africa and were worth about $25 in Brazil. Toward the end of the 17th century, before the great placers of Minhas Geraes were systematically opened, the price of negros on the coast of Africa was $15, and in Rio Janeiro about $65 to $80. After the placers became productive, that is to say, in 1705, negros were worth $75 in Africa and $250 in Rio. The christianised Indians kidnapped from the Paraguay missions by the San Paulo bandits, were sold to the mines at prices varying from $50 to $150 each. It is grievous to be obliged to record that the benevolent example of the Jesuits was not followed by all the clergy. On the contrary, many took part in entrapping and enslaving the natives, some of whom they deliberately trained as decoys to entrap others. The following affecting narrative appears in Bayard Taylor's work:

"Three years before the arrival of the travellers (Humboldt and Bonpland) the Missionary of San Fernando led his converted Indians to the banks of the Rio Guaviare, on one of these hostile incursions. They found in an Indian hut a Guahiba woman with her three children, two of whom were still infants, occupied in preparing the flour of cassava. Resistance was impossible; the father was gone to fish, and the mother tried in vain to flee with her children. Scarcely had she reached the savannah when she was seized by the Indians of the mission. The mother and her children were bound, and dragged to the bank of the river. The monk, seated in his boat, waited the issue of an expedition of which he shared not the danger. Had the mother made too violent a resistance the Indians would have killed her, for everything was permitted for the sake of the conquest of souls, and it was particularly desirable to capture children, who might be treated in the missions as slaves of the Christians. The prisoners were carried to San Fernando, in the hope that the mother would be unable to find her way back to her home by land. Separated from her other children, who had accompanied their father on the day on which she had been carried off, the unhappy woman showed signs of the deepest despair. She attempted to take back to her home the children who had been seized by the Missionary; and she fled with them repeatedly from the village of San Fernando. But the Indians never failed to recapture her; and the Missionary, after having caused her to be mercilessly beaten, took the cruel resolution of separating the mother from the two children who had been carried off with her. She was conveyed alone to the missions of the Rio Negro, going up the Atabapo. Slightly bound, she was seated at the bow of the boat, ignorant of the fate that awaited her; but she judged by the direction of the sun, that she was removing farther and farther from her hut and her native country. She succeeded in breaking her bonds, threw herself into the water, and swam to the left bank of the Atabapo. The current carried her to a shelf of rock, which bears her name to this day—The Mother's Rock. She landed and took shelter in the woods, but the President of the Missions ordered the Indians to row to the shore, and follow the traces of the Guahiba. In the evening she was brought back. Stretched upon the rock, a cruel punishment was inflicted upon her with straps of manati leather, which serve for whips in that country, and with which the Alcaldes were always furnished. The unhappy woman, her hands tied behind her back, was then dragged to the mission of Javita.

She was there thrown into one of the caravanserais. It was the rainy season, and the night was profoundly dark. Forests, till then believed to be impenetrable, separated the mission of Javita from that of San Fernando, which was twenty-five leagues distant in a straight line. No other route was known than that by the rivers; no man ever attempted to go by land from one village to another. But such difficulties could not deter a mother separated from her children. The Guahiba was carelessly guarded in the caravansary. Her arms being wounded, the Indians of Javita had loosened her bonds, unknown to the Missionary and the Alcalde. Having succeeded by the help of her teeth in breaking them entirely, she disappeared during the night; and at the fourth sunrise was seen at the mission of San Fernando, hovering around the hut where her children were confined. 'What that woman performed,' added the Missionary, who gave the travellers this sad narrative, 'the most robust Indian would not have ventured to undertake.' She traversed the woods when the sky was constantly covered with clouds, and the sun during whole days appeared but for a few minutes. Did the course of the waters direct her way? The inundations of the rivers forced her to go far from the banks of the main stream, through the midst of woods where the movement of the water was almost imperceptible. How often must she have stopped by the thorny lianas that formed a network around the trunks they entwined! How often must she have swum across the rivulets that ran into the Atabapo! This unfortunate woman was asked how she had sustained herself during the four days. She said that, exhausted with fatigue, she could find no other nourishment than black ants. The travellers pressed the Missionary to tell them whether the Guahiba had peacefully enjoyed the happiness of remaining with her children; and if any repentance had followed this excess of cruelty. He would not satisfy their curiosity; but at their return from the Rio Negro they learned that the Indian mother was again separated from her children, and sent to one of the missions of the Upper Orinoco. She there died, refusing all kinds of nourishment."

Something of this melancholy story Humboldt himself narrated in the hearing of the writer. The name of the mother was Lonira.

San Paulo was built about the year 1570 by the malefactors whom Portugal had cast upon the coasts of the New World. No sooner did these villains perceive that it was intended to subject them to some sort of surveillance than they fled from the coasts and buried themselves in places so remote or inaccessible that the power of the government could no longer reach them. Here they were recruited by other bandits and by the progeny of the Indian women whom they captured and enslaved, until from these vile beginnings a petty state arose whose laws were both piratical and sanguinary. These were primarily that: No person should be admitted into the community who was not prepared to remain with it and undergo a trial of his fidelity, courage and contempt of all obligations save those of supporting the band to the death; no member of the band should ever leave it; no travellers to be tolerated; all rejected applicants and all attempted deserters to be put to death. "They (the Paulistas) overran the inland parts of the Brazils from one extremity to the other. All the Indians who resisted them were put to death; fetters were the portion of cowards; the inhabitants (natives) hid themselves in the mountains to avoid slavery or death. It would be impossible to enumerate the devastations, cruelties, and enormities of which these atrocious men were guilty." Raynal, IV, 466. These were the men who first discovered gold in Brazil, who doomed its native population to destruction and supplied their places with negro slaves, in order that this gold might be dug out and carried to Europe

An attempt will now be made to follow the course of mining discovery in Brazil.

Chronology of Gold Mining in Brazil.

1549. The natives are observed to possess gold ornaments.

1573. Minhas Geraes. Sebastian Fernandes Tourinho discovers gold placers and emeralds in the valley of the Rio Doce. Almanach Sul Miniero, Campanha, 1874, p. 19.

1577. Minhas. Placers of Jaragụa discovered. Raynal, IV, 469.

1588. Bahia. Placers of Jacobino on the river Das Velhas. Raynal.

1594. French expedition up the Amazon found no gold.

1595 to 1605. Minhas. Gold discovered at Ouro Preto. When the news reached San Paulo a rush occurred of Mameluccos, who sought to dispossess the original discoverers by force. This gave rise to numerous deeds of blood, the Portuguese being the actors and the enslaved natives the bystanders. A fresh stream of gold-hunters soon afterwards appeared from Portugal, who dispossessed both parties and appropriated the richest washings for themselves and their slaves. Antonio Vieyra, and Dr. Guimaraes. When Pombal entered office in 1750 the population of Portugal had been reduced to two millions. "Famishing Portugal," in Lippincott's, 1876.

1625. Goyaz. Expedition of Fr. Christovao de Lisboa from Para to the highlands of the Tocatin river, which, after flowing 1,500 miles, empties into the Amazon near its mouth. Castelnau, 108.

1640. Goyaz. Another expedition to the Tocatin hills.

1662. Minhas. Expedition of Fernando Dias Paes Leme. Alm. Min. 19.

1669. Goyaz. Expedition of Gonzalo Paes and Manoel Brandon to the Rio Araguaya, an affluent of the Tocatin. Castelnau, 108.

1670. Minhas. Expedition of the Paulistas to Jaragua, carrying with them slaves to work the placers.

1670. Goyaz. Expedition of Manoel Correa with another band of Paulistas and slaves to work the placers. Castelnau, 123.

1672. Goyaz. Expedition of Pascoal Paes de Araujo with a band of Paulistas to Piauhy and Para, to capture Indian slaves, tributes, and gold. Castelnau, 108, 123.

1693. Minhas. Antonio Rodriguez Arzanou and Carlos Pedroso da Silveira with 50 men raid an Indian village, which the Portuguese named Cathay, on the Rio Doce, and from which they got three oitavas of gold, "the first product of Minhas Geraes," says the Almanach Sul Miniero; but this is a mistake, because Minhas began to be productive so early as 1670. It may have been the first registered product of Minhas.

1694. Minhas. Expedition of Bartholomeo Bueno. He presents 12 oitavas of gold to the governor of Rio. Alm. Min., 20.

1694. Minhas. Numerous expeditions of the Paulistas, who scour the country for Indian captives and carry them to the mines of Minhas, which are thus rendered very productive. Alm. Min. 20. New placers were discovered almost every day, so that to this period is usually ascribed the beginning of that vast production of gold which in the course of half a century exerted so remarkable an influence upon the social condition of Europe as to give rise to demands for the abolition of the feudal system, the ecclesiastical system with which it was connected, and the colonial system which had grown out of them and had now outgrown them.

1694. Goyaz. The Paulistas capture a number of Indians and set them to work in the mines of this province.

1699. Minhas. The mines of Sabara, Rio das Mortes, Cashoeira,

Sta. Lucia, Pousso Alto, Carmo, Campanha, Paracutu, Ouro Preto, Rio Doce, Rio Das Velhas, etc., are worked with negro slaves. Raynal. During the productive period, the following prices ruled at the mines: an alqueire of milho, 40 octaves of gold and of beans 80 octaves. Alm. Min. 20. The alqueire varied from half a bushel to 1¼ bushels; the octave of gold contained 55⅓ English grains, about $2.50. Beans, therefore, cost about $200 a bushel. A heavily paved road to the mines, constructed during the 18th century, at vast expense, is now in ruins.

1707. Minhas. Placers of Villa Rica discovered. Southey, III, 56.

1715. Minhas. Placers of Villa do Principe. Mawe, 222.

1718. Matto Grosso. New placers discovered. Macgregor, IV, 142. These were probably at Cuxipómirim, on a river of the same name, one league from Cuiabá, or Cuyabá, in the center of the province. Biblioteca Nacional, 1877, p. 269.

1721. Matto Grosso. Placers of Cuyabá and Rozario discovered by Miguel Subtil, probably the same as those mentioned under 1718. Rozario yielded 400 arrobas (about $3,500,000) in one month.

1722. Goyaz. Expedition of B. Bueno discovers gold placers on the Rio Vermelho.

1726. Goyaz. Placers of San Felix, Meia Ponta, O Fanado, Mocambo and Natividade discovered. Town of Goyaz founded. In 1830 there were 232 abandoned placers in this province, only 41 working. They are now all abandoned. The physical devastation wrought by the miners is frightful.

1727. Minhas. Sebastiao Leme do Prado, with a band of Paulistas, discovers gold in the river Bom Successo, at a place afterwards called Minas Novas, also on the Rio Capivary, near the Arraial da Chapada. These placers were extensively mined, and 300 arrobas (over $2,500,000) were sent to Bahia alone. In the Lavrado Batatal a single nugget weighted 28 pounds. Lock.

1730. San Paulo. Pompéo says that the gold mines of this province produced, from the period of their discovery to the beginning of the 19th century, 4,650 arrobas, about $40,000,000, while Scully says they only produced 70 tons, about $35,000,000. Lock, 219, 228. The difference in this instance is comparatively small. In the case of the other provinces, Scully's estimates are far too small to entitle them to consideration. It would probably be nearer the truth to credit Minhas with 550, Matto Grosso 230, Goyaz 150, San Paulo 40, and the other provinces with 30 million dollars; altogether about 1,000 millions.

The quartz mines of Brazil began to be worked after the placers showed signs of exhaustion. Among the most productive were Morro Velho, Maquiné, St. John del Rey (opened about 1725), Cata Branca and Gongo Soco, the last of which, according to Moraes, yielded in thirty-odd years, best period 1825–35, over $6,500,000. Hartt says that the first two of these were the only ones that were profitable.

1735. Matto Grosso. Placers of St. Vincent, Chapada, Araès, and St. Anne; also some new ones near Cuyabá. Raynal, IV, 470.

1736. Goyaz. An improved route (paved road) opened to the placers. Macgreggor, IV, 142. This road has now gone to ruin.

1738. Matto Grosso. Placers of S. José de Cocaes, six leagues from Cuyabá, discovered. Biblio. Nac. 269.

1739. Matto Grosso. Placers of Arinos, ten leagues north of Diamantino, discovered. 1741, Arraial Velho, one and a-half leagues north of Diamantino; and between 1740 and 1750, Brumado, six leagues southwest of Diamantino and Parí, three and a-half leagues south of Diamantico, were discovered. Biblio. Nac., 270–1.

1805. Matto Grosso. Placers of Diamantino and Ribierão de Ouro, whose sluiceways were filled up with tailings (entulhos), opened a second time. Biblio. Nac. 271. They have since filled up again and are now altogether out of grade and abandoned.

1808. Matto Grosso. Placers of S. Francisco, nine leagues northwest of Diamantino, discovered. Abandoned four years later in consequence of richer discoveries at Araés.

1812. Matto Grosso. Although the placers of Araès were discovered in 1735, some new and richer ground was opened this year, causing the abandonment of the poorer mines in the vicinity, an evidence that the mines generally were not paying. After the rich ground was worked out, the unhealthiness of the place caused it to be quickly deserted. Biblio. Nac. 271. Many other mines in Goyaz were opened at this period, but they failed to pay, and were abandoned.

1820. Goyaz, which formerly had a vast population of Indian encomienderos and 100,000 negro slaves in the mines, now has but 62,518 inhabitants, including 12,000 Indians. Castelnau, 121.

1821. In this year there were deposited at the mint of Rio only 39,286 oitavas of gold. Castelnau, 121. This was almost the last of the placer mines.

The first caza de fundiçaon, or smelting-house, where the gold dust was melted, refined, and cast into bars, appears to have been established in 1694. After the royal Quinto was deducted, these bars were stamped with the royal device, the date of their fabrication and their weight, fineness and mint-value, in Portuguese gold coins. Besides the Quinto on the production, the Crown levied a seigniorage of 6⅔ per cent., on the coinage of gold. To evade as much as possible the payment of so heavy an exaction, the colonists, though at some inconvenience to themselves, employed the gold bars as money, whenever the magnitude or peculiarity of their transactions permitted. Unwilling to lose its rightful dues of seigniorage, the Crown now forbade the use of dust or bullion as money, in any case; and although such an interdict was often difficult to enforce, yet, as a matter of fact, most of the bullion went to the mint and submitted to the seigniorage. But this was not enough for the Crown, which, in order to extort an additional profit, had recourse to a new expedient. As this was probably similar to the one employed nearly twenty centuries before by the Romans in Spain, it is worthy of a detailed description. The expedient was to place a higher value upon its silver and copper coins in Brazil than at home: it being borne in mind that the fabrication of coins is necessarily a state prerogative, and that at that period it was also a Crown monopoly.

Bars of gold bullion made a very inconvenient sort of money. Their value could only be definitely expressed in coined money. If they had been regarded merely as pieces of merchandise, a separate bargain would have had to be made in respect of each piece, and in fixing its value, very great allowances would have had to be made for such changes of value as each piece may have undergone, unknown to the buyer, since the last bargain of like nature took place. But nothing of this sort occurred. It was well known how many "crusados" each bar contained, and as well known how much merchandise, or labour, each crusado would buy, either in Rio or Lisbon. The principal inconvenience of using bars arose from their great and uneven weights. It was seldom that a bar weighed less than several ounces, and as it was necessary for reasons of prudence to keep each miner's lot of gold distinct from the other lots deposited at the caza de fundiçaon (the same practice is pursued in all mints today), it followed that the bars were of uneven weights, and therefore worth uneven sums of money. For example, the smallest bar probably had stamped upon it a certificate of somewhat the following character: "This bar of gold weighs 7 onzas, 7 oitavas, and 71 granos [1] Its fineness is 10 dinhieros, 19 granos." Weighing it against the crusado, its value after the payment of seigniorage, was 1,000$190. Such a piece as this one, of uneven weight and requiring an intricate calculation to determine its value in money, could not have been itself used as money, except in rare instances, and the miners must have been willing to submit to any reasonable exaction of seigniorage, rather than sell their bars, as they would otherwise have had to do, to the bullion-brokers. Another inconvenience arose from the legal requirement that each bar should be accompanied by "a printed ticket, stating the weight of the gold, its value in reis, and the quantity deducted for the royal treasure." Without this it was liable to be seized as contraband. Malte-Brun, III, 393.

Even after all this, the miners' difficulties were only half removed. By submitting to a tax of 20 per cent. on production and 6⅔ per

[1] 72 graos=1 oitava; 8 octavas=1 onza; 8 onzas=1 marco. The marco was equal to 3541½ grains Troy, or .6489 Troy pounds, or 229.5 metric grams. For the Spanish mark, see the chapter of the present work on "Spain." The Spanish mark was equal to 3550½ grains Troy.

cent. on mintage, together equal to about 25 per cent. on production, they could exchange refined gold bullion for coins containing an equal weight of fine metal; but such coins being of gold were too valuable for the ordinary transactions of life. What they wanted was minor coins, say of silver and copper, with which "to make change." The Crown being fully aware of this requirement, took advantage of it, by requiring the miners to pay more gold than before, for a given sum of silver coins. This system was commenced by Dom Pedro II., some time between 1690 and 1700. All the small coins—even the smaller gold ones—were valued higher in the colonies than in Portugal.[2] Thus, the gold piece of 3$800 in Portugal was ordered to, and did, pass for 4$000 or four milreis in Brazil; the pataca of silver, valued at 0$240 in Portugal, legally went for 0$320 in Brazil; and the vintem of copper, worth 20 reis in Portugal, was valued at 40 reis in Brazil.[3] The idea of the Crown was that if the producers of gold wanted small change, they must pay for it. Instances have since occurred in California and Australia, where there were no despotic governments to coerce them, when the miners voluntarily paid for silver coins, in gold dust, twice, four times, "and, in one case, sixteen times," their mint value.[4] These examples serve to measure the difference between the usefulness of bullion and money.

Having in this way sold its silver and copper coins at as high a rate as practicable to the miners in Brazil, the Crown of Portugal turned around and sold its gold at an enhanced rate to the merchants in Europe. This was done, by raising (for at that time it had the power to do so) the ratio between silver and gold to 16, which but a few years before had been but 13⅓ for 1. The aggregate profits on these various operations amounted in some cases to no less than 70 per cent. on the miner's product of gold. Thus, supposing a miner produced 100 pounds weight of gold, the quinto would have left him but 80, and the seigniorage but 74⅔ pounds in gold coins. Upon exchanging this gold—as in many cases would have had to be done—for small coins, he would lose one-half of its European purchasing power, leaving him with coins that, even if not over-valued in Por-

[2] John VI., April 18, 1809, reaffirmed this same measure, and as the currency was then wholly of copper, it completely changed the milreis of Brazil, which thenceforth became worth only one-half that of Portugal—a relation that was respected in the subsequent gold and silver coinages of the Brazilian Empire. Armitage, "History Brazil," ed. 1836, II, 54.

[3] A similar system had been pursued in England. From 5 Edward IV. to the reign of Elizabeth it was usual to legally value English coins one-third higher in Ireland than in England, so that what passed for three shillings in England was legal-tender for four shillings in Ireland. Case of the Mixt Moneys.

[4] "Hist. Prec. Met.," ed. 1880, pp. 221*n*., 320-21.

tugal, would only purchase as much in that country as 37⅓ pounds of gold bullion would have purchased. But by altering the ratio from 13⅓ to 16 silver for one gold the Crown had recently enhanced the value of gold coins 20 per cent.; so that supposing the miner's venture to have been made on the basis of the previous ratio, the silver coins, even in Portugal, were overvalued to this extent. On the whole, then, the miner ultimately received the equivalent of scarcely more than 30 pounds weight of gold for the hundred pounds weight which he may have committed all sorts of crimes to obtain; the Crown got the rest.

Statements concerning the population of Brazil at various periods since its discovery have been made by the following authorities: Humboldt, Malte-Brun, Sousa, Soares, Scully, Eubank, Armitage, Southey, Murray, Seaman, Weimer, McCulloch, Balbi, Cannabich, Midden, Behm and Wagner, Velloso, the National Censuses and the Encyclopedists. Upon a comparison of these authorities, and after a residence of two years in the country itself, the writer ventures upon the following estimates. At the period of the discovery Brazil probably contained a native Indian population of two or three millions. These are now reduced to about one million, of whom about three-fourths are classified as "domesticated," and one-half of the latter as "converted." African negros began to be imported into the Brazils so early as 1570, but it was not until after the accession of the Braganza family (A. D. 1640), that this heinous traffic assumed great proportions. In 1750 the negro slaves, free negros, mulatos and mestizos numbered about 1,500,000; 1800, same classes, 3,000,000; 1816–18, 2,500,000; 1830, 3,500,000; 1845, 4,400,000; 1855, 5,000,000; 1865, 5,500,000; 1875, 6,000,000; 1895, 6,500,000. During the 18th century the white population was about half a million; at the beginning of the 19th it was about three-fourths of a million; in 1816–18, about 850,000; 1830, 900,000; 1845, 1,250,000; 1855, 1,500,000; 1865, 1,750,000; 1875, 2,000,000; 1895, 2,500,000. Owing to the admixture of races, Indians, whites and negros, forming such classes as Mamalucos, Zambos, etc., as well as to the unreliability of the official censuses, it is exceedingly difficult to get at the truth. As a matter of fact, the population is chiefly African; but comparatively few white people being seen away from the coast cities. The persons actually employed in mining for the precious metals, who consisted almost exclusively of Indian cautivas (captives) and negro slaves, probably numbered during the 17th century, from 50,000 to 100,000; during the first half of the 18th century, from 100,000 to 200,000[5]; during the last half, from 200,000 down to 50,000[6]; and during the first half of the 19th century, from 50,000 down to 10,000; when the placers were abandoned, the slaves were emancipated and only those remained who were employed in the quartz mines,[7] or earned with the batea their scant 12 cents a day by panning gravel on their own account. If the whole number of captives and slaves forced to labour in the gold mines of Brazil be multiplied by the years they worked, and the quotient divided into the entire product of the mines, it will be found that they produced on the average less than $40 a year per man. Of this amount one-fifth went to the Crown at the outset, leaving but $32 per man. By the time the mine proprietor turned this into coins it would he reduced to about $26, thus reducing the average product to about half an American gold dollar per week per man, less than the solitary gleaner still wins in freedom from the neglected corners of the abandoned catas. The imagination can scarcely realize the cruelty and privation to which the mining slaves were formerly subjected in order to "make them pay" the cost of their maintenance.[8] But for all this, except in newly discovered and unusually rich catas, they never did pay, so that little by little, though very reluctantly, their masters permitted such of them as survived—and among these were few or no Indians—to be drafted into the less alluring, but, as it has turned out, far more profitable, fields of agriculture. The economical cost of the billion dollars worth of gold taken from Brazil was practically a fortnight's work of a slave for each dollar's worth of bullion. According to Mr. Wm. Jacob's Appendix No. 9, there were working

[5] Raynal estimated the "domesticated" Indian population for about the year 1750 at 245,000 males. This implies for a mining country at least 300,000 of both sexes; of whom it is estimated that two-thirds were employed in the mines.

[6] The correspondent of the London *Times*, writing from Rio in September, 1880, and evidently referring to the latter portion of this period, said that the diggings formerly employed 80,000 men. Captain Cook, who visited Brazil in 1768, said that "near forty thousand negros are annually imported to dig in these mines." Mavor.

[7] "It recently transpired that over 150 negros who had been emancipated twenty years ago, had been kept all that time in ignorance of their freedom at work upon the St. John del Rey gold mine!" Money and Civilization, p. 162. This mine was and is still owned entirely in England. The resident manager was an Englishman, who committed suicide when his crime was exposed.

[8] "What fools you are," said a Brazilian mine-owner to some run-away slaves who had been re-captured and duly whipped. "Out of the total product of these mines you get two-thirds, for that's what it costs to support you; the King gets the other third; and I, who furnish the enterprise and the capital, get nothing. Yet you are not satisfied, and ungratefully wish to abandon me!" The slaves were so struck by the justice of these remarks that they all resigned themselves to work, until they died from privation. A similar effect seems to have been recently produced upon the cardrivers of New York, when it was shown that they got 40 per cent. of the road earnings, while the proprietors or shareholders only got ⅞ of one per cent.

in all the lavras or placer mines of Minhas Geraes in the years 1812 and 1813, freemen and slaves to the number of 17,008, of whom 10,603 were "slaves of the miners," and 2,148 "slave washers." The total average annual produce of gold was 310,084 milreis, which at that period were roughly equal in value to Spanish silver dollars. Assuming that the men all worked in or about the mines, which is evidently what Mr. Jacob means, it follows that the average produce of each man was less than $20 per annum! In the lavras of Campanha, taken by themselves, the average produce per man was $27 per annum. From the writer's personal knowledge of this district, he is able to say that the statement appears to be entirely correct.

It is now time to collate the statistics of production. Except as to the gross sums derived from the placers during the first sixty years of their productiveness these are far from satisfactory. The best account, so far as it goes, is afforded by the Abbé Raynal. Says this distinguished author: "It is demonstrable, *from the registers of the fleets*, that in the space of sixty years, that is, from the discovery of the mines to the year 1756, two thousand four hundred million livres (about $480,000,000) worth of gold have been brought away from Brazil." Raynal admits that the registers of the fleets did not contain a full account of the production: for much of this found its way to market in a surreptitious manner to evade the Quinto, which, for all except a brief portion of the whole period of production, was fixed at 20 per cent., also the haberias or convoy duties, the seigniorage of the mints and other exactions.

During the earlier period of the placer development, the number of vessels annually dispatched from Portugal to the Brazils, did not exceed twelve, but as the mines grew into importance these amounted to a hundred. No ship was allowed to sail except with the fleets. Of these, one sailed for Rio in January, one for Bahia in February, and one for Pernambuco in March. Sometimes they sailed twice in a year. These fleets were convoyed by men-of-war, to protect them from enemies and pirates, and for this duty there was levied a charge called *haberia*. The haberia was established by Spain in 1522, and consisted, at first, of one per cent. ad valorem upon the freight. It was, however, shortly afterwards raised to 5 per cent. What the rate was in Portugal cannot be determined; but it was probably no less than the one last named. Insurance on the voyage out and return was eleven per cent. Interest 8 per cent. per annum. There were other exactions upon gold which will presently be treated more at length.

Taking into account these various exactions, and the extent to which the official returns were vitiated by evasions, the Baron von Humboldt in his "New Spain," estimated thc total product of the gold placers of Brazil from 1680 to 1803 at $855,500,000. Mr. Danson, in an elaborate paper on the subject which he read before the Statistical Society of London, going over the same ground, and with all of the authorities before him, estimated the product from 1680 to 1803 at 922 millions. Finally, Dr. Southey estimated the Quinto (or royal duty of one-fifth upon the production) of gold during the same period, and down to 1807 at 225 millions; which would make the total product more than 1,125 millions. "History of Brazil," III, p. 820. Southey's estimate makes an allowance of one-fifth for smuggling and is called the Product of the mines; not the Quinto. But from the estimates of separate years in other parts ofhis work it it is evidently the Quinto that is meant.

Guided by these estimates and by the accounts of the product in various years, obtained by consulting a number of other authorities, we have ventured to make the estimates given below. While the scantiness and uncertainty of the attainable data does not admit of any pretension to exactness in the table, so far as one particular decade or another is concerned, it is believed to be reliable, first, in respect to the period at which the placers of Brazil commenced to be noticeably productive; second, as to the total sum of the product from first to last, and even as to the total sum in any one period consisting of not less than two or three decades; and, third, as to the period of greatest productiveness, viz.: 1730 to 1750. With this qualified assertion of its reliability, the table will now be adduced.

Estimated total Gold product of Brazil, from the period of the discovery of the placers to the present time, collated from the following authorities: Raynal, Humboldt, Jacob, Danson, Mawe, Birkmyre, Southey, Phillips, Kelly, Beauchamp and others.

SUMS IN MILLIONS OF AMERICAN GOLD DOLLARS.

Period.	Product.	Period.	Product.	Period.	Product.	Period.	Product.
Before 1680	— 25.0	1720–29	100.0	1780–89	50.0	1840–49	13.0
		1730–39	200.0	1790–99	43.0	1850–59	15.0
1680–89	10.0	1740–49	150.0	1800–09	25.0	1860–69	13.5
1690–99	15.0	1750–59	75.0	1810–19	12.5	1870–99	12.5
1700–09	30.0	1760–69	50.0	1820–29	15.0	1880–89	13.5
1710–19	50.0	1770–79	50.0	1830–39	20.0	1890–99	15.0

According to this table the total product, from 1700 to 1755 inclusive, the period covered by Raynal's statistics, was about 580 millions; or about one-fourth more than was shown by the registers of the fleets. From 1680 to 1803 it was 833 millions compared with 863 millions estimated by the Baron von Humboldt, and 922 millions, estimated by Mr. Danson; a substantial agreement in both cases. The estimate of Dr. Southey is rejected as excessive.

Until within recent years the extent of the Brazilian gold product was scarcely less than had been that of either the Californian or Australian. When it is considered how much less gold there was in the world's stock of the precious metals, at the period when Brazil threw her auriferous product into Europe, than there was when California and Australia were productive, the importance of the Brazilian placers, will be seen to have been even greater than that of the great placers of the present century.

The impositions levied by the Portuguese government upon the product of the Brazilian gold fields now demand some further mention. In the absence of legal authorities upon the subject, there is considerable uncertainty with regard to the royalties, seigniorages, convoy-duties, and other impositions which the Portuguese monarchs levied upon the gold product of Brazil. The historical authorities have not paid that attention to the subject which its importance demands, and the most that can be hoped for with respect to this matter is substantial correctness.

First, as to the Quinto or Royal fifth on production. From the medieval workings of the auriferous sands at the estuary of the Tagus, the Portuguese monarchs had exacted one-half of the gold found in the lower layer or malhada, leaving all of the inferior quantities obtainable from the upper layer, or medaon, to the washers, who were, however, subject to a capitation tax; the adiceiros de malhada, to two coronas, and the adiceiros de medaon to one corona per annum. After this, the King, by compelling them to sell their gold to the mint for less than its weight in coin, managed to extort something more out of them.

In the proportion of production thus demanded from the gold miner, there appears no likeness to the Quinto or fifth, which afterwards so universally constituted the tax throughout the American possessions of Spain and Portugal, and which derived its origin from the Mahometan Koran. The Quinto was the tax imposed by the King of Portugal upon the production of gold in Brazil.

Previous to 1714, the Quinto appears to have been levied and paid pretty fairly. At that period the difficulties and excessive cost of production had become apparent to the miners. These considerations, together with the magnitude of the Quinto, which now amounted upon an average to a million dollars yearly, led to great dissatisfaction among the colonists. The upshot of this was that in 1714 it was "commuted" by some other tax, of whose character and weight we have no account. From the nature of the subsequent commutation, made in 1730, it may be surmised to have amounted to at least one-half of the Quinto, *i.e.*, to ten per cent., instead of twenty.

In 1720 the Quinto was resumed. Dr. Southey says of this period that the Quinto was not half paid; but Southey, though an industrious, was not always a careful writer, and in view of the discordance of opinion on this subject, presently to be shown, and the fact that generally, and as to the whole product of the mines from beginning to end, about two-thirds of the Quinto is known to have been collected, his assertion is to be received with caution.

In 1730 the Quinto, now amounting (if all paid) to upwards of $4,000,000 a year, to say nothing of the seigniorage of 6⅔ per cent. upon coinage, became so great and oppressive as to lead to open insurrection; followed in the end by a compromise. This consisted of a temporary reduction of the Quinto in 1730, and in 1736 by a capitation tax on the miners of two octaves and twelve vinteims weight of gold per capita per annum. The product of the imposts during both the reduction and the capitation tax was about one-half of the Quinto. In other words, the tax on production was practically lowered to ten per cent., plus a capitation tax.

In 1751 the Quinto was renewed. The government by this time had established military posts and such a system of supervision over the mines as rendered evasion difficult. The proportion of the product that escaped the tax, continued, however, to increase; until, towards the beginning of the 19th century, it probably amounted to one-half, as against one-fourth or one-fifth during the most productive periods. The general tendency of the evasions of excessive impositions is to increase in proportion as the impost grows old, and the tax-payer becomes more familiar with the devices of evasion. At the outset almost all imposts, however excessive, are generally pretty fully met.

This tendency of an evasion of duty to increase with the lapse of time, not usually being familiar to historical writers, and each one generalizing from the facts before him covering only a limited period, has occasioned the utmost discordance of opinion, with regard to the general proportion of the Crown taxes on gold, to the whole product of that metal. Dr. Southey's opinion was, that one-half of the gold escaped taxation; but he himself bears the strongest testimony to the vigilance of the Crown officials, and the severe penalties visited upon all who attempted to evade the Quinto. On the other hand, he avers that the officials themselves were dishonest, and that the collectors sometimes mixed copper with the gold composing the Quinto. As for the opinion of Comara, a Portuguese authority quoted by Southey, that the Crown only got one-twentieth of its dues, it is too extravagant to merit any consideration. Baron von Humboldt, with that more comprehensive and certain grasp of large facts, which formed the distinguishing trait of his ability, estimated the evasions on the whole at one-fourth, and this was probably very near the truth.

The colonial mint established in 1694, was located successively at Bahia, Rio de Janeiro and Pernambuco, finally settling at Rio in 1702, where it has remained ever since. Another mint was established at Bahia in 1714, and was in operation until 1834. A melting-house, caza de fundiçaon, was also established in Minhas Geraes in 1721. After being closed for some time, this establishment was reopened in 1825, when certain English capitalists commenced quartz-mining in Brazil; but it was closed again in 1835.

The colonial coinage previous to 1703, only amounted to about $700,000; from 1703 to 1833 $216,300,000; from 1834 to 1847, $600,000, and from 1847 to 1873 $44,600,000; total, 262 millions. It has since been very trifling, the currency being entirely of paper, with nickel and copper coins for small change. The remainder of the bullion product was formerly carried chiefly in English vessels to Portugal and England; a small portion finding its way across the Rio de la Plata. It now goes entirely to England. With regard to the haberia or convoy duty, this, as before stated, amounted to about 5 per cent. ad valorem on the registered product. As to the further imposts which the Crown of Portugal managed to extort from this product through its colonial policy, a policy whereby the gold miners were obliged to pay customs-duties upon the food, clothing, tools, quicksilver and other supplies needed for the prosecution of their industry, and to suffer other exactions—these may be passed over as subjects too remote from the main topic of this work to merit further consideration.

Another matter in this connection appears, however, to deserve mention here. The gold product of Brazil gave rise to a measure which, to this day, exercises an important bearing upon commercial affairs. In 1688 Portugal changed the legal ratio of silver to gold in her coinages to 16 for one. This at the time, as compared with the valuation at other mints, was an over-valuation of gold, yet from the magnitude of her gold coinages, it tended to change the ratio of the entire commercial world. In 1747, when the maximum yield of her Brazilian placers had declined, Portugal over-valued her silver coins.

CHAPTER XXIII.

NORTH AMERICA.

Marvellous story of Cabeza de Vaca, 1527—Expedition of Father Marcos, 1539—It catches sight of Cibola and returns with great news to Mexico—Expedition of Coronado, 1540—He reports the country a desert and the Seven Cities as worthless—Other expeditions to various parts of North America.

STRANGE reports reached the City of Mexico, about fifteen years after its conquest by the Spaniards, respecting the unknown countries which lay to the north and northwest. Those as yet undiscovered regions were supposed to abut upon the kingdoms of India, and were said to contain not only rich and populous nations and splendid cities, but also mountains of gold, silver, and precious stones, oceans of pearls, islands of Amazons, mermaids, unicorns, and all the marvels which for centuries had played a part in the fables and romances of Europe. The conquerors, even though in the presence of the glories of Tenochtitlan, believed they had entered merely the *threshold* of the wealth and splendour of the New World, and that the true Dorado lay in the far North beyond.

Among these stories the strangest were concerning the Seven Cities of Cibola, or the *Septem Civitates*, as they were called by the Latin-speaking priesthood of the day. The exact situation of these famous cities was not pointed out; but in all the ancient maps, however general and defective in other respects, they were invariably designated, and given "a local habitation and a name." In some, they were represented as rearing their giant towers where the then unknown Bay of San Francisco ought to have been; in others, as lying at the head of the California Gulf, and in others as more nearly in the centre of the great sandy wastes, like Palmyra in the desert. However erroneous, and at whatever times these maps may have been made, they all exhibited the Seven Cities, or the *Septem Civitates*, as if they were as well known as the cities around the Lake of Tescuco.

The bringers of these stories were Alvaro Nuñez Cabeza de Vaca and his companions, Alonzo de Castillo, Andres de Orantes, and a negro called Estevanico, the last of whom, by the way, is one of the first of his race named in American annals. These persons, according to the reports they gave of themselves, were of the unfortunate expedition, conducted by Panfilo de Narvaez, into Florida, in the year 1527. Managing to escape the death which their leader and comrades suffered, they found means, by persuading the Indians that they possessed miraculous powers for healing sicknesses, to subsist. Several fortunate recoveries under their hands gave colour to their pretensions. They passed from tribe to tribe, and gradually, after wandering for nine long years, reached the Pacific and at last made their way to Mexico—being thus the first Europeans who crossed the continent north of the tropics. In narrating their adventures they assured their wondering listeners that the interior of the country through which they had passed was full of various nations; that they themselves had seen much wealth in the shape of arrow-heads of the finest emerald, and big bags of silver; and that they had heard of many peoples, living further north, who possessed great cities and abundant riches. These reports, sustained as they were by the credit of Cabeza de Vaca, confirmed the Spaniards in their previously somewhat vague belief in the wealth of the northwest, and not only induced Cortes to continue his exertions, but attracted the enterprise of others, who, it might be supposed, would have been the last to engage in visionary schemes or mere romantic adventures.

One of these latter was Father Marcos de Niza, a Franciscan priest and provincial of his order. He was regarded as one of the most solid and substantial men in the New World, but he became so much animated by the reports of Cabeza de Vaca, that, without considering personal risk and inconvenience, he determined at once and almost alone to explore those wonderful countries, and reap the early harvest of uncounted wealth, as well as of regenerated souls, which they promised. Accordingly, having secured the services of the negro Estevanico as a guide, and a number of Indian interpreters, he set out for Culiacan, the most northern of the Spanish settlements on the Pacific, in March, 1539. He travelled first a hundred leagues northwestward along the eastern coast of the Gulf of California, and reached, and in four days crossed, a desert. This brought him to a country where the natives had no knowledge whatever of the Christians, and believed him a Man come from Heaven. They placed before him provisions in great quantities, and touched his priestly robes with reverence. He was conducted to an eminence near the Seven Cities which he saw, but was not permitted to enter. He obtained no gold, but returned to Mexico with an excited account of the

Septem Civitates; to which it was now resolved to send a strong expedition under Coronado, who was ordered to plunder them.

Coronado set out from Culiacan on April 22nd, 1540, with the express design of conquering the Seven Cities, and all the countries in that part of the world. When he reached the neighborhood of the valley among the mountains, which Marcos de Niza had reported to be full of gold, he sent off a detachment of horsemen to reconnoitre it. They did so, but brought back cold comfort. He assures us they found neither cities, nor gold, nor anything, but a few Indians, who lived upon maize, beans, and calabashes, and in warfare used poisoned arrows, with which they killed several of his soldiers. He thereupon continued his march; taking, however, a somewhat different route from that pursued by Marcos de Niza; for he crossed several mountain chains and two rivers, one of which he called the San Juan, and the other the Balsas. In the course of a month or more, having passed over countries diversified with deserts, fruitful valleys, mountains and plains, in all of which there was nothing to attract his attention, he at last stood, with his army, before Cibola—Cibola the famous, Cibola the renowned! The imaginations of all had been raised to the highest pitch by Marcos de Niza's account of his reconnaisance of this renowned locality from the mountains. But all that Coronado could now see was a few small towns, consisting of houses built, indeed, of stone, and having flat roofs, but peopled with only a few hundred miserable inhabitants. He admits that the country was delightful, and the soil fruitful; but he intimates, and indeed virtually declares, that the narrative of Marcos was a fable. He, Coronado, could finding nothing worthy of conquest, nothing to attract emigration, nothing to justify settlement. The country was remote; and there was in it neither civilization, nor splendour, nor wealth, nor turquoises, nor precious stones, nor silver, nor gold.

In 1496 John Cabota, a Venetian navigator and adventurer, living at Bristol, England, organized a private expedition to obtain treasure in Cathay and the Indies. He obtained authority from Henry VII. to search for islands or provinces and to take possession of the same as a vassal of the Crown, upon condition that the latter should receive one-fifth of the proceeds of the expedition. Under this charter Cabota, or Cabot, embarked in a single vessel and rediscovered, in 1497, the continent of North America, which the Goths had discovered five centuries previously. He sailed along its coasts, which he mistook or pretended to mistake for those of Cathay, or China, for a distance of 300 leagues, and claimed them for England and Venice.

Cabot returned to England without having seen a single inhabitant or obtained a piece of gold. This ill-fortune rendered abortive all his efforts to organize a second expedition; so he disappeared from the lists of American discoverers and died in obscurity.

In 1512 Juan, Ponce de Leon, governor of Porto Rico, allured by tales which were invented by the Indians of Cuba and Hispaniola—perhaps in the hope of getting rid of him and his rapacious crew—sailed with three brigantines and a band of adventurers from Porto Rico to the island of Bimini. In this island, one of the Lucayas, the Cuban Indians informed him, would be found abundance of gold and beautiful women, and a Fountain, to bathe in which, was to remain forever youthful. In the contract with the Crown of Spain, which formed the basis of De Leon's expedition, the adventurer was to be entitled to encomiendas of Indians and women, and in return he was to pay to the king, at first a tenth, afterwards a fifth, of the gold obtained.[1] De Leon found the island and the women, but neither the Fountain nor the gold; and as the latter was the principal object of the expedition, he resumed his voyage, this time guided, or rather misled by the Lucayan Indios, and steering westward, discovered an unknown land, which he named Florida, and proclaimed a possession of Spain. Believing his force inadequate to explore the interior, he followed the coast to Key West and the Tortugas, whence he returned to Porto Rico. After obtaining a royal charter in Spain, he devoted his energies to the project of conquering Florida; and, nine years later, he attempted to carry it into execution; but the natives fiercely opposed his advance, mortally wounded him in person, and drove him and his party away.[2]

In 1520, Lucas Vasquez de Ayllon, auditor and judge of appeals at San Domingo, landed in Florida, but found no gold. After enslaving some Indians, he returned to San Domingo, where he sold them. Upon a second expedition of a similar kind he was killed.

In 1523, Francis I. of France authorized John Verrazzani, a Florentine navigator, to explore the coasts of North America for gold and other spoil. He landed near the present harbour of Wilmington, N. C., then sailed into what are now the harbours of New York and Newport, but found no gold nor signs of any, and returned to France without farther adventure.

In 1524, De Geray, a Spaniard, landed in Florida, but like his predecessors, found no gold in that country.

In 1527, Pamfilo de Narvaez, fired by the riches and fame which

[1] Parkman's narrative. [2] For another Ponce de Leon, in Mexico, see *ante*.

Cortes had acquired in Mexico, attempted the conquest of Florida with 400 men and 80 horses. Their objective point was Appalache, a native city, said to be filled with gold, but which, in fact, proved to be a village of a few poor wigwams. It was evident that the Indios had misled him, in order that he and his command might perish. Surrounded by a difficult country, destitute of provisions, and harrowed by a sleepless enemy, the expedition entirely failed. After a terrible march from Tampa Bay to the Apalachicola, the remnants of his band attempted to reach Mexico. In this, only four of them succeeded, among them his second in command, Alvaro Nuñez Cabeça de Vaca. Narvaez and all the rest perished, while Cabeça de Vaca, with his three companions, only accomplished the journey after eight years of extraordinary hardship and suffering. They crossed the Mississippi and traversed the western plains, travelling as Children of the Sun, and officiating among the Indians as wizards. They arrived at Culiacan, on the Pacific coast, in 1536.

Amongst the band of adventurers whom Pizarro had led to Peru was one Hernando de Soto, whose share of the Inca's spoil was so great, that upon his return to Spain he was enabled to make an envious display of his riches. Thirsting for renown, he applied for and obtained permission from the Crown to subjugate Florida, which at that period meant all that was known of the entire continent of North America.[3] While the expedition was being organized, Cabeça de Vaca returned to Spain, and to advance his own fortunes, he spread abroad a report that Florida was filled with gold and silver mines, and far exceeded in riches either Mexico or Peru. Nothing else was wanting to recruit the expedition of de Soto. Nobles and gentlemen flocked to his standard. Every man saw himself in a flowery land, surrounded by beauteous concubines, with nothing to do but enjoy himself and count his wealth.

Such was the rush to join this expedition that one Gallego sold his houses, vineyards and cornfields near Seville, and determined to take his wife with him to the new world. Many others committed similar imprudences; only the married ones wisely left their spouses at Havana. A band of Portuguese fidalgos, completely armoured in steel, joined the expedition under one Vasconcelos. Ripartimientos of Indians were to be conferred upon all of them, and they looked forward to holding high wassail in the halls of the Floridian caciques. To further the objects of this expedition and secure his own

[3] Parkman, 14. Barcia, in 1611, speaks of Quebec as a part of Florida; in the time of Henry II., of France, North America was termed Terra Florida.

Fifth of the expected riches of North America, the King of Spain conferred upon De Soto a marquisate and the title of Adelantado, with an estate thirty leagues in length by fifteen in breadth, in any part of the conquered country which he might choose. Moreover, he was created Governor and Captain-General of Florida and of Cuba for life. The command of Cuba was added at the especial request of De Soto, who foresaw its usefulness in fitting out and provisioning his armaments.

In 1539 De Soto landed at the Bay of Espirito Santo, now Tampa Bay, with 644 chosen men and 223 horses. The former were armed cap-a-pie, and they brought bloodhounds to hunt the Indians and fetters to bind them. Shortly after landing they picked up Juan Ortiz, a survivor of Narvaez's expedition, who had lived ten years among the Indians. Guided by him and the account of Cabeça de Vaca, they marched westward through the present states of Florida, Georgia, Alabama and Mississippi, reducing the hapless Indians to slavery, ravishing their women, destroying their offspring and laying waste their wretched homes. They passed over many of the gold-fields or mines which were brought to light during the nineteenth century, but failed to discover them. In fact, they did not come to delve for gold, but to rob it from the Indians. Their eyes were not upon the ground to seek for either placer-grains or quartz-croppings; they were lifted afar off, to the Seven cities of Cibola, whose countless wealth they fancied lay ready to fall before their arms; they were fixed upon the South Sea, between which and them lay an empire filled with treasures. In sober truth, Cibola was but a few rude villages, 2,000 miles away in the desert, near the Rio Gila, and with scarcely gold enough to pay for one of their horses; whilst the South Sea was yet another thousand miles farther westward, over parched and solitary deserts, which still remain valueless to man.

In the third year of their journey they crossed the Mississippi river, at a point above the mouth of the Arkansas. Advancing westward, as far as the highlands of the White river, in the present Indian Territory, and finding neither gold nor the South Sea, but only hostile Indians, and the vast plains which fringe the desert, they abandoned all hope, and returned to the Mississippi. Here De Soto died of dejection and fever, and they buried him in the turbid waters of the great stream. Selecting a more southerly course, a portion of his command, under Alvarado, attempted to reach Mexico, but were foiled by the rigours of the march and the hostility of the Indians. Returning to the Mississippi, they constructed seven brigantines, upon which they embarked themselves and floated down the stream. Escaping many perils, they finally reached the Gulf of Mexico, which they coasted to the Rio Panuco, where there was a Spanish settlement. The survivors numbered but 311, half their original complement having succumbed to the hardships of the journey. Between the expedition of De Soto and that of Ribaut, next to be mentioned, there were two other Spanish expeditions. In 1547 Balbasteo, and in 1559 Arellano, attempted the conquest of Florida, the latter with 1,500 soldiers and a large complement of friars; but neither arms nor religion could subdue the indomitable natives.

As yet, the Spaniards had not planted a colony in America north of the Gulf of Mexico; indeed, their expeditions had really no such object in view. Whatever may have been the promises, expressed or implied, in their contracts with the Crown, their real object was to scour the country for such gold as the Indians already possessed. This is established beyond a doubt, both from their arrangements with one another, their mode of procedure, the enquiries they made of the Indians, their conduct towards them, and their resolute contempt for agriculture.

As with the Spanish Catholics, so was it with the French Huguenots, who next took up the task of exploring North America. Religion had nothing whatever to do with their want of success. They went to America, not to found a religion, nor to plant colonies, but to find gold. Their leaders are represented as mere adventurers, who cared nothing for religion, except to use it as a cloak.[4] For the most part, they consisted of men "on whom Faith sat but lightly and whose real advantage lay in adventure, commotion and change."[5] Among them were the younger sons and other neglected kinsmen of the Huguenot seigneurs, whose triumph in France, under whatever name, it is contended would have involved the redistribution of all the wealth of that country.[6] There were no stock-herders, no agriculturists, no mechanics among them, except such few of the latter class as were needed to forge their arms, or patch their boots. Says Parkman, 43: "There were no tillers of the soil. Indeed, agricul-

[4] Many of them renounced their faith when it suited their purposes ; amongst them Villegagnon who commanded the Huguenots in Brazil. [5] Parkman, 29.

[6] "The adventurers had found not conquest and gold, but a dull exile in a petty fort. . . . The young nobles, of whom there were many, were volunteers who had paid their own expenses in expectation of a golden harvest, and they chafed in impatience and disgust. . . . The religious element in the colony was evidently subordinate; the adventurers thought more of their fortune than their faith." Parkman, 53, 54, 59, 60.

turists were rare among the Huguenots, for the dull peasants who guided the plough clung with blind tenacity to the ancient faith. Adventurous gentlemen, reckless soldiers, discontented tradesmen, all keen for novelty and heated with dreams of wealth," composed their numbers.

An expedition of this character, under command of Jean Ribaut of Dieppe, sailed from Havre and reached the coast of Florida in 1562. On a beautiful morning in May they landed upon the banks of the St. John's river. The land was fertile, the climate mild, the river teemed with fish, the woods were filled with deer, and the Indians were friendly and hospitable. Did the Huguenots forthwith break the earth and plant seeds? Not at all. Their thought was not of subsistence, but of gold. They asked one Indian whence he had obtained a great pearl that hung suspended from his neck; and of another they enquired for "the wonderful land of Cibola with its Seven Cities and its untold riches," and were facetiously informed that it was distant but twenty days' journey by water. In June, Ribaut returned to France, leaving thirty adventurers behind him, to establish a colony. In the course of a few months their forays and cruelties had alienated the Indians, who left them to starve or return to their own land. They choosed the latter alternative, fed upon human flesh during the voyage, and finally fell prisoners to the English.

In 1564, three vessels and several hundred adventurers from France, under René de Laudonnière, reached Florida. After building a fort, which they named Caroline, at St. John's Bluff, on the river St. John, they set about hunting for gold. By this time the Indians had come to understand the Europeans so well, that, lacking weapons with which to destroy them at once, they employed artifice to wear them out. One chief named Satouriona, secured the alliance of the French against a rival named Outina, by representing that the latter barred the way to great deposits of gold. Arriving in the country of Outina, and learning from Mollua, a subordinate chief of the latter, that Outina's men wore armour of gold and silver plate, and that it was Potanou, a rival of Outina, and not Outina himself, who barred the way to the gold and gems of the Appalachian Mountains, the perfidious Huguenots deserted the cause of Satouriona, and embraced that of Outina. Lured by a promise of the wiley Mollua, of a heap of gold and silver, two feet high, the Frenchmen marched upon Potanou's town and put its whole population to the sword. They found nothing to satisfy their cupidity.[7] Disappointed in their hope of spoil, they

[7] Parkman, 53, 60.

now turned upon their allies, who, escaping into the woods, left them, like their predecessors, to starvation.

In September, 1564, a portion of Laudonnière's command stole his two pinnaces, and set forth on a piratical expedition to the West Indies. They took a small Spanish vessel off the coast of Cuba, but were soon compelled by famine to put into Havana, where, to gain favour with the authorities, they represented Florida to be a land full of gold, and treacherously denounced their comrades, who now formed its only European occupants. Meanwhile, the minority of Laudonnière's command, having sought in vain for the Appalachian gold-fields, mutinied, built two small vessels, abandoned Laudonnière, and started down the coast to plunder a Spanish church and obtain provisions by piracy. They captured and plundered a brigantine and a caravel near Cuba, and looted a village in Jamaica. Subsequently they captured another vessel with rich booty, but were soon afterwards attacked by the Spaniards, and all but twenty-six were killed. These escaped to Florida, where Laudonnière put them to death. While these events were transpiring, one La Roche Ferrière was sent by Laudonnière to spy out the country of the Indios. He pretended to have reached the mysterious mountains of Appalache, from which he brought, among other things, arrows tipped with gold and wedges of a curious green-stone, perhaps jade.[8] Another spy named Grotaut reached the dominions of Hostaqua, a chief who ruled three or four thousand warriors, and who promised with the aid of 100 arquebusiers to conquer all the tribes of the mountains and subject them and their gold-fields to the French. A third offshoot of this party, a man named Pierre Gambie, made his way to the island of Edelano, where he became the favourite of the chief and married his daughter. The Indians, however, afterwards killed him.

During the winter of 1564–5 certain Indians from the vicinity of Cape Canaveral brought in two Spaniards, wrecked in 1550 within the domains of King Calos, near the south-western extremity of the Floridian peninsular. They reported that "in one of his villages was a pit six feet deep and as wide as a hogshead, filled with treasure gathered from Spanish wrecks in adjacent reefs and keys."[9] Although, for a wonder, Laudonnière did not send a force to plunder this treasure, he again listened to the wiles of Outina and his promises of the gold deposites of Appalache, and sent another expedition to raid the villages of Potanou. The arquebuse did its work, producing the usual panic, slaughter, and plentiful harvest of scalps; but no gold

[8] Parkman, 62–8. [9] Ibid, 68–9.

was found, and the thirty Huguenot bravos, who composed the French part of the expedition, returned to the camp in disgust.

Famine reduced them to the necessity of eating snakes, until, on on the 3rd of August, 1565, they were relieved by an English squadron under Captain (afterwards Sir John) Hawkins, who put into the St. John's river for water. This adventurer, whose occupation at that time was capturing negros in Africa and selling them to the Spaniards in Hispaniola, offered the Frenchmen a free passage to France; but this offer, much as it pleased his men, Laudonnière felt constrained to decline. On the 28th of August, the adventurers were again succoured, this time by Ribaut, who had been to France and returned. But their joy at this relief was of short duration, for on the 4th September, 1565, a Spanish fleet, under Pedro Menendez de Avilés, entered the harbour and demanded the surrender of the entire French force.

The Spanish expedition had resulted, in part, from the account given of the riches of Florida by the Huguenots who had escaped to Havana during the previous year. It originally consisted of 34 vessels (one of 966 tons), and of 2,646 persons; and it cost a million of ducats.[10] Of this fleet about one-half reached the St. John's river. After chasing Ribaut's squadron and his succoured Frenchmen out of the harbour, Menendez landed and built a fort, around which afterwards grew the town of St. Augustine. Here he was about to be attacked by Ribaut, but a storm arose and caused the French to withdraw. Menendez then marched overland, captured Fort Caroline, and put its defenders to the sword, only a few escaping to a French vessel commanded by Ribaut's son. Returning to St. Augustine, Menendez learned that all of Ribaut's ships had been wrecked. Following the coast, Menendez came upon the crews of two of their vessels, numbering in all about 500 souls. These, with Ribaut at their head, were nearly all butchered by the Spaniards, Menendez alleging by way of extenuation, that had the French been spared, they would have made their way by the river St. Lawrence to the Mollucas and other parts of the East Indies; and thus robbed Spain of the valuable mines of Zacatecas and St. Martin![11] He seems to have forgotten that those whom he spared, on condition that they became Catholics, might equally well have carried out this fantastic design.[12]

[10] Menendez' dispatch to the King of Spain, cited in Parkman, 135–6.

[11] Those whom Menendez spared seemed to be of the third portion of Ribaut's shipwrecked command. Parkman, 137.

[12] Ibid, 144–6.

Turning from the French, to the Indians, Menendez endeavoured to follow up the scheme of conquest and plunder which had been planned by his predecessors; but the Indians, taught by a bitter experience that the whites were not to be trusted, retired before them wherever they went. "The Spaniards drove them from their cabins, ravished their wives and daughters, and killed their children." * * * "Friendship had changed to aversion, aversion to hatred, hatred to exterminating war. The forest paths were beset, stragglers were cut off, and woe to the Spaniard who should venture after nightfall beyond call of the outposts." Under such circumstances the conquest of North America by the Spaniards and French, like that of South America by the Portuguese, met with little success. Unlike the Indians of the West India Islands, Mexico and Peru, who were hemmed in by the sea, or the mountains, and could not escape from the fire-arms and the bloodhounds of the invaders, the Indians of what are now Brazil and the United States, had a vast open continent behind them, whose trackless forests afforded an ever-ready refuge. The gold-hunters could occupy the coasts of these countries but never the interior, and this fact has much to do not only with the history of the precious metals but with the conquest and civilization of America. Had the interior been accessible to the invaders it can scarcely be doubted—what with their thorough knowledge of mining and the one absorbing object of their numerous expeditions—that the vast subsequent production of gold in Brazil and in the Appalachian range and in California, and of silver in Nevada, Arizona, Utah, Colorado, etc., would in great measure have been thrown upon the world during a single century and long before Portugal and Brazil were subjected to the influence of England, or North America had beeen colonized by Anglo-Saxon communities.

In 1567 a French adventurer, named Dominique de Gourges, sailed from the mouth of the Charente, with 180 men, on a voyage of plunder.[13] He first visited the Rio del Oro, whose name probably indicates the object of his voyage. From thence he went to Cape Blanco, where the Portuguese beat him off. Cape Verde was his next stopping place; then the West India Islands, at many of which he landed in quest of plunder. Touching at Hispaniola, he bore away for Florida, where he landed and made an alliance with the Indian chief Satouriona, against the Spaniards. Attacking the the Spanish forts with his combined force of pirates and savages he took them all by

[13] Some authors say that his object was vengence upon the murderers of the French Huguenots in Florida; but it seems that De Gourges was a Catholic. Parkman, 140.

surprise or assault, put their defenders to death, and sacked the settlements of all they contained. Then impiously thanking the Creator for His part in this bloody work, he returned to France, hoping to gain pardon for his numerous offences by reason of the part he had taken in this last exploit But the king turned his back upon him, and he retired into obscurity, from which he emerged a few years later, only to die of a sudden illness at Tours.

Menendez, who was in Spain when De Gourges captured his forts, afterwards returned to Florida, and re-established the Spanish authority on that peninsular. With certain interruptions this government lasted until 1821. But "Florida" was no longer held as a gold-hunting province. The enmity between the Spaniards and Frenchmen, and the gold forays of both, had taught the Indians that the whites, whom they had at first believed to be demi-gods and Children of the Sun, were only a set of rapacious adventurers better armed than themselves. They had also learned how to avoid them. Left to their own devices, the Spaniards found it necessary to either cultivate the soil, or abandon the settlements. Choosing the former alternative, they gave up their vain efforts to plunder gold from the natives, and established missions and small colonies in the most favoured vales near the coast, beyond the strict precincts of which they seldom strayed. Their principal settlement was at Saint Augustine. The chronicles of this obscure colony during the two-and-a-half centuries which preceded its cession to the United States, furnishes an interesting chapter to the history of civilization.[14]

In 1578, Elizabeth, Queen of England, conferred upon Sir Humphrey Gilbert, the half brother to Sir Walter Raleigh, the right to make discoveries, and search for gold and silver in North America, upon condition of his paying to the Crown one-fifth of all of such metals as he might obtain. In 1583, Sir Humphrey sailed from England with five vessels and 260 men and landed in Nova Scotia. The ancient cavities on the seashore near Lunenberg, called the "Ovens," are believed to have been dug by these adventurers in search of gold;

[14] The following strange paragraph appeared in a Peoria (Illinois) newspaper, dated January 11, 1897: "The hull of an old Spanish gunboat has been uncovered on the farm of Chas. Brown, in Renville County, Minnesota, and hundreds of people have been flocking to the place. Joseph Bagne struck the vessel while digging a well, and has since completely uncovered it. Its armament comprised five cannon and two mortars. Cannon balls and bomb shells were found in large numbers. The impression is that the boat was run up into that region about the year 1600, when a much larger proportion of the country was under water than now. The gunboat was found directly on Birch Coolie Creek, a branch of the Minnesota River, which creek at that time was no doubt a navigable stream." Upon making enquiries in the vicinity of this alleged find, the author ascertained that it was a mere hoax.

with what results are not recorded. Mr. Wm. A'Beckett, in his Life of this hero, says: "He had carried out with him a Saxon miner, who pretended to have discovered a very rich silver mine on the coast and dug up some ore, which seems fully to have convinced Sir Humphrey that the means of wealth were within his reach. Misfortune, however, was impending over him. His largest ship was lost in a storm with all the crew, except twelve and his miner, and the ore perished with her." In their subsequent expeditions to North America, the English, profiting by the losses which it became evident Spain had sustained in her mad pursuit of the precious metals, turned their energies to the establishment of agricultural colonies and the monopolization of their trade. Says "The Constitution and Present State of Great Britain," London (anonymous), 1758, pp. 233–5: "While trade is thus carried on between Great Britain and her colonies, both parties must grow powerful; but should a mine of gold be found in such colonies, however desirable that metal may be, I should dread the consequences. . . . Spain, for instance, though she has possessed herself of the silver mines of Mexico and Peru, is evidently a loser by the bargain; for these provinces have drained the Mother Country of her children and left her plains uncultivated and her vines unpruned."

In 1606 Captain John Smith, adventurer, mercenary, and gold-hunter, was employed by a London Company which had obtained a charter to colonize Virginia and search it for gold. Smith, having been placed in command of the colony, failed of his mission, while his employers reprimanded him for not realizing their expectations of a golden harvest. The only outcome of this expedition contemplated by its projectors was the discovery of some gold in the Appalachian range. The hostility of the Indians, however, barred the way to the placers and these were left to be discovered and worked at a later date. Yet the enterprise was not wholly without fruit; for it gave rise to the beautiful romance of Pocahontas.

CHAPTER XXIV.

THE PLUNDER OF AFRICA.

Remote antiquity of the gold trade with Cephala or Sofala—Early voyages—Punic, Arabian, Greek and Roman navigators—Repeated openings of the Suez Canal connected with the gold trade—Necho—Xerxes—Voyages of Hanno, Eudoxus, Agatharcides and Arrian—Land expeditions of the Romans under Augustus, Nero and Antoninus Pius—Arabian expeditions to interior Africa—Their influence upon the natives, whom they refrained from enslaving—Gold trade of the Soudan—Mines of Guangaru—Kano—Timbuctoo—Ghanah—Bornu—Mines of Oungara—Congo "Free" State—Guinea—Gold Coast—Voyages of the Norsemen, Portuguese, Dutch and English—El Mina—Buncatee's Golden Stool—The British African Company—Ashanti—Gold product of the West Coast—Native aversion to gold mining—Voyage of De Gama—The Zambesi gold region, or Sofala—Mozambique—Melinda—The Arabians are driven away by the Portuguese, who seize upon the mines—Extraordinary preparations to exploit them—Operations of Alboquerque in the Red Sea—Barreto's Expedition to the Zambesi—Ancient towers of Zamboe—"King Solomon's Mines"—Produce of Sofala—Mining districts—Ruin of Portuguese Power—Recent discovery of gold in the Transvaal—Produce of the mines—Practical enslavement of the natives by the English mine-owners—Sacreligious coins—Desecration of the Sabbath—Boer discouragement of mining—The Boer War and Closure of the mines,

AFTER closely examining the narratives of Nearchus, Agatharcides and Arrian, concerning the ancient trade down the coasts of the Red Sea and Eastern Africa, Dr. William Vincent, Dean of Westminster, was of opinion that the Arabs, whom some of the Greek writers called Phœnicians, had been from time immemorial in possession of the entire coast down to the "Southern Horn," and that they carried on a traffic in gold-dust and ivory with Rhapta or Quiloa, Melinda, Mozambique and the great gold region of Cephala, or Sofala. An opinion from so eminent a source, and expressed with the deliberation evinced in his work on the Peripli of the ancient navigators, is entitled to the highest consideration. It also suggests that Ctesias, when he alluded to "one hundred and twenty thousand Cynocephalians" meant the population of the Arabian emporia of the south-eastern coasts, and not "dog-headed monkeys," as too literal an interpretation of the Greek term led some writers to suppose; for how could anyone pretend to number the monkeys in a wild and remote region! Be this as it may, the periods at which the voyages of the Greek navigators were made, too closely coincide with the various openings of the Suez Canal, to leave any doubt that they were connected; moreover, that such openings would hardly have been made at all had there not been a large and profitable trade with the eastern coasts of Africa to support then; for, except as to the single voyage of Eudoxus, neither the Greeks nor Romans are known to have ever attempted to trade directly with India or China. This trade was in fact always in the hands of the Phœnicians or Arabs; nor could it have been conducted by any nation not having control of the Arabian coasts and emporia. The trade with the eastern coasts of Africa could alone have been sufficient to warrant the opening, and the great expense of keeping open, from the drifting sands of Egypt, the water-way known to us as the Suez Canal; and such trade with the eastern coasts must have included the gold regions of Cephala; for without that it would have been too trivial in extent to warrant so mighty an undertaking. From the Gulf of Aden to Rhapta or Quiloa the coast was poor and offered but little inducement to trade; but at Melinda, Mozambique and Sofala, it was rich and lucrative. At a later æra the Arab dwellings at these places were made of hewn stone, their gardens filled with rare Indian plants, and their fields yielded a bounteous harvest of corn and fruits, while their pastures abounded with great numbers of horned cattle and other domestic animals. They traded with Aden and Ormuz, with Cambaye and with Calicut. The friendly natives of Africa brought to their marts abundant supplies of gold-dust which they gladly exchanged for Levantine implements, wares and ornaments. The iron works of Melinda and the gold mines of Cephala were at once the monuments of their art and the sources of their opulence. Such must also have been the condition of this trade when the Phœnicians conducted it; a trade too great to be wholly concealed and yet too valuable to be made public. Both the Phœnicians and Arabs kept the knowledge of it to themselves; even the channels by which it was reached were clothed in mystery and verbal perversion. Cephala was a sealed land, and the seal was only broken when Alexander destroyed Tyre and transferred the trade of Arabia and the eastern coasts of Africa to that New Emporium upon which he conferred his own name. It was broken again when De Gama rounded the Cape.

We have now to follow in detail such accounts of the Greek and

Roman expeditions into the gold regions of Africa as time and a jealous circumspection have permitted to reach us.

About the year B. C. 610, Necho, King of Egypt, re-opened the ancient Suez Canal of Rameses II., and admitted a Phœnician fleet from the Mediterranean to the Red Sea, bound probably for the gold regions of south-eastern Africa. Three years afterwards this fleet, or a portion of it, returned to Egypt by the Straits of Gibraltar, having meanwhile circumnavigated the entire continent of Africa. The Phœnicians said nothing about the gold regions, but reported that, having made a settlement on the coast, they ran short of provisions and were obliged to plant seed in order to secure a supply of corn. They added that this planting had to be done in autumn, a fact peculiar to the Southern hemisphere, where the seasons are the reverse of those in the Northern; and that as they afterwards proceeded on their voyage, they had seen the sun on their right hand, another fact peculiar to the course they claimed to have run, and which, together with the one previously mentioned, stamps this narrative with the strongest probability. Herodotus, Mel., 42, and Commentary of Dr. Larcher.

The grounds upon which this voyage have been discredited are unsound. These are "the successful agriculture of the strangers in unknown climates and new soils"; the improbability that "seeds of the temperate zone yield their increase between the tropics," *i. e.* between Northern Egypt and the Mozambique; and the improbabiity of "sailing round Africa in three summers to amuse the curiosity of a King of Egypt." Gibbon's Treatise on the Meridional Line. To these strictures it is to be replied that: considering the numbers of the fleet, the fact that it was conducted by Phœnicians, who were essentially a mining people and had already abundance of mining experience on the coasts of Greece, Italy and Spain, and their admission through so costly a channel as the canal of Suez, it is unlikely that the expedition was organized merely for purposes of curiosity; that a knowledge of the gold regions of south-eastern Africa could hardly fail to have reached Egypt overland, as, at a subsequent date, that of Peru reached Pizarro in Darien, and that the Phœnicians had doubtless been sent to work them; that, as Gibbon himself affirms in another part of the same Essay concerning Zanguebar, and as a subsequent experience overwhelmingly testifies, corn will and does grow upon immense regions "between the tropics." Says Macgreggor, III, 332: "Mozambique and Sofala (Cephala) have excellent soils which produce (where cultivated) indigo, sugar-cane, rice, wheat, potatoes, beans, maize, all kinds of fruits; and in a wild state, oranges, lemons, oleaginous plants and trees, the vine, mulberry and olive," etc. The experience of the Boers in the Transvaal and Orange River Republics abundantly confirms this opinion.

If the Phœnicians ever preserved a more minute account of this celebrated voyage, that account must have perished when all their literature was destroyed with the cities of Tyre and Sidon, by Alexander, in the 4th century B. C. But it is not likely that any record was kept of it, because they made it a practice, from motives of policy, to conceal the destination of their fleets and the sources of their commercial wealth. The Phœnicians have left us no records of their voyages to the coasts of Southern and Western Europe, and yet we admit upon precisely the same testimony as is herein adduced and offered, that between the 16th and 10th centuries B. C., they discovered, colonized and worked mines of gold and other metals all along the Mediterranean and Atlantic, from Thrace to Britain.

As going to show that the early maritime voyages were made rather for gold than motives of geographical discovery, and that where the former was not to be obtained, the latter was always neglected or abandoned, it is to be noticed that from the period of this Phœnician voyage to that of Vasco de Gama, the only portion of the African coast never visited was that which was non-auriferous, namely, from Guinea on the West around the Cape to Caffraria on the East. It can scarcely be doubted that if the gold placers had extended south of either Guinea on the West or Sofala on the East coasts, they would have been followed and colonized by the nation who first traced them from the North to those limits.

Some corroboration of the Phœnician voyages to Caffraria is found in the descriptions and representations of the unicorn (the one-horned rhinoceros or ndzoo-dzoo of Caffraria), which prevailed among the early Hebrews and Persians. The rocks of Cambebo and Bambo, in Caffraria, were also found covered with ancient representations of these animals.[1] Further corroboration of these voyages is found in the extensive ruins covered with inscriptions in an unknown language, which were discovered in Batua[2]; and in the narrative of Thomas Lopez, during his voyage to India, which records that the inhabitants of Sofala boasted of the possession of certain ancient books[3]—a significant circumstance when it is remembered that the art of writing, which was brought into Europe by the Phœnicians, was entirely unknown to the native African races.

[1] Malte-Brun, III, 70. [2] Ibid., III, 81; Morse's Geog., II, 537. [3] Morse, II, 537.

While it must be admitted that these evidences possess some value, we do not as yet insist upon their being regarded as conclusive. But there are still other evidences on the subject. The Caffres have a tradition that their ancestors came from some eastern country. The Batuans, a neighbouring nation, whose capital was Lattakoo, were acquainted with metallurgy. They could smelt copper and iron, and make brass-wire, steel, knives, needles, earrings and bracelets. They were also familiar with the art of tanning leather.[4] Their complexion is not black, but brown; their hair is straight, their manners are mild, and their language sweet and sonorous.[5] Above all, the gold-mines themselves exhibit evidences in every direction of having been systematically worked by some ancient people who had many more uses and a far more eager desire for gold than any of the native races.[6]

About B. C. 482, Sataspes, a Persian noble, who had been banished from the court of his kinsman Xerxes, B. C. 485–65, organized in Egypt, which at that period was a Persian province, a fleet of Phœnician ships, to circumnavigate Africa from West to East. This fleet passed through the Straits of Gibraltar and sailed down the African coast as far as Cape Bogador, where, finding no gold, and meeting with vast masses of floating seaweed, which threatened to interrupt the further progress of the vessels, Sataspes turned back and eventually reached Persia, where he was crucified, probably for failure to reach the gold regions. Like Lander, the valet of Mungo Park, who, after the latter's death, returned to Africa, and with slender means accomplished the very same voyage which his master had died in attempting, a priest of Sataspes' expedition claimed to have afterwards successfully circumnavigated the African Continent.

During the fourth century B. C., a Carthaginian expedition, under Hanno, sailed to the Western coast of Africa, made the river Senegal, the coast of Guinea, the volcanoes of Sierra Leone and the Cape of Tres Puntos, lat. 5 N., from which point it turned back.[7] At a subsequent date Hanno sailed down the entire African coast and

[4] McCulloch, Geog. Dic., I, 41.

[5] Malte-Brun, III, 75, 76, 77. See Dr. Bleeck and Custs' *Modern Languages of Africa* on the philological importance of the South African languages.

[6] "The negros are fully aware of the value of gold, but seldom search for it." Dr. Livingstone in Lock, 10. At Zanzibar, in 1799, the negros preferred brass buttons to guineas. Macgreggor, II, 339. Cameron states that in Katanga "the natives do not value gold; it is too soft." Lock, 25. In Coomassie, before the gold-hunters came, the natives patched their broken crockery with gold rivets. Lock, 30. In Madagascar "silver is, in general, preferred to gold." Macgreggor, II, 343. Batua is rich in gold and silver, but the natives appear to prefer mining for iron. Malte-Brun, III, 81.

[7] Periplus of Hanno, in Gibbon's Essay on the Meridonal Line.

finally made his way to the Red Sea.[8] It was at this period that the Suez Canal was re-opened for at least the fourth time.[9]

In the reign of Ptolemy Philopata, B. C. 177, Agatharcides sailed down the east coast of Africa, concerning whose trade he has left a scant account, in his De Mari Rubro, noting, among other matters, that the Sabæi valued silver ten times more than gold. This could not possibly have related to the Sabæans of Arabia, because that people had for ages traded between Arabia and India and well knew the relative value of the precious metals in both countries. It could only have related to some coast tribe of Africa, who were willing to pay ten weights of gold-dust, the produce of their own country, for one weight of silver coins, the produce of Egypt or of Rome; coins, the like of which they could not themselves fabricate, and which were valued rather for ornamental purposes than as money.

In the reign of Ptolemy Lathyrus, B. C. 117-81, Eudoxus of Cyzicus, a Greek navigator, made a voyage down the eastern coast of Africa, probably as far as the emporium of Rhapta or Quiloa, alluded to by Arrian, and situated in lat. 10 S., and thence across the sea to India, where he exchanged his cargo for the precious stones and spices of the Orient, and returned to Egypt. At a subsequent date he made a trading voyage down the western coast of Africa until he heard spoken the same native languages that he had heard on the eastern coast. This would carry him down to the Southern tropic and within hearing of the gold regions of Sofala. "Long before this, Caelius Antipater informs us that he had seen a person who had sailed from Spain to Ethiopia for the purposes of trade." Pliny, II, 67. The Eudoxus here mentioned as of Cyzicus, may have been the celebrated Eudoxus of Cnidos, who was a pupil of Plato, and flourished about B. C. 366, the confusion of dates arising from the alteration of 108 years in the calendar by Augustus Cæsar. Del Mar's "Worship of Augustus," p. 37.

Returning to Egypt, probably to urge the government to sustain the cost of an expedition to these regions and thus reap for itself great riches, Eudoxus fell under the displeasure of the court, and flying from its vengeance, yet determined to pursue the voyage he

[8] Pliny, II, 67; VI. 36.

[9] The Suez Canal was in existence during the reign of Rameses II., B. C. 1311, as the ruins on its banks attest. Rawlinson's Herodotus, *n.* to II, 58. This monarch is supposed to be the same with Sesostris, mentioned by Aristotle, Pliny and Strabo. The Canal was re-opened by Necho, B.C. 610. Herod., II, 158. Re-opened by Darius, B. C. 521-485. Herod., II, 158. Re-opened by the Ptolemies; by the Romans under Julius Cæsar, Augustus, and Trajan, or Hadrian; by the Arabs, A. D. 639; and lastly, by De Lesseps, A. D. 1869.

had planned, he embarked on the Red Sea, followed the entire coast down to the Cape, doubled that point, sailed up the western coast, and eventually arrived at Cadiz in Spain.[10]

During the Roman æra in Egypt, the Phœnician (Arabian[11]) navigators and merchants maintained an active and extensive trade for gold, slaves and ivory, with the eastern coasts of Africa, as far south as Cape Corrientes. Pliny possibly alludes to this trade where he regrets the efflux of the precious metals from the Roman Empire to India, because his words leave it to be inferred that the exports were from "our peninsular and Empire to India and Seres," the peninsular meaning Arabia, which, according to the Roman view, included the eastern coast of Africa. Pliny, XII, 41 (18). Arrian, who wrote about A. D. 140,[12] describes the coasts and the various emporia of the Arabians down to Rhapta, and concludes by saying that somewhere beyond that point "the ocean sweeps round to the westward, where it communicates with the great Western Ocean"—the Atlantic: a statement for which he could only have obtained the necessary basis from the Greek and Arabian navigators trading down to and past the gold regions of Sofala.[13]

McCulloch, in an able essay on Africa, says: "There is little room to hesitate in fixing Rhapta at Quiloa. . . . To the coast described in the Periplus of Arrian, Ptolemy adds an additional range stretching south-east from Rhapta to another promontory and port, called Prasum; considerably south-east from which lay a large island, Menuthias, evidently Madagascar. According to Gosselin, Prasum is Brava, while Dr. Vincent makes it Mozambique." With all this evidence before us, it is difficult to refrain from the conviction that during the Empire, to say nothing of still earlier periods, the Arabians traded with the gold regions of Sofala, and that some knowledge of this trade found its way into Roman geographical works.

The Greeks derived, or pretended to derive, the term cynocephalus from *kunos*, dog, and *cephale*, head, and they applied it to dog-headed monkeys; but this may be a mere piece of verbalism invented to account for the name of a remote people of Africa, who were mentioned by Æschylus, Herodotus, Ctesias, and other early writers.

[10] Cornelius Nepos, in Pliny, II, 67; Posidonius in Strabo, II, 155-160.

[11] Both Herodotus and Pliny in some places allude to the Phœnicians as "Arabians." Pliny, lib. XII.

[12] Dean Vincent assigns Arrian's Periplus to about the year A. D. 63.

[13] That the Arabians traded with the African coast down to and past Sofala, to the "Southern Horn," was the opinion of Dean Vincent. Consult his "Voyage of Nearchus and Periplus of the Erythrêan Sea," Oxford, 1809, 4to, pp. 45, 281.

Sofala is mentioned as a gold region by Edrisi, an Arabian author of the 12th century. "Omnium præstantissimum aurum in universa Sofala regione ibi repereri." Hartman's Latin Translation. Coviglium, a Portuguese commander, who voyaged from the Red Sea to Sofala about 1485, and Ulysses Aldrovandi, 1522–1607, who cites Johannes Hugo concerning the gold of Sofala, Mozambique and Melinda, both spell the name as Sofala; while Boisard, a French writer of the 17th century, and others, spell it Cephala, and describe it as a rich gold country of Africa, next to Monomotapa.

Of the Roman land expeditions into Africa we have several brief accounts, neither of which, however, contribute any essential information concerning the history of the precious metals.

It having been reported to Augustus that "Arabia" was rich in gold (eastern Egypt was then called Arabia), he equipped an expedition of ten thousand men and dispatched them to that distant region by way of Alexandria. They were commanded by Ælius Gallus, and included one thousand troops contributed by Obodas, King of the Nabatæ, and commanded by his lieutenant Syllæus, who undertook to act as guide to the expedition. It is also said that Herod, King of Judea, contributed five hundred of his body-guard to these forces. Be this as it may, the expedition began a six months' march from Egypt toward the sources of the Nile. It is only recorded that it entirely perished from hunger and disease. Nabathæa is described by the Rev. Dr. Lemprière as "a country of Arabia of which the capital was Petra, . . . and seems to be derived from Nabath, the son of Ishmael." This pious desire to connect the name with the son of Ishmael seems to have induced this author to remove the Nabathæ, who were none other than the "Nobatæ, or people of Nubia" (Gibbon, I, 440), from one side of the Red Sea to the other. Strabo, who accompanied this expedition to Ethiopia, has much to say of Egypt and Nubia, but scarcely anything of Arabia. According to Kramer this portion of his geography is strangely corrupted. Mire corrupta est hæc ultima libre pars; probably in order to drag in the name of "King Aretas."

A similar expedition, equipped by Nero, is mentioned by Pliny, Seneca and Dion Cassius. The first says it consisted of the prætorian troops; the last, that it only comprised two centuries, and that its object was to explore the sources of the Nile. It appears to have penetrated to Dongola, about a thousand miles beyond Syene; but without any results that seem to have been worth recording. A Roman expedition into Africa, under the Emperor Antoninus Pius, A. D.

138–61, is mentioned by Pausanias, in *Arcadics*, XLIII, who says that the Romans drove the Moors "entirely out of their country and compelled them to fly to the extremities of Libya, to the mountain Atlas, and to the people who dwell near Atlas." In the seventh century, when the Arabians advanced westward from Egypt to Morocco, they found the Romans of Byzantium strongly entrenched along the northern coasts, as well as in the possession of some of the interior parts of Africa. "The Byzantine legate reigned at Carthage; while Gregory, the patrician, governed at Sufetula." Del Mar's "Hist. Monetary Systems," chap. v. Indeed, Roman remains have been found of late years in the extreme parts of the Desert of Sahara; but whatever the objects of these expeditions and garrisons, whether conquest, plunder, geographical discovery, or trade, they all failed. Roman Africa was practically confined to the navigable parts of Egypt, Carthage and Mauritania; it never extended to the gold regions, either of Sofala on the south-eastern or of Guinea on the western coast.

Our knowledge of Africa, beyond the Roman or Byzantine pale, substantially begins with the explorations of the Moslem Arabs during the Middle Ages; yet so far as the precious metals are concerned it amounts to very little; for although the Arabs penetrated the Dark Continent in every direction—a fact attested by the presence of the Moslem religion or customs, among the remotest negro tribes—they always kept secret their knowledge of the districts where gold was to be obtained by mining, and they often even succeeded in concealing the channels of trade through which flowed the product of the mines. Captain Soule, whom the writer met at Sir Montague Nelson's mansion at Ealing, in 1886, and who had lived three years in the Congo country and seven years in the Ashanti country, besides many other parts of Africa, stated that traces of Arabian mining works had been encountered all the way between Zanguebar and Sofala on the east coast and across Central Africa to Morocco on the west coast. Cameron and Burton made similar statements. The writer himself has seen Arabian mining works in Senegambia. Notwithstanding the mystery which hung over interior Africa and rendered it to Europeans an unknown land, until it was explored by Livingstone, Stanley, Cameron, Strologo, Soule, Burton, and others, there are evidences that the Arabs had previously gone over the entire continent and skimmed it of all its easily-found gold. The rapid commercial development of the empire which the Moslems founded at Baghdad during the Middle Ages, though it began with plunder, was evidently sustained by mining and trading for gold. Their plunder of the precious metals was obtained chiefly in the Byzantine provinces of Asia and in Persia; their mining produce came for the most part from Africa. They began with re-working the Egyptian and Nubian mines, but they soon learnt that it was far more profitable to purchase gold than to dig for it.

Here it is to be observed that, unlike any of the gold-seeking nations who preceded or followed them, the Moslems refrained from procuring gold by slavery. No Moslem could be a slave. To accept the Moslem faith was to secure freedom; and this important fact not only explains the avidity with which Islam was embraced by the Roman, Persian, and Indian provinces; it also explains the conversion of Africa to the faith of the Moslems. Their traders knew the east coast down to Gaza, Swazi and Matabele Lands; they knew Madagascar; they knew the Soudan as far west as the Senegal and the Gambia; and they probably also knew Uganda and the Lake country. Their caravans from the Nile met those from Morocco; and in the great marts of Timbuctoo and Mogador, the salt of Senegambia, and the silks of Grenada were eagerly exchanged for the gold-dust which the dusky natives had learnt both to win from the earth and to despise. (Cada Mosto.) With the conquest of Spain by the Christians, and the breaking up of the Arabian Empire in Europe, this trade declined; and the knowledge of its sources died out.

The gold trade of the Soudan is described by the Arabian travellers Edrisi, Abulfeda, Ibn-al-Wardi, and Ibn-Batuta, the last named of whom has left us the most copious account. Ibn-Batuta left Segelmesa in 1353, and in a month's time reached Walet. Thence he travelled to Melli, in Eioucaten (it sounds like Yucatan), and afterwards to Timbuctoo. The numerous evidences of a traffic in gold which he observed upon the way put him upon the lookout for mines; but the only ones he found where work was being prosecuted were iron mines. The gold ones were evidently concealed from strangers. The travellers who next described this region were Hassan-ben-Mahomet, Cada Mosto, 1454, and Leo Africanus, 1511. Leo mentions the celebrated gold mines of Guangaru; yet he neglects to sufficiently describe their character or extent, or the manner of working them. Some gold ornaments which the King of Timbuctoo obtained from this region weighed as much as 1,300 ounces; but nuggets of this size, though uncommon, afford no criterion of the richness of a mining district. The fact that the mosque and palace of Timbuctoo were built by architects from Grenada in A. D. 1215, justly appeared to Leo of more interest than enquiring into mines. The people of

Guinea were the first to embrace Islam in Western Africa, and the seat of Mahometan power was originally at Gualata (Oualet) north of Guinea; a fact which indicates that Mahometan civilization was brought into this country from Spain or Morocco, rather than from Egypt. Upon the conquest of Guinea by Izchia, King of Timbuctoo, the seat of power was removed to that city, and Gualata lost its preeminence.

The principal Arabian kingdom in Central Africa was Ghanah, whose chief city was Kano. The Arabian travellers dwelt upon the splendour of its court, its throne of massive gold and other wonders. The next Arabian kingdom was Tocrur or Takrour, close to Sokato. Konkou or Bornu, celebrated for its manufactures, was the next kingdom to the westward. South of Ghanah was Oungara, where the richest gold placers were situated. At a later period, Belek-el-Soudan (Lordship of the Soudan) appears to have embraced all these kingdoms, extending from Darfur to Senegambia; but at the present time they are again broken up into separate principalities, or else absorbed into the newly-erected European empire facetiously styled the Congo Free State, where freedom appears to consist of enslaving and whipping men and women in order to compel them to produce gold, india-rubber, and ivory. (See Appendix to this Chapter.)

Gold was obtained, and is still obtained to a small extent, throughout all Equatorial Africa, but we have no quantitative accounts. This much, however, must again be remarked of this region: that so long as it remained under Moslem control or influence we hear of no cruelties, no sacrifice of native life; in short, no such tragedies as distinguished the career of the Spaniards and Portuguese in the East and West Indies.

Guinea, or the Gold Coast, which includes Ashanti, was visited for gold by the Spanish Arabs during the Middle Ages, and by the Norsemen from the coast of France so early as 1382. In 1442, Antonio Gonzales, returning from a voyage beyond Cape Bogador, brought the first gold and the first slaves from Senegambia to Portugal. Some of the latter were negros, whilst others were of a lighter colour, evidently half-breed Arabs. (Helps, I, 37, 52.) Of this spoil, both gold and slaves, the good Prince Henry reserved the Quinto, that is to say, a Fifth, for his own share. The success of the expedition led in 1444 to the formation of a Portuguese Company to capture gold and slaves. In 1469 the trade was farmed out to one Fernando Gomez for five years, at one thousand ducats a year, and it was this speculator, who, in 1471, by one of his ship-captains, Pedro de Escobar, "discovered" the Arabian occupation of the Gold Coast and built the fort of El Mina, a name which was also granted to the "discoverer," in a patent of nobility, which he paid for, with the gold obtained from this place. The "discovery" was followed by a "grant" from the Pope of Rome. Leo Africanus, a geographer of this period, stated that the King of Ghanah had a gold nugget weighing 30 pounds, which was bored through and fitted for a seat. In later times the King of Buncatoo had a solid gold stool, which proved to be the cause of his destruction. And in our own times King Prempeh, of the same region, had, or was reputed to have had, a stool of the same auriferous material, which, like Miss Kilmansegg's Golden Leg, became the means of his downfall; for it led to the invasion and conquest of Ashanti-land by the English. Yet the entire auriferous results of this war (1896), including the coveted stool, which was partly the occasion of it, did not exceed £2,000 in value.

In 1479 the rival claims of Spain and Portugal to lands which neither of them had discovered, conquered or colonized, were reconciled by a division of the spoil; Spain retaining the Canary Islands and Portugal the Coast of Guinea. In 1553 an English adventurer made his appearance on the coast, hungering for gold, ivory and slaves; but it was not until nine years afterwards that this trade was permanently established by another adventurer, John Hawkins. The gold and ivory trade soon dwindled away, but the slave trade to America, which began here, speedily grew to such vast proportions, that its profit to the British Government far exceeded the £50,000 a year, for which Queen Elizabeth pawned the Customs Revenues to Sir Thomas Smith. (Anderson's Hist. Com., II, 175.) For the honour and advantage thus secured for his native country, John Hawkins was advanced to knighthood, and returned to Parliament as Sir John. Meanwhile, the unhappy victims of his cupidity were crowded into the holds of slave-ships and hurried off to perish in the gold mines of San Domingo, there to replace the natives whom the Spaniards had already exterminated. In 1561 the Norsemen again made their dread appearance on the coast of Guinea, whereat the English became so fearful of losing their infamous traffic in mining slaves, that in 1572 they made an "arrangement" with Portugal, by which that country conceded to them, alone of all foreigners, the "freedom of trade" in negros.

This monopoly of the privilege to capture slaves was renewed by the treaty of 1642. It was in this same year that the Dutch commenced to compete in the gold and slave trade of Guinea, to forward which purpose they captured from the Portuguese several forts on the Senegal and Ashanti Coasts, including Acheen, or Axim, and El Mina; and commenced trading with the natives for gold-dust and slaves. In 1652 the Swedes also took a hand in this commerce; but their vessels being seized by the English, together with their lading of gold, they soon afterwards abandoned the trade. (Anderson, II, 117, 137, 395, 421.)

In 1664, "without a formal declaration of war, until some months afterwards, King Charles II., of England, made war upon the Netherlands," by treacherously seizing several of the Dutch forts on the coast of Guinea. The agent of this breach of faith was Admiral Sir Robert Holmes, while its beneficiary, according to Anderson, 480–81, was "the English African Company, at the head of which was the Duke of York." In 1665 the Dutch, under Admiral de Ruyter, retook all these forts, as well as "our own fort of Cormanteen, which they hold to this day by the name of Fort Amsterdam." The company mentioned by Anderson was formed in England to trade on the African coast for gold, ivory and slaves; and some of the large capitals, which since that time have been employed in England, are said to have had their origin from this source. In 1750, the company (indeed, there were several successive companies), was succeeded by the African Company of Merchants, incorporated under an act of Parliament, with power to erect forts and form settlements. These forts were held, "not as a territorial right, but at a rent from native governments"; that is to say, from the King of Fantee. (McCulloch.) In 1807, after the Ashanti conquest of Fantee, the rent was claimed by and paid to the King of Ashanti. This arrangement subsequently gave rise to difficulties, and negotiations; and in 1820, to a treaty, which, however, was not ratified by the Crown. In 1821, the entire property of the African Company was placed under the Crown. Upon the death of King Sai Quamina, in 1823, war was declared upon the English, who were charged with deceit and faithlessness. In 1824 Sir C. M'Carthy was defeated by the natives, but in 1826 the Ashantis were defeated in turn, and compelled to pay six thousand ounces of gold as indemnity. In 1874, the Gold Coast was erected into a British colony. The recent Ashanti expeditions have practically reduced the entire coast, from Liberia to the Niger, to a British possession.

"Gold is more abundant in Ashanti than in any other part of Africa, probably than in any other part of the world, not excepting even South America," wrote McCulloch in 1838. "The pits and washings in Tokoo alone, are reported to yield sometimes as much as 2,000 ounces per month. In 1726 Bosman gave the average annual exports of Ashanti at 7,000 French marks weight, or 4,590 pounds Troy, per annum, say, £250,000 sterling. (McCulloch, Com. Dic.) Captain (afterwards Sir Richard) Burton increased this estimate to £350,000; but the Captain, with whom the writer was well acquainted, was, like most prospectors, of a sanguine temperament, and in respect of gold mines, a confirmed optimist. His estimate must therefore be received with caution. The following letter on the subject of gold mining in Africa appeared in the San Francisco Daily Exchange of June 7th, 1880:

"An esteemed friend in England has lately sent me a marked copy of the African Times for April, 1880, a monthly newspaper relating to the West Coast of Africa, and published in London. The marked article relates to the gold mines and gold product of West Africa; and as the subject will doubtless interest your readers, I send you herewith a digest of the news.

Rich gold mines have recently been discovered near Tacquah, in Wassaw, which latter is in Ashanti-land, a portion of Guinea. The first information of the strike was conveyed from Cape Coast Castle to London by the steamer Biafra, on March 16th, 1880. Fifteen days later the steamer Benguela, from Axim, the seaport of Wassaw, brought confirmation of the news. Four mines have been located as follows: 1. The African Gold Coast Company, Limited. This Company was organized in Paris, and is known as "The French Company." It has 1,000 tons on the dump, worth £20 per ton. Stamp-mill ordered and expected out by May 6th. 2. The Effuenta Gold Mines Company, Limited, has a European staff at Effuenta and sixty native labourers; adjoins the south-west extremity of the French Company's claim. Tunnel commenced to intercept vein. 3. W. & A. Swanzy's mines. Have several shafts down. About March 15th they took out of a newly-sunk shaft, at a depth of 80 feet, a piece of quartz weighing 35 pounds, "one-quarter part of which was pure gold." Sent to England by the Biafra. 4. The Gold Coast Mining Company, Limited. The latest one organized. Will work the Abbontuyakoon property, adjoining the north-east extremity of the French Company's lands. Have a shaft 70 feet deep, from which rich ore has been extracted. A narrow-gauge railway from Chamah to Wassaw, 50 miles, is projected, and the British Government is being solicited to assist in its establishment.

So much for the news. I can see no good reason to doubt its substantial authenticity. That gold has long been obtained from this coast is a well-known fact. The coast of Guinea is believed to have been visited for gold by the Moors and afterwards by the Normans in the 14th century. It is certain that the Portuguese explored it in 1471, and called it the Gold Coast, while they called its port El Mina (The Mine.) Afterward this port was occupied by the Dutch West India Co., and known as Delmina. Bosman, Father Labat, and other African travellers have testified to the wealth of its placers.

The following table is from Bosman, pp. 61–97; the equivalents in Troy pounds weight and American dollars, having been added by myself:

Average Annual Export of Gold Dust from the Coast of Guinea during the early part of the eighteenth century.

Shippers.	Marks Weight.	Pounds Troy.	Value.
Dutch West Indian Co. . . .	1500	983.7	$243,997
English-African Co.	1200	786.9	195,198
Dutch Private Traders, . . .	1500	983.7	243,997
English Private Traders, . .	1000	655.8	162,665
Prussians and Danes, ...	1000	655.8	162,665
Portuguese and French, . . .	800	524.6	130,132
Total and Fractions, . .	7000	4590.7	$1,138,654

During the eleven years from 1808 to 1818 inclusive, the amount of gold-dust shipped from the West Coast of Africa by the English-African Company in men-of-war, was 81,905 ounces, equal to $692,-976, an average of $153,907 per annum. From 1819 to 1857 inclusive, the average annual shipments of gold from the British possessions on the West Coast were about $150,000. From 1858 to 1875 inclusive, the average annual shipments from the West Coast possessions (British and other, but not including French), were $118,000. From 1876 to 1895 inclusive, the average annual shipments from the British possessions on the West Coast fell to about $100,000. Since the conquest of Ashanti-land by the British forces in 1896, the negros have been replaced at the gold-mines and the output has increased.

The above is all the specific data I can find in the various works devoted to this subject. This data is no warrant for believing that the gold exports of the West Coast of Africa were ever important enough to excite attention. Mr. Jacob, in his "History of the Precious Metals," chapter XXII, says: "We are disposed to estimate at a very low rate the whole produce of gold from Africa, and in our estimate of the production of the world at large, we have not thought it necessary to take notice of it." Other recent writers on the subject have pursued a similar course. No estimate of the gold product of the world ever makes any allowance for the product of Guinea. It is too small.

Such being the case, what are we to think of the leading editorial in the African Times above alluded to, in which it is claimed that the shipment of gold by Europeans from the West coast of Africa began in 1471, and has continued ever since at a decreasing rate, which has lately sunk to half a million pounds sterling (say $2,500,-000) per annum; and that to estimate the entire shipments at one billion pounds sterling (say five billion dollars), is quite within the mark! Five thousand million dollars is four times as much as California ever produced; more than twice as much as California and Australia combined, and very nearly as much as all the gold coin and bullion now existing in Europe and America combined. In view of these facts, I am compelled to conclude that the West African editor is somewhat disposed to draw the long bow, and such being the case, would warn investors from putting too much faith in the great Ashanti Gold Boom.

The figures given by Soetbeer, who estimates the exports of gold from all Africa from 1493 to 1875, at about £100,000,000, are purely the result of conjecture. There is no warrant for them in either the commercial accounts, or the reports of travellers. If they were reduced to one fifth the amount, or say $100,000,000 they would still be ample enough to cover the truth. These accounts are re-published in Mr. Lock's bulky volume on Gold.

The source of the gold found in Senegal, Liberia, Ashanti and Dahomey, is the Kong Mountains, distant about three hundred miles from the Coast. In the foothills of this range are the quartz veins which furnish the gold, and, below them, the gravel beds, which receive it. The rivers, cutting through these alluvions, carry the gold in continually diminishing proportions to the seashore, where minute particles of it are still found in the sand, but not in quantities sufficient to pay for extracting it, even by the most economical processes known to man. The principal part of the country that has been "worked" consists of the gravel beds between the foothills and the Coast, especially these adjacent to the rivers. Most of these are dammed; and the water is run into small ditch lines for the purpose of washing gold. While the water season lasts the natives can "pan" on the average from ten to twenty cents worth of gold per diem, and, as next to planting a little maize, this is their principal occupation, there are probably several hundred thousand of them thus employed.

The native reef or quartz mines consist of narrow shafts, dug with adze and chisel, varying from 20 to 50 feet in depth. Two men can sink about one foot a day. The gold is extracted by pulverising the quartz and panning. One man can pulverize and pan a ton per month. Women and children are commonly employed in the work. At Tacquah the average earnings are ten cents a day, paid for, chiefly, in execrable gin, at 4s. 6d. (about one dollar) per bottle. The negro King's royalty is nominally one-half of the quartz from the reef mines and all large nuggets from the alluvions; but these exactions are largely evaded.

Whether arising from some ancient and forgotten ordinance of religion or from the cruelties to which the natives were subjected by the various races of gold-hunters—Phœnicians, Romans, Moslems, Normans or Christians—who have successively visited their coast, is not known; but there exists throughout all this country a superstitious horror of searching for gold. This aversion was once so common that after the eighteenth century and until quite recently, it was difficult to procure any gold on the West Coast of Africa. However, the insuperable attractions of gin and the elevating influences of total insensibility, have at length won the day; and for the sake of a thorough soaking in the Mountain Dew of European civilization, the native will now slave for a month and brave death in the most dangerous climate under the sun."

The aversion to gold-mining alluded to by this writer is not confined to the native races; it is shared very strongly by the Dutch Boers, whose opposition to it, evinced in many ways, led to a War in which their national existence is yet at stake.

The subject of aversion to mining, the causes that have led to it in various countries, and the Policy of Closing the Mines, which is connected with it, is treated at length in a another portion of the present work.

We now turn to the gold regions of south-eastern Africa. This embraced the eastern coast, from the mouths of the Zambesi, southward for about 800 or 1,000 miles, and in a general sense extended inland to the various placer mines whence the natives fetched gold in trade to the port of Sofala, in lat. 21 south. Though this region did not include the mining districts which have since become so productive, in the Transvaal Republic, it came close to them, and included the whole of the Monomatapa country, now known as Matabele Land, Swaziland, Rhodesia, etc. In this region the native word for gold is danama, probably a corruption of the Arabian dinar.

The opening of the Zambesi gold region by the Portuguese must not be regarded as either the result of accident or of geographical research. The captains of Prince Henry's ships were not seeking renown, but gold, and their entire course was governed by this consideration.[14] Gold was first met with by Henry's navigators at Rio d'Ouro (Gold River), under the tropic of Cancer; and the sight of a few ounces brought home by them was the immediate cause of establishing an African Company at Lagos.[15] So long as this Company continued to earn profits[16] from its nefarious traffic in slaves, little effort was made to push maritime discovery along the coast, but when these profits fell away, the timorous and unwilling navigators were urged on beyond Guinea, past the non-auriferous coast of the Continent, until, in 1498, they rounded the Cape and reached Caffraria, where, near Cape Corrientes, De Gama first saw in the hands of the natives those implements and ornaments of civilized manufacture which assured him that he was near the Arabian settlement at Sofala, of which Covigliam had advised the Portuguese Court.

De Gama's first actual meeting with the Arabians was at the Island

[14] "The views of Cada Mosta" (whose two voyages down the African coast were made in 1455 and 1456, and who has left us a journal of these voyages by his own hand) "do not seem to reach beyond the fame and profit of his immediate discoveries." Gibbon's Treatise, 498.

[15] Dean Vincent's Notes to Gibbon's Treatise. This Company, established 1444, enjoyed a monopoly in gold, slaves, ivory, etc., on the African coast. Prince Henry was a large shareholder. Gibbon, 497.

[16] These sometimes exceeded 1,000 per cent. Gibbon, 507.

of Mozambique; his first act was one of treachery. The Moslems greeted him with kindness; he repaid them with a cannonade. Similar hospitality was extended to him at Melinda, his northermost port on this voyage, but the Arabians at this place being better armed, De Gama refrained from assaulting them, and spread his sails for India. In the course of the next few years the Portuguese, either by treachery or force, succeeded in occupying all of the Arabian settlements on the eastern coast of Africa, and at once proceeded to work the coveted mines for their own profit.

The Arabian method of working the mines of the Zambesi was the commercial one, the method of their forefathers the Phœnicians, and of the Hindus, who were the forefathers of the Phœnicians.[17] The Portuguese method was that of their forefathers, the Romans of the Empire: it was the method of mine-slavery. The Arabians brought to the gold coast such merchandise as the native chiefs required, and exchanged these goods for gold-dust, ivory and slaves; slaves not intended to be exploited in mines, but reserved for a servitude but little harsher than patriarchal subjection. They never attempted to master the country or to sacrifice the natives to a ferocious or callous cupidity. This was the Portuguese method. After the latter had seized upon the Arabian coast settlements and slaughtered or driven away their masters, they penetrated into the coveted gold country, fortified a line of communications with the coast, hunted down the frightened inhabitants with fire-arms and blood-hounds, and thrust them into the mines, to work under the lash and with insufficient food and rest—leaving them no hope of escape but in death. Notwithstanding the stimulus to production which this atrocious system afforded, the produce of the Zambesi mines ceased to become important. As the cruelties of the Portuguese proceeded, the uncaptured natives removed farther and farther away from the mines, each one of which thus becoming the centre of a solitude which was only broken by the sound of the Portuguese mining picks (*cavadeiras*), and the despairing sighs of the slaves who were forced to wield them. Africa was neither a narrow island nor an isthmus, nor a strip of littoral, like Hispaniola, Darien or Peru. It was therefore impossible to keep the natives at work in the gold mines, as had been done in America. As the Africans fled before the Portuguese they drove their flocks before them, and thus rendered provisions more and more scarce and expensive to the miners. Years before the date of Barreto's expe-

[17] For the easy and indulgent methods of the Banian farmers of the customs, at Zanzibar, see the Reports of the U. S. Consul at that place.

dition, presently to be mentioned, it had become difficult to capture mining-slaves near the gold regions. As for purchasing slaves along the coast and conveying them to the mines, although such a policy was quite feasible, it is evident, because it was not pursued, that it would not have paid, and that the mines were too poor. In Brazil, although the average value of the auriferous gravel was comparatively low, yet, in many places, it was found to be rich enough to repay the expense of purchasing slaves in the marts of Western Africa, conveying them across the Atlantic Ocean to Rio Janeiro, and marching them from 300 to 500 miles into the interior of Minhas Geraes and Goyaz, a journey of one to two months. How poor the gravel of the Zambesi regions had proved to be is evinced by the fact that it did not pay the Portuguese to fetch a gang of negros from so comparatively near a mart as Zanguebar.

In spite of this poverty of the mines, a fact which a very slight research into their history, or else an ordinary exertion of reason would have been sufficient to establish, the fascination of gold mining was such that this region became the scene of one of the most extraordinary misadventures among the many which are to be found in the annals of mining. However, before describing the expedition of Barreto, it will be necessary to briefly allude to that of Alboquerque.

About the year 1510, this Portuguese hero coasted the Red Sea, plundering its maritime cities, and landing at Suakin, or some other port in the vicinity, from whence he sent a proposal to the christian King of Abyssinia to assist him in cutting a navigable canal to the upper Nile, with the object to divert the Indian trade from Alexandria, then in possession of the Moslems, and turn it to Abyssinia. It has been said that the real object of this scheme was to divert the trade by the sea route to Portugal, also to procure an army of labourers to re-open and work the gold mines of the Bisharee region. However this may be, it received no encouragement from the Abyssinian prince, and so came to nothing. Mr. Herbert Vivian, in a recent work, informs us that in Abyssinia "only the royal family are permitted to wear gold in any form"; an interdict that probably had its origin in the horror of those crimes by which gold was commonly procured, and in which the pious monarch of Abyssinia desired to take no part.

About the period 1552–1567, Francisco de Barreto was the Portuguese viceroy of India at Goa. In 1553, to punish the poet Camoens for reminding the grasping officials of his court that "honour and self-interest are never found in the same sack," he sent him to Macao.

Sixteen years later Barreto was in Lisbon at the head of a mining scheme. He solicited and obtained permission from the King Dom Sebastian to conduct an expedition to the gold regions of south-eastern Africa. The Arabian name of the country, Sofala, which means low country, or netherlands,[18] had been distorted by the managers of this scheme into Sophala and this into Ophira and Ophir. Hence they argued Sofala must have been the source of Solomon's and The Queen of Sheba's wealth. Even the particular mines which had furnished the gold to these scriptural heroes were made out; the queen having received her's from those of Massapa! A nugget valued at 12,000 crusados (about $6,000) had recently been found in Monomotapa, and this was exhibited as a sample of what was to be found in plenty throughout that highly favoured country. In a short time Barreto got together three ships, with a thousand adventurers (mostly of the gentleman or fidalgo class), and abundant supplies of horses, camels and provisions; and, in April, 1569, this expedition left the Tagus to make a descent upon Solomon's Ophir. It never seemed to have occurred to the adventurers that Solomon, in the plentitude of his wisdom, might have had enough of it to induce him to "clean up" the gold mines before he abandoned them to the modern world. The name Ophir and the 25-pound nugget were enough; so off sailed these one thousand lunatics, bound upon a voyage to the antipodes in search of a myth. Having successfully passed the Cape, they came to anchor at the Quilimane mouth of the Zambesi river, and at once began the ascent of that stream in boats. Upon reaching Senna (long. 35), where they were hospitably received by the Arabian traders, whose poverty should have admonished them of the futility of their errand, they left the river and pushed on at once for the gold-fields. And now their sufferings began. The tsetse fly, the marsh fever, the bites of reptiles and vermin, the heat of a tropical zone, the hostility of the natives, who knew from the experience of their race what gold-mining meant for them; all these perils had to be encountered, besides those of unaccustomed diet, brackish water and other privations. The result was that when the party reached the mines[19] their numbers were greatly thinned and the survivors were so reduced as to be scarcely able to move about. Their confidence in King Solomon must have been greatly shaken by what

[18] Malte-Brun, III, 80. "The coast is very low, and mariners discover their approach to it not so much by their sight as by the smell of its fragrant flowers." Morse II, 537.

[19] The mines they reached were those of Macoronga and Manica, the latter being in lat. 20, and the same as those discovered by Mauch and delineated in Wilmot's Map as Ramakuaban, Victoria Diggings, Mattoppo Mines, etc.

they found there. On the borders of a sluggish stream—the Tatti river—which had long been choked with mining gravel, were to be seen the pits, ground-sluices, and ruined ditches of some ancient miners, who, having exhausted the placers, or run them out of grade, had deserted them for the quartz-leads beyond, and worked these down to the water level. As for gold, there was none to be seen.[20] The negros could, indeed, wash out a few colours from almost any basin-full of selected earth, but this was an hour's work, and the result proved to be scarcely worth a vinteim. Moreover, negros were not to be obtained without violence, and when obtained, there were no means of supporting them.[21] There was but a single alternative, either to return to Portugal or capture and immolate the natives by dooming them to incessant labour and insufficient food. If these men had been human they would have returned to their ships at once; but being gold-hunters they remained and condemned the natives to mine-slavery, to torture and death. Notwithstanding this cruel resolve, and the energy which was applied to its execution, the gold proved so scarce, the natives so difficult to enslave, and their own privations so numerous that in the course of five years nearly the whole of Barreto's party, including the Ex-viceroy himself, miserably perished.[22] And so ended this search for King Solomon's mines. Yet, three centuries later it was revived by a German named Karl Mauch, and again a decade later by an Englishman named St. Vincent Erskine.[23] Six years after Barreto's death, the entire piratical Empire of the Portuguese in the East, whose dominion was the seas and coasts, and whose capital, the once palatial Goa, fell into the hands of Spain. From that moment the mines of the Zambesi were abandoned, save by those native labourers whose insignificant and desultory product

[20] The so-called gold mines "were found in no degree to correspond with the magnificent expectations formed of them, or of the labours and dangers through which they had been reached." Murray's "Africa," II, 362. (Edinburgh, 1817.)

[21] Murray tells an amusing story of a native who induced Barreto to give him a magnificent reward for leading him to the "silver mines of Chicova," and who, having "salted" a hole in the earth with a few pieces of silver, and obtained his reward, suddenly decamped, leaving Barreto to discover the fraud at his leisure.

[22] Barreto, who had not dared to return to Lisbon, died on the Zambesi in 1574. Two versions of this story are given by Lock, 10 and 16, and one by Wilmot, 6; and it is astonishing how many blunders their brief narratives manage to comprise. Lock gives the date of the expedition in one place in the middle of the 16th century, and in the other as 1650. In one account Barreto reaches the mines; in the other, he entirely fails to do so; and as to this, Wilmot says the same.

[23] While this chapter was being written the author was applied to professionally by an agent of the Portuguese government for particulars of King Solomon's mines, in order that he might transmit them to his sovereign and induce him to send out a gold-hunting expedition from Portuguese South Africa to work them! When advised of their mythical character, he became highly indignant, and said he intended to consult another engineer on the subject.

of gold continued to always form a portion of the shipments from the Coast. Notwithstanding the extravagant statements sometimes to be met with in reference to the auriferous wealth of this region,[24] no returns of its production since the Portuguese occupation have come under the author's observation which warrant an estimate to exceed twenty million dollars, as follows:

From 1507, the date of the Portuguese occupation of Sofala, to 1574, when Barreto perished, and gold-mining by or under the Portuguese, came to an end, about two million mithcals or metigals,[25] say	$6,000,000
From 1575 to 1808, a period of 234 years, the product of the natives could not have exceeded 100,000 crusados, or $50,000 a year,[26] say	11,700,000
From 1809 to 1836, 48 years, at about $10,000,[27] say .	480,000
From 1857 to 1866, 10 years, at $2,500,[28]	25,000
Since 1866, say for round figures[29]	1,795,000
Total, 1507–1897, 390 years, say	$20,000,000

Or scarcely more than one year's production of California, Australia, or Russia.

[24] "Gold was formerly so common in the interior that many of their (the Portuguese) household utensils were made of this metal." Macgreggor's Statistics, II, 336. "The gold mines yield more than five million dollars a year," Morse, Geog., II, 537, an opinion evidently derived from the Portuguese statement of the total product of 67 years: viz., two million meticals.

[25] The basis of this statement will be found in Macgreggor, II, 335, and Thos. Baines, in Lock, 16. The metigal, mithcal, or metical, is an Arabian weight which varies from 60.7 grains in Tunis to 73 grains in Aleppo and Algiers. That of Damascus is 69 grains. A metical of gold, nine-tenth fine, is equal to about $3. Lock converts the metical at about $2.50, and Macgreggor at $4.17.

[26] The following relates to the entire gold trade of the Mozambique Channel: "In 1593 the Governor of Mozambique, George Menzes, collected for himself and the Viceroy of India one hundred thousand crusados, and I do not believe that one-third of this amount is now (1808) annually produced." Salt's Voyage (1809) to Abyssinia, p. 68, quoted in Jacob, 373. Lock, 16, swells this sum by a blunder into one million crusados. In 1788, when Galvao da Silva visited these regions, "the supply (of gold) was very much reduced." Lock, 12.

[27] For trifling extent of the product in 1823, see Lock, 25, who quotes a statement made by the Arabs to Capt. Boteler.

[28] In 1808 the entire gold product of the Mozambique region was only 100,000 crusados. Lock, 13.

[29] Thos. Baines, in Lock, 17, says he was informed by B. N. Acutt, of the London & Limpopo Co., that down to 1872 the Tatti region had yielded from 1,500 to 2,000 lbs. of gold. In 1875, over 100,000 pounds sterling in gold were shipped from Cape Colony; but this may have been the product of more than one year. Lock, 22. The conjectural shipment by "private hands" is evidently a gross exaggeration. The *Colonies and India* stated that during the last six years (this was probably written in 1881) over 200,000 pounds sterling worth of gold had been sent through the Cape Commercial Bank alone. Lock, 24. It is not specified whether this gold was the product of the African fields or not. Probably none of it came from the Tatti region.

In 1857 Dr. Livingstone travelled through this country, but although a close observer familiar with mining and mine indications, and aware of the history of this country, he failed to notice the presence of any paying quantity of gold in either the gravels or quartz. He merely noticed that the formerly great gold trade had fallen to $2,000 or $2,500 a year. The discovery that this exhausted country was yet full of gold was reserved for the German traveller Karl Mauch, who is described variously as a "mineralogist," an emissary of the Prussian government, a geographer travelling for Dr. Petermann of the Gotha Geographical Journal, and a correspondent of the Natal Mercury. In 1864 this traveller discovered gold-fields in Moselikatze's country (lat. 20 S., long. 29 E.). This is a portion of the old Monomotapa, and is full of old works of the Portuguese. Mauch at once hastened to inform the world of his valuable find. By the year 1868 he had succeeded, with the aid of interested parties in Capetown, in getting up a boom concerning the South African gold-fields, the natural result of which was a "rush" to the fields from Capetown, and the organization of several mining companies in London. Among these was the London and Limpopo Company, 1870, and the Transvaal Gold Mining Company, 1872, capital $50,000, both of which enterprises failed.

In 1875, Major St. Vincent Erskine informed the Geographical Society of London that he had discovered the Ophir of Solomon in Sofala, between the Limpopo and Sabia rivers; that the Queen of Sheba's real name was Sabia, hence the name of the river; that in "a geography book, eighty years old, in his possession, it is stated that the Portuguese exported thence *annually*—(oh, Major!)—£3,000,-000 worth of gold"; and that he was anxious that everybody should share with him these important discoveries.

From this time to 1883, when Baron Grant undertook the difficult task of "getting up a new deal" in the South African mines, we are not aware of any further efforts of the kind; but from 1868 to nearly the present time placer mining on a small scale has been carried on by scattered parties of Boers, Australians, and Englishmen. The product has been small and the miners—on the whole—unsuccessful.

With the year 1884 commenced a new æra in the history of mining in Africa, due to the discovery of quartz or reef mines in the Witwatersrand district of the Transvaal Republic.

A letter written from the vicinity a year after the discovery of these mines will afford some evidence of what was thought of them at that period:

In the Witwatersrand district, the gold-bearing formation is what is known as "banket," a sort of conglomerate easy to reduce. It is composed of a sort of pebbly quartz cemented together by fine sillicious sand. A piece of it may easily be crushed beneath the foot, and a little water poured over a lump will cause it to crumble at once. Such is "banket," and the gold is contained chiefly in the cement. Veins of positive quartz are found in the same reefs as the banket, but the mills are working exclusively on the latter so far. Banket runs in lodes or veins varying from one to twenty feet wide. There is such an abundance of this material near the surface that it is practically inexhaustible, and as deep as shafts have thus far been sunk, it holds its own in width of vein and richness. Thousands of stamps may find remunerative employment night and day for years on banket now in plain sight.

At present nearly a thousand stamps are working steadily on the banket in the Witwaterstrandt district alone, and it is thought that by this time next year six times that number will be hard at it. The average clean-up yields about one and one-half ounces of gold to the stamp per day. At this rate some of the companies have commenced paying dividends at 50 per cent. a year on the capital invested, and shares are held at ten and twenty times their original cost a few months ago. The banket lodes were discovered a year ago by a Pretorian named Stuben. The district was totally wild and uninhabited, a barren plateau, considered fit for nothing. Then came the discovery and inevitable rush, and with mushroom spontaneity there has sprung into existence the town of Johannisberg already numbering six thousand inhabitants.

The country round about Johannisberg for miles contains no timber. Many of the houses are queer things, built entirely of movable iron sheets imported from England; others are of adobe, or mud and rock. It is a regular gold-field city, full of rowdyism and hard characters; dance-houses and saloons by the dozen are in full swing, and robbery and shooting affrays are of almost daily occurrence.

The following letter from the South African correspondent of the London Times, December, 1886, will evince the appreciation in which the mines were held a year or two later.

The development of the gold fields continues to absorb public attention. From every town and village in South Africa, during the past three weeks, a stream of fortune-seekers have wended their way to the DeKapp and Witwatersrand. The population of Barberton has more than doubled and is daily increasing; stands for building and business sites are being eagerly purchased there; new syndicates are being formed and new companies floated. The total capital of all the gold-mining companies is stated to be not far short of £2,000,000, while their value, as represented by the ruling share prices, is nearly double that amount. In many cases the realization of returns is a long way off, as there is no machinery immediately available for the development of the properties; in other instances some of the companies have been—even with very inadequate appliances—very successful; and it is this, together with the handsome return from "company promoting," which has been the incentive to so much speculation.

Of the Witwatersrand fields, between Pretoria and Hiedelberg, some of the capitalists of Kimberly have secured gold properties from which wonderful results are also expected. W. Knight, who has been long and favorably known in connection with successful mining enterprises at the diamond fields, has secured mining rights on the farm of Driefontein, where prospecting work has resulted in the discovery of four conglomerate gold-bearing reefs, giving together an average thickness of about twenty-three feet, extending over three and a-half miles in length, with a proved depth of about 100 feet. Adjacent to this, Messrs. Rhodes, Rudd and Caldecott have purchased for £10,000 the properties of Reitfontein and Witkoppies; and several other syndicates and individuals have secured mining privileges in the same neighbourhood. When companies are organized and mills set to work it is expected that these fields will give an average return of an ounce per ton, and that the total cost of extracting the gold, mining royalties, and other charges, will not be more than 15 shillings per ton.

Fifteen years' experience since this last letter was written has demonstrated that the quartz was estimated at thrice its real value, which is about seven dollars a ton. The DeKaap properties have proved of little worth, and Barberton has not become a metropolis.

According to the official returns issued by the Witwatersrand Chamber of Mines, the produce of the Transvaal mines has been as is shown in the first column of the following table, the valuation of the gold in the second column being that of the Chamber itself—namely, £3 10s. per ounce for mill gold, and £3 for cyanide gold, an average of about $16 per ounce. The figures in the third column are those of the United States Mint.

Produce of the Transvaal Gold Mines.

Year.	Gold. Ounces.	At $16 per ounce.	As given by U. S. Mint.	No. of Miners.
1887	35,125	$ 562,000	—	1,000
1888	222,122	3,553,952	—	5,000
1889	385,557	6,168,912	$7,888,37	10,000
1890	494,817	7,917,072	10,438,356	15,000
1891	729,268	11,668,288	14,885,639	20,000
1892	1,210,869	19,373,904	23,220,108	40,000
1893	1,478,477	23,655,632	28,293,831	45,000
1894	2,024,164	32,386,624	39,696,330	60,000
1895	2,277,640	36,442,240	43,893,300	60,000
1896	2,280,892	36,494,272	43,779,669	65,000
1897	3,034,678	48,154,848	57,633,861	80,000
1898	3,831,975	61,311,600	79,213,953	120,000
1899	3,500,000	56,000,000	71,691,163	100,000
1900	Est'd.	10,000,000	—	15,000

It should be remarked that the value of the gold, as returned by the mining companies to the Chamber of Mines, was a mean between the two extremes shown above, or about $18 per ounce.

From the discovery to the closure of the mines by the Transvaal government, who worked them on its own account from October, 1899, to May, 1900, and then closed them down, the entire product of gold was about 350 million dollars. As the Transvaal mines are owned chiefly in England, the gold for the most part has gone to that country, there to be coined into those beautiful "sovereigns," upon one side of which is stamped Saint George and upon the other the good Queen "Victoria by the Grace of God," etc. Some of the gold went to the United States, where it was stamped with the arms of the Republic and the name of God—"In God we Trust." In both of these instances, in every instance where the name of the Almighty is stamped upon a coin of Transvaal gold, the Boers regard the act as a sacrilege; for almost every ounce of it has been obtained through slavery. Not such slavery as the Spaniards and Portuguese inflicted upon the natives of America; but another sort of slavery: a slavery that is inflicted without remorse and yet which hypocritically pretends not to be slavery at all.

They say that the mines of the Transvaal have been worked from first to last by natives, who were entrapped and forced into them against their will; that they were bought from contractors at so much per man, bound, strapped, made drunk with rotten liquor, and thrust naked into pits which avarice had dug and hypocrisy has covered over. The following extracts from the carefully-guarded report of the Chamber of Mines (a British organization), for 1896, have been supplied to the author.

Towards the close of 1896 this Chamber applied to the Transvaal government for eight thousand additional native labourers. On the part of the government, Capt. Dahl "referred to the disinclination of the natives to work" in the mines, and although the government had issued orders that the mining companies should be assisted in the matter, he feared that natives could not be got without using "pressure." Turning to the Portuguese authorities at Lourenço Marques, the Chamber, in September, 1896, obtained leave "to *emigrate* natives from any part of the Portuguese territories, and the right to arrest any native similarly engaged who is not in the employ of the Chamber, or in possession of direct authority, in writing, from the government of Lourenço Marques." In the same month the Chamber resolved to reduce the wages paid to the natives, then £1 or $5 per month, from 20 to 25 per cent., and steps were taken to "prevent any disturbance which might arise from discontent" with this arbitary measure. It was further resolved that the miners should be taxed at the rate of three shillings per native employed, of whom there were then seventy thousand. The tax would yield ten thousand five hundred pounds, and the amount was to be paid over to the Rand

Native Labour Association, and if found necessary, a like sum of ten thousand five hundred pounds was to be raised and applied in a similar manner. Here were one hundred thousand dollars appropriated and confided to a "Native Labour Organization" for an unspecified purpose, which looks uncommonly like purchasing slaves. Meanwhile, the natives escaped from the mines in such great numbers as to call for new measures of repression. A Pass Law was obtained from the Transvaal government which it was expected would operate like the Fugitive Slave Act of 1850; it authorized the arrest and detention of any native without a Pass; he was then to be delivered to the mine proprietors. "The object of the new law was to bring the natives under effective control and reduce the risk of desertion to a minimum. . . . Many offers had been made to supply natives for fixed periods, at wages far below those ruling (at that time), but nothing could be done under the then existing conditions, and desertions remained as easy as ever." . . . The chief defect of the old law was that the penalty for desertion was insufficient. This was ten shillings, or a week, with hard labour and *lashes*; second offence, *double;* subsequent offences, *at discretion.* The new law, dated December 23, 1896, raised the penalty for first offence to £3, and second offence to £5, with imprisonment, hard labour and lashes; and other amendments to prevent desertions. However, as the Transvaal government refused to execute the law, "the question of regulating and controlling native labour still remained a difficult one." . . . "Besides the harm done by the indiscriminate sale of liquors to the Kaffirs, the evil is intensified by the vile quality of the drink generally supplied to them. . . . As already stated, special permission has been given to the Native Labour Commissioner of the Chamber and some few others to engage natives in and *to emigrate them from* Portuguese territories. The Boers having insisted upon the observance of Articles 131 to 134 of the Mining Regulations, forbidding the running of mining mills on Sundays, this Chamber sent a delegation to the Valksraad to represent that such stoppage would mean ruin to the mine owners. A reduction of the gold output by one-seventh, or 14 per cent., would convert profitable mines into losing properties." The point was eventually settled in favour of the miners. . . . The scarcity of native labour checks the expansion of mining operations. . . . Success depends on the natives who work our mines. . . . The object of the Pass Law was to identify natives who deserted from their employers, that they might be brought back to the mine from which they had deserted." The Reports of 1897 and 1898 contain similar passages. The British mine-owners continually pressed the Boer government for harsher measures to the natives, and although the Boers, anxious not to arouse resentment, seemingly yielded to this pressure, they did all they could to defeat it, by neglecting to execute the Pass Law. The running of mills on the Sabbath day was also offensive to them, and in this respect they likewise made concessions which were very unpalateable to their people. But all in vain. The mining companies desired to be free of all control, so that they could work the "niggers" their own way. The result was antagonism and war.

Without expressing any further opinion of his own concerning the merits of this contention, the author gladly turns from the unsavoury subjects of mine-slavery and Sabbath-day gold to the economical aspects of the Witwatersrand mines. It will be observed that they have consumed the labour of about 640,000 men for a year. From 10 to 15 per cent. of this number were English, Australian and American miners, whose salaries ranged, as mining superintendents, from £10,000 a year down to £1 a day, and who, therefore, absorbed a very considerable proportion of the entire yield of this ill-gotten metal. Leaving this out of consideration for the present, if we divide the numbers of miners employed into the whole product, it would only amount to $547 per annum, or about $1.50 per diem! So it is quite evident that without forced labour, without slavery, and without working the mills on the Sabbath, the industry could not have been made to pay. Its profits have therefore been the profits of forced labour.

THE CONGO FREE STATE.

The Rev. John B. Murphy, of the American Baptist Mission, who has just reached England, comes from Equatorville, a very important centre of the State Government, situated 800 miles in the interior, and has lived at various stations on the Congo for the past nine years. He left Equatorville on May 9, and remained on the upper river until the middle of August, 1896. Mr. Murphy said:—

The attitude of the natives of the Congo Free State is everywhere unfriendly, and if the people do not universally rebel against authority, it is because they are reduced to a state of despair. If possible they leave the territory. Two of the most flourishing towns in Mr. H. M. Stanley's time, situated at Stanley Pool—viz., Kintamo and Kinchassa—are now no more, and the people have gone over to the French Congo. Besides the natives of towns I have named, many people have left the main river and gone into the interior, in order to escape the arbitary demands of the State. Difficulties have arisen, too, between the State and the porters, and if the requisite number of carriers were not forthcoming, detatchments of soldiers were sent, with orders to capture all the women they could find. Several Christians were arrested in this manner. The natives and missionaries remonstrated and presented a letter to the governor without getting any redress. The people were so enraged at these outrages that they took matters into their own hands. They killed three white men and met and defeated the State forces in more than one pitched engagement.

I have seen these things done and have remonstrated with the State in the years 1888, 1889, and 1894, but never got satisfaction. I have been in the interior and have seen the ravages made by the State in pursuit of this iniquitous trade. In one place I stood by the side of the river and heard a little boy describe how he had seen the Belgians shoot people for not fetching rubber, and at the same time he pointed to the flag-staff to which the poor victims had been tied, and which still bore the bullet and bloodmarks. Let me give an incident to show how this unrighteous trade affects the people. One day a State corporal who was in charge of the post of Lolifa was going round the town collecting rubber; meeting a poor woman whose husband was away fishing, he said, "Where is your husband?" She answered by pointing to the river, and he then said, "Where is his rubber?" She answered, "It is ready for you," whereupon he said, "You lie," and lifting his gun, shot her dead. Shortly afterwards the husband returned and was told of the murder of his wife. He went straight to the corporal, taking with him his rubber, and asked why he had shot his wife. The wretched man then raised his gun and killed the corporal. The soldiers ran away to the headquarters of the State, made misrepresentations of the case, with the result that the commissaire sent a large force to support the authority of the soldiers: the town was looted, burned, and many people killed and wounded.

In November last there was heavy fighting upon the Bosira, because the people refused to give rubber, and I was told upon the authority of a State officer, that no less than 1,890 people were killed.

Upon another occasion, in the month of November, 1894, some soldiers ran away from a State steamer, and, it was said, went to the town of Bompanga. The State sent a message telling the chief of the town to give them up. He answered that he could not, as the fugitives had not been in his town. The State sent the messenger a second time with the order, "Come to me at once, or war in the morning." The next morning the old chief went to meet the Belgians, and was attacked without provocation. He himself was wounded, his wife killed before his eyes, and her head cut off in order that they might possess the brass necklet that she wore. Twenty-four of the chief's people were also killed, and all for the paltry reason given above.

Again, the people of Lake Muntumba ran away on account of the cruelty of the State, and the latter sent some soldiers in charge of a coloured corporal to treat with them and induce them to return. On the way, the troops met a canoe containing seven of the fugitives. Under some paltry pretext they made the people land, shot them, cut off their hands, and took them to the commissaire. The Muntumba people complained to the missionary at Irebu, and he went down to see if the story was true. He ascertained the case to be just as they had narrated, and found that one of the seven was a little girl who was not quite dead. The child recovered, and she lives to-day, the stump of the handless arm witnessing against this horrible practice.

The people of the district of Lake Muntumba, of Irebu, and Lokolala, and all the Mobangi towns have crossed over to the French side. The French treat them kindly, and are glad to get people. The people did not run away without great provocation, for it meant starvation to them, all their gardens and their homes being upon the State side. Even then the State could not leave them in peace; they heard that they came over at night to their old homes to get food, and they stationed canoes and laid in wait for them, with orders that the soldiers should shoot as many as they caught; and to my knowledge they shot seven people in one night.

The white officers do not know the language of the people that they govern, and trust too much to their native soldiers, who are, as a rule, men belonging to a hostile tribe, whose chief aim in life is to plunder. These men are sent out to fight very often without any responsible officer being with them, with the result that many cruelties are being perpetrated which might have been avoided. It is impossible for the Governor of Boma—four weeks' journey from Stanley Pool, which ought to be the real seat of Government—to manage his vast and unwieldy territory; so that the commissiares and petty governors of the interior districts have almost unlimited power. The officers of the State are young and inexperienced; they do not come out as colonists to develop the country, but in order that they may receive quick promotion, the Congo decoration, and, above all, to get money. I have been told by naval and other officers of the State, that a certain sum per head is paid by the Government to the commissaires of districts from which slaves are received, and to the naval officers who bring them into camp. These wretched slaves soon find that they have only changed masters. About fifty per cent. are in a starving condition. Let me (said Mr. Murphy in conclusion), again just revert to the rubber question, which is by far the most pressing, being accountable for most of the horrors perpetrated on the Congo. It has reduced the people to a state of utter despair. Each town in the district is forced to bring a certain quantity to the headquarters of the commissaire every Sunday. It is collected by force, the soldiers drive the people into the bush. If they will not go, they are shot down, their left hands cut off and taken as trophies to the commissaire. The soldiers do not care who they shoot down, and they more often shoot poor helpless women and harmless children. These hands, the hands of men, women and children, are placed in rows before the commissaire, who counts them to see that the soldiers have not wasted the cartridges. The commissaire is paid a commission of about 1d. a pound upon all the rubber he gets, therefore he gets as much as he can.

The King of the Belgians is our very good friend. Great Britain is the effective protector of his realm. However, the atrocities in the Congo State, of which he is Sovereign, have reached a point at which interference with his sovereignty could not have been far off, but for his appointment in September of a Commission charged with the protection of the natives. Even now we have no assurance that the worthy priests and ministers composing the Commission will be able to influence the judicial authorities to whom they are to report. Those authorities would have probably interfered on their own motion if they had been minded that way. And the Aborigines' Protection Society aptly shows the King, in a long memorial it has sent him and Lord Salisbury, that the proposed measures are quite inadequate. The ordinances made previously, were flagrantly neglected. The free trade provisions of the Berlin General Act have been contravened, and the country has been worked for all it is worth "in utter disregard of the early pledges that the interests of the natives should be cared for, their customs and usages respected, and nothing done to deprive them of their lands and means of living."

A modified slavery has been established; the people are compelled to sell their rubber or ivory to the State for what it chooses to give; and their lands have been added to the public domain. The case of Mr. Stokes, who was foully done to death by Major Lothaire, as yet with impunity, shows that even white men cannot look for justice from the administrators of the Congo State, or King Leopold himself, and of course the natives' appeals are never heard. They "are at the mercy of Europeans, who consider themselves entitled to all the ivory, rubber and other produce that the natives can be induced, by any process, to bring them." The regulations do not provide properly for the protection of non-military natives. And this one fact goes a long way to show that offences, even when proved, are practically winked at. In May, 1890, M. Deghilage, chief of the Manyanga Station, flogged two of his black servants to death, was tried for it, and "in consideration of his having been drunk at the time," was fined 20*l.* But in December, 1893, this flogger was decorated, being given the "Star of Service."

London Chronicle, Dec. 14, 1896.

CHAPTER XXV.

JAPAN.

Enormous treasure obtained from Japan—Parallel between the inhabitants of New Spain and Japan—Many circumstances common to both—The principal differences consisted in the American belief in a coming Messiah and the Japanese familiarity with horses and with steel weapons—The precious metals in Japan—Operations of the Portuguese to acquire them—The means employed—Slavery, priestcraft, and intrigue—Penances and Indulgences—Introduction of the Inquisition—Sedition and civil war—Plot to overthrow the native government—The plot discovered and the Portuguese banished from the country—Closure of the ports for over two centuries—Curious memorial of a Japanese Finance Minister to his government—The mines of Japan—Production of the precious metals.

THAT during the seventeenth century, Europe obtained from Japan so great a sum in the precious metals as 360 million dollars, is a statement that will probably be received with surprise, for it is not to be found in the usual works of reference. Yet there is little room to doubt its truth. It is supported by the testimony of both Japanese and European writers.

From 1545 to 1597 the exports of the precious metals from Japan, chiefly by the Spaniards and Portuguese to Manilla and Macao, were about 30 million dollars, nearly all gold. Martin's "China," I, 292, 317, and Sir Stamford Raffles' "Hist. of Java." From 1598 to 1610 the exports were about 16 millions gold and 49 millions silver. Hildreth. From 1611 to 1646 the exports were about 25 millions gold and 113½ millions silver. Arrai Tsikugono-Kamisama in "The Riches of Japan," quoted in Klaproth's Noveau Journal Asiatique, Vol. II. The electrum Koban (that of 1605–22) is reckoned at $6.68 gold and $0.32 silver; the silver tael of that period at $1.50. The exports of this period, as given by Hildreth, are evidently copied from Arrai's statement. From 1647 to 1671 the exports were 16 millions gold and 57 millions silver. Arrai. From 1672 to 1705 we have no separate data. From 1706 to 1840 the exports were about five millions gold and about one million silver.

In addition to these sums, Raffles states that Japan exported unlisted gold and silver to China. There were also considerable quantities of gold and silver mingled with the exported copper. It will therefore not be far wrong to estimate the total exports during the course of 250 years at about 360 million dollars, of which perhaps two-thirds were of silver and one-third gold.

Mr. Lock (p. 349) says that between 1649 and 1671 the annual exportations of the precious metals from Japan by the Dutch, until it was stopped by the imperial edict of the last named year, "averaged nearly three million dollars; but little less than that of the Portuguese in the previous century." This statement is fragmentary and defective. It omits from view the exports of the 18th and 19th centuries, while as to the 17th, it fails to observe that the edict did not really put an end to the exportations. It was an edict of the Shogunate, and owing to various reasons, was largely evaded. Nevertheless, Mr. Lock's statement admits an average exportation of over three millions a year during the 16th century, and adds more than sixty millions to the 17th. It is evidently drawn from the same sources as the account more fully given above.

The circumstances under which this treasure was obtained are not only similar to those which attended the despoilment of Spanish America, they are connected with it in other ways. Columbus' voyage across the western ocean was to discover this very Japan which Polo had described two centuries before, and which Pinto was to stumble upon half a century later. The object of the adventurers who followed in the wake of Columbus to America, and in that of Pinto to Japan, was the same—to obtain from the inhabitants of those countries their accumulations of the precious metals. The internal condition of New Spain, and of Japan, at the period of the arrival of the strangers was much alike, and from the few respects in which they differed, we gather a world of historical meaning.

The first arrivals of Europeans in Japan were those of two Portuguese who were wrecked on the coast in 1542. In 1545, Mendez Pinto, a Portuguese adventurer and pirate, was driven to Japan (so he claimed) by stress of weather. Within a year after this event, and attracted by the glowing reports which Pinto carried to Ningpo (in China) of the great wealth and magnificence of Japan, the Portuguese at Ningpo fitted out nine ships to trade with or conquer the newly-found islands, as opportunity offered. Of these vessels only one survived the perils of the voyage, and the intention to piratically conquer Japan was necessarily abandoned. The voyage was, however, not without important fruits. The Portuguese fell to trading with the natives, and the profits were so great—amounting to 1,200 per cent. upon the value of their cargo, the avails of which they carried to Ningpo in gold—that upon their return new expeditions were fitted out, and Japan was practically opened to the commerce of Europe.

In outward appearance the natives of New Spain (that is to say, Mexico, Central America, the Isthmus, and the northern and western shores of South America) and those of Japan bore at that period no little resemblance to one another. They were dark of visage, short of stature, lightly clad, brave in action, intelligent, amiable, and, as their visitors regarded them, partially civilized, though we of to-day may well doubt if in this respect they fell a whit behind their visitors. The condition of society in Mexico, Peru, etc., and in Japan was much the same. All these countries were in a feudal condition, and it was owing to this circumstance that Cortes was enabled to make allies of the Tlascalans, Pizarro to divide the rival Incas, and Xavier to make a dupe of Bingo.[1]

Both the Mexicans, Peruvians and Japanese were ignorant of the use of firearms, though the latter were familiar with gunpowder. The character of their European visitors was the same. Cortes and his desperate followers had stolen away from Havana without authority, and as a band of pirates. When Pizarro set forth to discover and ravage Peru, associating himself with Pedrarias, De Luque, and Almagro, as co-partners, he was a pirate. Pinto was on board of a Chinese corsair when he was driven into Japan. The subsequent expedition from Ningpo was of a piratical character, and but for the loss of eight of the ships, its promoters might with rude hands have at once grasped those riches for which, in the end, they were content to trade and intrigue.

But here the parellel ceases. The Aztecs believed in a Messiah who might appear at any time, and when Cortes craftily announced himself as the emissary of this heavenly personage, Montezuma was credulous enough to believe him, to yield him possession of his person, and to advise his people to surrender themselves to the strangers. The Japanese, who were Shintos and Buddhists, looked for their Messiah only at an appointed time, and in the bigotry of an ancient creed they regarded tne strangers more with disdain than fear.

The Aztecs looked upon the Spaniards as children of heaven; the

[1] Bingo was a Japanese daimio whom the Portuguese converted to Christianity, and whom they employed to betray and enslave his countrymen in order to further the sordid aims of the strangers.

Japanese treated the Portuguese as a parcel of barbarians, shrewd and acquainted with many useful arts, but not at all to be held in reverence. The Spaniards appeared off the shores of America in great ships with lofty sails, the like of which the natives had never seen before. The Portuguese visited Japan in Chinese junks, with whose appearance the natives were quite familiar. The Spaniards landed in Mexico and Peru with horses, animals whose wonderful strength and speed rendered them a greater source of dread to the natives than their masters. The Portuguese carried no horses to Japan; nor, had they done so, would they have created any surprise, for the horse was common in that country. The most formidable weapon possessed by the Mexicans and Peruvians was a wooden club, studded with flints. The Japanese, on the contrary, were as well acquainted with iron and steel as their visitors; and whereas the latter carried each a sword about their persons, the former carried two.

The features of resemblance between the Aztecs and Japanese, which most nearly concerned their future relations with the Europeans, were their common feudal condition and lack of firearms. The features of dissimilarity were, first, the American belief in an immediate Messiah, and the familiarity of the Japanese with horses and steel weapons. But for the superiority of the Japanese over the American aborigines in these respects there is little reason to doubt that the fate of Mexico and Peru would have also been that of Japan —to fall beneath the arms of a few marauders whose single purpose was the acquisition of gold, no matter at what cost of life, or of lingering misery in the hideous subterranean prisons which they called mines.

At the period of the arrival of the Portuguese in Japan, the currency of the country substantially consisted of rice for agricultural rents, public taxes, and many other kinds of payments; together with gold, silver, and iron or copper coins, issued by the Shogun. It seems to be pretty well established from the narratives of Marco Polo in the thirteenth, and Mendez Pinto in the sixteenth century, and the ease with which the first foreigners were enabled to obtain returns in silver and gold for their cargoes, that at the time of their arrival these metals existed in Japan in considerable quantities. It is probable that the Shogun, and perhaps also the Mikado, as well as each of the great daimios, possessed a reserve of these metals, chiefly gold, in slugs or rude hammered forms. After the arrival of the Portuguese these slugs were coined into kobans, boos, etc., and exchanged for European weapons, dress-stuffs, medicinal drugs, and other commodities. It was when this stock gave signs of becoming exhausted that the Portuguese entered upon those transactions which, on being exposed, led to their expulsion from the country, and its virtual closure to the world for more than two centuries.

These operations first assumed importance during the shogunate of Iyé-Iasu (posthumous title, Gongen or Gangin), which began in the year 1603. They consisted first of winning over Gongen and many of the great daimios to the project of opening the native mines of gold and silver, and of working them with their own subjects, who were to be condemned as slaves for this purpose; and next, of obtaining the metals thus produced from their owners by means of ecclesiastical dues, penances, and indulgences. The instruments of this vile scheme were slavery and a sordid priestcraft, both of which the Portuguese were the first to introduce into Japan. Says Griffis: "The arrival of these foreigners was the seed of troubles innumerable. The crop was priesthood of the worst type, political intrigue, religious persecution, the Inquisition, the slave trade, the propagation of Christianity by the sword, sedition, rebellion, and civil war. . . . All foreigners, especially Portuguese, then, were slave-traders; and thousands of Japanese were bought and sold and shipped to Macao in China and to the Philippines. . . . Even the Malay and negro servants of the Portuguese speculated in the bodies of Japanese slaves, who were bought and sold and transported. . . . Such a picture of foreign influence and of Christianity, which is here drawn in mild colours, as the Japanese saw it, was not calculated to make a permanently favourable impression on their minds." In another place Griffis states that Okubo, a Christian convert and Governor of Sado, promised by the Christian priests, in view of the treasonable plot of 1611, to be made hereditary Emperor of Japan, and really a catspaw in their deeper design connected with the abstraction of the precious metals, had thousands of other "Christian converts," his countrymen, working for him in the mines of the province over which he ruled, and was about to betray.

The lowest servitude known to the Japanese before the arrival of Europeans was a mild form of feudal villeinage, which kept the peasant in his own province and at work in the fields. Religion sat lightly upon these happy islanders, and the bonzes were rather their welcome companions than their gloomy masters and inquisitors. The Portuguese changed all this: they converted villeinage into chattel slavery, and religion into an instrument of robbery, torture and oppression. The Jesuits acquired their ascendancy over the natives by preaching a religion artfully filled with liberal promises for all classes. This they called Christianity: it was in reality mere demagogy, which was propagated the more readily by employing the forms and paraphernalia of the religion it was destined to supplant. "The very idols of Buddha served, after a little alteration with the chisel, for images of Christ." When these devices failed, others were resorted to. Fire and sword, as well as preaching, were employed as instruments of conversion. By these and other means they rapidly made a million of "communicants," including many of the daimios, military leaders, officers of the fleet, and other persons of influence, and at the head of these communicants they placed themselves.

Through the control thus obtained they induced the "christianized" daimios to consign their vassals to the mines, where thousands of them miserably perished, though not before they had extracted, for the benefit of the instigators of this tyranny and their countrymen and abettors, the Portuguese traders, a portion of those vast sums of treasure which were shipped to Europe. Another portion of these sums was obtained directly by the Portuguese priests, under the pretence of releasing the souls of the natives and of their ancestors from a purgatory which the priests themselves had invented.[2]

These measures could not long be practiced upon so intelligent a race as the Japanese without exciting suspicion of their true character; but the country, notwithstanding its numerous mines, was poor; the people were separated into feuds, and the daimios longed for those European luxuries which the foreigners offered them in exchange for a metal, whose only cost to them was the sweat and blood of their vassals. Nevertheless, a reckoning day came at last. In 1611 the Portuguese were convicted, by documentary proof, of a design to seize the Shogun and enthrone a usurper, one of their own creatures, in his place; whereupon the Shogun made war upon them and their native allies, until in 1615 they were defeated, and in 1624 every Portuguese was banished the land. In this war, the last in which, until very recently, the Japanese were engaged, more than one hundred thousand lives were sacrificed.[3]

The Dutch had arrived in Japan previous to these occurrences, their first voyage having been made in the year 1600. Down to the outbreak of the war against the Portuguese the former had acquired but little influence, and accomplished but little trade with the natives. Actuated by jealousy of the Portuguese, a jealousy which was sharp-

[2] "The poor natives at home often pawned or sold themselves as slaves to the Spaniards and Portuguese." Griffis, 244. [3] Griffis, pp. 252–6, 292.

ened by sectarian and patriotic hatred, for the Dutch were Protestants and Holland was a rebellious province of Spain,[4] they opened the eyes of the Japanese to the arts of the Portuguese, and thus hastened their expulsion.

The Dutch trade commenced from the time when Iyé-Iasu declared war upon the Portuguese and their adherents, viz., in 1611. From 1624 to 1853 the Dutch were the only Europeans permitted to trade with Japan, and from 1639 they were confined to an islet (Deshima) close, in fact attached, to the port of Nangasaki, limited to a few vessels a year, and restricted in their intercourse with the natives by the most exacting and insulting regulations. For the sake of the enormous profits which the trade yielded at first, they bore these conditions with fortitude, and for their reward managed to obtain about one-half of the total exports of the precious metals from Japan. Very little of this was procured before 1640, or after 1706, when through the medium of the artificial prices (in zenni) fixed by the Japanese upon their merchandise, including the precious metals, the trade became unprofitable.

The Japanese were probably moved to this policy of restriction by the desire of terminating that slavery of the mines which had been greatly extended, if not, indeed, at first introduced by the Portuguese; but in carrying it out, the feudal state of the country, and respect for its powerful lords, obliged the native statesmen to assign a different reason in its behalf. This was the fear of exhausting the deposits of the precious metals. Says a native memorial of the period: "A thousand years ago, gold, silver, and copper were unknown in Japan, yet there was no lack of necessaries. The earth was fertile, and produced the best sort of wealth. Gongen[5] was the first prince who caused the mines to be diligently worked, and during his reign an inconceivably great quantity of gold and silver was extracted from them. These metals greatly resemble the bones of the human body, inasmuch as what is once extracted from the earth is not reproduced, if the mines continue to be thus rapidly wrought they will ere long be exhausted.

"Since these metals were exhumed, the heart of man has become

[4] In 1555 the crown of the Netherlands was inherited by Philip II. of Spain, who soon after his accession commenced those persecutions which eventually resulted in the revolt and freedom of the Low Countries. In 1567 Alva was sent to Holland; in 1598 the crown fell to Philip III. of Spain and Portugal (Portugal belonged to the Spanish Crown from 1580 to 1640), and thus at the period of the arrival of the Dutch in Japan, both Holland and Portugal belonged to Spain.

[5] This was either, as previously stated, the Shogun Iyé-Iasu, who was entitled "Sho ichi i To sho Dai *Gongen*," or his son Hidétada. The regnal period of the first was 1603–4, and of the second 1605–22. Griffis, pp. 273 and 285.

more and more depraved. With the exception of medicines (European drugs) we can dispense with everything that is brought to us from abroad. The stuffs (cloths) and other articles are of no real advantage to us. If we squander our treasure in exchange for them what shall we trade upon? Let the successors of Gongen reflect upon this matter, and the wealth of Japan will last as long as the heavens and the earth.[6]"

The gold of Japan has been largely obtained from placers; the quartz mines having been chiefly opened in comparatively recent times. The placer deposits occur in about one-half of all the civil districts into which Japan is divided, and lie either in the present valleys, or else in the former (parallel) valleys, of the existing rivers; not in transverse valleys, beds, or benches, as in California and Chile. For the most part the auriferous gravel of Japan is comparatively poor, the average yield not exceeding two or three cents worth of gold to the cubic yard of material. Numerous tests yielded 1 to 2 grains of gold per cubic yard. The average amount of gold in large fields did not exceed 3¾ cents in value to the cubic yard of gravel. The richest gravel found in Yesso yielded less than 7 cents per cubic yard, while the average assay value of the best fields was only 5½ cents. The first native washings usually yield about 65 per cent. of these amounts; the second washings add 15 to 20 per cent.; the third washings usually bring the total yield up to about 90 per cent. of the assay value. The extraction of probably 120 to 150 million dollars worth of gold from gravel thus poor bespeaks great privation, suffering, and loss of life to the feudal peasants, who were forced to this work by the exactions of their lords and the sordid intrigues of the Europeans, of whom these nobles had become the dupes.

Mr. Isaac Titsingh, who for fourteen years was the agent of the Dutch East India Company at Deshima, made a collection of Japanese moneys, which began with issues assigned to the seventh century before the christian æra. In 1818 this collection was sent to Paris, and there entrusted to M. Klaproth for arrangement. The first issues appear to have been entirely of bronze. Those which purport to be earlier than the 7th century A. D. are probably false. In A. D. 203 the Japanese plundered Corea of eighty shiploads of gold, silver and other precious merchandise; a circumstance that evinces the Japanese appreciation of the precious metals at an early period of their progressive career. In A. D. 645 (9th Jurukia), says

[6] Memorial presented to the government of the Shogun by a Japanese Finance Minister in 1710, quoted in Martin's "China," I, 292.

Titsingh, occurred the first native coinage of bronze zenni, a statement which implies that his earlier issues were slugs, ingots, or other uncouth pieces, not coins. The Japanese zenni were modelled after the Chinese chuen—round, with a square hole in the middle. In A. D. 604 (12th Suiko), the Empress of Japan ordered the bronze image of Siaka (the Buddhic pope) to be cast into coins and a plaster image to be set up in its place. In the same year a supply of gold was obtained from Corea, whether by plunder or trade is not mentioned. In 669 or 670 silver mines were opened in Japan, and in 675 other silver mines in the island of Tsushima, which lies midway between Japan and Corea. From the silver thus obtained there appears to have been struck an issue of coins, all of which were retired by the Emperor in 682, and replaced with cast bronze zenni. In 708 and 715 new copper mines were opened; and according to Mr. William Bramsen's "Japanese Chronological Tables," Tokio, 1880, this was the year, 708, when the first bronze zenni were cast. But here occurs a conflict of evidence. Mr. Tinsingh fixes this event in 465; Mr. Bramsen says 708; while the Dai-nihon-Kuwahei-Shi, a native work, says 698, adding that these coins were valued at one-fourth of their weight in silver. It goes on to say that in 721 this valuation was lowered, or that of the silver coins raised, until the proportion was 1 to 25. The zenni of this period were struck in the provinces of Chikugen, Harima, Suwo, Nagato and elsewhere. In 722 the relative value of bronze and silver in the coins was altered to 1 to 50; and in 760 gold coins were added to the previous issues. In the coins of 760 the ratios of value between the metals were 1 gold=10 silver; and 1 silver=10 copper.[7] These ratios are evidently arbitrary and of Brahminical sacerdotal origin, with no relation whatever to the market value of the metals at the time, either in Japan or any country with which it had commercial intercourse.

In connection with the sacerdotal attitude of the Mikado should be mentioned another circumstance: The date of the first coinage of zenni, A. D. 698, is the result of a computation based upon the Greek calendar, while the date A. D. 708, which not only appears in Bramsen, but also in the Encyclopedia Britannica, is based upon the Roman calendar. Mr. Bramsen adds that "the first coinage of copper in Japan, A. D. 708, was adopted for a new Nengo" (æra). The same discrepancy of ten years exists between the Greek and Roman Anno

[7] From the native work quoted in the text, as translated by W. Walanobe for Mr. C. Netto, Professor of Mining at the University of Tokio (Yeddo), and in 1880 kindly furnished by the latter to the author of the present work.

Mundi.[8] The two Japanese dates corresponding to A.D. 698 and 708 must therefore be regarded as one.

In 749–50 the gold-and-copper (quartz) mines of Suruga and the gold mines of Oshiou were opened. It is from this date that began the official annals of mining in Japan. In the Kanaba mines are to be seen ancient tunnels, over two miles in length, all cut with hand-tools, evidences that when these works were constructed the use of gunpowder was unknown. "One of these tunnels is said to have been the scene of a desperate fight between the rival miners of two provinces, whose workings, driven from opposite sides of the mountain, met in this place." A similar story is told of one of the ancient Roman mines; probably invented in both cases in order to emphasize the precision with which the galleries were driven by the engineers of the underground surveys.

The electrum or gold and silver mines of the island of Sado are among the oldest and richest in Japan. They have been opened for more than a thousand years and are still worked in some places, the produce, however, not equalling the cost. Yet, the women who carry the ores on their backs from the mines to the works receive but four cents a day for their services. Lock. Many of the mines of this district are water-logged, and will need a vast outlay to render them again workable. In the first quarter of the last century 300 miners lost their lives by the flooding of one of them.

In 1205 several hundred miners from the Oshiou, or Koshiou, province of Nippon, crossed over to the island of Yesso and attacked the native Ainos, at that time in possession of the gold placers of Musa and Toshibets, drove the natives away and jumped their locations In thirteen years' time the usurpers took out some $21,000 worth of gold—an average of about $5 per year per man—then the Ainos rallied, attacked and massacred them all! Lock. Less than seventy years after this, to wit, in 1274, Kublai-Khan's envoys reported to Marco Polo, then living in China, that the Japanese possessed "gold in the greatest abundance, its sources being inexhaustible; but as the King (the Mikado) does not permit of its being exported, few merchants visit the country, nor is it frequented by much shipping from other ports. To this circumstance are we to attribute the extraordinary richness of the sovereign's palace, according to what we are told by those who have had access to the place. The entire roof is covered with a plating of gold, in the same manner as we cover houses, or more properly churches, with lead. The ceilings of the halls are

[8] Del Mar's "Worship of Augustus Cæsar," chap. VII, sub anno B. C. 5492.

of the same precious metal; many of the apartments have small tables of pure gold of considerable thickness; while the windows also have gold ornaments. So vast, indeed, are the riches of the palace that it is impossible to convey an idea of them." Travels of Marco Polo, p. 569.

Such is the immortality of a well-told Lie, that this second-hand story of Kublai-Khan's envoys travelled the whole length of Asia and Europe to Venice, thence to Spain, thence with Columbus to America and with De Soto, half-way across America, until it prematurely perished in the arid wastes of Colorado; but not before it had enjoyed a triumphant career of more than 270 years.

According to the more veracious Kaempfer, "gold is dug up in several provinces of the Japanese Empire. The Emperor (Mikado) claims the supreme jurisdiction over all the mines in his dominions, and demands two-thirds of all that is procured; but of late, as I am informed (writing in 1698), the veins not only run scarcer, they do not yield nearly the same quantity of gold as formerly." Hist. of Japan, I, 107.

The prerogative of gold here adverted to, appears to have arisen from the sacred character attached to the metal by that venerable religion of Asia, from which all its other religions have either sprung, or else by which they have become modified. The religion of Japan is nominally Buddhism; Shinto meaning not a religion but a Way of Life. On the other hand, the establishment of a sovereign-pontiff, in the person of the Mikado, is not Buddhic, but Brahminical; and so is the ordinance of Sacred Gold. However, this ordinance, that is to say, so far as the production of gold is concerned, like many others pertaining to the Japanese pontificate, was usurped by the Shogun, whether before or during the Portuguese intrigues for gold, is not determined. At all events, the date of the Shogun's ursupation of the coinage of gold is fixed in 1598, when Taiko Sama struck temporal coins of that metal. Hildreth, 213*n*. The usurpation of the prerogatives by the Shogunate was accompanied by so radical an abatement of the duty on production that it amounted to virtual abandonment.

In 1610 new discoveries of electrum mines were made in the island of Sado, which thereupon became the centre of the Portuguese intrigue with the daimio Okubo, and the immediate cause of the civil war and the exclusion of foreigners which followed.[9] In 1613 still

[9] The Abbé Raynal, I, 204, also accuses the Portuguese adventurers of marrying the Japanese heiresses, merely in order to possess themselves of their wives' fortunes.

other mines were opened in the same vicinity, and these were worked at intervals and by manual labour down to 1869, when machinery was introduced. In 1880 the principal main shaft was down 600 feet and two of the mines were connected by a gallery 3,000 feet long. The ores yielded about $7 gold and $15 silver to the ton; but at an annual loss to the government (which worked them) of about $25,000 a year. The works employed 1,080 persons, including 120 females, and produced about 20 tons of ore per diem; facts which bespeak a large proportion of dead work, waste rock, and fruitless exploration.

The product of the precious metals in Japan has never been, and will probably never become important, unless, indeed, the newly-discovered leads in Formosa should develop a bonanza. During the past fourteen years the average annual product of gold has been 25,000 fine ounces, say $500,000, and of silver about 1,500,000 ounces, say $1,000,000. The details are shown in the following table:

Production of Gold and Silver in Japan.

Year.	Gold, oz.	Silver, oz.	Year.	Gold, oz.	Silver, oz.
1886	8,827	767,656	1893	21,540	1,916,549
1887	14,963	1,084,852	1894	24,150	2,229,906
1888	16,768	1,147,113	1895	25,553	2,338,229
1889	19,057	1,376,436	1896	30,981	2,078,396
1890	24,709	1,382,611	1897	33,385	1,748,609
1891	23,632	1,703,878	1898	38,253	1,660,200
1892	22,548	1,890,010	1899	*65,000	1,500,000

* The production for the year 1899 is estimated. New and productive gold mines have recently been opened in the islands of Yesso and Formosa, the two extremities of the Empire.

About one-third of the gold and one-sixth of the silver shown in the table were obtained from government mines; the remainder from private properties. Since the demonetisation of silver by the Western world the miners of America and Australia have abandoned the silver mines and have prospected for gold, with the result of decreasing the product of the white metal and increasing the product of the yellow. A similar movement is to be discerned in Japan.

Owing to the long-continued isolation of Japan from the commercial world, the history of the precious metals in that empire is closely and instructively connected with its monetary systems, and these with the progress of its civilization, a pretty full account of which monetary systems is printed in the author's work on "Money and Civilization," chapter XX. This account ends with the following language:

The Japanese are a singularly energetic and intelligent race, their country is insular and free from the dangers of foreign interference, its topography is favourable to industry, the natural resources are abundant, the soil is fertile, the rains are perennial, the climate is temperate, the government is paternal, and the religion tolerant, and of a character that offers no obstacle to progress. Yet, until quite recently, Japan had made no progress for centuries. The land is only sparsely inhabited and partially cultivated, the mechanical arts are in their infancy, the fine arts have enjoyed but scant development, and science is almost unknown. As for recent progress, it is merely the result of European intercourse, and is not known as yet to possess inherent force. Let any of the causes that have been assigned by writers on civilization, as a reason for social retardation or decay, be applied to Japan, and it will fail to explain the backwardness of this country. The sun shines there as brightly as elsewhere, the rain falls as favourably to man, the soil is as rich, the government is as mild; yet Japan from the societary point of view, has been a petrifaction, except at rare intervals, ever since its history began.

These rare intervals of progress offer the only solution to an otherwise unaccountable phenomenon. In every instance these were intervals of increasing moneys and rising prices. The same correspondence between these occurrences, which Hume observed in the affairs of England and Alison in those of Rome, is to be found in the history of Japan. Not high, but rising prices, have invariably been followed by progress; and not low, but falling ones, by decay. The stationary condition which has characterized Japan is to be imputed neither to the influence of nature nor the operations of individual men, but rather to those governmental arrangements, foremost among which stands Money, which, instead of promoting the development of civilization, has proved to be an obstacle and a drag.

CHAPTER XXVI.

PLUNDER OF ASIA.

Darius—Alexander—The Romans—The Huns—Moslem raids—Mahmoud of Ghazni—Mahomet Ghor—Sack of Hindu temples—Sack of Guzerat, Delhi and Benares—Genghis Khan—Kublai Khan—Timur, the Tartar—Second sack of Delhi—Baber—Third sack of Delhi—Sevaji—Sack of Surat—Nadir Shah—Fourth sack of Delhi—Ravages of the Portuguese—Plunder of the sea-ports and islands—The Dutch—The French—The British—Enormous wealth drained from India and remitted to Europe—Present poverty of the country—Its exposure to famine—Degradation and desperation of the people—Plunder of India and China.

ABOUT B. C. 500 Darius of Persia undertook that series of military expeditions, which, like the later ones of the Spaniards in Mexico and Peru, appear to have had for their principal object the acquisition of the precious metals; for after having attained this object their other fruits seem gradually to have been abandoned. Darius acquired and plundered the Punjab, Bactria, Asia Minor, including Phœnicia (at that time the sole depot of the metallic products of Spain), the Greek islands, Thrace, Egypt, etc., and carried away to Persia probably the bulk of the gold of placer origin which was then to be found in Hither India, Europe, and Africa. Of this gold there are numerous coined specimens, known as staters or darics, still in existence in the various numismatic cabinets of Europe.

Besides obtaining this spoil, Darius levied tributes in gold and silver upon the conquered countries, amounting annually to 13,710 Euboic talents of gold, a sum which has been computed as equivalent to about three millions sterling. According to the inference of Gibbon, this revenue was the surplus after discharging the expense of maintaining the army and provincial administrations. Whatever the amount was, the tribute was probably paid for only a few years, because not long afterwards the conquered countries recovered their independence.

Cyrus, in his plunder of Asia Minor, secured a booty of 24,000 Roman pounds weight of gold, the total value of which in our present American coinage would be about five million dollars, "besides vessels and other articles of wrought gold, as well as thrones (solia) of gold, a plane-tree and a vine, all made of that metal. . . . He also carried off 500,000 talents of silver, and a silver vase called the Vase of Semiramis, of 15 talents weight, the Egyptian talent being equal, according to Varro, to eighty of our pounds." Pliny, XXXIII, 15. If the half million "talents" of silver are also meant as weights the statement is incredible; but if they are meant as talents of account their value was five gold staters, or about $25, each, in American coin. The half-million silver talents would therefore be worth about twelve to fifteen million dollars in American coin. Hist. Mon. Systems, American edition, p. 95*n.*

Alexander was the next conqueror who plundered Asia, and it is likely that a considerable portion of the precious metals which Darius had acquired, now found their way back to the Occident. In addition to this, a prodigious spoil was acquired by Alexander from India. Strabo, XV, III, 9, sums up the result of his expeditions at 180,000 talents, but does not specify either the metal or the talents. The rendition of these weights, or sums, whichever they were, has engaged the industry without determining the doubts of Letronne, Boeckh, Arbuthnot, De Vienne, and numerous other metrologists. We shall refrain from entering upon this perplexing subject and content ourselves with quoting the conclusion of Athenæus, who says, VI, 19: "When Alexander had brought into Greece all the treasures of Asia then really there did shine forth, to use the language of Pindar, 'wealth predominating far and wide.'" Not only were the treasuries of Babylon, Susa, Persopolis, and Pasagarda,[1] sacked, the mines of Bactria were plundered and the graves of Persia and Hither India despoiled. Gmelin, who visited the mines of Bactria in the early part of the 18th century, and inspected what remained of the old workings, states that the ancients excavated the passages underground with incredible pains, and in numerous instances lost their lives by "caves." The underground passages were so narrow that the workmen must have crawled and wriggled through them like worms, in order to get at the quartz, from which they picked the gold (as he believed) with the sharpened fangs of boars. How they ob-

[1] Exped. Alex., vol. III, pp. 16–18. Consult also note to Rawlinson's Herod., 396. Garda, grad, and gorod, are variants of a Gothic word, meaning city. The name of a city on the Gulf of Persia, ending with *garda*, is an indication of its settlement by some Gothic tribe. Traces of a Getic or Gothic invasion of this portion of Asia are to be found in many ancient authors. As worshippers of Buddha the Getæ were called Budini, by Herodotus; from their custom of drinking out of their enemies' skulls (still commemorated in the toast of "Skoll!") they were classed as Anthropopagi, by Pliny, VII, 2, etc. Herodotus and Strabo call the Buddhists Musioi.

tained light and air or removed the worthless rock and debris, is not described. The writer himself having visited many ancient quartz mines, found that the passages, though narrow and irregular, were large enough to enable the miner to swing a hammer, and that as for picking out gold from quartz with boars' teeth, the marks on the rocks very plainly indicated that such teeth were made of steel. What was the origin of the Asian gold or how originally procured, the followers of Alexander did not stop to enquire. The form in which they obtained it was Oriental coins, plate and jewelry. Much of this appears to have been coined anew in the field, probably as fast as it was plundered. To enable him to pursue this policy, Alexander must have carried in his train a corps of expert mintners. The same practice is known to have been afterwards followed by Hannibal in Italy.

Between the expeditions of the Greeks and those of the Romans, Asia does not appear to have been despoiled by any people who carried its treasures to distant countries. As the Roman plunder of Asia is mentioned in another portion of this work, it only remains to describe the various plunderings which took place in later times. Among these the first in order was the invasion and plunder of Northern India, by the White Huns, in the fifth century of our æra. Of this expedition we have only the most meagre account. It is mentioned by the historian Priscus, who was in the camp of Attila, A. D. 448. It is also alluded to by the Chinese writer, Sung-yun, A. D. 520, who calls the conquering people the "Yesha," or Iesha. In the Skanda Gupta inscription, A. D. 450 to 480, they are called "Hunas." All we know of this conquest, is that immediately afterwards the Huns commenced to strike gold and silver coins, of which some examples are still extant. From these coins we gather the names of about a score of Hunnish kings, whose reigns are believed to have extended from A. D. 402 to A. D. 554. On the coins of one of them, by name Toramana, appears the date "52," which is supposed to commemorate the year of the Indian Invasion. This conqueror is said to have built or dedicated a temple to the Sun at Mooltan, on the Ravi, an affluent of the Chenaub. (Num. Chron. 1894, III.)

The Moslem raids in India began as early as the eighth century, but they exercised no noticeable influence upon the history of the precious metals until the eleventh century, when the celebrated Mahmoud of Ghazni, or Ghuzni (in Afghanistan) conducted that series of predatory expeditions which eventually transferred a considerable portion of the metallic wealth of India to the Arabian Empire in Asia Minor and Europe. These expeditions began soon after Mahmoud's elevation to his father's throne, A. D. 997, and continued at intervals until the period of his death, in 1030.[2] This conqueror successively plundered Batinda, about 1000; Nagercot, 1008, with its rich temples and immense treasures; Tanesar, 1010, with its temple and treasure; Muttra, 1017; Lahore, 1023; and Guzerat, in 1024, with the enormously rich temple of Somnath.

Professor Wilson thinks that the wealth of this famous temple has been greatly exaggerated, places no confidence in Ferishta's account of the hollow idol filled with gold and precious stones, and ascribes the "crores" and "millions" of treasure to the mistakes of Dow and Gibbon; whilst Raynal, (II, 341) believes that the Afghans gathered immense spoils which they "buried under ground in their wretched and barren deserts." It is difficult to discern the truth amid these conflicting accounts. Certain is it that after the return of these plundering expeditions to Ghazni, Mahmoud celebrated his success in "triumphant feasts, alms to the poor and presents to persons distinguished for supposed merit or sanctity," (Garrett, 37,) and equally certain is it that a vast current of the precious metals began to flow from Afghanistan and India to the Arabian Empire. This flow was interrupted in 1152, when the Ghorians captured and sacked Ghazni itself, but it was followed a quarter of a century later by another influx of spoil, when Mahomet Ghor captured and sacked the splendid city of Delhi. In 1195 this same chieftain sacked Benares and loaded four thousand camels with the wealth of that sacred city. Following a similar career to that of Mahmoud, he made nine predatory expeditions into India and accumulated treasures which almost rivalled those of his great predecessor.

India was next plundered in A. D. 1217, by the Mongol Tartars, under Genghis Khan. This conqueror, who claimed, like Alexander, to be an incarnation of the Deity, marched, in 1214, from the vicinity of Lake Baikal to the conquest of China. After capturing and sacking upwards of ninety cities of the Empire, including Pekin, his forces only rested when they had penetrated to the Shan-tung peninsular and ravaged Corea. At Pekin, the besieged, "when their ammunition was spent, discharged ingots of gold and silver from their engines." (Gibbon, VI, 294.) In 1219 Genghis organized four great Mongol armies, which overran Lahore, Peshawur, Melikpur, Turkestan,

[2] In Briggs' "Mahometan Power in India," I, 33, quoted by Ball, 209, it is related that in the first month of Mahmoud Ghazni's reign, his followers discovered a mine in Seistan, containing a gold streak "three cubits in depth" and shaped like a tree. This streak was of pure metal, and the mine continued to be worked until the reign of Musaud, when it was destroyed and lost by an earthquake.

Transoxania, Chorasmia (to Khiva), Persia, and all the intervening countries. The cities of Otrar, Samarcand, Khogend, Balkh, and Herat, all fell to his arms. In the precincts of Herat alone no less than 1,600,000 persons were slaughtered; the total number of lives destroyed by the Mongols being estimated at between five and six millions. The plunder was enormous. The entire metallic wealth of the countries named, except what the terrified inhabitants saved by burying it, would seem to have been carried away by this scourge of the earth, who, in one day, distributed 500 wagon loads of gold and silver to his followers. (Gibbon, vi, 305.) Yet all that remains of the vast Empire which Genghis founded is a granite tablet erected near the silver mines of Nertschinsk, dug up in recent years and deciphered by Prof. Schmidt of Petersburg. It commemorated in the Mongol language the conquest of the kingdom of Kara-Khatai.[3]

In 1290, Ala, or Alau 'a din, the nephew of Feroze I., Patan king of Delhi, after having assassinated his benefactor and uncle, invaded the Dekkan, took Deogiri (now Dowlatabad) by artifice, and returned to Delhi with immense treasure. The ransom paid by the rajah of Deogiri, or Deoghar, alone amounted to 17,500 pounds weight of gold Lock, 306. Ala successively plundered Guzerat, Telingana, Carnata, and Malabar. It was this usurper who affected the names of Isskander and Jul-al-addin (Divine Glory) and meditated setting himself up for an incarnation of the Deity. He died from poison in 1316, the twentieth year of his reign.

In 1252 Kublai, the grandson of Genghis, conducted a marauding expedition through the heart of China to Yunnan, and in 1259 he ascended the throne of his ancestors as Grand Khan of Mongolia, or Tartary. In 1264 Kublai removed the seat of his Empire from Karakorum to Yenking (now Pekin), which he thoroughly rebuilt and fortified. From this centre, and after repeated campaigns, covering nearly half a century of time, the whole of China and the Western steppes were subdued, until Kublai's sway extended northward to the Arctic Ocean, westward to Poland, southward to the Strait of Malacca, and by the family of his brother Hulaku, south-westward from the Oxus to the Arabian desert, a greater extent of territory than was ever previously governed by a single monarch. It was the court of Kublai Khan that was visited by the Venetian merchant Marco Polo. Kublai conquered Corea and even threatened Japan, but here

[3] At Karakorum, the original capital of Genghis, there was found by the monk Rubruguis, in 1254, a Paris silversmith, Guillaume Boucher, who had executed for the Khan, "a silver tree supported by four lions, and ejecting four different liquors." Gibbon, vi. 306*n*.

his good fortune failed him, and his expedition of 100,000 men was entirely destroyed. Chagrined at this and the loss of his son, he died soon after, in 1294, at the age of 78: Du Halde says 80. The Mongolian name for Pekin was Ham Balu, or King's Court, written Kam Balu, or Cambalu, by Marco Polo. Kublai destroyed the Tao books, altered the calendar of China, and built the great canal 300 leagues long, chiefly for the purpose of conveying the tributes of the Empire to Pekin. But little mention is made of the plunder which he accumulated, and it is reasonable to believe, both from this circumstance and the honourable character of his sway, that so far as the precious metals are concerned, it was comparatively small. Kublai's Chinese titles were Chi Tsou, Chi Yuen, and Ho-pi-ley.

Lock computes the accumulations of Kales Dewar, rajah of Mabar in 1309, at 1,200 crores of gold, which he converts into "1,200 millions sterling, or 90,800,000 pounds" weight of gold! The accumulations were, in fact, measured in crores of billon coin, and therefore amounted to infinitely less. The same author gives as "part of the spoil of Devara Samudra (Halebid, Mysore) presented, in 1310, by Kafur to the Emperor of Delhi, 2,400,000 pounds (weight) of gold," another preposterous computation, the result of a similar blunder. The reader who desires to make the proper calculations for himself will find the materials in the "History of Monetary Systems," London edition, chapter on "India." The original amount from Ferishta is 96,000 maunds of gold, besides 20,000 horses and 312 elephants, all pillaged from Halebid. Taylor, 110.

In 1398 India was again plundered, this time by Timur, or Tamerlane, a Tartar chieftain, who, under a solemn promise of protection, induced Delhi to surrender to his arms, when he resigned it to the fury of a savage host, who condemned its hapless people to every horror, in order to extort from them the secret of their hoards of gold and silver. In this design his followers succeeded; and loaded with immense booty, they left the city to desolation and ruin. A famine and pestilence afflicted the few inhabitants whom the sword had spared; so that many generations elapsed before Delhi assumed its wonted appearance.

In 1441 Abdul Razzak, the ambassador from Persia to the court of Delhi, mentions the "enormous quantities of gold which were to be seen in the King's treasury at Vijayanagar, or Hampe, in the Bellary district"; but he gives no specific sum. It is evident from these references that the Mahometan conquest of India kept a large number of natives forcibly at work washing out gold from the attenuated gravels of the river-beds; with what outcome of privation, misery and death we have no account. The means employed, though not literally chattel slavery, was a system of grinding taxation, which had a similar effect upon the natives. It is probably this form of tyranny which gave rise to that aversion to dig for, or even to use, the precious metals, which Mr. Ball mentions in connection with the Buddhists of Laddak.

In 1526, Baber, the sixth in descent from Timur, invaded India with an immense army, captured Delhi and distributed its treasures among his soldiers. He successively reduced and plundered the Punjab, Oude and Rajpootana, fixing his capital at Agra and establishing that Mogul empire which eventually embraced nearly the whole of India and continued to exist down to the period of the European invasions.

In 1664 (Raynal), or else in 1670 (Garrett), Sevaji,[4] a Mahratta chief, captured the city of Surat, the principal seaport of the Grand Mogul's dominions, with £1,200,000 ($6,000,000) in treasure. At that period the British had a fortified factory in Surat, which they had been permitted to occupy since 1612. This they defended with so much courage that Sevaji abandoned his attempt to reduce it. In recognition of this act, the Grand Mogul (Auranzebe) remitted the duties on all British goods. (Raynal, II, 257; Thornton, 948.)

In 1736 (Macgregor), 1738 (Raynal), 1740 (Garrett), or 1749 (Haydn), Nadir Shah, an usurper of the throne of Persia, invaded Afghanistan, reduced Cabul and Candahar and then marched upon Delhi with such rapidity that it had not sufficient time to prepare for defence. "Till mid-day the streets of Delhi streamed with blood; after which the conqueror suffered himself to be appeased—and so complete a control did he exercise over his rude followers—that at his mandate, the sword was immediately sheathed." . . . The imperial repositories were now ransacked and found to contain specie, rich robes, and above all, jewels to an incredible value. The Mogul emperors, since the first accession of their dynasty had been indefatigable in the collection of these objects from every quarter, by presents, purchase, or forfeiture; and the store had been continually augmented without suffering any alienation, or being exposed to foreign plunder. The invaders continued during thirty-five days to extort by threats, torture and every severity, the hidden treasures of that splendid capital. Historians estimate the spoil carried off by

[4] Sevaji's name, like most Oriental ones, is spelt in many different ways, the last and most approved Anglo-Indian form being Shivaji. The Madras "Indian Review," April, 1901, p. 184.

the Iranian monarch and his officers at £32,000,000 ($160,000,000), of which at least one-half was in diamonds and other jewels." (Garrett, 136). Included in this spoil was the Peacock Throne and the Kohinoor diamond. Macgregor estimates the value of this plunder at £125,000,000; Haydn, not to be outdone, swells it to the preposterous figure of £625,000,000!

We now come to the ravages of the Europeans. Before the close of the 16th century the Portuguese had virtually sacked the ports of the Orient of their accumulations of gold, besides acquiring such other riches and advantages as enabled Portugal to rapidly rise from the condition of a petty kingdom in Europe to that of a first-class power. They began by ransacking the coasts of Africa, east of the Cape. They established an empire extending from Sofala to Melinda, of which the island of Mozambique was made the centre. The principal object of this establishment was the search for gold. They explored all the rivers, opened a trade for gold with the negros, and delved for gold at Tatti. But this last was an exceptional case: their usual device for obtaining gold was not mining, but piracy.

"They supposed that the Pope of Rome, in bestowing the Kingdom of Asia upon the Portuguese monarchs, had even included the property of individuals. Being absolute masters of the Eastern seas, they extorted a tribute from the ships of every country; they ravaged the coasts, insulted the princes, and became a terror and a scourge. The King of Tidor was carried off from his own palace and murdered with his children, whom he had entrusted to the care of the Portuguese. At Ceylon the people were not suffered to cultivate the earth, except for the benefit of their new masters. At Goa they established the Inquisition, and whoever was rich, became a prey to that infamous tribunal. At the island of Calampui, Faria plundered the sepulchres of the Chinese emperors. In Malabar, Souza plundered and destroyed all the pagodas and massacred the wretched Indians who went to weep over the ruins of their temples. The Portuguese Correa made a treaty with the King of Pegu, and swore to it on a book of songs, with the view to disregard his oath and betray his friend. Nuno D'Acunha, after sacking Diu, and other places in Guzerat, descended upon the island of Daman, on the coast of Cambaya, where the inhabitants offered to surrender to him, if he would permit them to remove their effects. He refused, put them all to the sword, and carried off their treasures. One of the commanders, for a consideration, granted a passport to a Moorish ship richly laden. Another commander, Diego de Silveira, cruising in the Red Sea, overhauled the

Moor, and after examining the passport, put the bearers to death. The passport read: 'I desire the captains of ships belonging to the King of Portugal to seize upon this Moorish vessel as a lawful prize.' In a short time the Portuguese preserved no more humanity or good faith with each other than with the Moors or Indians. Almost all the States where they had command, were divided into (Portuguese) factions. Avarice, debauchery, cruelty and devotion marked their manners. Most of them had seven or eight concubines whom they kept to work with the utmost rigour and forced from them the money they earnt by their labour. . . . The King of Portugal no longer received the tributes which a few years back had been paid to him by more than one hundred and fifty eastern princes." It was embezzled by his servants. (The Abbé Raynal, I, 207.)

In 1507–11 the Duke of Alboquerque plundered all the Arabian and Persian and many of the Indian ports, from the Red Sea to the Straits of Singapore. Among these were Muscat, Ormuz, Goa and Malacca, the latter with precious metals to the value of one million gold crusados, each of about the weight of a guinea, or half eagle. The atrocities committed by Alboquerque, were they not avowed by the man himself and corroborated by several respectable authors, would exceed belief. Accounts of them will be found in the "Commentaries of Afonso D'Alboquerque," published in Lisbon, 1774, translated into English by Walter De Gray Birch, and published in London, 1875, by the Hakluyt Society; also in the "Toh fut-al Mujahidden," an Arabian MS. work in the British Museum (Addl. MS. No. 22375), and translated into English by Lieut. M. J. Rowlandson, London, 1883. The limits of the present work will only admit of some brief abstracts; but the reader who desires further light concerning the cost of gold to Europe in the 16th century, and the influence of such cost upon its present value, will do well to consult the original works for himself.

Sailing with a warrant from the King of Portugal to take "free booty" everywhere in the Eastern Seas, Alboquerque began his piratical proceedings at Braboa, or Brava, an Arabian port on the Zangueban coast, which carried on a trade for gold with Sofala and Monomotapa. Practically his demand was: "Surrender everything, especially gold, silver and women, or perish by the arquebus and sword!" Refusal was answered by assault and followed by butchery. No quarter was given to any but such a number of maidens as fulfilled the requirements of the murderers. All else, man, woman and child, were slaughtered, and the Hebrew bible was impiously cited as authority for the atrocious deed. The houses were razed in the search for treasure, the mosques were robbed of their sacramental vessels and "reconsecrated" as christian churches, together with their lands and appurtenances, and the graves were opened and despoiled before "the honourable cavaliers," as they regarded themselves, set sail for the Indies; where Goa, Malacca, and several other places fell to their treachery or their arms. It was on his return voyage that Alboquerque attacked and destroyed the Arabian emporia at Ormuz, Muscat, etc. At Muscat the order was to put all the Arabs to the sword, both men, women and children, to spare none, save for the purpose of torture, in order to extort the secret of such hoards of gold and silver as the natives might possess. The noses and ears of the Arabs were cut off, and their bowels torn out. A Portuguese cabin-boy, armed with a match-lock, boasted of having killed in a day eighty Arabs, the limit of either his ammunition, or his strength to reload. The hoards and sepulchres were then robbed and the pirates sailed away, leaving only death and desolation behind them. At Ormuz the resistance of the Arabs led to a parley, in which the native chief or governor was required to acknowledge fealty to the King of Portugal, and pay a tribute of 30,000 xerafins per annum.[5] This being refused, a siege was commenced, in which the city suf- all the horrors of thirst. Beginning at 10, the jar of water rose to 200 xerafins, and finally was not to be procured at any price. With the women and children around them dying from thirst, the stout hearts of the Arabs finally gave way, and Ormuz capitulated, only to suffer extinction from the sword. No quarter was given. The order was to kill all, except the cattle; and with their lances and swords dripping innocent blood, and their wallets filled with gold and silver, whose cost of production is claimed by the modern political economist to be a mere arithmetical problem, the Portuguese again hoisted their sails. From Kalhat (Calayate) they only carried away 100,000 xerafins: a sum which seemed to men who were surfeited with plunder, contemptibly small.

It would fatigue the reader to continue an account of the atrocities which Alboquerque committed to acquire the gold and silver of the Orient. The pretence which appears in one portion of his "Commentaries," (Birch. p. 119), that these crimes were committed "for the Glory of the Passion of Jesus Christ," is falsified by the fact that where the "infidels" were found too strong to be murdered and de-

[5] Mr. Birch regards the silver xerafin of Ormuz as equivalent to about 15 pence sterling, while according to Kelley's Cambist, it was worth as much as the Spanish silver dollar.

spoiled, they were invited to make treaties of amity and trade. No quantitative account of Alboquerque's plunder has fallen under the observation of the writer, but judging from the spoil obtained at certain places and the affluence of the victors, it could hardly have amounted to less in value than fifty million dollars. One-half of this sum was obtained from Malacca and Muscat alone. De Villars informs us that Alboquerque exhibited among his own private property upon his return to Portugal, 120 dozen silver plates, 500 large silver dishes, 500 small silver dishes, and 40 silver ladders, with which to mount to the silver repositories, besides numerous other articles of the same precious metal. ("Memoirs," ed. 1861, p. 7; Edin. Rev. Jan. 1861.) His countrymen, in admiration of his military prowess, dubbed him the Portuguese Mars, and are never tired of singing his praises. In fact, he was a coward who murdered inoffensive women and innocent children, and a thief who robbed peaceful communities and dug into the graves of the dead for spoil with which to augment the splendour of his own palace. This palace is now a ruin; the Crown which he dishonoured has long been the football of British politicians; while Portugal itself, naturally one of the most beautiful and fruitful countries of Europe, is abandoned by its people, until many parts of it are a mere desert, in which the indigent survivors exist upon the pods of the carob tree. Lippincott's Mag., 1871.

Of the enormities which the Portuguese committed in Japan, by means of which they obtained from that country gold and silver to the amount of about thirty millions sterling ($150,000,000), we have written elsewhere. The decline of the Portuguese power in the Orient antedates the period, A. D. 1580, when Portugal was annexed to the Crown of Spain. Says Raynal: "The original conquerors of India were none of them now in being; and their country, exhausted by too many enterprises and colonies, was not in a condition to replace them. At the same time that they gave themselves up to all those excesses which make men hated, they had not courage enough left to inspire the people with terror. They were monsters: poison, fire, assassination, every sort of crime was become familiar to them; nor were only private individuals guilty of such practices; men in office set them the example. They massacred the natives, they destroyed one another. The newly-arrived governor often loaded his predecessor with chains, that he might deprive him of his wealth. The distance of the scene (from Europe), false witnesses and heavy bribes, secured every crime from punishment.

"The island of Amboyna was the first to avenge itself. A Portuguese had at a public festival seized upon a very beautiful woman; and regardless of all decency had proceeded to the greatest of outrages. Under the leadership of Genulio, the islanders flew to arms, and before attacking the Portuguese, thus addressed them: 'To revenge affronts such as this one calls for action, not for words: yet listen! You preach to us a Deity whom you say delights in generous actions; yet theft, murder, obscenity and drunkenness are your common practices. Your hearts are inflamed with every vice. . . . Go, fix your habitations among those who are as brutal as yourselves. . . . The Itons are from this day your enemies; fly from their country and beware how you approach it again!"

It is hardly necessary to turn from the Portuguese to the proceedings of the Spaniards in the Orient. They were simply the counterpart of previous doings in America. The Moluccas or Spice Islands, the Philippines—indeed every place conquered by them was robbed of its gold, enslaved, and subjected to the most revolting cruelties. If the reader is not tired of these horrors let him read Mr. Horace St. John's "Indian Archipelago," where he can get his fill. In a single day the Spaniards betrayed, entrapped and massacred 20,000 Chinese in the Philippines, and on another occasion 33,000, of both sexes and all ages; the motive in these cases being simply gold. St. John, pp. 238, 327. The object of the present work is not to prove how cruel man becomes in his search for the precious metals, but to place before the reader the evidence that our Institution of Money, so far as it is based upon the theory that the value of gold conforms to the economical cost of its production, is built upon a fatal error; that the cost of gold is not to be measured in economical effort, but in blood, in tears, in hellish passions and in beastiality; and that the economical cost is unknown and unknowable.

We now turn to the transactions of the Dutch. The early pecuniary success of the Dutch East India Company has remained a mystery, which the laboured efforts of De Luzac never succeeded in explaining. The exports from Europe were comparatively small; the imports from Asia were immense. Whence arose the difference? From commercial profits? Impossible: the difference amounted to some thousands per cent. The East Indians are among the shrewdest traders in the world. The honest Abbé Raynal explains the whole matter in a few words: whilst the Portuguese robbed the Indians, the Dutch robbed the Portuguese. "In less than half a century the ships of the Dutch East India Company took more than three hundred Portuguese vessels . . . laden with the spoils of Asia. These brought the Company immense returns." Thus we find that much of the Eastern gold, which found its way to Amsterdam, was the proceeds of a double robbery.

One of the earliest of the Dutch acquisitions in the Orient was a fort which the natives permitted them to erect near the present city of Batavia. Here they conducted an establishment which was called a trading factory, but which was little more than a trap for slaves, who were captured in the Moluccas, the Celebes, and other islands, and brought hither to cultivate spices or else shipped to Sumatra, to dig gold for their Dutch masters. Raynal (1,337) estimated these slaves at 150,000. They were not liberated until 1850, before which time spices had ceased to be a monopoly, and the gold of Russia and California had rendered the mines of Sumatra practically worthless.

Of these mines (drift mines in banket), near which the Dutch built their factory in 1649, Marsden (1812), says that in the domain of Menangkabau alone, to say nothing of others, there were 1,200; and that the annual receipts of gold at the ports, whither only a portion of the product found its way, were about 15,000 ounces. Crawford (1820), estimated the total annual product of Sumatra at 35,530 ounces, or £151,000, or $671,125. According to Verbeck (1879), this rate of production continued throughout a long period. (Lock, 460.)

In 1682 the Dutch East India Company sent out a party of miners from Saxony under an expert named Benj. Olitzsch, to open up a quartz mine at Sileda, but the undertaking was opposed by the natives and several of the miners were killed. Attempts on the part of the Company to work the gravel mines beyond the protection of their forts, met with little better success; whereupon it was shrewdly resolved to resign the actual working of the mines to the natives and obtain the gold by exchanging for it the iron, opium, and piece-goods of trans-marine origin, which the natives coveted. The process by which the gold was obtained was therefore as follows: One hundred and fifty thousand slaves are captured in the Moluccas, Celebes and other islands to cultivate spices in Java; the spices thus raised are exchanged for European goods brought to Batavia, and these goods are exchanged for the gold washed out by the natives of Sumatra. It would be interesting to learn, from those sciolists whose clamour and false doctrine on this subject fill the commercial world, what they regard as the economical cost of the production of gold in a case of this sort.[6]

[6] In the ancient Indian mines of the Wynaad, in Mysore, a large number of skeletons were found lying in such positions as to establish the inference that they were those

The Dutch, who knew the importance of Malacca, used their utmost efforts to conquer it for the sake of its trade in gold, tin, opium, ivory and linen. Having failed in two attempts, they had recourse at last to artifice. "They endeavoured to bribe the Portuguese governor, whom they knew to be covetous. The bargain was concluded, and he admitted the Dutch into the city in 1641. The Dutch hastened to his house and murdered him, to save the payment of the half million of livres ($100,000) which they had promised him. . . . The Dutch commander then ironically asked the leader of the Portuguese forces when he thought he would be able to recover the place. 'When your crimes are greater than ours,' gravely replied the Portuguese." (Raynal.)

In 1658 the Dutch obtained possession of Ceylon by a similar act of treachery. This island was valuable for its trade in precious stones, pearls and gold. The Cinghalese name for this metal is rang. Amurang means native gold; ratrang is bullion. The names of many places in Ceylon indicate the presence of gold, as Rangboda (gold-town), Ruanwelle (golden-sand), and Rang-galle (the rock of gold). Ptolemy and Solinus both mention the gold of Ceylon. It is referred to in the Mahawanso. The natives are well aware of its presence, and frequently wash out small quantities of it from the river gravels. Spilberg, the Dutch admiral, persuaded the King of Ceylon that the Portuguese intended to subvert the government and religion of his country, and he offered his disinterested assistance to drive out those wicked people, stipulating, however, that he should be permitted to build a fort on the island. The offer was joyfully accepted, the grateful monarch offering both himself, his wives, and his children, to bring together the necessary materials for this work. It was thus that in 1658 the virtuous Hollanders made themselves masters of the island.[7]

In 1663, after the Portuguese had been driven from most of their ill-gotten possessions in India, they still maintained their ground in Malabar, whereupon they were suddenly attacked by the Dutch, who took from them Culan, Cananor, Grandganor, and Cochin. "The victorious general had but just invested this last place, the most important of all, when he received intelligence of a peace that had been concluded between Holland and Portugal. This news was kept secret. The operations were carried on with vigour, and the be-

of workmen who had perished from the caving of the mines. Even at the present day, when mines are opened systematically and supported in weak places by heavy timbers skilfully adjusted, the caving in of a mine is not an infrequent occurrence, and in some instances is attended by great loss of life.

[7] Mavor's Voyages.

sieged, harrassed by continual assaults, surrendered on the eighth day of the siege. The next day a frigate arrived from Goa with the Articles of Peace. The conquerors gave themselves no trouble to justify their treachery, further than by alleging that those who complained had followed the same course in Brazil a few years before." (Raynal, I, 303.)

"For two hundred years the Dutch enriched themselves by the sale of cloves and nutmegs. To secure themselves the exclusive trade in these articles they destroyed and enslaved the nations who were in possession of these spices; and lest the price of them should fall, even in their own hands, they rooted up most of the trees and have frequently burnt the fruit of those they possessed." (Raynal, II, 415.) Their policy with respect to gold was of a similar character. Gold they could not destroy, but gold coins they rendered scarce by abrogating the customary seigniorage and removing all impediments to the melting down or exportation of coins. This policy, ruinous to any State but that one which happens to command the carrying trade of the world, was followed by Philip of Spain, in 1608, and by Charles II. of England, in 1666, and thoughtlessly copied into the Mint Act of the United States, adopted in 1790.

After the depredations of the Portuguese and Dutch, those of the French seem light by comparison. Like their predecessors, they plundered many of the sea-coast towns, and, under pretence of defending the native princes against their neighbours, managed to gain from them valuable concessions and monopolies of trade, which savour more of blackmail than commerce. Among the places thus acquired was a territory near Pondicherry, the territory of Kurical, and the city of Mausulipatam with its dependencies, with annual revenues computed at £43,250, together with £100,000 in cash out of the £2,500,000 plundered from the treasury of Nazir-jing, soubah of the Dekkan, whom the French assisted to dethrone in favour of his rival, Muzaphe-jing. For defending the nabobship of Arcot in favour of the same house, the French obtained from Salabut-jing the provinces of Mustaphanagar, Ellore, Rujamundy, and Chicacole in full sovereighty. "The acquisitions, added to Mausulipatam, rendered the French masters of the sea-coast of Coromandel and Orissa for 600 miles, from Moatapillo to Jaggernaut. The revenues of these territories were computed at 42,87,000 rupees; and the French now ruled over a greater dominion in extent and value than had, up to this period, ever been possessed by Europeans in India." (Macgregor, IV, 301.) In the Anglo-French convention of 1755, the French were to be undisturbed in the enjoyment of these revenues, which were then computed at 68,42,000 rupees, or £855,250 sterling.

The Abbé Raynal claims that in the war between France and England, which followed the deposition of James II., the French privateers took no less than 4,200 English merchantmen, valued at 675 million livres (about 135 million dollars), and that these vessels included "the greatest part of the ships returning from India" laden with the plunder and other merchandise which their rivals had amassed in the Orient. (Vol. II, p. 35.) On the other hand, Macgregor, vol. IV, estimates the loss to England at not more than 3 or 4 per cent. upon the entire trade, including losses from wrecks. This proportion, however, appears to be derived from the rates of insurance covering the entire period of British commerce with India, and is not confined to that of tne French privateering expeditions.

The account which Raynal gives of the misconduct of his countrymen in India does not differ much from what we are told of the other Europeans in those distant regions: "The misfortunes which befell the French in Asia had been foreseen by all considerate men who reflected on the corruption of the nation. Their morals had degenerated in the voluptuous climate of India. The wars which Dupleix carried on in the inland parts had laid the foundation of many fortunes. They were increased and multiplied by the gifts which Salabatjing lavished on those who conducted him in triumph to his capital, and fixed him on the throne. The officers who had not shared the dangers, the glory, and the benefits of those brilliant expeditions, found out an expedient to comfort themselves under their misfortunes; which was, to reduce the sipahis (Sepoys) to half the number they were required to maintain, and to apply their pay to their own benefit; which they could easily do, as the money passed through their hands. The trade agents, who had not these resources, accounted to the Company but for a very small part of the profits made upon the European goods they sold, though such profits ought to have been all paid to the Company, for whom they also bought merchandise in India at a very high price, which the Company ought to have had at prime cost. Those who were entrusted in collecting the revenue of any particular place, farmed it to themselves under Indian names, or let such farms for a trifle, upon receiving a corrupt gratuity; they even frequently kept back the whole income of such estates, under pretence of some imaginary robbery or devastation, which had made it impossible to collect it. All undertakings, of what nature soever, were clandestinely agreed upon and became the prey of persons employed in them, who had found means to make them formidable, or of such as were most in favour, or were the richest. The authorized abuse which prevails in India, of giving and receiving presents on the conclusion of every treaty, has multiplied these transactions without necessity. The navigators who landed in those parts, dazzled with the fortunes which they saw increased fourfold from one voyage to another, no longer regarded their ships but as means of carrying on private trade and acquiring clandestine wealth. Corruption was brought to its greatest height by people of rank, who had been disgraced and ruined at home; but who, being encouraged by what they saw, and impelled by the reports that were brought to them, resolved to go themselves into Asia, in hopes of retrieving their shattered fortunes, or of being able to continue their irregularities with impunity. The personal conduct of the directors made it necessary for them to connive at all these disorders. They were reproached with attending to nothing in their office but credit and money, and the power it gave them; with conferring the most important posts upon their own relations, men of no morals, industry, or capacity; with multiplying the number of factors, without necessity and without bounds, in order to secure friends in the city and at court; and, lastly, they were accused of palming upon the public, commodities which might have been bought cheaper and better in other places. Whether the French government was ignorant of these excesses, or lacked the resolution to put a stop to them, it was, by such blindness or weakness, in some measure, accessory to the ruin of the affairs of the nation in India." (Raynal, II, 372.)

These crimes and offences committed by the Portuguese, Spaniards, Hollanders, etc., in their search for the gold of Asia, remained to be eclipsed by the British. In 1612 permission was granted to them by the Grand Mogul, Shah Jehan (Jehangir) to establish factories at Surat and at several minor places in Guzerat. In 1616 this was extended to all the ports of the Mogul Empire. In 1620 permission to trade and to coin "Lakshmi" pagodas was given to the Company by the Rajah of Madras. In 1622 the English plundered Ormuz, in the Persian Gulf, which had been previously plundered by the Portuguese. In 1628 the British erected a factory and fort at Armegon, 66 miles north of Madras, and in 1639 a similar establishment (Fort St. George) at Madras itself. In 1668 the island of Bombay, which Charles II. had acquired from his wife, and she from the Crown of Portugal, was ceded by him for a pecuniary consideration to the East India Company, soon after which transaction the headquarters of the

Company in the Indies were removed to this place, which was strongly fortified, provided with dock-yards and filled with commodious storehouses, residences and other buildings. Next to Surat, the principal mart had been at Bantam, on the island of Java, then "the greatest place of trade in the Indian Seas." This is included in the list of factories which the English Company possessed in 1617. (Macgregor, IV, 311.) They were expelled from Bantam by the Dutch in 1680 whereupon the Company fitted out a fleet of twenty-three ships and 8,000 regular troops. To retrieve the loss of Bantam this force was ready to sail, when Charles II., for a bribe of £100,000, which, according to Raynal, "was paid to him by the Dutch," countermanded the order, and the fleet never sailed. The Company had been to such great expense in this matter that it could not command the funds necessary to purchase the next year's cargoes in Surat, and was therefore obliged to purchase them from the native merchants, on credit. These purchases amounted to so large a sum that Sir Josiah Child, the tyrant of the Company, resolved, if possible, to avoid its payment, by trumping up a series of fictitious claims against the governor of Surat. With this object Sir Josiah sent out his brother John, as governor of Bombay, with instructions, in case his claims were not allowed, to sack every ship departing from that city. "This terrible pillage, which lasted the whole year 1688, occasioned incredible losses throughout all Hindostan"; losses and outrages which were soon afterwards avenged by Auranzib. In 1689 he sent 20,000 troops to Bombay, who carried the outworks, captured the cannon, besieged Child in the citadel and induced him to sue for pardon. His deputies were admitted to the Grand Mogul's court, with their hands tied and their faces toward the ground. The conditions of the pardon were the dismissal of Child and restoration of the plunder. These being arranged, Bombay was released. It afterwards became the principal base of British operations in the Indies.

While Child was plundering the ships of Guzerat, other agents of the Company were busy at the same sort of work in Sumatra. A squadron of the Company's ships had been dispatched to Madras, with orders to set up a factory at Indrapoura, "the part of the country most abounding in gold." Contrary winds having driven the vessels to Sumatra, which was also reputed to be rich in gold, the adventurers resolved to fix a factory at Bencoolen. "Their trade with the natives was made at first with a great deal of frankness and confidence, but this harmony did not last long. The agents of the Company soon gave themselves to that spirit of rapine and tyranny which

Europeans so universally carry into Asia. Animosity was at its height when the natives saw rising out of the ground, at a distance of but two leagues from their city, the foundations of a fortress," the future Fort York. This led to a conflict, which was only accommodated after much loss and difficulty. In 1720 Fort York was abandoned; but Fort Marlborough was built a league distant.

It was at this place that a large proportion of the gold product of Sumatra was concentrated: for the English, following the example of the Dutch, had found it more profitable to kidnap slaves and exchange slave-products for native gold, than to work the mines in the face of native opposition.[6] When Fort St. George (Madras) was taken by the French in 1759 it contained only a small amount of booty, the English having shrewdly sent their gold to England as fast as it was acquired. Among the sixty-odd factories which the Company claimed to possess in 1710, were Aden, Mocha, Bassora, Ispahan, Bombay, Calcutta, Madras, Hooghly, Patna, Siam, Pegu, Cochin, Acheen, Bantam, Macassar, and Amboyna. (Macgregor, IV, 342.)

In 1723 the "Emperor" Charles VI., of Germany, granted a charter to the "Ostend Company," with license to trade with the Indies, erect factories, build forts, etc., under the protection of the Holy Roman Empire. The capital of this Company, six million florins, was subscribed in one day, and ships were at once dispatched, laden with merchandise and silver bullion, to take part in this lucrative commerce. But the remonstrances of the English and Dutch against this assumption of authority were so effective that a few years later, this monarch was induced to withdraw his license; when the Ostend Company came to an inglorious end. Several other attempts of a similar character were afterwards made, but they were defeated by the same influence and ended the same way. (Macgregor, IV, 344–8.) From these circumstances it is evident that the Holy Roman Empire was practically dead before Napoleon pretended to kill it in 1806.

Down to the middle of the 18th century, if we except the factories already mentioned and alluded to, the British East India Company had acquired no territory in India. They were nominally a trading Company, rich, bold, and covetous, but presumably governed by British law and British principles of honour, justice and national polity. In their haste to snatch Bengal from its rightful owners they

[6] In the evidence given by Mr. F. H. Barber, before the Committee on Indian Affairs of the English House of Lords, printed April 2, 1830 (cited in Ball, 181), it is stated that the gold mines of Malabar were crowded with slaves, who cost the English proprietors from 5 to 20 rupees—say, $2 to $8 each; that their sole clothing was a plantain leaf; and that their wretched appearance suggested baboons, rather than men.

threw off all disguise and trampled upon every principle of justice. The moment for action was propitious. The Portuguese had been virtually driven from Hindostan by the Dutch, and the Dutch by the English, the former retaining only Malacca, Cochin and the Sunda Islands. The French still governed the Carnatic, but they were hardly to be feared in Bengal. This was the time when Clive, Omichund and Meer Jaffier put their wicked heads together and plotted to rob the viceroy Surajah Dowlah of his master's empire. Bribery, treachery and forgery were the instruments employed by Clive. He condoned the Black Hole of Calcutta, by the most friendly professions toward the guilty viceroy; he conspired with Meer Jaffier to betray the viceroy; he agreed to pay Omichund, the go-between, £300,000 for his vile services; and, by a trick with white and red paper, afterwards cheated him of his guilty reward. The "battle" of Plassy was fought by 900 Europeans with a loss of 22 killed and 50 wounded. It was little more than a skirmish, employed as a blind to cover Clive's treacherous bargain with Meer Jaffier, who at the critical moment drew off his forces from the field and left the viceroy to his fate. The immediate results of this episode were that Meer Jaffier enjoyed for a brief interval the empty viceroyship of Bengal; Omichund, upon finding that he was duped, sank into idiocy; and Clive entered the treasury of Moorshedabad to find, like another Pizarro, an apartment filled with gold and silver. Reserving for himself what he afterwards declared was altogether too modest a proportion, namely $1,500,000, he turned over the remainder to the Company. In addition to his modest reservation, he compelled Meer Jaffier to grant him an annual revenue from the Bengal treasury of $140,000. Such is the account of the plunder. The larger results of the battle of Plassy were the conquest of Bengal and afterwards of all India.

In 1759 the English took Mausulipatam, and gained the ear of Salubat-jing, with whom they entered into a treaty to attack the French and compel them to evacuate the Peninsular. This convention was literally carried out. During the year 1760 the French were driven from the Dekkan, from Orissa, and, in 1761, from Pondicherry and the entire Coromandel Coast. By the Anglo-French treaty of 1763 the French factories thus captured were to be restored, but not the French sovereignty of the territories of which they had been dispossessed. In 1778 Pondicherry and the other French factories in Coromandel and Orissa were again taken, and in 1793 the French settlements in Bengal; with them the French power in the Indies was completely extinguished.

In 1773 Clive was impeached by the House of Commons, but aquitted with honour. In the following year he committed suicide. The plunder acquired by Clive was enormous. In 1755 he sacked Gheriah of £150,000 in the precious metals; in 1756 he took Calcutta; in 1759 Hoogly with £600,000, and Moorshedabad with several millions. With part of these treasures he corrupted the commanders of the native armies and thus acquired Benares, upon which he levied a contribution of eight million livres. (Raynal, II, 172.) The English agreed to serve the Grand Mogul on condition that he surrendered to them the absolute sovereignty of Bengal; yet, after he had ceded this vast realm, they repudiated the agreement. (Raynal, II, 173.) Nay, after they had plundered the country of its treasures and acquired the right to govern and tax it, they stooped to the debasement of its coin and altered its metallic basis. They sold exchange on India in mohurs, then they debased the mohurs with over 40 per cent. alloy. Raynal states that the amount struck of these over-valued pieces was about £625,000; that they were nominally rated at 10½ rupees, while the Company only paid 6 for them. Finally, they refused to accept tributes in the billon coins of the Moguls —all that were left to the miserable inhabitants—and insisted upon payment in those silver rupees which they now declared to be the only lawful money of India.

"We find the agents of the Company almost everywhere exacting their tribute with extreme rigour and raising contributions with the utmost cruelty. We see them carrying a kind of inquisition into every family, sitting in judgment upon every fortune, robbing indiscriminately the artizan and the labourer. . . . In consequence of these measures we find despair seizing every heart, and a universal dejection oppressing every mind and uniting to arrest the progress of commerce, agriculture, and population.

"In consequence of these measures, when the famine of 1770 occurred, the people were impoverished to the last degree and totally unable to meet it. The unhappy Indians were every day perishing by thousands in this famine, without any means of help and without any resource, not being able to procure themselves the least nourishment. They were to be seen in the villages, along the public ways, in the midst of our European colonies, pale, meagre, fainting, emaciated, consumed by famine, some stretched upon the ground in expectation of dying, others scarce able to drag themselves on to seek for any food, and throwing themselves at the feet of the Europeans intreating them to take them in as their slaves.

"To this description, which makes humanity shudder, let us add other objects equally shocking; let imagination enlarge upon them if possible; let us represent to ourselves infants deserted, some expiring on the breasts of their mothers; everywhere the dying and the dead mingled together; on all sides the groans of sorrow, and the tears of despair; and we shall then have some faint idea of the horrible spectacle Bengal presented." (Raynal.)

This picture was drawn a century ago. Has the situation improved? Let England herself reply.

"The known conditions of the Indian ryot's daily life were his bare subsistence in times of plenty, his constant hand-to-mouth struggle with want, his miserable physique, and the narrowness of the margin between what he ordinarily can obtain and that which is the irreducible requirement of life itself, let alone health. It is not easy to bring home to the English people—save by such shocking pictures as are now appearing in the illustrated papers—the utter destitution of the great mass of the Indian people. Perhaps some faint idea may be formed by a consideration of the amount of the wages paid on relief works. The average rate may be taken to be two annas (1½ pence, or 3 cents gold) a day for a man, one and a-half annas for a woman, and one anna for a child above seven. Without making any elaborate calculations, we may say that it is actually possible to maintain life in the people in times of famine at an average daily rate of one and a-half annas per head of the population. This means three rupees a month, or thirty-six rupees a year. This is in times of famine, when the staple foods of the people average double the normal price. Now, what does this mean? Thirty-six rupees are equal to about £2 6s. 8d. (a fraction over $10) a year; yet such is the poverty of the people of India that when a scarcity threatens them they have not even this small amount of reserve resources upon which to fall back. . . . Their whole existence is literally one of hand-to-mouth, and that on such a low plane of vitality, that with the extra pinch of the slightest scarcity they die like rotten sheep." London Daily Chronicle, Jan. 6th, 1897.

It is not for the idle purpose of offering a cup of shame to England's lips that these passages are adduced: it is to exhibit the cost of gold in India; gold which has been obtained by conquest, by piracy, by treachery, by slavery, and, worst of all, by exactions which condemn a vast population to hopeless indigence and famine.

This colossal crime cannot be palliated by shouldering it upon the East India Company. The gold of India went sooner or later to the Crown.

"By a development, the steps of which need not here be retraced, it became recognized that prize captured on land or at sea was the King's property. At an early date our courts took the view thus concisely stated by Lord Brougham; 'That prize is clearly and distinctly the property of the Crown; that the sovereign in this country, the executive Government in all countries, in whom is vested the power of levying the forces of the state and of making peace and war, is alone possessed of all property in prize, is a principle not to be disputed.' That is the principle applied too in modern times. In the wars with France, but especially in our Indian wars, the matter of booty assumed great importance. Prize of enormous value, in the shape of treasure, stores, and munitions of war, fell into the hands of the conquerors, in the first Mysore war, at the capture of Seringapatam, Pondicherry, and Serampore, on the conquest of the Dekkan, and in the war against the Maharajah of Bhurtpore in 1825. The catches were prodigious, the net was hauled often, and the conquerors became versed in the technique of pillage. Between 1779 and 1858 are recorded more than 130 important captures of booty. That taken at Banda and Kirwee was valued at £750,000. In the campaign of 1825-6 in Burmah, booty to the value of 517,683 rupees fell into the hands of Sir Archibald Campbell. In the fort at Bhurtpore was found nearly half a million sterling, and at Hyderabad Sir Charles Napier, the conqueror of Scinde, captured more than half a million sterling. To the Indian soldier, the chances of his share in booty counted for much. 'You and I,' writes Wellington to a correspondent, 'know well that there is nothing respecting which an army is so anxious, as its prize money.'"—London Contemporary Review, April, 1901.

THE PEACOCK THRONE OF DELHI.

In a smaller building adjoining we inspected the still more famous Dewanikhas, or hall of private audience. This proved to be a square pavilion of marble, resting on massive square pillars and moresque circles of the same material, all highly polished. The ornamentation is simple rather than elaborate. The apartment is small and oblong, opening into a court one one side, to the river Jumna on another, to the palace gardens on a third, and to the apartments of the Zenana on the fourth. The outside pillars are connected by superb marble balustrades, carved with graceful filigree work. At each corner of the roof is a marble kiosk with a small gilt dome. The ceiling of this Dewanikhas was one entirely composed of gold and silver screen work, wrought by the famous Delhi goldsmiths. Here stood the famous Peacock Throne, unparalleled in rare beauty and costliness by any other royal court in the world. Tavernier, a skillful French jeweler, saw it and made a computation of its value. He calmly estimated that its metallic worth, aside from its claims in an artistic way, was £6,000,000, or about $30,000,000 in American money.

This magnificent throne, six feet long and four feet broad, was composed of solid gold inlaid with rare Oriental gems. It received its name from the jeweled images of peacocks which adorned its canopy. This canopy was also of gold, supported on twelve gold pillars, and hung all around with a fringe of pearls. On each side of the throne stood two symbolical chuttas, or umbrellas. They were made of crimson velvet, embroidered with gold thread and pearls, and were equipped with solid gold handles eight feet long, studded with diamonds! This throne was constructed by Austin de Bordeaux.

The Peacock Throne was carried away by the Persian Nadir Shah, while the jeweled ceiling was melted down by the Mahrattas at the time of their invasion in 1759.—Delhi Corr. NewOrleans Picayune.

PLUNDER OF INDIA.

1505. Lopé Soarez takes Calicut and Cannamore; also seventeen large Arabian ships, with great booty. Taylor, Hist. Ind. 221.

1511. Sir Henry Middleton cruises the Red Sea with three ships and piratically plunders both Portuguese and Arabian ships of treasure, which he and Capt. Saris safely convey to England. Taylor, 289.

1530. Antonio de Silveira sacks the rich city of Surat. Taylor.

1537. The Portuguese sack many Indian cities, which they afterwards burn, condemning men, women and children to slavery. Taylor, 282.

1543. Prince Mulloo Khan of Beejapoor takes refuge with the Portuguese at Goa. His brother, Ibrahim Adil Shah, pays the Portuguese ten million ducats for his ransom. The latter accept the money, but refuse to surrender the refugee. Taylor, 282.

1553. Mahomet Shah Soor, of the Afghan dynasty, flings away his treasures, even shooting golden arrows among the populace, for his amusement. Taylor.

1554. The Portuguese play Mulloo Khan for the sovereignty of the Dekkan, but not succeeding, they abandon him to his enemies, who capture and execute him. Taylor, 283.

"The Portuguese chiefs and principal officers admitted to their table a multitude of those singing and dancing women with whom India abounds. Effeminacy introduced itself into their houses and armies. The officers marched to meet the enemy in palanquins. That brilliant courage which had subdued so many (Oriental) nations, existed no longer among them. The Portuguese soldiers were with difficulty brought to fight, except where there was a prospect of plunder. Such corruption prevailed in the Portuguese finances, that the tributes of sovereigns, the revenues of provinces, which ought to have been immense, the taxes levied in gold, silver and spices on the inhabitants of Hindostan, were not sufficient to keep up even a few citadels, and to fit out the shipping that was necessary for the protection of trade.

"It would be a melancholy circumstance to fix our attention upon the decline of a nation that should have signalized itself by exploits useful to mankind, that should have enlightened the world, or increased its own splendour and happiness, without being the scourge of its neighbours or of distant regions. But we should consider there is a great difference between the hero who spills his blood in the defence of his country and a set of robbers, who exploit a foreign land and put its innocent and wretched inhabitants to the sword. 'Serve or die,' the Portuguese used insolently to say to every people they met in their rapid progress, marked with blood. It is a grateful thing to behold the downfall of such tyranny, and a consolation to expect the punishment of those treacheries, murders and cruelties with which it has been preceded or followed." The Abbé Raynal, I, 207--9.

1557. King Ibrahim Adil Shah of Beejapoor pays 12 lacs of hoons as indemnity to Ramraj, the Hindu rajah of Beejanugger. His son, Ali Adil Shah, makes an alliance with the rajah, and together they march to the conquest of Ahmednugger. Taylor.

1564. Asof Khan Azbek, a predatory Mongol chieftain, and a subject of Akbar, the Grand Mogul, attacks the Hindu State of Guna, capturing enormous booty in jewels, bullion and coin. He also captures Oude. For these offences he was punished by Akbar, who was determined to protect his Hindu subjects. Taylor.

1564. The alliance of Ali Adil Shah and Ramraj is resented by the Mahometan princes, who combine against them and advance to the river Chrishna. In a great battle, which decided the control of India, the Moslems fired bags of copper money from their cannons, and the Hindu allies were cut to pieces with the flying coin. Ramraj was defeated, taken prisoner and beheaded. An immense treasure fell to the conquerors. Taylor.

1586–88. Cavendish returns to England from a piratical voyage in the Orient, with great booty in gold and silver. Taylor.

1605. Akbar's revenue was £30,000,000. At his death, his treasury contained £10,000,000. Taylor, 229.

1611. The Grand Mogul, Jehangir, marries the beautiful widow, Noor Jehan, and a new coinage is struck in her name; an act unprecedented in the annals of Islam. Taylor.

1623. The British and Persian forces unite in besieging Ormuz, then in possession of the Portuguese. After a contest of two months, the place is taken, with booty amounting to two millions sterling, which is duly divided between the conquerors. It is then destroyed. McCulloch.

1648. Sevaji, the Mahratta chieftain, whose miraculous birth and "divine mission to deliver India from Moslem rule," had been revealed to his Virgin mother by the goddess Bhowanee, intercepts a convoy of treasure from the governor of Kulliam to the Emperor Shah Jehan, and conveys it to his capital city of Chrishna. Taylor.

1653. Sevaji captures Surat and the Moslem districts of Dowlatabad, and carries off five or six million dollars in treasure. Raynal.

1656. Shah Jehan orders his son Auranzib, viceroy of the Dekkan, to plunder the King of Golconda, Abdulla Kutub Shah, for his disobedience. Auranzib surprises and plunders Hyderabad, and levies a fine equal to one million pounds sterling. Taylor.

1657. Sevaji takes Joonair and captures the Moslem treasury with £120,000. Taylor, 337.

1657. Auranzib marches to Beejapoor, at that time the finest and most populous city of India. Ali Adil Shah II. offers a ransom equal to one million pounds sterling. Pending the negotiation, Shah Jehan dies, 1658, and Auranzib returns to Delhi to claim his father's throne. He finds 24 million pounds sterling in the treasury, besides bullion and jewels, rich beyond precedent. Taylor.

1668. Josiah Child, now at the head of the British East India Company, together with his brother, John Child, governor of Bombay, piratically seize the merchant ships of the Grand Mogul, especially the rich ones from Surat, until the plunder amounts to immense sums, and attracts Auranzib's army to Bombay to demand restitution, the terms of which, as described by Raynal, are extremely humiliating to British pride.

1672. Auranzib levies the hated jezia, or capitation tax, upon the Hindus, exciting a general commotion, which Sevaji turns to account.

1675. Sevaji plunders Khandesh and Berar of great booty.

1679. Sevaji plunders the Carnatic, annexes the territory west and south of the river Chrishna, and levies "choath." He dies in the following year. Choath is a tax of one-fourth of the produce.

1686. Prince Muazzim plunders Hyderabad and marches upon Golconda, whose king, Abu Hassam Kootub, purchases peace with the sum of two millions sterling in money and jewels.

1687. Auranzib repudiates a treaty which he had made with the king of Golconda, whose treasury he plunders and whose dynasty (the Kootub Shahy) he overthrows.

1702. The Mahrattas under Pám Naik, plunders portions of Guzerat and Khandesh.

1705. The Mahrattas plunder Malwah. Taylor.

1739. Nadir, Shah of Persia, after expelling the Afghan monarch of that empire and defeating the Turks, takes Cabul, 1738, Bokhara, Candahar, Chorasmia (Kaurezm), and many other cities and provinces, with great booty. In 1739 he ravages the northern provinces of India and captures Delhi, killing 100,000 of its inhabitants, regardless of sex or age, and capturing treasure to the value of £125,000,000.

1740. Serefraz Khan, an unfaithful viceroy of the Grand Mogul, is defeated by the Afghan, Ali Verdi Khan, a rival viceroy, when the conqueror enters Moorshedabad, in Bengal, which was tributary to Delhi, captures one million pounds sterling in coin and nearly an equal sum in plate, jewels, etc., all of which is transmitted to Delhi. Taylor, 398.

1742. The Mahrattas plunder the bank belonging to Juggut Sett, of Moorshedabad, obtaining two millions sterling.

1748. Through the defeat and assassination of Nazirjing, Soubah of the Dekkan, his treasure of £2,000,000 and jewels valued at £500,000 is acquired by the victors, the French troops receiving £50,000 and the French East India Company an equal sum; besides Pondicherry, with an annual revenue of 96,000 rupees, Karical 106,-000, and Mausulipatam 144,000 rupees, in all a revenue of £43,250 sterling, and later on the provinces of Mustaphanagur, Ellore, Rajahmundy, and Chicacole in full sovereignty, whose revenues amounted to 42,87,000 rupees. Macgregor, IV, 351.

1749. The Mahratta Brahmins, under Sanhojee, offer the English immense sums and vast territorial concessions for their aid against the Moslems. This offer being readily accepted, Major Lawrence and Capt. Cope advance to Dévicotta, which they capture from the

Moslems, with great booty. Dévicotta and the adjacent territory are ceded to the English by the rajah of Tanjore. This is practically the beginnning of the British conquest of India.

1750. The alliance of Joseph Francis, afterwards Marquis of Dupleix, commander of the French forces in India, is secured by Mozuffer Jung, the newly-crowned Soubah of the Dekkan, who appoints Dupleix governor of all the Mogul territory south of the Chrishna river, with capital at Pondicherry and the prerogative of coinage. Mausulipatam and its dependencies, with its annual revenues of five lacs (translated £50.000) are also ceded to France. Twenty lacs in cash, besides gratuities to his officers, are granted to him from the treasury of the Soubah.

In 1754 Ahmed Shah Abdalla, king of Afghanistan, had marched from Kandahar and plundered the oft-plundered Delhi, "extorting a vast sum of money from the people by torture and massacre." Then he proceeded to the rich city of Muttra, where, on the occasion of a Hindu religious festival, he slaughtered and despoiled thousands of the defenceless inhabitants. In consequence of the disturbed condition of his feudalised and distracted empire, the Grand Mogul was so far unable to resent these outrages that he permitted the Afghan to name a commander-in-chief for his army. In 1759 Ahmed Shah again set forth from Kandahar, this time to punish the Mahrattas for plundering the Moslem city of Delhi. After much manouvering and delay, the two armies met on the field of Paniput, January 6, 1761, when the Mahrattas, 70,000 strong, with 300 pieces of cannon, were utterly destroyed, the prisoners being murdered in cold blood and the women and children sold as slaves. The spoil consisted of 27 lacs of gold mohurs, and "of the silver and copper (money) the total cannot be computed." Grant Duff. The remains of twelve contending powers now devastated India from end to end: the Moguls, Rohillas, Oudhs, Bengalese (under Clive), Rajpoots, Mahrattas, Jats, Dekkanese, Mysoreans, Carnatics, Tanjorese, and Cochinese; to say nothing of the Portuguese, French and English The plunder was immense. Its cost was simply blood: blood and slavery. Says Taylor, "The shameless demands of my countrymen upon Meer Jaffier for private presents and losses were pursued with more than usually stringent rapacity," until no less than £350,000 were disbursed to private claimants, and Meer Jaffier was harried into his his grave. Mr. James Mill, taking a more comprehensive view of Clive's operations in Bengal, computes the booty at £5,940,498 in money alone, to say nothing of other property. Hist. British India, III, 326–9.

1757. Robert (afterwards Lord) Clive, together with Watson, Drake, Watts and other British commanders, swear "on the Holy Gospels" to lend all their forces to Meer Jaffier in his design to usurp the throne of his master, Suráj-ud-Dowlab, the Nawab, Nabob, or Soubah, of Bengal, Bahar and Orissa: Meer Jaffier being at that time in command of the Nawab's army. Having by this device practically disarmed his ally, Colonel Clive, who had secretly determined to conquer these rich provinces for himself, called a meeting of his officers at Cutwah, on the 21st of June, 1756; when he threw off the mask and imparted his treacherous design to them. The enormous booty which he held in view quieted their consciences, and the next day, under his orders, the British army crossed the Ganges, and with a loss of 72 killed and wounded, gained the "battle" of Plassy, and practically the paramountship of India. After permitting the defeated Suráj-ud-Dowlah to be assassinated, and placing Meer Jaffier upon a puppet throne, which Clive held completely under control, the labours of the army were devoted to the main object of the campaign. This was plunder—plunder of the richest and most populace provinces in the world. Part of the nefarious transaction with Meer Jaffier consisted of a promised bribe of 30 lacs of rupees (£300,000) to Omichund the banker, for betraying Suráj-ud Dowlah. This sum was regarded as one-twentieth portion of the Nawab's treasure, which therefore consisted of 600 lacs, or six million pounds, about thirty million dollars. The promise to Omichund, after being reduced to 20 lacs, was written upon a red paper, which was signed by Clive and all his officers, except Admiral Watson, who refused to become a party to the fraud. With this promise in his possession, and relying upon the good faith of the British, Omichund, who might have warned his master in time to summon forces capable of defeating the British, kept silence until after the conflict at Plassy, when he claimed his reward, and was cooly told by Clive and Scrafton that the promise, on account of its being on red paper, was invalid, and that he was to have nothing: an announcement which upset the traitor's reason, so that shortly afterwards he died a drivelling idiot. This act, says Taylor, himself an Englishman, was one of "deliberate and unworthy treachery" on the part of Clive. In addition to the concessions specified in the treaty of alliance with Clive, Meer Jaffier paid from the Nawab's treasury 50 lacs to the British army and navy, 28 lacs to Clive as a member of the "Committee of Calcutta," 24 lacs to each of the other members, and in addition, 160 lacs to Clive himself: altogether about 400 lacs. "The first installment of 80 lacs was brought to Calcutta in a triumphal procession of boats from Moorshedabad." The exactions of Clive did not rest until he had completely exhausted the usurper's treasury. Taylor, 433–37. He afterwards made a treaty with the Grand Mogul, Meer Jaffier's suzerain, by which he secured a jahgire, or appanage, worth three lacs, or £30,000 per annum. Taylor, 433–9. After this he deposed Meer Jaffier, and for an immense bribe proclaimed Cossim Ali Khan in his stead. Raynal, II, 171.

In 1757 the Mahrattas, under Sudasheo Rao, bombarded Seringgapatam, and ransomed it for 32 lacs. After defeating the Nizam Ali in the field, and exacting as indemnity the perpetual jahgire of provinces yielding 62 lacs annual revenue, he marched to Delhi, the capital of the Grand Mogul, which he captured and plundered, the silver ceiling of the imperial palace alone yielding 17 lacs, or £170,-000. With this crowning misfortune the Mogul Empire of India practically ended.

1765, August 12. On this day the Grand Mogul was seated on a throne constructed of the dining-table and an armchair in Lord Clive's tent, and made to sign a firman which conferred upon the English in perpetuity the three provinces of Bengal, which possessed a popularion of 25,000,000, and a revenue of £4,000,000 sterling, less 26 lacs pension conceded to the Emperor, who was no longer an Emperor, and 50 lacs to the Nawáb, who was no longer a Nawáb. Clive also obtained a formal grant of the whole of the Circars down to the river Chrishna. Substantially, these domains included the entire north-eastern half of Hindostan. Taylor, 463. In 1768 the pension to the Emperor was repudiated, and the crown domains of Corah and Allahabad, which did not pay, were sold to his so-called Vizier for 50 lacs. Rohilkhund, a province belonging to the Indian allies of the British, was delivered over to the same minister to be plundered, the consideration being 40 lacs, an immediate gain and an eternal infamy. Says Taylor: "It was treacherous, because the Rohillas had already professed their attachment to the English and high trust in their good faith." The Vizier commenced his operations in Rohilkhund by demanding "two crores, or two millions sterling," by way of immunity. This demand being refused, a detachment of the British troops were sent in 1774 to assist the Vizier in plundering the province. Upon paying the Vizier a sum of 15 lacs, hostilities were suspended. The next operation of the British was to "support the claims of the Begums, the mother and widow of Suráj-ud-Dowlah, to the whole of the treasure amassed by him, which was about two millions sterling." These demands occasioned a mutiny of the Nawáb's army, "which was only quelled by the slaughter of many thousands."

In January, 1775, the British authorities hired out 3,000 of their soldiers to Rughoba, a rebellious chief of the Mahratta Brahmins, in consideration of the cession of Salsette and Bassein, in perpetuity, and the assignment of jahgires, amounting to 24 lacs per annum. Then they turned around and for a payment of 12 lacs deserted Rughoba and agreed to support the legitimate chief of the Mahrattas, by name Mahdoo Narrain, a posthumous child forty days old, offering Rughoba a pension of 25,000 rupees a month to abandon his pretensions. Upon his refusal to accept this illusive grant, they again changed sides and undertook to support him, with the result that the allies were defeated, Rughoba made a prisoner and the British obliged to yield up the concessions which they had so unworthily acquired. This was on the 13th January, 1779.

While the British forces were employed in these sordid and dishonourable transactions, the sole object of which was to plunder the Indies by dividing its distracted factions and nibbling from each in turn, the American colonies on the opposite side of the world seized the opportunity to assert their independence. On the 19th of April, 1775, the battle of Lexington was fought; on the 19th October, 1781, Lord Cornwallis laid down his arms at Yorktown. The cost of the Asiatic spoil, with which Clive and Hastings filled the British treasury, was no less than the empire of the earth; for who can doubt that with America still subject to its sway, the British crown would have attained that universal dominion of the civilized world which Cæsar once established at Rome and for which the nations of Europe have ever since contended?

When Hyder Ali, Sultan of Mysore A. D. 1760-82, took Bednur, or Nagur, in Mysore, he captured £12,000,000 in treasure, and in order to coin it, established a mint there. Lock, 307. The sum is deduced from an erroneous valuation of the billon money of the period.

"During the whole of this time (the spoliation of India), the Ganges was covered with carcasses; the fields and highways were choked with them; infectious vapours filled the air and diseases multiplied; and one evil succeeding to another, it was likely to happen that the plague (of 1770-71) might have carried off the remainder of the inhabitants of that unfortunate kingdom (Bengal). It appears by calculations pretty generally admitted, that the famine carried off a fourth part, that is to say about three millions. . . . Yet, although all the Europeans, especially the English, possessed magazines, these even were respected as well as private houses; no revolt, no massacre, nor the least violence prevailed. The unhappy Hindus, resigned to despair, confined themselves to begging for succour which they did not obtain, and peaceably awaited the relief of death. What disorder, what fury, what atrocious acts, what crimes would have been committed had a like calamity visited Europe!" Raynal, II, 189.

In 1773 the revenues of Bengal, exacted by the British through the medium of a puppet throne, amounted to $14,250,000, of which about $12,275,000 were "absorbed by plunder or the necessary expenses" (of the British East India Company). The agents of the Company exacted the tributes with extreme rigour and raised contributions with the utmost severity. They carried a kind of inquisition into every family; robbing indiscriminately the artizan and the labourer and punished others whom they suspected of being rich. They sold favour and credit as well to oppress the innocent as to screen the guilty. Despair seized every heart, dejection oppressed every mind, and these united to arrest progress and discourage commerce, agriculture, and population." Raynal, II, 177-185.

1773. The revenues of Bengal in 1773 rose to 71,004,465 livres, but 61,379,437 livres were absorbed in plunder, or the necessary expenses. Raynal, III, 177.

In 1775 Cossim Ali Khan, tired with being plundered by Clive, united with the nabob of Benares in a campaign against the foreigners, with the result that he and his coadjutor were defeated and forced to pay the victors a sum of about $1,600,000. Raynal, II, 172.

In 1779 Hyder Ali, a Mahometan chieftain of Mysore, who, during the decline of the Mogul dynasty, had gradually risen to power, until as Nawab of Mysore he claimed for himself the throne which the British had overturned, established his capital at Seringapatam, declared a Holy War against the infidels (British), and in 1780 defeated them in an action between Arcot and Madras, and afterwards besieged and captured and plundered Arcot. While this contest was waged for dominions to which neither party had the least right, Mysore suffered from the devastations of both of them, with the result that a famine afflicted the province, which carried off myriads of its unfortunate inhabitants: the price of the gold, which meanwhile was accumulating at Madras and Seringapatam. In 1782 Hyder Ali died and was succeeded by his son, Tippoo Saib, in whose service, as a sergeant of artillery, was the celebrated Bernadotte, afterwards King of Sweden. After capturing Bednore, where the British are charged with having violated the text of their capitulation by secretly conveying away the treasure, Tippoo took Mangalore, and there dictated those insulting terms of surrender to the hitherto invincible foreigners which cannot be read without indignation.

Sultan Tippoo Saib, 1785–99, appears to have worked the mines of Mysore with great activity, and for this purpose he condemned to practical slavery a large portion of its lower castes. A mine at Marcurpam, in Mysore, which he placed in charge of a Brahmin, was worked for a while, then abandoned as unprofitable. (Ball's Geology of India, p. 185.) A silver mine, eight miles north of Kadapah, in Madras, was treated the same way. (Ball, 233.) The gold mines of the Coimbatore district appear to have been profitable, for upon the approach of Gen. Cornwallis, in 1792, Tippoo took the trouble to fill the shafts with waste rock in order to discourage the British from working the mines. Ball, 310.

In 1781 the British authorities in Bengal suddenly demanded from the rajah of Benares, who was paying them a rental of 22½ lacs per annum, a further cash payment of 5 lacs and the service of 2,000 horsemen. This being refused, the British threatened hostilities, whereupon the rajah offered 20 lacs as indemnity. The British demand was now raised to 50 lacs, and upon this being refused, they besieged his capital of Bidgeghur, and captured it with plunder amounting to 50 lacs, which were distributed to the soldiers. The rajah's rental was afterwards raised from 22½ to 40 lacs.

In 1789 the rajah of Travancore ransomed from the Dutch two coast towns which Tippoo Saib declared belonged to his ally the rajah of Cochin, but which the British claimed as within their sphere of influence. Lord Cornwallis, who, after his disastrous campaign in Virginia, was in command of the British forces at Madras, was prepared to resist Tippoo's claim, and wrote a letter to him to that effect. This letter was withheld by Mr. Holland, the governor of Madras, who then secretly approached the rajah and offered to sell him the service of those British troops who were designed by Cornwallis to be employed in the public interest. This plot being exposed, Holland fled to England, and a campaign was begun against Tippoo with 15,000 troops, who, however, were forced by the indomitable Moslem to retreat.

In 1790 Lord Cornwallis seized the revenues of the Carnatic, which had been assigned to the Nawab, and in 1791, with the assistance of the Mahrattas, he opened a campaign against Tippoo, in which Bangalore, Nundidroog, Savandroog, Ráyacotta, and other places were successively captured and plundered. On the 5th of February he besieged Seringapatam with so vast a force that on March 19th Tippoo was induced to offer terms. This consisted of one-half of his dominions, the English share being Malabar, Coorg, and the Báráh Mahál and £3,300,000 sterling in cash, one-half down and the remainder within a year.

The treaty of Seringapatam having virtually established the British paramountship of India, Lord Cornwallis at once proceeded to secure its fruits. The feudal revenues exacted by the Zemindars were based upon a valuation of land fixed in the 16th century during the reign of Akbar the Great. These were now so greatly increased that "the cultivators were depressed to an extent hardly realisable." The civil service was closed to all natives, and Mr. Wilberforce's motion in Parliament, to permit English missionaries, schoolmasters, and other instructors in religion and knowledge, to enter India, was rejected. Parker, 529.

In 1795 a war broke out concerning a division of plunder between the Mahrattas and the Nizam of Hyderabad, in other words, between the Brahmins and Moslems, in which each party brought into the field of Khurdlah upwards of 125,000 men, when the Nizam was defeated and obliged to submit to a payment of 300 lacs, one-third down, and the surrender of districts yielding 35 lacs per annum. Although the Nizam had a brigade of British troops in his pay, they were instructed by Sir John Shore to take no part in the fray, it being the policy of the administration to let the natives destroy each other as rapidly as possible.

In 1797 a dynastic war broke out in Oudh, in which, for an annual payment of 76 lacs, Sir John Shore took sides with one of the contestants, and seated Saadut Ali on his pasteboard throne. In the following year, in order to raise 200 lacs, the Peshwar, or sovereign of the Mahrattas, gave up Poona to pillage, the native bankers being his principal victims. Unspeakable tortures were inflicted by his orders upon these unfortunate men, until at length the required sum was extorted from them.

Upon the arrival of Lord Morington, afterwards the Duke of Wellington, the new governor-general, in 1798, it was determined to make an attempt upon the capital of Tippoo Saib, in which vast treasures were believed to be deposited. The movement began by inducing the Nizam of Hyderabad to withhold the pay of and dismiss his force of 40,000 sepoys, who had been organized and drilled by French officers. Upon assurances by the British authorities that the sepoys would be paid, this force of mercenaries was either re-enlisted or disbanded; and thus one of Tippoo's resources (for he was the Nizam's suzerain), was destroyed. War was then declared, and the British, under Lord Harris, advanced upon Seringapatam with 21,000 European and native troops, 10,000 of the rebellious Nizam's contingent and 4,000 of the purchased sepoys, altogether 35,000 men, besides a reserve of 6,000 Europeans under Generals Stuart and Hartley. On the 6th of April the capital was invested. The British demands were £2,000,000 sterling in cash and one-half of the Sultan's dominions. This being refused, the city was bombarded, and on the 3rd of May it was carried by assault and devoted to pillage. The scenes of carnage beggar all description. The brave Tippoo died sword in hand fighting to the last. The plunder of gold and silver distributed to the army amounted to $7,500,000, to say nothing of the jewels, arms and furniture that fell to the victors. The entire force of Tippoo Saib, consisting of 22,000 men, was cut to pieces. The British loss in killed and wounded was 1,164 men, many of whom were sepoys. This was the cost of the gold.

If these details of carnage, intrigue and plunder seem to have been followed to a greater extent than the general character of the present work would warrant, it has not been without a purpose which, to the reflecting reader, will carry its own apology. We have no details of the circumstances under which Asia was plundered by Darius, Alexander, or Seleucus. We have but few details of the circumstances under which it was plundered by the Moslems. The barbarities of the Portuguese reach us chiefly through Portuguese writers, whose partiality has left us only glimpses of what took place. The conquest and plunder of Asia by the British, as told by their own historians, comes to us in detail, and it is the only one which enables us to compute with any degree of precision the cost of that gold which forms the bulk of the world's monetary symbols of the present day. We have been taught that this cost was the economical value of the labour devoted to its production. We now know that this teaching is false; that the cost was blood and slavery, and that with such cost its value has no relation. There is no known connection between them.

PLUNDER OF CHINA.

In consequence of a famine in 1342, which destroyed 13 millions of people, a general revolution broke out in China, 1358, headed by a Buddhist monk, Chu-yuen-chang, who overthrew the Mongol dynasty and established the Ming, a native Chinese dynasty, which lasted 276 years (1368–1644). The progress of the revolution was facilitated by the issuance of government paper notes, some few specimens of which are still extant.[1] It was under the reign of one of the Ming Emperors that commenced those inroads of the Mantchu Tartars, 1522–67, which eventually overthrew the native dynasty and established the present (Mantchu) sovereignty; and it was during this period of trouble and civil strife in China that the Portuguese, Spanish, Dutch and English, successively attacked and plundered its coasts.

"It was in the year 1498 that the Portuguese, under Vasco de Gama, made their way around the Cape. In 1510, under Alboquerque, they treacherously seized the East Indian city of Goa, and leaving a garrison in it, sailed away to Malacca, which they had seen and coveted in 1508. This great city they treacherously and piratically captured. The superiority of their arms will be understood when it is stated that this act was committed by only eight Portuguese, assisted by two hundred Malabar natives. They plundered Malacca of 'a booty so enormous that the Quinto, or fifth, of the King of Portugal amounted to 200,000 gold cruzados, a sum equivalent to $5,000,000,' exclusive of ships, naval stores, artillery, and other property. Malacca was at that time a vassal state of the Chinese Empire, and our first acquaintance with maritime Europe was, therefore, begun on its part by the greatest act of piracy the world has ever witnessed. Pizarro's plundering of Peru, committed a few years later, was nothing compared with it. Hearing at Malacca of the great Chinese cities to the northeast, and hoping, no doubt, to pillage them as his companions had pillaged Goa and Malacca, one of the Portuguese, Rafael Perestralo, sailed away in a junk to view our coast. Finding the Chinese better prepared for pirates than he expected, he returned to Malacca.

"We have our own theory concerning the sources of your present riches. We ascribe it, in part, to your gains from the piratical conquest, enslavement, and murderous extinction of the American races, but chiefly to the profitable trade with the Orient. From the opening of this trade to 1640, when the Portuguese were driven from Japan, and the British first acquired territory in Hindostan, three of your European nations alone took a thousand million dollars in gold and silver from Asia; two-thirds as much as they wrung from all

[1] Only three specimens of the Ming paper notes are known to be extant. One of these is in the possession of the writer. It is about eight inches long and four inches wide, printed in black ink with a red countermark: the seal of the Chinese treasury.

America during the same period. From Malacca, alone, they took twenty-five millions; from Japan, up to the date mentioned, four hundred millions; from India and China still greater sums. Add the interest on these sums for three hundred years and subtract the total from your present wealth, then how much would there be left?

"In 1565 the Spaniards in Mexico sent a fleet to the Philippine Islands, which they captured and occupied. Under assurances of protection from these marauders, a considerable number of Chinamen were induced to reside upon the the islands, which, under the effects of their industry and enterprise, became as rich and productive as before they had been poor and barren. In 1602 there were upward of twenty thousand Chinese in Manila, whilst the number of Spaniards did not exceed eight hundred. There never had been the slightest disturbance between them. The Chinese were hard workers, who meddled with nobody. The Spaniards rode about on horseback, enjoying the fruits of the Chinamen's labour and living like lords; and yet they were not satisfied. They wanted to rob the Chinamen of the gold and silver which they had managed to save under the hard conditions of their life. The Spaniards met together in secret, planned a massacre of the Chinese, and carried out this atrocious design with such expedition, that, in the course of a few months, but few of the twenty thousand victims were left alive. The marauders then divided the spoils they had gained, and rejoiced in the name of civilization and religion. Thirty-seven years later, a new generation of Chinamen having arisen, who were ignorant or careless of what had occurred before, some thirty-three thousand of my countrymen gradually found their way to Manila. Precisely the same thing happened as before. The Spaniards, coveting the gains of the Chinese, planned their massacre, and slaughtered twenty-two thousand of them in four months, with a loss on their own side of but three hundred and thirty. (Martin's History of China, I, 378.) From that moment the Philippines decayed and sunk to nothing. In 1762, when Sir William Draper captured Manila from Spain, his most numerous and eager allies were the Chinese. It was a punishment and retribution to the Spaniards." —The "Letters" of Quang Chang Ling, 1878.

1596. Queen Elizabeth despatches a fleet to China, with instructions to obtain a footing in that country by treaty if possible, by force if necessary. The vessels are wrecked on their way to the Orient, and the design is temporarily abandoned.

1680. The East India Company open a factory at Canton and introduce the sale of Indian opium in China. The development of this traffic leads to frequent conflicts between the traders and the officers of the imperial government, who have instructions to restrict or stop it. These conflicts lead to threatening expeditions on the part of the English, in each of which some Chinese coast towns or forts are destroyed, and some indemnity demanded, by special embassy. Such were the objects of Lord Macartney, 1793, and Lord Amherst, 1816.

1828. The imperial laws against the use of opium exasperated the British, who sent fleets to Canton in 1831 and 1834. Although they failed to intimidate the government, they fostered and protected a contraband trade of opium in armed ships. In 1838 the Chinese government, desiring to destroy a traffic which corrupted the morals and promoted the degeneracy of its people, made the use of opium a capital crime, and destroyed the British stock of opium at Canton, worth twenty million dollars (March, 1839). On June 28, 1840, the British fleet, under Admiral Elliott, occupied Chusan, destroyed the fortifications of Amoy, and in 1841 the forts at Boca Tigris, which defended Canton. This city, together with Chusan, Amoy, Ningpo, Cha-pu, and Shanghai, were successively bombarded, occupied and plundered, when a peace was patched up by a concession of the Island of Hong Kong and the payment of $6,000,000 indemnity to the British, May 27, 1841. In the same year Chin-pae, Yuyao, Tsikee, Funghwa, Cha-pu, Woosung, Chin-keang-foo, and many other towns, are captured and plundered by the British, who now raised their demands for indemnity to $21,000,000. This sum was conceded by the treaty of August 29, 1842, and Chusan given as security.

The inferiority of the Chinese arms in these attacks led to disgust of their Mantchu rulers on the part of the Chinese, and to internal commotions, which developed into the Taiping Rebellion of 1850. This, being aided by the foreigners, secured them a further footing in China. Foreign adventurers enlisted in the ranks of the rebels, and foreign ministers entered into negotiations with them. The rebels plundered Han-Yang, Kan-Keu, Nanking, Chin-Kiang, Kua-chow, Yan-chow, Tung-yang, and many other places, while the foreigners alternately aided one side or the other, apparently with the view to prolong the rebellion and weaken the government. The rebellion was no sooner crushed by the imperial government than a new pretext was found for war and plunder.

In 1856 the Chinese authorities at Canton seized the Chinese crew of a Chinese smuggler sailing under the British flag. This being regarded as an insult to Great Britain, was resented by the bombardment of Canton, the French fleet, recently released from operations in the Black Sea, by the conclusion of the Crimean war, taking part with the British. The American frigate Portsmouth also took a hand in this cowardly attack, though the act was not approved by the American government. On January 1, 1858, Canton was taken by the British and French forces, who committed many wanton cruelties and plundered the city right and left. In May, the forts of the Pei-ho were destroyed, and Tsin-tsin threatened, when the Chinese government procured peace by the payment of $22,000,000 to the allies (16 to England and 6 to France), the surrender of the left bank of the Amur river to Russia, and numerous concessions to the other Christian nations who were the unmoved spectators of these acts of brigandage.

In dealing with Chinese affairs, it must not be forgotten that ever since 1728 there has been maintained at Pekin a convent or college for educating Russians in the Chinese language, who go there and leave the place by rotation when instructed. Russia also enjoys an immense moral advantage over Great Britain in the circumstance of her abstention from the degrading and destructive opium trade. When the events of 1842 blocked the British opium trade at Canton, they tried to force the deadly drug into China by way of Smyrna, Moscow and Kiakchta; but the Russians stopped it by imposing a prohibitive duty of 40 silver roubles per pood, equal to 3s. 6d. per pound English. Macgregor, II, 621, 644.

In 1899 Baron Ketteler, of the German legation at Pekin was killed by a Chinese fanatic. This incident, and the pretence of protecting their own legations, was made the occasion of an attack upon China by the combined forces of Germany, Great Britain, France, Austria, Italy, the United States, Russia, and Japan, in which upwards of one hundred Chinese cities were captured and plundered, upwards of 50,000 innocent people destroyed, and 450 million taels demanded as indemnity by the destroyers; the Chinese armies and fleet being the cowed and silent spectators of this ruin, wholly unprepared to prevent, modify, or resent it. The German Emperor's command to his forces were to "Kill all; spare none!" and this Scriptural injunction appears to have been implicitly followed by all of the allies, with, perhaps, the exception of Japan and the United States, both of which powers limited themselves chiefly to plunder and "indemnity."

July 19, 1900. Brig.-Gen. Heywood, of the U. S. Marine Corps, on October, 29, 1900, quotes the report of Col. Meade, dated July 19, which states that the Treasury of Tien-Tsin had been looted, and that Major Waller, searching for further treasures, recovered $376,-300 from the ruins of the Salt Company's Yamen. This plunder was sold for drafts on J. P. Morgan & Co., of New York, and the latter were forwarded to the Secretary of the Navy.

Aug. 18. The London Times says that Pekin is being systematically looted. The French and Russians are searching for the Imperial treasure.

Aug. 24. The Elberfeld Journal publishes a letter from one of the German soldiers at Tien-Tsin, which states that the German troops killed only the Chinese prisoners; while the Russians killed everybody, men, women and children. The Crefeld Zeitung has a letter saying that the Russians and Japanese spared none. Leslie's Weekly says that every city from Taku to Pekin was looted. The allies took the silver, jade, silks and furs, everything of value. The English gathered the furs, ornaments and furniture from every house in their quarter and sold them at auction. The Japanese devoted their energies to gold, silver, and munitions of war, which they shipped to Japan. Every Russian helped himself. Much loot was shipped to Shanghai and the Philippines and there sold at auction. The French took all they could find. The Germans came late, so they organized "punitive expeditions." A Pekin despatch says that the Chinese women, to escape the nameless bestiality of the Russians, drowned themselves by tens of thousands. The scenes of wanton carnage, outrage and spoliation, defy description. The women, owing to their feet being cramped by bandages and tight shoes—and this bespeaks women of the higher classes—could not escape from the brutal ruffians who pursued them; and thousands of them committed suicide. The poor innocent children were slain without mercy. Gold, silver and lust were not the only incentives. Wanton murder was added to the other horrors of war upon a defenceless people, against whom war had never been declared.

Sept. 1. Joaquin Miller, in the Chicago American, tells of horrible scenes witnessed by him in China. At Tong-Ku, which the Russians burned, they beat out the brains of a fleeing Chinaman. They looted the place and massacred women and children. The people, robbed of their food, were reduced to cannibalism. Many women and children drowned themselves to escape the Russians. He estimates that from 10,000 to 50,000 bodies of dead Chinese floated down the Peiho river.

Pekin, Sept. 1. A private letter from Pekin says that a German cavalry officer, having translated the United States Mint Reports, which estimate that there are seven hundred million dollars of silver in China, his countrymen are resolved to capture the treasure if they have to tear out the hearts of the Chinese to get at it. Upon being told that the estimate was the wild conjecture of an ignorant official, and that almost the only money of China was copper "cash," he replied that the Germans intended to find that out for themselves. I fear that our preposterous Mint estimates will cost many thousand innocent lives.

Sept. 6. The London Globe, of Nov. 13, publishes a letter from a Belgian gentleman who travelled to Pekin via the Trans-Siberian Railroad. He describes, under date of Sept. 6, what he saw in the Amur River. His account surpasses in horror those previously published.

"The scenes I have witnessed during the three days since the steamer left Blagovetchensk," he says, "are horrible beyond the powers of description. It is the closing tableau of a fearful human tragedy. Two thousand were deliberately drowned at Moro, 2,000 at Rabe, and 8,000 around Blagovetchensk, a total of 12,000 corpses encumbering the river, among which were thousands of women and children. Navigation was all but impossible."

Pekin, Sept. 25. This Chinese expedition is the biggest looting affair since the days of Pizarro. Men's fur coats, women's jewels and hair-pins, and children's baubles, are auctioned off every day, Sundays excepted, in front of the British legation. Sir Claude MacDonald is frequently in the crowd. Royal Marines and engineers, Welsh Fusileers, Americans, Germans, Sikhs; in short, everybody except the Russians and French enjoy the sport. The wealthy houses and pawn-shops were first gutted. Many thousands of houses, after being plundered, were wantonly destroyed, the occupants killed or beaten, the women grossly treated, mules and other animals bayoneted or shot. Even the civilians who accompanied the army took part in the plunder.—Dispatch to the London Express.

Berlin, Nov. 13. A new batch of letters from soldiers in China give horrible details of the wholesale murders of Chinese at Liang-Sian and Ta-King, which have been suppressed by the newspapers.

November. 15. Dr. Morrison, the Pekin correspondent of the London Times, considers that the plunder of the Chinese tombs was a just punishment for the murder of the missionaries.

December 1. The Kobi (Japan) Chronicle reports the silver plundered from Pekin by the Japanese troops at 2,637,700 taels, valued at $1,895,500.

December 5. In the Kalgan expedition, the German cavalry exacted tolls of skins and silver. At Swen-hwa-Foo they overtook the fugitive Chinese, killed thirty of them, and bagged twenty thousand taels of silver. This is but a small part of their doings. The Pall Mall Gazette regards the plundering expedition to Kalgan, and the looting and burning of Chinese villages, as barbarism and brigandage.

December 7. The London Morning Post reports that the Germans at Cho-Chow plundered the christian converts equally with the other Chinese.

December 12. A confession extorted from a former Court official has induced Col. Tullock, commander of the British forces, to dig for a large amount of gold and other valuables said to have been buried by the Chinese Emperor at a spot some twenty miles northwest of Pekin.

Dec. 21. The Pekin correspondent of the London Morning Post reports that Count Von Waldersee's command are engaged in looting. At Lung Ching, the Germans shot sixty Imperial troops who were engaged in supressing Boxers, killed thirty other Chinese, including three christians, and took two hundred prisoners, including thirty christians, whom they ransomed for twenty thousand taels of silver. A Reuter telegram from Pekin says that blackmail is levied upon the people by the native employés of the allies. Murder, pillage and burning are going on in Pe-chi-Li and Shan-Tung. We are making preparations to celebrate Christmas.

December 26. Mr. H. J. Whigham, correspondent of the London Morning Post, reports that the Germans, not content with looting Pekin, rob the poor peasants who bring their produce to the Pekin market. The result is that only the markets controlled by British, American and Japanese troops, are freely supplied.

Dec. 26. The St. Petersburg correspondent of the London Times reports that Vice-Admiral Alexieff has issued orders to his command to prepare full lists of the gold and silver, cattle and provisions, seized in China.

London, Jan. 4, 1901. The Daily Mail publishes a severe arraignment from Mr. Willard, a correspondent in Pekin, of the European and American missionaries in China. He accuses them of urging the military to send expeditions to different points of the country, ostensibly to protect native converts, but really to get an opportunity for wholesale looting.

Mr. Willard declares that the missionaries have had their share in all the phases of loot, and gives the evidence of American officers, whose names he withholds, in support of his charges, the whole going to show that in several cases American officers declined to sanction the urgings of the missionaries. In one case the "rebels," whom the missionaries called upon them to shoot, turned out to be a parcel of Chinese school boys brandishing sticks. The soldiers said it would be sheer murder to fire at them, so they fired over their heads and the boys ran away.

Jan. 13. M. de Witte, the Russian Minister of Finance, says in his annual budget, that: "Russia has accomplished all that it was her duty as a christian power to do in China."

Jan. 26. The Europeans, under the command of Field Marshal the Count Von Waldersee, in their various unprovoked and unpunished acts of murder, arson, robbery, and rape, do not shine by contrast with the Boxers. It is rather the Boxers who shine by contrast with them. . . . All the accounts that have reached us have represented the missionaries as having been as active in the looting of Chinese property as they had been in instigating the promiscuous taking of Chinese lives.—New York Times, editorial.

Jan., 1901. In the Contemporary Review for this month, under the title of *The Chinese Wolf and the European Lamb*, Mr. E. J. Dillon presents in harrowing details an account of the massacres of the Chinese by the allied European soldiery. "The doings of some of these apostles of culture," writes Mr. Dillon, "were so heinous that even the plea of their having been perpetrated upon wild savages would not rid them from the nature of crimes." This is the description of a scene at the bar of the Taku, toward the mouth of the Pei-ho: "Dead bodies of Chinamen were floating seaward, some with eyes agape and aghast, others with brainless skulls and eyeless sockets, and nearly all of them wearing their blue blouses, baggy trousers and black, glossy pigtails. Many of them looked as if they were merely swimming on their backs. Hovering over each was a dense cloud of flies, and higher still, in the hot, heavy air, unclean birds of prey wheeling round and round. . . . Fire and sword had put their marks upon this entire country. The untrampled corn was rotting in the fields, the pastures were herdless, roofless the ruins of houses, the hamlets devoid of inhabitants. In all the villages we passed, the desolation was the same. Day after day, hour after hour, sometimes minute after minute, bloated corpses, pillowed on the crass ooze, drifted down the current, now getting entangled in the ropes, now caught by an obstacle near the shore."

The terrible conditions in Tung-Tschau are thus described: "I speak as an eye-witness when I say, for example, that over and over and over again the gutters of Tung-Tschau ran red with blood, and I sometimes found it impossible to go my way without getting my boots bespattered with human gore. There were few shops, private houses, and courtyards without dead bodies and pools of dark blood. Amid a native population whose very souls quaked with fear at sight of a rifle, revolver, or military uniform, a reign of red terror was inaugurated for which there seems no adequate motive. Even if all the Chinese within the city walls had risen in revolt against the foreigners, the latter could have quelled it almost without an effort. . . . To compare nationalities in respect of the guilt of their representatives, in the violation of Chinese women, would be at once misleading to the historian and prejudicial to the cause of humanity. It is enough to know that outrages against female honour were heinous and many; together with the taking of unprotected lives and property, they were the crimes most frequently committed by the allied troops. . . . China has never meddled in European affairs, never given the powers any just cause of complaint. In fact, her chief sin consists in her obstinate refusal to put herself in a state to do either. She is not encroaching upon the territory of others, although her population has become too numerous for her own. Her only desire is to be left, as she leaves others, in peace. She has a right to this isolation. Russia allows no foreign missionaries to convert her people. To induce a Russian subject to abandon his Church for Protestantism or Catholicism is a crime, punishable by law. Why should a similar act not be similarly labeled and treated in China?"

Pekin, Jan. 30. In October last, while men of the Fifteenth U. S. Infantry were on guard duty along the river, they blackmailed several villagers, demanding from 100 to 200 taels protection money, and in some cases their demands were complied with. They were captured by the French and turned over to the American authorities. Dickson escaped after the court-martial began, and is still at liberty. Seamons was sentenced to twenty years imprisonment, and Dickson to twenty-one years.

Berlin, Jan. 30. A dispatch to the Cologne Volks Zeitung from China relates horrible details about the warfare in that country, and says: "We hope these awful conditions will soon cease. The depravity and bestiality among our troops are enormously on the increase. Large numbers of old soldiers are sentenced to long terms in the penitentiary and jail for murder, burglary, and ill-treatment of women."

March 14. Lieut. Chas. Kiburn, of the Fourteenth U.S. Infantry, has arrived at San Francisco from China, on sick leave. He says that the reports of looting in China are untrue. All that he got was an embroidered coat belonging to the Chinese Emperor, Kwang Su.

March 15. A large force of mixed allies, under no command, are plundering every home between Pekin and Tien-Tsin. Having defeated a German squad, a force of British cavalry has been sent to disperse them.

March 15. Rev. W. S. Ament, a missionary and member of the American Board of Foreign Missions, charges the Germans with plundering the city of Man-Ming, sixty miles from Pekin. He said that they ransacked and desecrated a *Christian* chapel and despoiled the women of their trinkets, even tearing the rings out of their ears and ill-treating them in other ways. The Germans replied by charging the missionaries themselves with partaking of the plunder.

April 6. The Kobe (Japan) Herald of this date publishes an interview with Rev. W. S. Ament, an American missionary on his way home from Pekin. He admits that the American and English missionaries looted the premises of Prince Yu and other Chinese magnates and sold the plunder for the benefit of the missions, the sale lasting about two weeks. After disposing of all the loot, and finding that the demand for sables and other valuable things was undiminished, they purchased other plundered articles from the Russian and East Indian soldiers and sold these at a profit. Some of these furs were worth $600 cash.

April 13. Germany, which had fewer troops in China before Pekin fell than any other great power, has presented a bill which has not been definitely stated, but which is understood to amount to not less than $80,000,000, or more than the amount claimed by Great Britain, Japan, and the United States together. The reason of this is that after the fighting was all over Germany poured into China thousands of troops who have been there ever since, and she is now saddling the expense of their keep on China. The Chinese indemnities for war expenses, exclusive of the claims of private individuals and missions, have been thus far fixed as follows:

Russia, $90,000,000; France, $65,000,000; Germany, $80,000,000; Great Britain, $22,500,000; United States, $20,000,000; Japan, $25,000,000. France will also present claims on behalf of Italy. The indemnity is ten times the amount paid to Japan at the close of the Chinese-Japanese war, which was $37,000,000.

April 14. Dr. L. L. Seaman, surgeon of the Seventeenth U. S. Infantry in China, returned to the United States and reported that Gen. Chaffee was careful to prevent his soldiers "from taking part in the looting that has been going on at such a fearful rate in China."

"It is an outrage upon humanity and upon civilization that the Chinese have been treated as they have by the foreign troops who have gone there since Pekin fell. China has suffered a thousand times more than she ever deserved. Her homes, her cities, her sacred places have been looted and devastated, her women have been outraged, and her defenceless citizens have been murdered.

"The German troops have for months been engaged in what they term punitive expeditions. Money is demanded of villages, and proceedings go on which are not warranted. The war has been prolonged months beyond what it ever should have been. When Pekin fell the spirit of China drooped. The war was over. Peace could have been made then and there. The moral effect of the war on China has been rendered nil by the conduct of the allies since Pekin was taken. By this conduct the Chinese hatred of the foreigner will be very justly intensified, and the Chinaman now hates the foreigner a thousand times more than he did when the Boxer troubles began.

"The Germans have now recaptured a large quantity of the best guns they had sold to China but a short time before. These have been sent down to Kiau-Chau, which is the centre of the German sphere of influence.

"It should be borne in mind that nearly all the piliaging that is being done is in territory which formerly gave its trade to the United States. All American trade with China is being interfered with. Whole sections of country are been devastated.

"I witnessed the execution of one of the leading men of China in Pekin on February 26. He was one of two whose heads the allies had demanded. Not even the famous speech of Nathan Hale, 'I regret that I have but one life to give up for my country,' was finer than the patriotic utterance of old Hsi-Chang Yu just before his decapitation on that day. He was President of the Board of Pardons in Pekin, a most important position, and was of noble birth, his family having been of the ruling class since the Ming Dynasty. Just before he was led to the block he was brought, arrayed in the magnificent sable and silk robes of his rank, before the members of the Board for identification. He said to them:

"'I know there have been mistakes made in China for which I have been in part responsible, but if my death will bring peace and prosperity to my country, I cheerfully surrender my life.'

"At the command of Gen. Grodekow, now Governor-General of Manchuria, between 12,000 and 13,000 men, women and children, unarmed and defenceless, were driven into the Amur river at Blagovetchensk because an attack had been made on the Manchurian railway, further up the river. Homes were ferreted out, and the old and decrepit were driven with the women, and children to their death—all to strike terror to the heart of the Chinese. One of the officers in command of the troops which executed that order is a friend of mine, and he told me that his heart grew sick as he assisted in carrying out this command.

"The only Chinese to escape from that terrible massacre were sixteen employes of the firm of Kunst & Albers. This firm protested strongly against the murder of their shroffs, compradores, and clerks. 'Well,' said Grodekow, 'if you don't like it give me 40,000 rubles.' And this amount was paid, then and there, to save the lives of their men.

"In one of my talks with Li Hung Chang we were discussing this subject of the treatment of the Chinese by the foreigner. I said to him: 'I hope, your Excellency, that some day China will adopt some of the methods of Western civilization, so that the time may come when she can regain her ancient possessions, that she may come to be so powerful and so capable of caring for herself that every person who shall reside within her borders shall be there only by the courtesy and right of international law. I hope she may rise, phœnix-like, and reassert her proud position among the nations of the world, and the world looks to you, your Excellency, to begin this rejuvenation of your venerable empire.'

"'Ah,' responded the old man, pathetically, 'the bird is very weak, and is not able again to spread her wings. Like me, she, too, is old.'"

April. 24. United States Minister Conger, returning from China, reports that the missionaries did not join the Germans in plundering the Chinese; they only "appropriated their property for justifiable ends." Dr. Ament only received $75 from the British Loot Committee. "Previous to this, a sale of garments and curios was held, and the $400 netted was given to the American Board of Foreign Missions," with which Dr. Ament is connected. Dr. Ament explained that from the sale of goods plundered from the Mongol Prince's house only $4,000 was realized, and this was devoted to the needs of the native Christians. These possibly included spiritual needs. It will be observed that the amount of plunder admitted to have been acquired by the American Board of Missions increased gradually from $75 to $4,000. No doubt it can be proved to have been all "justifiable."

Pekin, April 25. Very serious famine spreads over the whole province of Shan-Si. Over 11,000,000 population affected. Urgent relief necessary. Conditions warrant immediate appeal.—LI HUNG CHANG.

To this piteous appeal christianity has made no response.

May 1. "The Russian troops were soldiers of fortune from the borderland of civilization; they were fighters, murderers and plunderers by nature and profession. . . . The French were a lazy, cruel and plundering lot. Their actions in China disgraced the name of France. . . . The German campaign of vengeance is absolutely inexcusable. . . . British (chiefly Indian troops) looted systematically. They sold their loot to the highest bidder, and distributed the money among their officers and privates. The auction of loot occurred daily in the British legation. The Japanese took silver from the public yamens, and sent to Japan shiploads of rifles and ammunition from the arsenals. If the veneer of civilization gave way where it was thinnest, and brutal vengeance inspired men's hearts, human nature must be held partially responsible."—Ex-U. S. Vice-Consul General at Hong Kong, Edwin Wildman, in Munsey's for May, 1901.

London, May 1. Dispatches from Pekin report that the German expedition throngh Shan-Si to the Ku-Kwan Pass is returning, after having plundered the country and having inflicted terrible bloodshed. The expedition had produced a very bad effect. Details are not permitted. Every man is laden with plunder. The gold and silver alone will amount to millions. Famine prevails in Shan-Si.

May 6. The New York Sun publishes from Dr. Tewksbury the conditions of the fines levied upon Chinese villages by the missionaries backed by the military. These are one-and-a-third times the value of all property destroyed by the Boxers, the proceeds first to satisfy Christain claimants; "any balance after claims are paid to be used as designated by the Church." All villages where disturbances had occurred shall devote to the Church suitable lands whereon to erect chapels.

Pekin, May 8. The foreign Ministers to-day decided to address a collective note to the Chinese government, informing it that a joint indemnity of 450,000,000 taels will be demanded, and asking what method of payment is proposed.

CHAPTER XXVII.

THE SPANISH-AMERICAN REVOLUTION.

Opening of the American ports to private traders—Tax on quicksilver—Overthrow of Feudalism in France—Feudalism, the conservator of metallic specie systems—Modern freedom and bank notes developed together—The Napoleonic Empire conserved the legitimate fruits of the French revolution—Bank of France—Development of the bank note system opportunely made good the scarcity of money occasioned by the Spanish-American revolution—French invasion of Spain—The injury thus indirectly caused to the mines by the French revolution was repaired by the extension to which it gave rise of the bank note system—Effects of civil war on the mines—Desertion of mining towns and localities—Abandonment of local agriculture—Neglect and ruin of farm buildings and implements—Sinking down of the mines—They are choked by debris—Filled with water—Destruction of machinery—Mischievous work of the *buscones*—Withdrawal of capital from the mines—Permanent blow to production—Immediate effects—Produce of mining reduced nearly one-half—Diminution of the world's stock of precious metals—Development of the bank note system—Beginning of free mining due to Spanish-American revolution.

IN 1778 Señor Galvez, the Minister of the Indies (America), declared thirteen of the American ports open to private traders carrying the Spanish flag, a measure that swelled the exports from Spain five-fold, doubled those from America, and increased the import and export duties levied in Spain from 6¾ million reals in 1778 to 55½ million reals in 1788. Not only this, by affording a ready market for American productions, this measure imparted a marked stimulus to the growth of sugar, tobacco and coffee, and in promoting agriculture and creating a class of yoemen, it materially hastened the Revolution. Mining and the produce of mining may be taxed so insidiously as to furnish no pretext for revolution. Very few people understand the intricacies of coinage, seigniorage, ratios of value between the precious metals, exchange, arbitrage, etc., while everybody is more or less familiar with the cultivation of the earth and the bearing of taxes upon its products. When Galvez's reforms had borne due fruit he attempted, by increasing the taxes, to augment the revenues of the Crown. In 1781 this produced a revolution in the vice-royalty of Sante Fé, essentially a gold-mining district.

The same causes afterwards occasioned a more serious revolt in Peru, where it was only quelled by the most sanguinary measures.

Meanwhile, the news from British North America was not without its due effect. If the Americans would only recall the tyranny, injustice, and wretchedness from which their brave example redeemed the Roman world, they would deport themselves rather as a nation of priests than as a parcel of money-seekers! An accident assisted the yearning of the Mexicans for liberty. The American mines annually consumed 1¼ million pounds weight of quicksilver for the purposes of amalgamation. Of this amount one million pounds came from Almaden, in La Mancha. In 1782 the galleries of Almaden caved in, crushing a number of workmen to death and temporarily diminishing the supply of quicksilver. Although, to make good the deficiency, the king of Spain arranged with the emperor of Austria for an annual supply of 600,000 pounds of quicksilver from the mines of Carniola, the price rose from $41 to $52 per quintal, and many of the mine proprietors of Mexico, wearied with the exactions of the Crown, and irritated by this additional discouragement, turned their peons into the fields and their own thoughts to freedom from the thraldom of the Spanish monarchy.

These inspirations were materially assisted by the overthrow of the feudal system in France. This system was the especial conservator of metallic money, and whenever, and so long as that system lasted, all extensions of the monetary circulation by means of bank notes or any substitute for coins were impracticable. When these devices were employed by the Italian republics they were free from Roman control: when banks of issue were established in Catalonia, Sweden, Holland, England, &c., those countries had removed their feudal yokes. When France arose in turn to throw off this weight, which in that country had attained uncommon and crushing proportions, her long smothered resentment overleaped its legitimate bounds, and the freedom of the Revolution quickly degenerated into licence and the intolerable rule of a mob. It was in harmony with these social changes that in that country money was represented by the illimitable assignat and mandat.[1]

[1] The assignat was at first a communal note, bearing interest and acceptable for purchases of the forfeited lands of the condemned nobles. It was decreed legal tender by the Republic, and became a numerary whose value was independent of the influence of either accumulations of interest or promises of land, and depended solely upon its acceptability for taxes and the quantity of notes emitted.

The emissions becoming excessive, and the notes extensively counterfeited, their value eventually fell to a mere nothing.

The mandat, a subsequent form of note, was payable in, and formed a claim upon, certain lands at fixed prices; but being also issued in excessive numbers, and being also widely counterfeited, it fell in value like its predecessor, until it became worthless.

The Empire of Napoleon was at once a protest against the ancienne régime and against a premature republic in France. It was the conservator of all the legitimate fruits of the Revolution, and marked not only the termination of feudalism in France, but also its termination throughout Europe generally. For the suppression of the feudal system at its centre did not fail to evoke the resistance of surrounding countries, and to bring about those wars which a narrower view of events has attributed too exclusively to the personal ambition of Napoleon. Wherever the French arms prevailed, the feudal system was swept away and monetary systems were changed to conform to the new order of affairs.[2] Banks were chartered in nearly every country of Europe,[3] and their promissory notes for the first time became an essential portion of the circulation. But for this timely and general introduction of an acceptable substitute for coins, it would be difficult to estimate the calamitous consequences which would have followed Napoleon's plunder of continental Europe and the stoppage of the Mexican supplies of silver which took place upon the breaking out of the Spanish American revolution in 1810. Even as it was, these consequences were of a magnitude that seriously threatened the march of European civilization.[4]

The French Revolution and the subsequent invasion of Spain by the French imperial army, were the immediate causes of the overthrow of the royal government in Mexico; thus the very same events which led to the closing of the American mines had provided beforehand the only mitigation, that the times rendered practical, for the evils which that closing was about to entail upon the world. The French Revolution and Empire was about to stop the supplies of silver from America; but it had already provided a partial substitute for it in the paper-note systems which it had rendered possible in Europe.

The news of the invasion of Spain, and the proclamation of Joseph Bonaparte as king, was received with great indignation by the European Spaniards in the American colonies, and regarded with equal pleasure by the natives, to whom it held out hopes of liberty. Down to this period the former had monopolized the public offices and enjoyed every privilege in the colonies; the latter, when they were not held in actual slavery, had been treated as inferiors. Education was

[2] The feudal tenures of Germany were abolished during the French occupation, and restored after the downfall of Napoleon. Hodgskin's Travels in North Germany, II, 89. The author gives an interesting sketch of these tenures and the base services they entailed. [3] The present Bank of France was chartered in 1803.

[4] J. R. McCulloch, Com. Dic. art. "Precious Metals."

forbidden to them; they were not allowed to cultivate flax, hemp, saffron, the olive, the vine, or the mulberry; nor to manufacture any article which could be supplied by the mother country. For a long time foreign trade had been prohibited on pain of death. Taxation had always been kept at a point which rendered its consequences a mere euphemism for slavery [5]; justice was corrupt, partial, and venal; life itself hung upon the words of men, who, to their self-assumed title of Spanish hidalgo, could add none of the virtues which that title represents.

Aware of the discontent generated by this state of affairs, the viceroy, Don José Iturrigaray, upon hearing of the ominous change which had taken place in Spain, sought to conciliate the Mexicans by affording them an opportunity to take part in the administration of colonial affairs: a wise and generous scheme which was thwarted by the resident Spaniards, who seized the viceroy in his own palace and sent him a prisoner to Spain. This proceeding and the conduct of the new viceroy, Venegas—who brought from Spain [6] rewards and distinctions for the leaders of the revolt against his predecessor, and who persecuted those who had supported the plan of admitting Mexicans to the vice-regal council—alienated and incensed the natives.

In September, 1810, a revolt broke out in the province of Guanajuato, headed by a priest, one Miguel Hidalgo, who possessed considerable talent and had much influence with the native population. Such was the practical beginning of the Revolution, which, after having swept through Mexico, eventually liberated every state of Spanish America.

This liberation was, however, not accomplished for many years; in Mexico, for example, not until 1824. During this time the mines of the country became the prey of neglect and vandalism. Among the first consequences of the civil war was the desertion of the mining towns and localities; the inhabitants preferring for greater security to live in the larger and better settled towns, and in districts where supplies of food and other necessaries were more to be depended upon. Ward alludes to this depopulation of the mining towns with reference

[5] For government prices of quicksilver, blasting-powder and other monopolised adjuncts of mining, see the author's "History of Monetary Systems," and U. S. Mint Reports, 1868, p. 617.

[6] The royalists in Spain had demurred to the authority of Joseph Bonaparte and organized a National Regency. This was followed by a second Regency which assembled in the Isla de Leon in September, 1810, and was declared by the Cortes, "the only legitimate source of power during the captivity of the sovereign." "Mexico in 1827," by Sir H. G. Ward. London, 1828, I, 125.

to the unreasonable expectations of the Anglo-Mexican mining adventurers of 1825. He says: "No allowances were made for the moral, as well as physical effects of fourteen years of civil war: the dispersion of the most valuable portion of the mining labourers; the deterioration of landed property; the wanton destruction of stock; and the difficulty of reorganizing a branch of industry so extensive in all its ramifications as mining, and so dependent upon other branches not immediately connected with the mines themselves, and consequently not under the control of their directors. All this was to be effected, too, in a country in many parts of which it was necessary to create a population before a single step could be taken towards repairing the ruin which the revolution had occasioned. And yet nine-tenths of those who engaged in the arduous task did so under the conviction that *water* was the only obstacle which they had to overcome, and that the possibility of surmounting this by aid of English machinery, was unquestionable."

The desertion of the mining districts had been followed by the abandonment of agriculture in their vicinity, the neglect and ruin of farm dwellings, and the destruction of farm implements, stock, etc., so that no present resumption of mining could possibly occur, however peaceful a turn political affairs might take, or however well preserved the mines might be. In fact, neither of these favourable conditions happened. The civil war went on and the mines were abandoned to the elements, the *buscone*, and the bandit.

Rock mines are liable to sag, cave in and fill up with superincumbent rock and earth, especially in countries like Mexico, where perturbations of the earth's crust are of common occurrence.[7] Even without these, the danger of caving is very great, and it can only be averted by incessant watchfulness and continual repairs to the timbers.

At Virginia city, Nevada, a mining town built directly upon the great Comstock lode, there are places where, from the operations of the miners beneath, the surface of the ground has sunk many feet, and in one instance an entire house was engulfed in the "cave." In the galleries of the lode below, though these are supported by the most massive timbers, the signs of overwhelming pressure, twisting, distortion, and sagging, meet the eye on all sides. The shafts have frequently to be re-excavated and re-timbered, and the lines of partition between adjoining properties are so prone to shift, that they require to be frequently surveyed and re-located.

[7] In 1887 an earthquake destroyed Bapispe, while shortly afterward, July 1887, another one nearly destroyed Bacariac, a village of 1,200 inhabitants, some 20 miles distant.

All these circumstances occurred in the silver mines of Mexico. The galleries not only became distorted from the perturbations of the earth and the pressure of the incumbent masses of rock, the timbers rotted and the galleries caved in and became choked with débris. Being now entirely neglected, many of them became filled with water, and, except in some instances where the excavations had been tapped by adits or tunnels at low levels, so as to afford drainage for the water, the mines were rendered entirely unapproachable and unworkable.[8]

Coincident with these processes of destruction and decay occurred the rotting, rusting, and plunder of machinery. According to Ward, this was all destroyed during the war, and "it became necessary to erect anew horse-whims (*malacates*), magazines, stamps, crushing-mills (*arastras*), and washing-vats; to purchase hundreds of horses for the drainage (of each mine) and mules for the conveyance of the ore from the the mine to the haciendas, where the process of reduction was carried on; to make roads in order to facilitate the communication between them; to wall in the *patios* or courts in which amalgamation was at last effected; and to construct water-wheels, wherever water-power could be applied.

However, all the mines were not abandoned during the civil war. A few of them in favourable locations were not disturbed at all, and were worked as usual; others were worked by the peasants, who crept into the deserted galleries and gleaned from their more accessible levels such remains of ores as the proprietors had not deemed it worth while to remove, or such as had served as pillars to support the excavations. These peasants were called *buscones*, *i.e.*, searchers or gleaners. The mines also became the hiding-place and refuge for tramps and bandits, who, in common with the *buscones*, played such havoc with the works as to still further unfit them for a return to systematic mining.[9]

But perhaps the most serious, because the most permanent, source of misfortune to the mining industry of Mexico, was the removal of the large capitals which had formerly been invested in its prosecution. Some classes of theorists and politicians have supposed that capital has but little to do with the progress of mining; but practical men, and others, familiar with the requirements of mining countries,

[8] "In other mining districts water was allowed to accumulate to an immense extent in consequence of the suspension of the usual labours." Ward, II. 56, 119.

[9] In alluding to the depredations committed by the *buscones*, Ward says they were still at work in many mining districts of Mexico when he visited them in 1827, and that he could not help believing that they managed during the civil war to produce as much as $5,000,000 a year, in addition to the $6,000,000 obtained from regular mining. Ward's "Mexico," II, 21, 23, 24.

have abundant reason to believe otherwise. For example, there are numerous mines throughout the world to-day which are not paying dividends; yet, through the power of capital, the workings continue and thousands of miners are employed, with the hope that before long some new ore body will be found to cover the hazardous outlay, and reward the adventurers.[10] If this capital were withdrawn the mines would have to be abandoned; for the fruit of the present outlay may not appear, if at all, for months or years, and the unaided miner could not afford to wait for it. The withdrawal of capital from the Mexican mines is frequently alluded to by Ward as the principal cause of their abandonment and destruction. "This was the real evil of the Revolution. It was not the destruction of the *materiel* of the mines, however severe the loss, that could have prevented them from recovering the shock, so soon as the first fury of the civil war had subsided; but the want of confidence, and the constant risk to which capitals were exposed, which, from being in so very tangible a shape, were peculiarly objects of attraction to all parties, led to the gradual dissolution of a system which it had required three centuries to bring to the state of perfection in which it existed at the commencement of the War of Independence."[11]

The accounts of the mineral riches of Mexico given by Baron von Humboldt in his celebrated Essay on New Spain, so excited the enterprise of British capitalists that after the close of the civil war a vast capital was invested in the mines,[12] and half the population of Cornwall[13] was sent to reopen them; yet the previous destruction had been so great that several years after these new aids had begun to be applied, Ward, who was a careful observer, regarded the works as nothing in comparison with what must formerly have existed.

The remarks here applied to Mexico alone, may with equal propriety be applied to the rest of Spanish America. The withdrawal of capital from the mines, the depopulation of mining districts, the neglect of those industries, agricultural and other, upon which the successful working of the mines depended, the abandonment of the works and machinery to the elements, the vandal and the thief, were common throughout the whole of Central and South America.

The immediate effect of these circumstances was very decided, and full of important consequences to the world. Taking Mexico

[10] Pliny, in his account of the Spanish mines, made a similar observation nearly two thousand years ago. It seems a pity to be obliged to repeat it.
[11] Ward's "Mexico," II, 56, 57, 116, etc. See also Jacob, chap. XXV.
[12] For a particular account of the capital thus invested, see Ward, II, 64, *et seq.*
[13] Ward's expression, II, 76.

by itself and following Ward, the production from 1796 to 1810 inclusive amounted to an annual average of $24,000,000, of which about five per cent. was gold, the remainder silver. From 1811 to 1825 the production fell to an annual average of less than $11,000,000, of which about eight and a-half per cent. was gold, the remainder silver. In 1826 the production did not exceed $8,000,000; in 1829 it was still less.[14] By the year 1842 it had again risen to about $23,000,000,[15] and during the five years ending with 1875 it averaged $22,500,000 a year.[16]

If the view be extended to the whole of Spanish America, including Mexico, the decline in the production of the precious metals was still more pronounced. Previous to the Revolution, the average annual produce of gold and silver—chiefly the latter—amounted to $36,000,000. From this sum it fell during the Revolution, that is to say, from 1810 to 1825, to $25,000,000, and in 1829, which seems to have been the lowest point of the depression, it probably did not exceed $20,000,000.[17]

The significance of this decline in production will be better understood when it is remembered that before the produce of the Russian gold mines began to be important (in 1841 they amounted for the first time to so much as $5,000,000 a year) the silver mines of Spanish America constituted the principal, almost the only source, from which for a long period, the European world had been accustomed to derive its supplies of the precious metals.

With the falling off of the customary supplies began a diminution of the European stock of coin, which, it is estimated, fell from $1,900,000,000 in 1809, to $1,557,000,000 in 1829.[18] The civilized world was thus threatened with a calamity similar to that which had befallen it upon the decline of the Roman mines. The customary supplies of the precious metals were arrested, the general stock had begun to diminish, and prices threatened to fall, not, indeed, simultaneously, but in that irregular order which would subvert all the existing relations of society, and plunge them into a confusion as profound as that of the Dark Ages.

[14] Jacob, 327, allows eleven and a-half million dollars for each of the years 1826 and 1827, and twelve millions for 1829, and 1830; but, as he himself admits, these estimates were based on very imperfect data.

[15] Thompson's "Recollections of Mexico," p. 203, quoted in De Bow's "Resources," I, 294. [16] "Report U. S. Monetary Commission," Appendix p. 383.

[17] Jacob, 342, and elsewhere throughout his XXVth chapter, where the details by countries and the delapidated condition of the mines is described.

[18] Jacob. This calculation is supported by Gerboux, Storch, Tooke, and several other writers of eminence. They all admit an important decline of the European stock of coin during the period of the Spanish-American revolution.

It is to the effort to avert this calamitous fall of prices that we are indebted to that extension, which amounts almost to a creation, of the modern banking system, and which took place between the years 1810 and 1830. As with many new systems, this one at first was employed without due precautions. Banks were established on too slender a basis, with too many and too important privileges, and the result was that, until better systems were devised, failures were common, and financial crises followed one another in rapid succession. Improved systems were, however, eventually established, and society at length learned how to conduct its exchanges, and sustain the accustomed level of prices even after the customary supports of such level—namely, the European stock of the precious metals as maintained by the annual supplies from Spanish America—had fallen away. The perturbations in the commercial world during the interval when this experience was being acquired, viz., from the breaking out of the Spanish American Wars of Independence to the period when the produce of the Russian mines became important—say from 1810 to 1842—will long be remembered for their frequency and violence.

Moreover, the Spanish American Revolution, and therefore its precursor the French Revolution, bequeathed to us something even more important than the modern banking system. They emancipated the slave and peon, and substituted the institution of mining for the precious metals by free labourers, an institution which since the days of the Roman commonwealth had been unknown.[19] When the tocsin sounded that proclaimed the revolution in New Spain, the fetters fell from off the person of every mining slave in the world. Thenceforth the precious metals were to be obtained by FREE LABOUR; and although the stock which conquest and slavery had amassed might last the world long enough to influence the value of whatever metal might be mined by freemen for centuries yet to come; yet that influence was to continually grow less, and, at some time or another, would and will assuredly come to an end.

[19] Ward's work contains an interesting and comprehensive summary of the mining laws and regulations of New Spain under the Spanish crown; but so far as it leaves the impression, which to some extent it certainly does, that the degree of freedom enjoyed by the mine owners (as to the workmen they were virtually slaves down to the period of the revolution) was due to any liberality on the part of the government, it is misleading. The reason why the Spanish government accorded the privileges which he mentions, is due to the fact, and it is one of the proofs, that gold and silver mining was unprofitable. Had this been otherwise, there is no room to doubt, from its proneness to monopolise every profitable industry upon which it could lay its hands, that the government would have conducted the mines itself.

As it was, it shrewdly left the miner to enjoy or suffer all the hazards of discovery, and encounter all the obstacles to production; and when this was effected, it taxed the product right and left, in every conceivable way, so as to leave as little of it to the producer as he was content to take without rebelling against its authority,

CHAPTER XXVIII.

THE APPALACHIAN MINES.

Circumstances which influenced the opening of these mines—Scarcity of gold—Æra of low prices—Humboldt's opinion—Product of the mines—Number of slaves employed—Cost of their subsistence—Unprofitable nature of the Industry—Closure of the mines—Revival in 1859—Second closure—Recent condition of the mines—Poverty of the neighbourhood—Wealth of the surrounding agricultural country.

THE great mountain chain which separates the Atlantic States from the Mississippi Valley, and extends from Eastern Canada to the northern portion of Alabama, was called Appalache by the southern Indians, and Alleghany by the northern tribes. Gold and silver have been found at intervals all along the chain, from the Chaudiere Valley, below Quebec, to the foot-hills of South Carolina. The greatest development has been of gold on the eastern flanks of the chain, south of the Potomac river.

Indications that the Spanish gold-hunters of the 16th and 17th centuries were aware of the auriferous character of the Appalachian range have been met with near Abbeville, South Carolina, and elsewhere. In 1606, under a charter of James I., the right to explore and settle the North American continent, from the 34th to the 45th parallel (say from South Carolina to Canada), was granted to the London and Plymouth Companies upon condition that one-fifth of the gold and silver and one-fifteenth of the copper discovered should go to the Crown. The only results of these grants were Captain John Smith's expedition to the Pacific Ocean, by way of the Chickahominy River, and the shipment to London of a cargo of pyrites, which that hero mistook for gold-dust. Lock, 123.

In North Carolina the Anson Mine consisted of the bed of Richardson's Creek, which ran dry in summer and disclosed a gravel deposit containing gold. Read's Mine, in Cabarras county, occupied the bed of another affluent of Rocky river. It was here that in 1799 a slave had taken out a single nugget weighing 34 pounds Troy; but although the vicinity was afterwards thoroughly dug over, no other rich find was encountered. Twenty miles northward, near the Yadkin river, a nugget of 600 pennyweights was found. Parker's Mine, four miles south of the Yadkin, yielded a nugget of 4¾ lbs. weight, and several of 100 pennyweights and upwards. Before the mint was erected at Charlotteville gold dust in quills was commonly passed in the vicinity for money at the rate of 90 cents a pennyweight.

After the placers were worked out, the quartz was located and developed. A mine in Rowan county, N.C., with a small vein 4 to 12 inches wide, and working only seven hands, produced $12,000 in a single month. The Washington silver mine, near Lexington, costing for ground and buildings about $500,000, and yielding $40,000 during the first two years, was opened in 1842; but abandoned a few years later as hopelessly unprofitable.

In Georgia the principal mines were in Habersham, Hall and Cherokee counties, and at one time so many as 5,000 miners were employed on these alone; but with the rise in the value of slaves, they were gradually abandoned. Early in 1899 a deposit of quicksilver was discovered in Grant Park, Atlanta, but as yet it has not been worked.

Yielding to the clamours of the gold-producing localities for free coinage facilities, the Federal government erected branch mints at Charlotte, N.C., and Dahlonega, Geo., neither of which States ever produced enough gold to pay for these establishments. Their practical uselessness may be gathered from the cost of coinage as compared with that at the principal mint in Philadelphia.

Cost of coining $100 worth of coins at the various Mints of the United States as reported to Congress, March 31, 1842:

Year.	*Philadelphia.*	*New Orleans.*	*Charlotte.*	*Dahlonega.*
1838	$1.52	$154.06	$17.82	$12.43
1839	2.07	19.72	9.03	10.78
1840	2.48	5.68	9.44	9.32
1841	4.34	8.12	9.02	6.09
Average,	$2.60	*	$11.33	$9.65

* In 1838, owing to the limited coinage, it cost more to fabricate the small coins at New Orleans than the coins were worth; for this reason an average of four years would be misleading. An average of the two years 1840 and 1841 is $6.90.

In short, while the cost of coinage in Philadelphia was about 2½ per cent. ad valorem, it was 7 per cent. in New Orleans, 9⅔ per cent. in Dahlonega and 11⅓ per cent. in Charlotte. These expenses did not cover the rent and repairs of buildings and plant, nor interest on the cost. They constituted in effect a gratuity to the owners of bullion for which the government received no consideration whatever.

Reverting to the period about 1824, when the Appalachian mines first became noticeably productive, the commercial world, for a quarter of a century, had been drained of the precious metals, to fill the void occasioned by the change made by the East Indian Company, from copper and billon to silver money in India. Coincidently with this drain to the East occurred the failure of the Brazilian mines, the Spanish-American revolutions, and the closure of the remaining mines of the West. The metallic stock of the Occident began to diminish; commerce halted; prices fell; bankruptcies multiplied; sharp cries of distress arose on all sides; and the auriferous deposits which, before the fall of prices, were not worth working, came again into play. This is what occurred in North Carolina and Virginia; the same thing occurred at the same time in Siberia. The mines had been worked long before; all at once they became unusually attractive and productive. Why? Because gold had become scarce and dear, and because now that mine-slavery was largely over, it paid better than before to search for it, to take the chances of finding it, and when found, to carry it to the eleemosynary mints, which the policy of "free coinage" had erected.

Baron von Humboldt, than whom there is no higher authority on the subject, expresses a somewhat similar opinion: "Almost at the same moment in which the Ural opened its golden treasures and began to replace what the diminished alluvions of Brazil were no longer in a condition to supply, strata containing gold were discovered in the southern part of the Alleghanies: in Virginia, North and South Carolina, Georgia, Tennessee and Alabama. The most flourishing period of these American gold washings was from 1830 to 1835."[1]

The produce of the Appalachian gold mines is indicated by the deposits of American bullion by private individuals in the mints of the United States, the larger portion of the produce having been thus deposited: Year 1824, $5,000; 1825, $17,000; 1826, $20,000; 1827, $21,000; 1828, $46,000; 1829, $140,000; 1830, $466,000; 1831, $520,000; 1832, $678,000; 1833, $868,000; 1834, $898,000; 1835, $698,000; 1836, $467,000; 1837, $282,000; total 1824 to 1837 inclusive, $5,126,500. From 1838 to 1847 inclusive the amount was $2,623,641; total from 1824 to 1847 inclusive $7,750,141; total from 1804 to 1866 inclusive, $19,375,890; and to the present time about twenty-five million dollars. Allowing for gold not deposited in the mints of the United States, the entire product has been estimated in

[1] Humboldt's "Fluctations of Gold," p. 62.

round figures at forty million dollars.[2] When the California mines were discovered in 1848 the Appalachian mines were abandoned and the produce declined. Of the total amount, North Carolina produced about 50 per cent.; Georgia, 33 per cent.; Virginia, 8 per cent.; South Carolina, 6 per cent.; Tennessee and Alabama, 3 per cent.

The work of the mines was done by about 2,500 negro slaves, who, being valuable workmen and servants fit for other employment, were well clothed and fed; so that their labour cost to their owners not less than half a dollar per day, say $180 a year, or for the whole number, about $450,000 a year. In seventeen years (the productive period of the mines) this amounted to $7,650,000, to say nothing of the first cost of the mines, machinery and other plant, quicksilver and mining supplies, expense of superintendence, office expenses, legal fees and numerous other charges.

After the hydraulic system of working placer mines was invented in California (1851), a feeble revival of gold placer mining took place on the Appalachian range in 1859. This saved the mints of Charlotte and Dahlonega from being closed. The Civil War, which broke out in 1861, reserved the mints for the service of the Confederate States; but the Appalachian mines contributed no metal to them.

The writer visited the Appalachian mines of North Carolina in his youth and again some fifteen years ago. The alluvions (mostly shallow deposits in gulches) had been worked off; the quartz mines were for the most part abandoned and filled with water; there were no fences or railings around the pits, which therefore were exceedingly dangerous to persons not familiar with the locality; pieces of machinery and the remains of dismantled structures strewed the ground; and the people of the vicinity were poor and needy. A more desolate picture than that presented by the immediate neighbourhood it would be difficult to conceive. Yet, once away from this scene of of devastation and neglect, the signs of prosperity multiplied on all sides. North Carolina is a rich State, but its riches do not consist in gold mines. They are to be found in her farm-lands and factories. These surround the mining neighbourhoods on all sides and evince, by striking contrast of appearance, their superior ability to maintain the general prosperity.

[2] The deposits of domestic gold at the mints down to 1836 inclusive, were furnished to Humboldt by Albert Gallatin, formerly Secretary of the Treasury. For the subsequent period, see Macgregor's Statistics, III, 1029; Putnam's Cyc., p. 760; Ure's Dic. Arts, Mines, etc., II, 647; Lock, p. 123, and the official report on the "Mineral Resources of the United States," 1868, Part II, p. 20; which is the authority for the "forty millions" of total produce down to that year. We very much doubt if the whole produce, even to the present time has been as much.

CHAPTER XXIX.

RUSSIA AND SIBERIA.

Ancient Scythia—Norse conquest of Novgorod—Mongol invasion—Establishment of the Muscovite Empire—Money—Precious Metals—Mines—The conquest of Siberia was effected by gold-hunters—Cossack expeditions under Yermak—Expedition of Polish exiles to the Amur—Expedition of Bukholtz—Scarcity of precious metals in Russia—Expeditions of the Zaparog brigands—Survey of early mining—The Scythian Mongols—Huns—Tschudi—Norsemen—Tartar Mongols—Gold mines of Petchora—Silver mines of Olkusz—Silver and gold mines of Nertschinsk—Gold mines of Berezovsk—Valley of the Yenesei—Ancient remains—Little Bokara—The Oxus—Kolyvan—Voetsk — Ekaterinburg — Berezovsk — The Altai—Miask—Employment of discharged soldiers—Tschudi remains—Gold panic—Humboldt's work—Tungousa range—The tundras—Remains of tropical animals—The Yenesei tundras—Convict system—Atchinsk—Minusinsk—Olekminsk—The Amur valley—Mantchuria—Produce of the Silver mines—Total gold produce—Mania for gold mining—Second gold panic—Demonetisation of gold in Holland and Belgium—Proposal to demonetise it in France and England—Failure of Banks of England and Austria—Californian and Australian produce—Chevalier's "Fall of Gold"—Russian tax on gold and regulation of the mines.

RUSSIA, before the Mongol Tartar occupation of 1223–1462, is so little known to history, and that little is so perverted and mixed with fable, that the merest outline will be sufficient to prepare the reader for the more authentic history which begins only when that occupation quite ended.

What is known to-day as Russia was to the Greeks known as Scythia, the ancient accounts of which are to be found in the pages of Herodotus, Strabo, and Pliny. In the ninth century of our æra, about the year 862, the northwestern part of Russia was invaded by the Norsemen, under Ruric, who established at Novgorod a Gothic republic, whose arms and commerce maintained a jealous supremacy of the Hyperborean world. At length these were destroyed by that Byzantine influence which made itself felt in the military career of Ivan (John) III. Vasilovich, who captured the city of Novgorod and with it three hundred cartloads of gold and silver. Macgregor, II, 407, dates this event in 1450, while other historians prefer 1462 or

1471.[1] Novgorod at that period contained 400,000 inhabitants; its territorial possessions extended to the White Sea and the Ural Mountains, in which it worked the gold mines of Petchora; it was the capital of that pagan Hansa, whose trade, previous to the establishmant of the christian Hansa, embraced the coasts of Europe and extended to farthest India and China; and it was justly called Novgorod the Great. The Norse history of Russia is lost in the ruin wrought by the Muscovites and in the fables erected upon that ruin. Novgorod appears to have been ransomed by, or else it defended itself successfully against, the Tartars, only to fall at last to the arms of Byzantium; when the fanes of Woden, which had escaped the Mongols, were destroyed, the graves plundered and the Norse runes effaced; until all that now remains of the once famous Gothic republic is the proud blood of the Vikings and the familiar name by which they were known. This was *Routsi*, the *rowers*, a name still preserved in the title of the Empire.

From the conquest of Ruric, A. D. 847 or 862, possibly even from the remoter period of Woden, or Woote, B. C. 90, down probably to to the period of the Mongol invasion, the money of Novgorod had consisted of parchment notes, which, having no currency beyond the the jurisdiction of the republic, obliged the clearings (permutations) of the great Fair to be conducted in metallic moneys. These consisted of the coins, baugs and slugs of various nations—East Indian, Chinese, Arabian, Roman, Scandinavian and English, the Arabian largely predominating and the universal integers being the gold Mark and the silver Ora, each of 240 grains, and valued at 8 for 1.[2] Commerce probably supplied all the silver that was needed for the clearances of the Fair, while the more precious metal was obtained in limited quantities from the western flanks of the Ural.

Between the period of the Mongol invasion and the fall of the Gothic republic we have no account of the moneys of Novgorod. The decline of its trade and the general confusion ot the times render it all but certain that the parchment system had to be abandoned in favour of metallic slugs and coins; chiefly silver. Indeed, some specimens of both classes of these moneys are still extant, and though they are without dates, they have been from other evidences ascribed by numismatists to this period of turmoil.[3]

[1] In many medieval dates relating to Russia a difference occurs of fifteen years, which may be due to those changes in the Roman calendar which are discussed in the author's "Worship of Augustus Cæsar."

[2] A full account of these moneys will be found in the "History of Monetary Systems."

[3] "Money and Civilization," chapter on Russia, pp. 294-5.

Practically, therefore, the history of the precious metals in Russia begins with the establishment of the monarchy by Ivan III., in 1462. The first coins were struck by Basil IV., 1425–61; the title of Czar or Cæsar was assumed by Ivan III., in 1482; and the mines once worked, but afterwards neglected by the Norsemen, were opened in the same reign. "The mines of Petchora were discovered (?) in 1491, and Russia for the first time saw silver and copper money, the produce of its own territory, coined in the capital."[4] This capital was now Moscow.

The monetary history of Russia, from the establishment of the monarchy to the conquest of Siberia, proves that the precious metals were scarce and little used. The earlier czars of Russia had to accomplish the almost insuperable task of cementing together a vast empire, composed of divers races and nationalities, without the incentive of that spoil of cities which had so powerfully assisted the establishment of the Moslem Empire, nor that spoil of the mines which at a subsequent period built up the empire of the Spanish monarchs in America. The only resource left to the Russians was the old Roman one of an overvalued copper currency, which, notwithstanding its many disadventages, sufficiently served its purpose, until the Mongolian placers east of the Ural were prospected, invaded, conquered and exploited. So late as 1663, the inhabitants of Moscow broke into revolt on account of the overvalued copper coinage. Meanwhile, brigands, miners and Cossacks were scouring the Urals and prospecting for gold even beyond the Khirgis steppes. Ample relief was to come within half a century, but the unpatriotic and rebellious Muscovites could not have seen so far ahead.

It has been often remarked that every step in the extension of the Russian Empire eastward, from the Ural chain to the Okhotsk Sea, was preceded by the discovery of gold; in other words, that Siberia was wrested from the Kirghis, Tschudis, Tungouses, and other natives, not, as has been pretended by some Russian historians, to confer upon it the fruits of civilization, but to exploit it for the precious metals. Says Humboldt: "Upon first taking possession of Siberia, the tombs were despoiled, and so great a quantity of gold was plundered from the graves (karganui) of Krasnojarsk, that, according to Muller, a reliable historian, it occasioned a remarkable (local) fall in the value of that metal." "Fluctuations of gold," p. 38. Says Lock:

[4] "Russia," by Karamsin, Tooke and Segur, edited by Kelly, 1840, I, 127. These writers could hardly have been acquainted with the Norse moneys, or the Tartar coinages, of which an interesting collection is to be found in the archæological museum of the Hermitage, at St. Petersburg.

"Lured onward by reports of fabulous wealth, a band of Cossack adventurers, led by that 'illustrious brigand,' Yermak (1580–4), laid the foundations of a new empire for the czars, beyond the Urals." "Gold," p. 38. Says Atkinson: "The wars by which Russia conquered Siberia were entirely carried on by private adventurers at their own cost, who were incited to the undertaking by their love of plunder. (In an almost desert country this could not mean plunder of the inhabitants, but plunder of the graves and the gold placers). A Pole and some other exiles escaped from Yeneseisk and built a small fort on the Amur; but having quarrelled with the Tungouses, they offered the conquest to the emperor of Russia, and begged forgiveness for their former offences; while the Tungouses about the same time applied for assistance to the emperor of China. This led to disputes between the two governments, but war was prevented and the boundary between China and Siberia was established by a treaty concluded at Pekin in 1689." "Oriental and Western Siberia," 1858, cited in Appleton's Encyc., XVI, 624. Says Lock, again: "In the early part of the 18th century, Bukholtz led his Cossacks, ever the vanguard of Russian explorers, to the Semipalansk gold region next to Altai. . . . The chief object of the Trans-Baikalian expedition of 1849, ostensibly sent to define the boundary with China, was to discover the gold placers in the Amur country (Chinese-Mantchuria). Two mining engineers and a corps of miners accompanied the expedition." "Gold," pp. 374, 415.

Says Tatarinof: "Hardly had our troops, under Major-General Chernaief, entered Kokand than reports of its prodigious wealth flew from ear to ear. Sévertsef, the geologist, was present with the active forces in the field during the summer of 1864. He was told of a gold mine on the Kukréú, and discovered a whole system of alluvial deposits near the sources of the Tersi. Thither the gold-miners rushed in quest of a new Dorado."

Nor were the forays into Siberia confined to the strictly criminal classes: during the last half of the 18th century they were largely reinforced by the ferocious Zaparogs, or Cossacks of the Dnieper (Boristhenes), whose territory extended from that river to the Bug (Hypanis). These people had been the terror of the Ukrane and the adjacent provinces, through which they rode headlong, plundering corn and cattle and setting fire to the villages, to force the women out of their hiding-places, in order to subject them to the worst of indignities. The dispersion of these brigands began in 1741–62, when many of them escaped to Siberia, where they took part in those forays and gold-hunting expeditions which gradually brought that immense territory under the dominion of the czar. The remainder were surrounded by a Russian army in 1798, and those who did not surrender, fled to the Danube or to Asia Minor.

Mining for the precious metals within the boundaries of Asiatic Russia began at a remote period. To say nothing of the Argonautic expedition to Colchis, which may be either fabulous or distorted,[5] nor of the mines of Scythia alluded to by Herodotus and Strabo, the Tschudi, or Judi, worked mines in the Ural and Altai ranges long before our æra. Their stone and copper implements and many traces of their workings are still to be seen. In 1723, Demidof's miners, while prospecting for minerals, came upon old Tschudi mines at the southwest end of Lake Kolyvan. In 1733–43 John George Gmelin, the botanist, examined others. No iron tools were found in them; but as iron oxidises and disappears in the course of ages, this is not positive proof that the Tschudi were unacquainted with that useful metal.[6]

In the reign of James I. of England, 1603–25, Walter Busbee, assayer in the Mint, was sent to Russia to give advice as to the gold mines of the Ural. He was taken prisoner by the Tartars. Calvert's "Gold Rocks of Great Britain," p. 10. From this passage it appears that the Ural as a source of gold was not being neglected in Russia, though, so far as the Russian annals are concerned, the only allusion to mining during the interval between the working of the mines of Petchora and the mission of Walter Busbee, relates to the piratical expeditions of Yermak and his Cossacks.

The lead-silver mines of Olkusz, northwest of Cracow, were worked early in the 17th century. In 1655 the royal tithe or production tax of 10 per cent. amounted to 1,225 marks of silver and 1,514 quintals of lead. . . . It appears from the lowest calculation that the ore extracted from the mines must have been worth 476,773 florins, each of which was worth four florins of the present day. Malte-Brun, IV, 373. The distinguished writer does not specify the interval which this production filled, though it is presumed that he means a year. Nor does he determine what proportion of the total valuation was derived from silver and what from lead. The mines could not have

[5] The gold deposites of the Caucasus, though immortalized in the tradition of Jason and the Argonauts, are now entirely abandoned, the last attempt at working them having been suspended in 1875. Am. Encyc. Brit. art. "Gold."

[6] Baron Alex. Von Humboldt, who was a mining engineer, and who alone of all the writers on the mines of antiquity, recognized the essential difference between a quartz, vein, or reef, and a placer, gravel or drift mine, is not disposed to admit that the ancient mines examined by Gmelin were quartz mines. In his courtly manner of never

been profitable, because when Malte-Brun wrote, about 1820, they were abandoned and Olkusz was deserted and in ruins.

But in fact, west of the Ural chain, the discoveries of either gold or silver within the boundaries of Russia were too inconsiderable to deserve notice. Practically, mining for the precious metals began with the forays of Yermak and the opening of Siberia. During the whole of the 17th century bands of predatory Cossacks roamed through these vast domains, prospecting, conquering, plundering the inhabitants, ransacking the Tschudi sepulchres, prospecting again, washing the river sands for gold, digging through the turf or tundra (tchundra), into the frozen placers below, or pointing out to the government where quartz mines might be located (for they themselves cared nothing for quartz), and where serfs or convicts might be led to delve for gold and silver, and incidently to establish those needful posts and centres of supply and trade, the demands for which increased with the wealth which these brigands had obtained through plunder and placer mining. Foremost among such quartz workings were the silver mines of Nertchinsk, opened in 1704. These mines are in the eastern part of Siberia, in the same longitude as, and 720 miles due north from Pekin, China. Nertchinsk has often been alluded to as a quicksilver mine. In fact it is a group or series of about fifty silver-lead mines, in which no quicksilver or cinnabar occurs. The sufferings of those victims to Russian tyranny who were supposed to be rotting in its dark recesses, have therefore been greatly exaggerated. The truth is bad enough. Quartz mining at great depths is not especially exhilarating work, even when done voluntarily and at good wages. To the convicts of Nertchinsk it may not have been altogether congenial; but, as the Rev. Henry Lansdell, an eye-witness, shows in a comparatively recent work, it was not marked by the exercise of any noticeable degree of hardship or tyranny. No women or children were to be seen in the mines. The author also declares that there are no quicksilver mines at or near Nertchinsk, and that the narratives of Mr. Lenke, Baron Rosen, and others, were exaggerated. "Through Siberia," 1882. The writer of the present work, who visited Russia ten years previously, and had abundant opportunities of observation, can offer corroborative testimony: the horrors of the Siberian penal system were greatly magnified. To these distorted

directly contradicting anybody, he merely alludes to them as mines known by the "ambiguous denomination of Tschudisher Tchurfe." This is equal to saying that they were Tschudi gravel (drift) mines, covered with a layer of turf or peat, like those recently discovered in the Klondyke, and worked in a similar manner, by removing the moss-covered turf, thawing the frozen gravel beneath and then drifting. Such work as this can be done without iron tools.

views even Mr. Lansdell contributed a share, when he repeated the tumid story of a British sea captain who had visited "the mines of Kara" (possibly Berezovsk), in 1859:

"He saw 2,000 men branded and chained to their barrows by night and by day. The overseer of the gold mines, a German, told him he had shot four men who had killed others when at work. And I (Lansdell) have heard since my return that some of the predecessors of Col. Trononovitch (1879) were so cruel that the mere mention of their names made convicts tremble."

The ores of Nertchinsk contained from five to forty ounces of silver to the ton, and had to be smelted. They produced: 1704–20, 4,732 Russian pounds of silver; 1721–30, 1,498 lbs.; 1731–40, 1,333 lbs.; 1741–50, 15,657 lbs.; 1751–60, 43,631 lbs.; 1761–70, 126,247 lbs.; 1771–80, about 160,000 lbs.; 1781–1800, about 140,000 lbs.; 1801–10, about 120,000 lbs.; 1811–20, about 100,000 lbs.; and 1821–30, about 95,000 lbs. From this date the silver produce of Nertchinsk rapidly declined, until it lost all historical importance. During the latter part of the 18th century these mines employed about 2,000 convicts, as many "free colonists," and 11,000 peasants—in all about 15,000 persons. The average production did not amount to the value of over $6 per capita per annum: not enough to pay for victualling the miners. When after many years these facts were fully realized by the administration, a portion of the convicts, to their great relief, were set to work upon the gold placers of Nertchinsk, which, though well guarded by military, were at all events above ground, where there was plenty of fresh air and a possible chance of escape. These placers were opened in 1830, but, though very productive and worked with convicts and serfs until 1863, they proved to be scarcely more remunerative than the subterranean galenas. The product is included in that of the Amur district mentioned further on. A few exiles annually escaped from Nertchinsk to China, where they were promptly caught and religiously surrendered, as polluters of its sacred soil. In 1863 the government threw open the placers of Nertchinsk to free labourers, thus affording a stimulus to mining that during the following quinquennial period augmented the production about 50 per cent.; but it afterwards fell to the old figure of about a ton and a-half of gold per annum. The returns bespeak a considerable amount of gold-stealing by the workmen.

The progress of mining discovery in Siberia next leads to the exceedingly interesting narrative of Baron Philip Johan von Strahlenberg, a Swedish officer who was captured by the Russians at the battle of Pultova, in 1709, and sent a prisoner to Siberia, where he enjoyed the advantage of personal intercourse with the governor of the territory, Prince Gargarin, who, like himself, was a student of history, and whose pages they greatly enriched by their learned and unprejudiced observations.

The part of Siberia to which Von Strahlenberg was consigned, and where he spent thirteen years of his life, extended from the Obi eastward to the valley of the Yenesei, in which were found numerous remains of an ancient mining race. About twenty or thirty years before his time, a former governor of Siberia had authorized the plunder of the native graves upon condition of his receiving ten per cent. of the proceeds. In these graves were found gold and silver coins, jewelry, chessmen, images, bronze (cast) swords and bronze metal plates with inscriptions in an unknown language. Many of the tombs were of stone, and lofty, similar, as Von Strahlenberg remembered, to the tombs of the ancient Cimbri, in Holstein. Some of the educated Chinese (Mongols), who visited the town of Yenesei, asked permismission to visit these tombs as those of their "ancestors"; but as the tombs had been rifled, such permission was not accorded.

The Tschudic tombs of the Irtisch, an affluent of the Obi, which flows into the Arctic Sea between the Yenesei river and the Ural range, are of a different period. Some of them, indeed, contain "Mongolian" remains, but these may be regarded as having been captured by, or relinquished to, the Tschudi, perhaps stolen by them from the tombs of the races who preceded them: for Von Strahlenberg says that the existing Tschudi are too ignorant and unskillful to have fabricated such articles. The remains include Mongol and Kalmuck inscriptions printed on cotton or silk paper. Of a far later date are the Arabian relics of the ninth, tenth and eleventh centuries of our æra, acquired by the Tschudi probably from the Arabian traders at Novgorod. Among these was a copper medal containing a Moslem religious inscription, which the Tschudi piously venerated or worshipped.[7]

In the Yenesei valley the rocks through which the river in some places has cut its way, are covered with pictorial illustrations, which appear to be similar to those on the coasts of Sweden, facsimiles of which will be found in Du Challu's "Viking Age." Von Strahlenberg's observations lead to the impression that the Mongols, whose

[7] The inscription was as follows: "To him who seeks God there will be given blessings, increase, abundance, prosperity, plenty, wealth, joy, grace, assistance, favour, honour, dignity, hospitality, stability, welfare, eternal life, power, authority, strength, applause, and immortality!" Von Strahlenberg, 327.

religion seems to have been largely tinctured with Buddhism,[8] made five great conquering expeditions to the West: first, under Ogus or Oochus Khan, B.C. 670 and 632; second, under Woden, or Woote, about B C. 90; third, under Attila, A. D. 440, and Genseric, A. D. 445, both of whom led vast armies of Mongols into Europe; fourth, under Genghis Khan A.D. 1238, and his nephew Hulaku, or Halagou Khan, A. D. 1257; and fifth, under Tamerlane, A. D. 1401. In all of these expeditions a vast spoil of the precious metals was accumulated and sent into Asia.[9]

In 1713 Prince Gagarin reported to the government at Moscow that placer gold was obtained in Little Bokhara, by the inhabitants of Yarkand. At the same period the Turkoman elder, Hádjá Neffez, from the eastern shore of the Caspian, reported gold washing on the Oxus; but neither of these sources of the precious metals turned out to be prolific, and the receipts of gold at the smelting houses of Russia continued without material augmentation until the discovery of the silver mines of Kolyvan, between the rivers Obi and Irtisch, near the Altai mountains, rivers which separate Siberia from the country of the Kalmuck Chinese. These mines were first discovered in 1723 by Akinfi Nikitich Demidof, who worked them secretly for his private emolument until 1744, when he made a merit of a necessity by turning them over to the government. They produced from 1749 to 1762 between 8,000 and 16,000 pounds[10] of silver; from 1763 to 1769 between 20,000 and 32,000; and after that period to 1778, between 40,000 and 48,000 pounds per annum. The silver contain upwards of three per cent. of gold. The whole produce from the outset amounted in 1771 to 400,000 pounds of silver, with 12,720 of gold. From 1772 to 1778 inclusive, the produce amounted to about 44,000 pounds of silver, and 1,200 pounds of gold per annum. "These mines employed nearly 40,000 colonists, besides the peasants in the district of Tomsk and Kusnetz, who, in lieu of paying the poll tax in money, cut wood, make charcoal, and transport the ore to the foundries." Bishop Coxe's "Travels," ed. 1792, III, 435. In 1838 the Kolyvan mines

[8] Witness the giant Foot of Hercules (Buddha) the legend of Xalmosis, the name Budini, the remarkably tall tombs, the drinking from skulls, and many curious customs related of the Scythians by Herodotus in Melpomene, LXXI, and elsewhere. In after ages Attila's capital in Pannonia was called Buddha, the modern Buda-Pesth.

[9] According to Von Strahlenberg, the Mongols who invaded Scythia during the seventh century before our æra were from the Volga, on one side of which dwelt the Unn (Hun) Oigurs or Ogus, and on the other the Nokos Oigurs—the latter being the generic name. There is no racial affinity between the ancient Huns and the modern Hungarians or Magyars. The former were Tartars; the latter are Finns.

[10] The pounds herein mentioned are Russian pounds, which are about one-tenth lighter than avoirdupois, and one-tenth heavier than Troy pounds.

were still producing 34,200 pounds of silver and 100 pounds of gold per annum, besides a considerable quantity of copper. McCulloch.

In 1737 the gold mines of Voetsk, near Olonetz, between Lake Onega and the White Sea, were discovered. In 1738 they were opened. From 1744 to 1776 this district yielded only 57 pounds of gold, and about 9,000 poods of copper. As the expenses amounted to $80,000 more than the income, the district was abandoned until 1772, when it was again worked. After that time it only yielded two or three pounds of gold per annum. In 1794 the district was exhausted and again abandoned. Lock, 369 and 720.

The gold mines near Ekatherinburg, east of the Ural, were discovered in 1743. The annual produce never exceeded 200 pounds of gold, and it was commonly much less: in 1772 it was only 101 pounds.

The celebrated mines of Berezovsk, east of the Ural, were discovered in 1745, and first worked in 1754. They consisted of quartz veins with iron pyrites containing gold. From 1754 to 1788 they yielded about one million dollars in gold, an average of about $28,000 a year. Towards the end of the century this rose to upwards of $150,000 a year. From this point to the year 1850 the produce of the quartz mines gradually fell to about $25,000 a year. In 1823, out of 66 quartz locations only eight were worked. But meanwhile, commencing nominally in 1814, and practically in 1819,[11] the placers or alluvions of the district had been opened, and for a while the combined product of quartz and placer mines rose to an average of $1,000 a day. From 1814 to 1874, a period of 60 years, the product was 2,201 poods, or 88,040 pounds of gold. During the 18th century the number of workmen employed was upwards of 3,000, of whom about 1,250 were daily engaged. During the 19th century the number of miners in this district was about 4,000. The wages were a mere pittance. The gold produced was of low grade, a considerable proportion of it not being over 0.85 fine. On the other hand, gold-stealing by the workmen was common; so that if the official returns of the product are valued as though the gold were nearly fine, the computation will bring the value of the real product to very nearly the truth. From first to last this could hardly have been less than

[11] The Ukase of May 28, 1812, forbade the working of the placers or alluvions by private individuals. In 1814 some relaxation of this edict was made. In 1819 the owners of quartz mining, or smelting, works, in the Ural were permitted to work the placers in their districts. In 1826 this right was extended to such Russian subjects as might be recommended by the Minister of Finance. In 1828 a similar right was extended to districts beyond the Ural. Lock, 431, 439.

twenty-five million dollars. Yet, after the share of the Crown was deducted, there did not remain enough of the entire product (even had there been no other outlay save wages) to pay the men ten cents a day. Malte-Brun; Hermann; Phillips; Lock; London Mining Journal.

The first returns of gold from the Altai (the Schlangenberg) were received at St. Petersburg in 1745. In 1774 the smelting ores of Miask were first worked. This was at Petropavlofsk, almost due south from Ekaterinburg. In 1798 these mines and works were taken over by the government. In 1823 the alluvions of this district were opened, producing about a ton of gold annually. In 1876 the alluvions of Miask were thrown open to individuals and their productiveness was increased to about two and a-half tons of gold per annum, giving them the reputation of being the richest placers in the Ural. Many large nuggets have been found in them: one of them, of very irregular shape, from the Tsarevo Alexandrovsk placer, containing no less than 79 pounds of metal, valued at about $17,500. A facsimile of this nugget, made of plaster of Paris, gilded, was courteously sent to the writer by the superintendent of the School of Mines at St. Petersburg.

The liberation of those vast armies which, until 1813, Russia deemed it necessary to maintain in Europe, marks an epoch in the production of the precious metals. The soldiery had nothing to do; why not open the placer mines to their activity? It will scarcely be denied that views of this character had some consideration in the Russian regulations of this period. From 1704, when Nertschinsk was opened, to 1813, when the armies returned from Europe, the entire precious metals product of Russia had been about 1,800 poods of gold, value about 18 million dollars, and 68,000 poods of silver, value about 42½ million dollars. A glance at the General Table, printed elsewhere in this work, will show how rapidly this rate of production afterwards advanced.

The stoppage of the Spanish-American supplies also began to be felt at the period mentioned. On the one hand, a deficiency of the precious metals, on the other, a redundancy of idle men capable of supplying the deficiency. The not altogether unnatural consequence was that they were made to supply it. The new source of supply was the Altai diggings. Herodotus had written them up twenty-two centuries before. In 1723 old Tschudi workings and remains had been found in them. In 1830, after a few years of prospecting, the Altai placer district began to be noticeably productive. It then assumed a political importance. Baron von Humboldt's expedition to Siberia, at the request of the Emperor Nicholas, was confined to the Ural. It was only after his return to Berlin that the Altai placers became so productive as to prompt the publication of his "Fluctuations of Gold." This work was designed to allay the apprehensions of those capitalists who feared that Russia was about to dump upon an unprepared world a mass of the yellow metal so great as to unsettle prices and destroy the stability of investments. Coming from so illustrious a source, who can doubt that it successfully performed its mission? [12]

Altai, or Altyn, is an old Mongol word for gold: also for money. It appeared in the name of the medieval *altynnik*, or one-kopeck piece, the *altyn*, or three-kopeck piece, and the *piath-altyn*, or five-kopeck piece. The same word occupies a conspicuous place in the nomenclature of Asiatic mountains, passes, gullies, etc., for example, the Altyn-Immel, or Golden Saddle, a pass in the Semirétchian Ala-tau, where Genghis Khan is said to have buried his saddle. The smelting works for silver ores, originally erected at Kolyvan, by Demidof, had been in operation more than a century before the gold placers of the Altai were unearthed. The discovery no sooner reached the government than a capitation tax was clapped upon 180,000 idle men, who, in default of payment, were set to work upon the new diggings. These were situated in a northwestern spur of the Altai, the eminences of which were distinguished by the names of Abakans-Chrish, Kusney-Chrish, and Ala-tau mountains. According to Von Helmerson, the geology bore a striking resemblance to that of the Ural. Placers were first discovered on the western slope, and here, as in the Ural, the Crown was advised to make its locations, throwing the eastern slope open to miners. It was the eastern slope that proved to be the richer. Humboldt, 35.

A year or two later another auriferous district was discovered in the Altai. This was on the western slopes of the Kuznetsk Ala-tau in the Salair ridge, a northwestern spur of the Altai, which separates the headwaters of the Obi and Tomsk rivers. The contribution of the Altai to the metallic produce of Russia became so important that in ten years' time the produce of the whole empire rose from 300 to 600 poods, or from three to six million dollars per annum. Taken in connection with the gold yield of the Appalachian range at the same period, these results afforded so great a relief to the commercial world, that, as before stated, some fears of its being "rather too much of a good thing" began to be put forth in the organs of financial

[12] The English translations of Humboldt's "Fluctations of Gold," and Grimaudet's "Law of Payment," both in one volume, are published by the Cambridge Encyclopedia Company, of New York.

opinion, until Humboldt's work and the panic of 1837 admonished these tremblers that gold had not quite yet become worthless.

The Russian government, having by this time learnt how easy it was to find attractive employment for its idle and adventurous classes, now entered upon a sort of Socialistic experiment; it turned its attention to the discovery and development of gold mines, opening them with assisted labourers and relinquishing them to private "enterprise" and the tax gatherer whenever they ceased to pay. The experiment pleased many parties and proved to be quite successful. It found employment for idle peasants; it relieved the apprehensions of the bourgeoisie; it developed Siberia; it occupied the gamblers; it increased the imperial revenues; it rewarded the courtiers; and upon the coins stamped from the product it bore the arms and language of Russia to the most distant climes.

The auriferous district, which next engaged the interest of the administration, was situated in the foot-hills and valleys of the rivers which rise in the Tungousa or Tungusska mountains, and flow northwestward into the Yenesei; this last being the Mississippi River of Central Siberia. The Tungusska range, running north and south with numerous lateral spurs, stretches from the Sayan range (Mt. Goletz is in the vicinity of Lake Baikal), to the Northern Ocean, separating the systems of the Yenesei and Lena. From the western spurs of the Tungusska range flow four remarkable rivers, all of which empty into the Yenesei. Beginning at the north, these are the Nijni (Lower) Tungousa, the Podkamenaia (Middle) Tungousa, the Pite, or Great Pite, and the Verknai (Upper) Tungousa. Near the source of the Verknai, and close to Lake Baikal, stands Irkoutsk: near where the Verknai empties into the Yenesei, stands Yeneiseisk. The auriferous districts north of the Pite were called the Northern gold regions; those south of it, the Southern. The latter, discovered in 1838, were first systematically worked in 1841; the former, discovered in 1839, were first worked in 1845. The auriferous strata consisted of brecciated and gravelly detritus from the mountains, such detritus varying from a few feet to upwards of 150 feet in depth, and containing enough gold to pay at first for hand-work, next for rude machine work, and finally not enough to pay at all. The detritus was almost invariably topped by a heavy overlay of moss-covered peat, usually ten to twelve feet thick, though in some places it exceeds fifty feet. It was in this peat and the frozen tundra northward of this region that were found those remains of the woolly elephant, the rhinoceros, and other tropical and semi-tropical animals, which to-day adorn the museums of St. Petersburg and Copenhagen, and excite with-satisfying our curiosity, concerning the antiquity and climatic viccisitudes of our planet.[13]

The Yenesei gold regions were first worked by the government, employing convict labour, including women and even children. This practically meant cruelty, revolts, desertions, the knout, and the stealing or embezzlement of gold. (Lock, 395–6.) In 1847 this system was abandoned and the mines were relinquished to private individuals, usually court favourites, who were, nevertheless, required to pay a heavy tax on the produce. Owing to the systematic evasion of this tax, the government, in 1850, resumed control of the mines. In 1853, due to a falling off in the returns, it again abandoned the mines to "private enterprise,"[14] a system, which, combined with police lettres de cachet and the hiring of convict labour, increased for a time the productiveness of the mines. Under this system, for which the courtiers rather than the government must be held responsible, the cost to Russia of the 806,630 pounds of gold obtained from the Yenesei region during the forty years, 1834–74, may be briefly summed up as follows: Twenty thousand Russian lives, consigned to exile, neglect and oblivion: plus the trifling cost of keeping these unhappy creatures alive until they were worn out. The entire quantity of alluvium and detritus washed was 130 million tons. The average value of the gravel was about 12 cents (6d. sterling) per ton. There are hundreds of square miles of far richer gravel in California, which has never been worked and which can be had to-day for the asking, but from which American labourers turn with disdain. This computation may serve to measure the difference between a free man's ration and the meagre one upon which the Russian convicts were condemned to die.

After 1846 the productiveness of the Yenesei region declined. At that period the annual produce of the district was about 40,000 pounds weight of gold, employing 25,000 labourers. In 1874, 13,000 labourers produced 12,000 pounds of gold. In 1894 the average annual

[13] The organic remains are mostly those of the mammoth. By the year 1840 more than 2,000 of these animals had been found; at the present time the finds are exceedingly numerous. Some with portions of their bodies still soft were discovered by the miners in the frozen tundras. Remains of the woolly rhinoceros (*Rhinocerus tichorinus*) are also frequently encountered. As all these animals required abundant vegetation, the present treeless state of the tundras and the long intervening glacial period indicate for them an age far more remote than is covered by our cosmogony. Human remains were found in 1860 in the Chtogolef mine at a depth of ten feet. Ancient hearths have also been found; and a stone slab with inscriptions in a mine in the basin of the Kigas. All these last may have been the remains of very ancient miners, but not necessarily, of a former geological period. Cf. U. S. Consular Reports, October, 1900.

[14] Among the principal grantees were ladies of the court: Princess Madatova, Princess Gortchakoff, Troubetskoÿ, Patkul, Maksimovitch, Madame Rodstvennaia, etc.

yield of each labourer was less than three-fourths of a pound weight of gold. From this the government exacted about one-fourth in production taxes, seigniorages and other charges; while, after "paying" the labourers, the proprietors of the mines and the vodka distillers, are described as living in luxury. The word *pay* in Eastern Siberia must have an entirely local meaning.

The history of the Atchinsk and Minusinsk gold districts is that of the Yenesei on a small scale. They lie in the foothills and valleys of the Sayan range and were discovered and first worked in 1835. During the Fifties they employed 20,000 labourers; in 1874 only 5,000. During the forty years between these dates they produced 75,000 pounds weight of gold, from 30 million tons weight of gravel washed.

The Olekminsk district has a similar history. It lies from 60 to 75 miles north of Nertchinsk, in the basins of the Vitim and Oletma, streams that empty into the Lena. Between 1849 and 1874 this district employed about 5,000 to 6,000 labourers, who, from 22 million tons of gravel obtained 245,000 pounds weight of gold. In the year 1874 the produce was 22,252 pounds weight of gold; in 1877 this rose to 37,120 pounds (928 poods), with 15,000 labourers; down to 1880 the total yield from the outset was 416,206 pounds weight. At the present writing (1900) these placers are declining; nevertheless, Olekminsk remains easily the richest and most productive gold district in Siberia. It is surprising that though so near Nertchinsk, which was opened in 1704, the Olekminsk deposits remained for nearly a century and a-half unknown. This is attributed to the circumstance that the auriferous strata lay buried beneath a heavy covering of peat, through which there was no occasion to dig. The mode of working the tundra placers is similar to that which is practised in Alaska: thawing the ground, sinking, drifting, hoisting, and washing in sluice-boxes, or else barrel-machines, together with mercury-amalgamation and blankets.[15]

The basin of the Amur, from the vicinity of Nertchinsk through Chinese (now Russian) Mantchuria to the Pacific Ocean, was explored in 1832, but until 1849 the mining operations were few and unimportant. In the year last named a more systematic exploration was made by government mining engineers, who reported the region as being highly auriferous; when operations on a larger scale were begun. The

[15] "It is a notorious fact that gold mining in Siberia has left behind it not the least trace of civilization, the sites of the abandoned diggings present nothing but howling wildernesses, and there is not a sign of a habitation. In the towns it has left no memorials except ruined pleasure gardens, in which the miners used to drink away their earnings; and the laments of proprietors, the latter totally bankrupt and begging almost for bread."—London Mining Journal.

full development of the region, however, awaited the acquisition and organization of such resources as were deemed needful to successfully accomplish this object. These embraced the possession of northern Mantchuria and the extension of Russian military posts eastward from the Nertchinsk region to the ocean. This time the outlawed (not necessarily the criminal) classes led the way. The discovery of gold in California, and the knowledge that under the American government there was freedom for the refugee and immediate employment and reward for the labourer, floated many a fugitive canoe down the Amur in the vain hope to reach the Aleutian Islands and the ships of the fur-dealers who traded with San Francisco. Whenever the adventurers succeeded in escaping the double vigilance of the Russian and Chinese authorities, they commonly took refuge with the Humaris and Chiliaks, whose solitudes they roamed in search of food and gold. Few or none reached the ocean. It was the researches of these outlawed prospectors that confirmed the conjectures of the Russian mining engineers and stimulated the Russian government to acquire, in 1854, that large portion of Mantchuria which lay north of the Amur. In 1857–64 (Anasof), and in 1870 (Boguliubsky), this region was again and again examined by mining engineers for the Russian government. Meanwhile, in 1866, prospecting by private individuals was permitted by the authorities, and in 1868 the Upper Amur Company commenced systematic operations. The detritus of the Amur region is usually about 18 feet thick, with an equal overlay of peat and a light vegetable mould on top. From 1832 to 1850 about 9,000 pounds weight of gold were secured by all the miners of the Nertchinsk circuit (estates of the czar), who made returns to the government. From 1851 to 1874 the returns amounted to 97,500 lbs. weight; altogether 106,560 lbs. of schlichs gold from 16½ million tons of gravels washed. This is equal to about $1.25 per ton: far richer gravel than what is now washed. From private estates the returns were 48,054 lbs.; from the Amur district proper, 34,500 lbs.; from the sea coast district, 400 lbs; altogether, 189,514 lbs. of schlichs gold.[16] The subsequent returns are included in the general Table of Production printed elsewhere. It may be stated comprehensively that of the whole amount of gold produced in the Russian empire from 1814 to 1880 the Ural contributed 27.6 per cent., Western Siberia, 6.4 per cent.; Finland, 0.1 per cent.; and Eastern Siberia, 66 per cent., and that at the present time Eastern Siberia contributes

[16] Schlichs is impure gold, just as it is found in the mines. It commonly runs about 0.888 fine. A Russian pound will coin into about $250, or a pood into $10,000.

more than 80 per cent. of the entire auriferous produce of the Russian empire.

Turning for a moment from gold to silver, it may be said in a few words that the silver produce of Russia has been small and singularly uniform. From 1830 to 1846 inclusive, the produce only varied between 1,300 and 1,200 poods per annum; from 1847 to 1868, between 1,200 and 1,000 poods and from 1869 to the present time, between 1,000 and 600 poods: not enough to make good the wear and tear of silver coins in the empire. In order to keep up her old and worn silver coinage and acquire the metal for such additional coinage as the growing population and expanding commerce of the empire demanded, Russia has had to import large quantities of silver. It has become the permanent policy of the government to sell its surplus gold and with the proceeds to purchase silver. A glance at any table of Russian commerce will show that this movement has been going on for half a century. For example, in the year 1896 Russia purchased the silver for, and caused to be struck at the Mint of Paris, no less than 40 million roubles in pieces of whole roubles, half-roubles and quarter-roubles. It is obviously the interest of Russia to purchase the silver needed for her coinages at as low a price as possible. Therefore, to permit her delegates to vote upon the proposal to demonetise one of the precious metals, so that the other might be purchased at half price, as was done at International Monetary Congresses, may have been an act of courtesy, but was hardly one of policy. It has cost the silver-producing states, represented in those Congresses, one-half the value of their entire silver product: nearly the whole of which has been gained by Russia and British India.[17]

It is now time to tabulate the returns of gold-mining in Russia. According to Jacob, the whole amount of gold obtained from the mines of the Russian empire from 1704 to 1810 was 1,726 poods, which, assuming it to have been "schlichs" or unrefined gold, amounted in value to about $17,260,000. From 1811 to a very recent date, we have the following returns from Russian official sources.

[17] Now that the subject is politically dead, and the disclosure can only be of interest in future movements of like nature, it may as well be stated in this place that Mr. Samuel B. Ruggles, who misled the New York Chamber of Commerce, and Senator John Sherman, who misled the United States Senate, were themselves the dupes of the far cleverer men who sustained and managed the Conventions of the Latin Monetary Union, at Paris. Among these were Artur Raffalovich, who represented the silver requirements of Russia, and Charles Rivers Wilson and Charles Fremantle, who represented those of British India. The two gentlemen last named, in recognition of the part they took in these intrigues, were rewarded with titles of honour. It is not known in what manner Mr. Raffalovich has been rewarded for his services to the Russian government.

General Table of the Production of Gold in Russia. (Schlichs, or unrefined gold, about eight-ninths fine, worth about $10,000 per pood.) Fractions of poods omitted.

Year.	Poods.	Year.	Poods.	Year.	Poods.	Year.	Poods.
1704		1831	402	1854	1,597	1877	2,502
to	1,726	32	423	55	1,649	78	2,572
1810		33	410	56	1,655	99	2,632
11	16	34	405	57	1,734	1880	2,642
12	15	35	393	58	1,711	81	2,444
13	15	36	406	59	1,539	82	2,207
14	16	37	443	1860	1,491	83	2,183
15	14	38	443	61	1,456	84	2,178
16	16	39	496	62	1,461	85	2,016
17	18	1840	458	63	1,459	86	2,042
18	17	41	646	64	1,398	87	2,128
19	14	42	909	65	1,584	88	2,147
1820	20	43	1,241	66	1,659	89	2,274
21	28	44	1,280	67	1,650	1890	2,404
22	54	45	1,307	68	1,711	91	2,386
23	106	46	1,612	69	2,007	92	2,625
24	206	47	1,757	1870	2,157	93	2,739
25	237	48	1,685	71	2,400	94	2,622
26	231	49	1,589	72	2,331	95	2,510
27	282	1850	1,454	73	2,025	96	2,272
28	291	51	1,474	74	2,027	97	2,326
29	290	52	1,367	75	1,929	98	2,346
1830	383	53	1,464	76	2,055	1899	2,291

It was not until after the year 1842 that the annual produce of gold reached a thousand poods or ten million dollars. Then ensued what Shuroffsky and other writers termed the Gold Mania. Burghers and aristocrats tumbled over one another to secure mines and mining privileges. In 1845 the produce rose to thirteen millions. As the currency of Russia then consisted almost entirely of paper notes, all this gold went abroad, carrying alarm wherever it went, to the conservative classes, who saw in its ample production a rise of prices and ruin to their vested interests. In Holland the necessity of couching all contracts in silver florins was openly discussed. It had become evident that the world was about to be "deluged" with a metal which would soon become "worthless," because it was becoming plentiful. Financiers and politicians forgot all about the cost of production theory, or, if reminded of it, they declared it was false.[18] The yellow metal was only fit for gold-beaters and jewelers. Money to be "honest" should only be made of silver. These views, repeated throughout Europe, resulted in a panic, to allay which it was everywhere proposed to demonetise gold. This was actually affected by the Netherlands Mint Acts of September 26th, 1847, and June 23rd, 1850, whereby

[18] "It is the cardinal rule of commerce that quantity governs price. . . . There must be a fall in the value of gold in consequence of its greatly increased quantity." Cobden's Introduction to Chevalier's "Fall of Gold." It is a misfortune that such gleams of wisdom only appear at those distant intervals when the pockets of the rich are threatened!

the silver florin, reduced to 145.85 grains fine, was declared the sole legal tender, and the gold coins were demonetised, melted down and sold at a loss of ten million florins. Belgium soon afterward followed suit. Chevalier; Tooke; Vrolik.

In 1847 a run occurred on the Bank of England, numerous failures followed, and on the 25th of October the Bank obtained leave from the Government to suspend coin payments. In England the crisis was ascribed to speculation; on the Continent it was imputed to be the over-production of Russian gold. In the following year the Bank of Austria stopped payment, and numerous provincial banks failed in England. These events were also attributed to the Russian gold. Still the tundras went on yielding up more and more of the despised metal. In the two years, 1847 and 1848, they furnished over thirty-four million dollars worth of new gold. When to this unprecedented and alarming supply California began, in 1849, to throw its auriferous treasures upon the world, the Gold Panic reached its height. Holland and Belgium had demonetised gold. Chevalier advised the government of France to demonetise it; Maclaren advised the British life assurance companies to adopt a "silver standard"; while Cobden thought it might be necessary to revert to corn-rents and even to the primitive practice of "paying in kind!" Cobden's Introduction to Chevalier's "Fall of Gold," p. ix.

It is not necessary to follow the history of this unreasonable alarm any further. Everybody knows what happened. It is true that gold became more and more plentiful; it is true that prices rose; but it is not true that anybody suffered from these occurrences: on the contrary, they inaugurated a period of unprecedented prosperity all over the commercial world—a prosperity that was only checked when the same shortsightedness deduced from the exuberance of the Comstock Lode a new and pressing reason for demonetising the other precious metal.

In 1869 the produce of the Russian gold mines rose to 20 million dollars, and in 1877 to 25 millions, at or about which figure it has since remained constant, varying from 27 millions in 1873 to about 23 millions at the present time: a constancy which is due less to any natural influences, such as the supply of water for washing the auriferous gravel, than to the restrictions imposed by taxation.

The tax on the produce of the mines and the other mining regulations of the Russian government, though deemed onerous and vexatious by the miners, are imposed with no oppressive or illiberal purpose. They are designed first to afford the State some return for yielding up its domains to private spoliation; to afford the State a revenue; and to keep the production of the precious metals within such prudent limits as shall not oblige the State upon a future occasion to depend upon commerce or foreign caprice for a supply of the materials of coinage. It must be remembered in this connection that Russia is not a new State. It is the successor of the Byzantine Empire and many of its governmental arrangements, among them its mining laws, are drawn from the great experience of its illustrious predecessor.

The oldest nations of the world deemed it expedient and just to claim in the form of taxation, a portion of the miners' produce. Though the surface of the land for agricultural, building, and traffic purposes might equitably become the property of private individuals, there could be no such basis for the claim of individual owners to the earth beneath, least of all in respect to the monopoly of those minerals, which, like the precious metals or coals, were the necessary instrumentalities of communal life. India, China, Persia, Babylon, Syria, Egypt, Greece and Rome successively taxed the produce of the precious metals, some to the extent of one-half, one-third, or one-quarter, but finally and generally to the extent of one-fifth, a tax in every age lighter than that which was imposed upon the produce of agriculture.

The tax of one-fifth, or the Quinto, as laid down in the Koran, eventually became the rule to the States which threw open the great mineral resources of the Indies, both East and West. This was the proportion levied by Spain, Portugal and even England, until the decline of slavery, the diminished profits of mining and the clamours of the mine proprietors, compelled it to be reduced to one-tenth, and lastly to one-twentieth. The American revolutions abolished the production tax altogether, because it had been levied by distant governments and presumably for the benefit of needy monarchs and rapacious revenue farmers. But now that it concerns the welfare of the American governments themselves, it may not be deemed out of place to calmly study the working of this impost as levied by the government of one of the most modern, and yet in certain aspects one of the most ancient, States now in existence.

Originally the mines were regarded as Crown property. In 1745 all of them were worked by the government for its own benefit. Under the Empress Anna (1730–40) a more liberal policy was adopted. By the Ukase of 1812, which went into effect in 1819, a portion of the mining lands were let to private individuals upon part produce; the government retaining and working the remainder. The proportion required to be paid by lessees was one-tenth of production and about 2½ per cent. for police duty and other charges.

The Crown mines were formerly worked chiefly by convicts or exiles; a few by paid labourers: both under supervision. A suspicious decrease in the produce of the Miask placers, about 1823, induced the government to permit the paid labourers to work them as they pleased, and without supervision, rewarding them with a small per centage of the product; but although this greatly increased the produce, it resulted in the gravel mines being worked unskilfully and in some cases practically ruined.

At this period, 1826, it was urged (it is believed by Count Cancrin, the Minister of Finance), that the government policy of keeping control, by limiting the lessees, of the mines, though sanctioned by antiquity and religiously observed by the Chinese and some other nations, was unwise; that the placers of Siberia had been formed by natural agencies, which were still at work; and therefore that the government should acquire all the gold it could, by permitting the mines to be exploited at once, because as time went on these natural agencies would furnish fresh supplies! Perhaps owing to financial necessity this fantastic argument prevailed, and permission was accorded to every Russian subject, *recommended by the Minister of Finance*, to prospect for gold. By the year 1838 there were issued 200 such permits. The concessionaires were taxed 5 to 35 per cent., an average of about 15 per cent. on the produce, plus about 1¼ per cent. for police duty, besides charges for smelting and for transporting the metal to St. Petersburg, where all the produce had to be sent for coinage.[19]

In 1840, after the opening of the Yenesei placers, the latter were taxed 24 per cent. on the produce, while all other mines in Siberia were taxed 20 per cent., besides the other charges above mentioned. In 1843 the mines in one of the Irkutsk districts were taxed 30 per cent. on production. In some cases the police duty was raised from 1¼ to 2½ per cent. In 1847, during the halcyon period of the Siberian placers, some individuals offered the government 50 per cent. on production for permission to drift for gold.[20] Although the offer was not accepted, it led, April 14, 1849, to the imposition of higher

[19] At the present time all the gold extracted is required to be sent to the smelting houses at Ekaterinburg (Ural), Tomsk (Western Siberia), or Irkutsk (Eastern Siberia), for refining, and afterwards to St. Petersburg for coinage; the entire charge of the double transportation and for refining and coining is required to be defrayed by the owner of the bullion.

[20] Among the wonders of Russian mining was the invention of Poetsch, in 1883. Some of the tundras were in loose quicksand, difficult and expensive to timber and

taxes on the more profitable and to lower taxes on the least profitable mines. The scale of imposts is too complex to find place in this work. Suffice it to say, that it averaged over 25 per cent. on production and still further increased the government revenues from this source. The tax may be taken to have averaged on the whole about one-third of the produce. Leon Faucher; Roscher; Chevalier.

In 1854 (Aug. 4,) it having been found that many of the mines which had formerly been justly classed as most productive, were so no longer, that the tax was oppressive and deterrent.and that the revenues, which in 1851 were over three million roubles, had fallen to 2.4 millions in 1853, the scale of taxation was raised; with the result that production increased, while the revenues rose, in 1855, to 2.7 millions; 1856, to 2.9 millions; and 1857 to nearly 3¼ million roubles.

In 1858 (April 14), a graduated tax, which amounted to from 15 to 33 per cent. on production, was substituted for the previous scale; and with small exception, this applied to all leased mining property, whether in Siberia or elsewhere. "Summarising the whole period from 1812 to 1847 Boguliubsky estimated that one-fifth of the gold raised belonged to the State, and four-fifths to individuals, . . . so that 32 per cent. of the production fell to the share of the government." Lock, 436.

If we may safely apply these proportions to the product down to the present time, it would appear that the Russian government has derived from the gold mines no less a sum than 450 million dollars, or nearly one hundred millions sterling. Other modifications of the mining taxes were made in 1876, effective from January 1, 1877; also in 1880. In the year last named the duty levied on Miask gold was 20 per cent. (Lock, 371.) Still other modifications were made in the laws of 1893. These are exceedingly complex. They include permits, passports, police supervision, fees of various kinds, taxes varying from 3 to 10 per cent on the produce of mines in freehold and imperial demains, and 5 to 15 per cent. on the produce of Crown lands; besides land taxes, a smelting tax of 3 to 5 per cent., transportation, and many other charges. Report U.S. Mint, 1899, pp. 344–5.

Rapacious foreigners, anxious to exploit the Siberian tundras with steam dredgers and electrical appliances, may view these various reg-

dangerous to the miner. By causing a current of cold brine to pass through a series of buried pipes, Poetsch froze the quicksands and rendered them hard enough to be excavated with safety. The proprietors of the Courrierres Mines published a formula and tables which enabled the safe thickness of frozen wall to be computed for round or square shafts of any dimensions. This appliance has been used for the canal-lift at Les Fontinettes and the construction of a tunnel at Stockholm.

ulations with contempt or derision: yet they are evidently the result of careful deliberation and an attentive consideration to the welfare of a great empire. It is not the writer's office either to recommend or deride them. Since the exchanges of the commercial world have now been made to rest entirely upon coins of a single metal, the magnitude of this subject is being perceived more and more clearly every day; and it must soon demand the very serious consideration of thoughtful men in every country. Among others, it proposes these questions: Shall the raising of gold, the exhaustion of mines, the coinage of gold, and the melting down and exportation of gold coins be left, as now, in America, practically without taxation or restraint, or shall they be as in Russia, subject to the regulation of government? Shall the mines belonging to the public domain be worked gradually and systematically, or shall they be exploited as rapidly as ingenuity can invent the necessary mechanical and metallurgical appliances? Have we a moral right to exhaust the supplies of a material, whether it be gold or silver, which Nature produces but once; but which posterity may need always?

Said Aristotle: The growth of society is impossible without exchange, while the development of exchange is impossible without money. These views were repeated by Paulus; they are embalmed in the Civil Law; they were enunciated from the Bench in the celebrated case of the Mixt Moneys; finally, they have been carefully analysed and fully elucidated by Frederick Bastiat, Destutt Tracy, and other philosophers and statesmen; all of whom have declared them sound and unimpugnable. If money must continue to be made of gold, it certainly becomes a very grave question whether or not we have the right to exploit it, and leave the world, as the Romans left it eighteen centuries ago, to Decay and to Dark Ages. And when these questions come to be answered, the example of Russia in taxing and restricting the produce of the mines may not be without its practical value.

GOLD STEALING.

"Gold stealing is a prosperous business in Siberia. It can hardly be called mere stealing, for it is conducted on such a large scale, so many men are engaged in it, it is so well organized and pays so well, that it fully deserves the name of an industry. True, the industry is not recognized by law; on the contrary, it is harshly punished when discovered. Discovered? But that is where all the difficulty lies. It is not easy to discover a new and unknown thing, but to discover what is known to every gold miner, to every proprietor of gold mines, to thousands of secret gold agents and to the czar's authorities, is exceedingly difficult, if it be possible at all. Here is room for another Columbus, but then he must be ready for Columbus' fate.

With the beginning of summer, when the work in the gold mines becomes lively, secret agents appear in the woods all around for miles. They set up their camps and open business, which lasts till the end of the gold season. At night a bright torch is seen on some hill and the gold miners understand the signal. With their gold dust they go to the camps in the woods and then begins a lively trade. The miners get vodka, provisions, or clothes, but chiefly vodka, and return to the mines. There follows in the miners' barracks an indescribable frolic.

Now, is such a transaction secret? Far from it. The guards were aware of it, but they remained silent, for they had their portion of the vodka and their share of the golden booty. It would be unjust, however, to say of the officers that they, too, take part in this retail business. No, no. As to the wholesale—well, we had better recollect that though speech may be silver, silence is golden.

The gold miners cannot resist the temptation to steal gold. Their wages are ridiculously small. They get from $25 to $100 for a season, which lasts nearly half a year. They are kept in wretched barracks, destitute of every sort of comfort. Their food is bad, even in the moujiks' opinion. If one of them wants something extra in the way of provisions, tools, clothes or vodka, he may get it at the store of the gold mine proprietor, the only store within reach of the labourers, and in that store the goods are usually sold at treble the ordinary prices. During the working season the miners remain in a state of SLAVERY. They can be fined, flogged and even imprisoned by their employer with impunity. In order to save their wages they steal gold, and thus procure vodka and provisions.

The numberof secret gold agents is legion, and they are excellently organized. There are the militant agents, armed to the teeth, who deal directly with the miners. There are the protectors, the spies, the station-keepers, the postmen and the foreign agents. They have their own secret routes through the czar's empire, and speak their own jargon. 'Grits,' for instance, means gold-dust; 'to reap' means to go for gold; 'to make pilgrimage to St. Innocent,' means to go to Irkutsk, in order to give gold to Chinese merchants for tea.

When the gold season opens the militant gold agent starts out in the woods with a large amount of provisions, as if going on legitimate business. The Siberian police and administration are too shrewd to interfere with these volunteer gold-hunters. The agents stay in the woods during the whole gold season and it is extremely dangerous to visit their camps. The gold they get they would not surrender even at the risk of their lives. They become regular desperadoes when interfered with. At the close of the gold season the agents carry their booty to the wholesale dealers.

The Siberian gold brokers are to all appearances perfect gentlemen. They are not armed and they do not tremble for their lives or for their gold, which they keep in piles on the counters. The authorities are on the best of terms with them. The transportation of the gold collected by the brokers is attended with danger. Formerly there were only two gold markets. At Nijni-Novgorod, during the fair, gold was disposed of to Russian merchants for home use. But a far larger part of the stolen gold went to Kovno, whence it was carried abroad, principally to Prussia, by Jews, who constitute a regular 'golden militia.' During the last few years two more markets have been opened, one in Odessa, where the gold is bought by Jews who exchange it for tobacco on which no customs duty has been paid, and for other smuggled goods; and the other in Kiakhta, on the Chinese-Siberian frontier, where the gold is exchanged for tea. To carry the gold over from Siberia to European Russia is the most difficult task of all. In this case customs, excise, and other officers, or ladies of high standing, for golden reasons, of course, join the 'golden militia.'

How much gold is stolen from the bowels of Siberia? Parties interested in the business affirm that annually there is disposed in this way gold worth at least 10,000,000 roubles. Annually Siberia produces about 100,000 pounds of gold. So it appears that a considerable portion of this precious metal goes abroad without bringing any return to the country. As to the demoralization produced by gold-stealing and dealing in and transporting stolen gold, who can estimate it at a money value?"—*Quoted from the Siberian Gazette by the San Francisco Chronicle, Sept. 18, 1883.*

CHAPTER XXX.

THE UNITED STATES OF AMERICA.

Accidental character of gold discoveries—Unavailing searches of the Spaniards in California—Suspicions of the Mission Fathers—Previsions of geologists—The annexation of California to the United States due to the chances of an election, and the great discovery of gold, to the building of a mill and the observation of a child—Statistics of production.

NOTHING more forcibly illustrates the part played by accident in the supply of the precious metals to the world than the history of California as a mining country. The Spaniards, ransacking America for the precious metals, had discovered and partly explored California so early as the sixteenth century. They afterwards colonized and held possession of it for nearly three centuries. They established missions; they pacified and converted the Indians, whom they used as domestic servants and workmen; they cultivated many of its arable valleys; they introduced the cereal grains, and even transplanted the grape, the fig, the pomegranate and the artichoke. Persuaded from the outset that it was a country rich in the precious metals they never ceased to look for them. Occasionally they found traces of gold, but not enough to attract a numerous colony from Spain or Mexico. The few adventurers who were drawn to its distant shores by stories of mineral finds, generally made their way home, disappointed and impoverished. A few remained to turn rancheros, and settle down with the mission fathers and the semi-christianized Indians into pastoral communities, which eventually bred a race of apathetic half-breeds.

In addition to these communities, there settled in the southern part of Alta California a number of Spanish and Mexican families of high culture and breeding, who entirely refrained from mining or prospecting for the precious metals, their sustenance being derived from the cultivation of the soil, chiefly with the aid of native labour, and their happiness consisting in the education of their children, the exchange of social courtesies and the pursuit of innocent diversions. All the early American travellers through this part of the country, especially the vicinity of Los Angeles, agree in describing these communities with admiration and envy. Theirs was an almost idyllic life, their manners the perfection of courtesy, their attitude toward strangers one or cordial welcome and unbounded hospitality. Among these early travellers was James Waters, who crossed the Plains from St. Louis in 1844, bringing several "cargas" of fine French boots, shoes and muslins, which, though past the fashion in Paris, were yet new to the señoritas of this remote region. Another of them was old Starkie, of San Bernardino, who, in 1849, left his ship at San Francisco, and pursuing the coast line on foot, entered the Paradise of Los Angeles during the next year. Both of these travellers, after witnessing the vicissitudes of California life for nearly half a century, united in the declaration that if there ever was a heaven on earth, it was Southern California previous to the gold discoveries of 1849.

Such were the economical and social conditions of California when the adventitious "success of the Democratic party (of the United States) in the close presidential election of 1844"[1] led to the independence of Texas, the war with Mexico, and the annexation of California. But for these events, which have no necessary connection with mining, or the search for the precious metals, Alta California might have remained as Baja (Lower) California has remained to this day, a province of Mexico; it might still have been the obscure and undeveloped country that it was under the Spanish mission fathers.

The existence of gold in Lower California was determined by Cortes, who fitted out an expedition in 1536, and returned with some small quantities of the precious metal. Its existence in Alta California was suspected by the chaplain of Drake's expedition, 1577–79, and noticed in Hakluyt's account of the region. The finding of gold was mentioned in 1690 in a work published in Spain by Loyola Cavello, a priest at the mission of San José, in the bay of San Francisco. Capt. Shelvocke, in 1721, noticed signs of gold in the soil. The historico-geographical dictionary of Antonio de Alcedo, 1786–89, affirms the abundance of gold, and speaks of lumps weighing from five to eight pounds. In 1837 a priest from California went to Guatemala, and induced Mr. Anderson, a Scotch gentleman, to endeavour to obtain English capital for the purpose of mining gold, which, he averred, was to be found not far from San Fernando. The favourable geological appearance of the country for gold was noticed by Professor J. D. Dana, in 1841, and recorded in his report on California. In April,

[1] "Report of the United States Monetary Commission" of 1876, p. 43.

1847, Com. John D. Sloat, U. S. N., made a very decided statement, concerning its auriferous character, in the New York "Merchants' Magazine." About this time the Mormons, in the Cajon Pass, together with some Mexicans and Indians, were engaged in washing gold gravel upon the banks of a small stream. In January, 1848, another party of Mormons gathered a quantity of gold on Mormon Island, near Sacramento, and within a few miles of the rich placers afterwards discovered.

And yet none of these observations or operations led to the great discovery, which was made by the daughter of James Marshall, the overseer of Captain Sutter's mill. While a race was being dug for this mill on the American fork of the Sacramento, the child found a lump of gold which she showed to her father as a pretty stone. That lump was the beginning of nearly one and a-half billions of dollars, which the California mines have yielded to the world.

During the first few years after the discovery of the placers the entire precious metal product of the coast was derived from this source. Hydraulic mining was commenced in a small way so early as 1851, but it did not assume material proportions until after the close of the American Civil War. Vein mining was begun in California about the year 1860. It has only recently become important. Meanwhile, it assumed enormous dimensions in the adjoining State of Nevada, where it was commenced at about the same time, and where the chief part of the product, in quantity, though scarcely in value, has been silver. The principal locality of this industry was the Comstock lode, upon which Virginia City was built, many of the mines being directly beneath the town.

From the following Table it will be seen that the California placers reached their maximum productiveness almost immediately after their discovery, the greatest yield having been in the year 1853, when it reached 65 millions, and in the opinion of some well-informed writers, 75 millions. From that time to 1893 it declined to 12 millions a year. Since 1893 it has slightly increased, the produce at the present time being about 16 millions a year. The produce of silver in California is too inconsiderable to merit further attention in this place. At the present time it is about a million a year, and for various reasons this sum is not likely to increase. It is only in San Bernardino County that silver has been found in California, and there the miners and prospectors, since the settlement of the Silver Question, have deserted silver and gone into gold mining and prospecting, which they now pursue exclusively.

PRODUCTION OF GOLD AND SILVER IN THE UNITED STATES.

Except in the first and last columns, the sums are in millions of gold dollars. (The silver is stated throughout at the coining value of $1.2929 per oz. Troy, though it ceased to retain that value after 1873.)

Year.	Gold. California only.	Gold. Rest of the U. S.	Gold. Total U. S.	Silver. Nevada only.	Silver. Rest of the U. S.	Silver. Total U. S.	Silver in London, per oz.
1848	10	—	10	—	—	—	$1.29
49	40	—	40	—	—	—	1.30
1850	50	—	50	—	—	—	1.32
51	55	—	55	—	—	—	1.34
52	60	—	60	—	—	—	1.33
53	65	—	65	—	—	—	1.35
54	60	—	60	—	—	—	1.35
55	55	—	55	—	—	—	1.34
56	55	—	55	—	—	—	1.34
57	55	—	55	—	—	—	1.35
58	50	—	50	—	—	—	1.34
59	50	—	50	—	—	—	1.36
1860	45	1	46	—	—	—	1.35
61	41	2	43	2	—	2	1.33
62	35	4	39	5	—	5	1.35
63	32	8	40	8	1	9	1.35
64	35	11	46	10	1	11	1.35
65	35	14	49	9	2	11	1.34
66	25	15	40	8	2	10	1.34
67	25	17	42	12	2	14	1.33
68	22	16	38	9	3	12	1.33
69	23	14	37	9	4	13	1.33
1870	25	12	37	12	4	16	1.33
71	20	15	35	14	4	18	1.33
72	19	19	38	17	3	20	1.31
73	18	21	39	19	6	25	1.29
74	20	18	38	15	10	25	1.28
75	18	15	33	21	9	30	1.24
76	19	24	43	27	11	38	1.16
77	15	30	45	25	15	40	1.20
78	17	21	38	19	20	39	1.15
79	18	18	36	20	21	41	1.12
1880	18	15	33	11	28	39	1.15
81	17	14	31	7	36	43	1.13
82	16	13	29	7	40	47	1.14
83	14	14	28	5	41	46	1.11
84	14	11	25	6	43	49	1.11
85	14	12	26	5	47	52	1.07
86	15	15	30	5	46	51	.99
87	14	19	33	5	49	54	.98
88	14	16	30	7	50	57	.94
89	13	20	33	6	54	60	.94
1890	13	13	26	6	58	64	1.05
91	13	12	25	4	56	60	.99
92	12	12	24	3	52	55	.87
93	12	13	25	2	50	52	.78
94	13	24	37	2	51	53	.63
95	15	32	47	1	53	54	.65
96	17	34	51	1	54	55	.68
97	16	42	58	2	53	55	.60
98	16	47	63	1	54	55	.59
99	15	57	72	1	54	55	.59
1900	17	60	77	1	59	60	.61

With reference to the foregoing Table, it is to be remarked that from 1863 to 1877 was the productive period of the Comstock Lode, and that from 1884 to 1893 the figures given for the produce of gold are probably from five to ten million dollars a year in excess of the fact. They are retained in the Table, first, because they are the actual figures forwarded by the author's correspondents; and second, because they are a concession toward the still higher figures included in the official Mint Reports

The Unites States Government coins and fully monetises gold bullion for individuals, and without limit, charging for these services and advantages merely a fraction of the cost of coinage. Practically it coins and monetises privately-owned gold-bullion gratuitously. As the owners of gold coins may have them minted and re-minted as often as they please, without loss, and may melt them down or export them at pleasure, the mint and market value of that metal is substantially identical. Such, however, since 1873, has not been the case with silver. This metal the government coins for itself only. A troy ounce of fine silver has always been and is still coined into $1.29; but none is now coined on individual account. The "coining value" of silver is therefore merely nominal; and, since 1873, a "market" value, the result of a competition between the mints and the silversmiths throughout the world, has taken its place. This market value is best determined in London, because that city is the centre of the world's exchanges. But though determined in London, it does not hold good elsewhere; freight, insurance and time (interest), all contributing to effect local deviations.

There are no taxes on the mining or production of the precious metals in the United States, except a statistical tax in the State of Nevada. There are no official or sworn returns of production from the mines, nor no laws by which such returns can be enforced. In the Report of the Monetary Commission of 1876 the defects of the various methods employed to ascertain the bullion product, namely, the Export and Consumption method, the Express and Railway Carriage method, the Bank method, and the Mint Estimates, were shown at length. For example, ores containing bullion are shipped from Montana, Idaho, Dakota, etc., to Utah, for treatment in the smelting works at Salt Lake City, and they often appear in the statistics of both States. Other ores are shipped for reduction from Nevada, Arizona, etc., to California. Here a similar duplication frequently occurs, that is to say, the bullion is returned as the product of the mines in one State, and of the reduction works in the other. The

Denver Mint returns include bullion from the ores of other States sent to that establishment for reduction, and deposited in the Mint as Colorado bullion. The produce of Mexico and British America, including the Yukon, finds its way into the American returns. Through these means the American Mint Estimates are greatly exaggerated.

After these defects and blunders were pointed out, the Mint Bureau adopted a method of collecting production statistics even worse than the previous ones. It made them up from items in the mining newspapers and from the reports of local correspondents, who obtained their news from others.[2] As it is known beforehand that the details of such information will be published in the Mint Reports as "official," this new system afforded an opportunity for mine-vendors and their friends to advertise their properties to the world. The result was that the statistics were replete with erroneous and exaggerated returns[3] from "prospects" and broken-down mines, which were thus prepared to be floated upon the Bourses of the world as productive or promising adventures. In this way not only was the American Administration misinformed in regard to the bullion product, but mine-promoters and the investing public, both in San Francisco, Chicago, New York and London were grossly misled with regard to the productiveness and value of such properties.

When the Silver Question assumed political importance, the Mint Bureau, going from bad to worse, adoped a most extraordinary method of guiding the public on this subject: it sent to Germany for its production statistics. These were generously supplied from the ample imagination of Dr. Adolf Soetbeer, who knew nothing about the American mines. In 1887 the State Department of the United States paid a considerable sum of money to one man to go to Germany and get a copy of Dr. Soetbeer's book, and other sums to another man, who merely translated the headings of this precious work. Then the results were handed to the United States Government Printer, in order that they might be published as a volume of the United States "Consular reports," and copied by the Mint Bureau as official. Being thus doubly stamped "official," first as part of the Consular reports, and second, as Estimates of the Mint Bureau, these figures, made

[2] For example, all that portion of the "Report of the Director of the Mint upon the Statistics of the Production of the Precious Metals" for 1882, which relates to California, was farmed out to an infirm person named R. W. Hardenberg for $600. This man sublet the job to a mine-owner named R. C. Donns, of Amador, and to several others, at $30 to $50 each, making a profit of $400 by the operation. *Information received by the writer from R. W. Hardenberg himself, Feb. 19, 1884.*

[3] All this and more is admitted in the Mint Report for the calendar year, 1883, than which there could scarcely be adduced a more piteous exhibition of official confusion, blundering and incompetence.

in Germany, have found their way into almost every work of reference printed in America. It is, perhaps, hardly necessary to say that, whatever their usefulness as political pointers, they are of little value as statistics of the precious metals.

Now that the Silver Question is settled and both Russia and India have been enabled, largely through these intrigues, to purchase their supplies of the white metal from the United States at half price, some more sober arrangement of the production statistics appears to be needful. Besides the information gathered by the present writer, almost the only returns which exhibit the Production of the Precious Metals in the United States are those of the bullion carried from the mines to the mints, or else to refiners or reduction works, by the expresses and other transportation companies.[4] One of the express companies, which, thirty years ago, enjoyed almost a monopoly of this carriage, but enjoys it no longer, has been in the habit of compiling an annual production table of the precious metals from its own waybills and such others as courtesy supplied. The tables thus compiled were published as business cards or circulars by the Superintendent of the Wells Fargo Express Co., Mr. Jno. J. Valentine. This gentleman received no official reward for his work. There was no responsibility attached to the figures which he gave out. Whether they were right or wrong made no difference to him or to anybody else and could do them neither harm nor good. It was only the public, the people at large, who were interested in the results and in the policy of government, or the legislation, which might flow from these statistics. Yet they were repeatedly shown to be grossly incorrect. The Commissioner of Mining Statistics, in his Report for 1873, p. 507, pointed out blunders in Mr. Valentine's tables amounting to many millions of dollars. In the evidence which Mr. Valentine himself gave to the Monetary Commission of 1876, he admitted errors, omissions and blunders of a still more startling character. For example, during the productive period of the Comstock Lode, the bullion from these mines consisted of doré bars, which were registered as doré or silver, whereas nearly one-half of their value was derived from the gold in the bars. Not

[4] While the Comstock Lode of Nevada was in bonanza, the official returns of the Nevada mines to the State Assessor (Nevada is the only State which levies a tax—in this instance a trifling statistical tax—upon the produce of mining) furnished a valuable guide to the production of the precious metals; but this halycon period has long since passed. The bullion deposited at the Mints affords no guide, because nobody is obliged to send his bullion to the Mints, and in fact much of it never goes there. On the other hand, the Mints receive large amounts of foreign bullion for coinage, which they are not always enabled to distinguish from native bullion. This inability is admitted in the Mint Report for 1893.

only were they thus registered in the bullion receipts given by the express companies (most of which were personally examined by the writer on behalf of the Monetary Commission), the bars were transported by the express companies and appeared in their way-bills as "doré," or else as "silver."[5] Moreover, when the bars were exported to foreign countries, as many of them were, they were registered in the official Commerce and Navigation Reports as "silver."[6] The importance of this blundering will perhaps be appreciated when it is remembered that the Comstock Lode produced over 350 millions, nearly half of which was gold.[7]

Such was the material which for more than a quarter of a century formed the only basis, apart from conjecture, for the production statistics of the Mining Commissioners and the Mint Bureau of the United States, and such to a great extent is the material upon which their estimates still rest; for in truth, for many years there has been little other.

As for the estimates of production compiled by the Mint Bureau from the letters of its correspondents in the Mining States, most of them are of little worth. Exception must, however, be made in favour of the returns by mines from California, Nevada, Washington and Colorado published in some of the later Mint reports. These are based on a proper system and may be regarded as reliable. Justice to the Mint Bureau compels us to declare that no comprehensive production statistics of the Precious Metals in the United States can be relied upon until a statistical tax is placed upon the produce of the mines. At present the mines are not taxed; they are not obliged to make any returns of production, and in fact they furnish no returns at all except when it pleases them, and this is chiefly when they are in bonanza. Such returns are grossly exaggerated, and they reduce the correspondents of the Mint Bureau to the necessity of accepting tumid statements, or else to fall back on Mr. Valentine's tables and

[5] "Your suggestion that the doré bars in our Tables are misleading to persons at a distance has been acted upon, and having remedied these features, I now take the liberty of enclosing you a corrected copy."—Jno. J. Valentine, Jan. 3rd, 1878.

[6] "United States Custom House, San Francisco, Collector's Office, Dec. 30, 1876. No doré bars are manifested as such (for export), but we assume that when the value of silver predominates, which is usually the case, they are manifested as 'silver bullion.'" J. A. Perkins.

[7] The Mint Report, 1899, p. 149, shows that the Comstock Lode produced, 1859–99, 203 millions silver, 146 millions gold, and 18 millions unclassified: total, 367 millions. We believe the proportion of gold to be underestimated. During the first few years of its activity the Comstock produced more gold than silver. In many of Mr. Valentine's annual statements, previous to his examination by the Monetary Commission of 1876, the whole produce of Nevada, averaging 25 to 30 millions a year, was credited to silver and nothing to gold. Yet it was upon this slipshod commercial circular, which had only a remote relation to fact, that the Mint laws of France, England, Germany and the United States were made to turn!

adopt these as the basis of conjectures, which in many instances have proved to be entirely misleading.

Table showing the Production of Gold in the United States of America, according to different authorities. Sums in millions of dollars.

Year.	Del Mar.	Valentine.	Mint Bureau.	Year.	Del Mar.	Valentine.	Mint Bureau.
1880	34	33	36	1890	26	32	33
81	32	31	35	91	25	32	33
82	27	29	33	92	24	30	33
83	25	28	30	93	25	34	36
84	23	25	31	94	37	46	40
85	26	26	32	95	47	48	47
86	25	30	35	96	51	53	53
87	26	33	33	97	58	64	57
88	24	30	33	98	63	66	64
89	26	33	33	99	72	73	71
Totals,	268	298	331	Totals,	428	478	467

The produce of gold during the twenty years, 1880–99, according to Del Mar, was 696; Valentine, 776; and the Mint Bureau, 798 million dollars; a difference between the first and third estimates of practically 100 millions, and between the second and third of 32 millions. The disagreement between these various authorities concerning the produce of silver is even greater; but in order not to confuse the subject, it is enough in this place to consider the discrepancies with reference to gold. Leaving the first estimate out of view for the present, here is a difference in twenty years of 32 millions between Mr. Valentine, who bases his computations on the Express carriage of bullion, and the Mint Bureau, which bases them partly on the academic figures of Dr. Soetbeer, a German author, who has never visited the mines, and partly on the gossip of mining newspapers and the letters of distant correspondents, some of whom are known to have sublet their commissions to irresponsible and ignorant persons. When the importance of these computations is held in view, the enormous discrepancies between them should surely suggest the necessity of imposing a statistical tax on the produce of the mines, so that the government may have some data upon which to form a judgment concerning the production of the precious metals. These metals furnish the basis of its monetary system and fix the value not only of all the property in the country, but what is of infinitely greater consequence, they determine the value of those exchanges of services and commodities which most deeply affect the welfare of the nation.

The indifference exhibited by the American government with regard to the production of the precious metals from its own domains finds expression in another direction. It is a remarkable fact that the United States have always made their coins of a material which was either altogether or largely under foreign control. Until the Civil War, when the Comstock Mines were opened, the United States produced no silver, yet the circulating coins which it adopted were chiefly silver ones: at first Spanish and Mexican silver dollars, halves, quarters, eighths, and sixteenths, together with a few French five-franc-pieces; afterwards American dollars and fractions. When the Comstock Lode began to be prolific of silver, the authority of the Mints to coin any more silver dollars was destroyed by Act of Congress. Neither the prolificity of the Appalachian nor of the Californian gold mines could influence the government to demonetise silver, nor could the Nevada mines influence it to demonetise gold. Such self-denial, such faith in the disinterestedness of those foreign states, whose emissaries suggested this policy, might be commendable, could it hope for an equally generous reciprocity; but, unhappily, national policy is not commonly based upon sentiment, so much as interest. And it is greatly to be feared that these strange instances of liberality on the part of the American government will go without reward.

CHRONOLOGY OF GOLD IN CALIFORNIA.

The discovery of gold at Coloma was not the beginning of gold mining in California. Gold was very well known to exist in the soil; but so long as the Missions lasted, mining was discouraged by the Jesuit fathers, and this was quite sufficient to put a stop to it. Opposition to gold and silver mining was the avowed policy of the church. The dreadful accounts of mine-slavery, which, during three centuries (from the time of La Casas to that of Miguel Hidalgo), the priests in charge of the American missions, had sent to Rome, furnished the ground for this policy. It was represented as hopeless for the priests to civilize the natives so long as they were systematically brutalized by mine-slavery; so the church was asked to oppose it.

After the Spanish-American Revolution, the Mexican government refused to support the Missions, and when the Missions fell, the opposition to mine-slavery (it was called peonage) fell with them. Whereupon gold-mining, long suppressed in California, was at once resumed. Of the many localities in which gold was believed to exist, the San Fernando Valley was the only one which was at that time practical to work, because it was there alone that peonage could be maintained. That part of the country was open and the Spanish horsemen could ride in any direction. In the northern foothills, where gold was afterwards found in still greater plenty, the Indian, owing to the ruggedness of the country, was secure from arrest and mine-slavery, and therefore mining was impracticable; and so it continued to be until the mines fell to a free flag and a free people.

1534. Cortes sends out an expedition from Mexico which discovers Baja Calfornia, and prospects for gold.

1536. Cortes himself sails from Mexico, and lands upon the coast of Baja California in search of mines.

1536. Cabeça de Vaca starts overland from Florida in 1528 and after great suffering and the loss of all his party, except two whites and a negro, he reaches Texas in 1535, where the adventurers are enslaved by the natives, and where they hear of the Seven Cities of Cibola filled with gold. In 1536 they escape to the Rio Fuerte in Mexico, and are brought into San Miguel, a Spanish settlement on the west coast. The account of Cabeça de Vaca gave rise to Coronado's fruitless expedition to Cibola in 1540. Del Mar's "Hist. Money in America," p. 32; Bancroft's "California," I, 6.

1539. Marco de Niza sails from Culiacan to the Colorado River, and returns without success in finding gold.

1540. Coronado goes from Culiacan to the Colorado River; sees the great Cañon; explores the country now known as Arizona, New Mexico and Kansas; discovers the Seven Cities of Cibola; and returns to Mexico disappointed as to gold.

1540. In a work published during the 16th century, it is asserted that some of the soldiers in the Expedition of Francisco Vasquez de Coronado to the Seven Cities of Cibola, not sharing the pessimist views of their commander, resolved to desert their party and push on to Quivera, *in California*, a country governed by Tatarrax, a hoary-headed, long-bearded king, who worshipped a golden cross and an image of the Queen of Heaven, and who was amply provided with gold from the mines of his country. Nothing is said concerning the result of this adventure, which, if it took place at all, probably ended in the entire loss of the party in the endeavour to cross the terrible desert of the Mohave. There is some blundering in this narrative. The Cities of Cibola, reached by Coronado, must have been in the Gila Valley. From this point he is said to have travelled "northwardly and eastwardly for nearly three weeks," when he saw immense herds of buffalos. This would bring him to the northern part of Arizona. It is here that he heard of Quivera and its King Tatarrax, who lived still further north, and therefore away from California. If any of Coronado's soldiers left him to seek Tatarrax, they must have wandered into the desert of Nevada, rather than that of the Mohave. However, one would have been quite as fatal as the other. Mineral Resources, 1884, p. 548; and Del Mar's History Money in America, p. 41, in which will be found a full account of Coronado's Expedition.

1542. Cabrillo discovers the bay of San Diego.

1570. "Very interesting is a paper in the National Archives in the Hague, a letter written from Japan in the year 1635 by a certain William Verstege and addressed to Governor General Henricq Brouwer, at Batavia. Many years ago, says the writer of the letter, a vessel sailed from Manila to Nova Hispania (Mexico). In the Pacific Ocean, at a northern latitude of 37½ degrees, and at a distance of from 380 to 390 miles to the east of Japan, this vessel was overtaken by a heavy storm, in which she lost her mast, and therefore was obliged either to return or put into the nearest land which she could reach. As the weather cleared up a little they saw a large and mountainous island, to which they took their course and landed. The people they met on the shore were fair and well proportioned and *extremely kind*.

After having fitted a new mast the voyage was continued. But the visit paid to this island had been satisfactory, and the crew said that gold and silver was so plentiful on the island that one could pick it up on the shore. Even the kettles and other cooking utensils which the people there used, were made from these metals.

The king of Spain, having heard of the occurrence, sent an order to the viceroy of Nova Hispania that he should make investigations after the island, and a vessel was accordingly fitted out at Acapulco for this purpose. The vessel was first to take her course to Japan, and having sold there the wares and merchandise which she carried, from their produce, a second one should be built. This expedition failed, and most probably the Spaniards, after the ill-success of the first attempt, had not renewed it. William Verstege, however, thought it his duty to communicate the information which he had obtained to the Governor-General, and thought it worth while that the matter should be taken into careful consideration. It would not be such an expensive undertaking to send out a couple of vessels from Japan in search of the island. It was mountainous and situated at a distance of from 380 to 390 and at the utmost 400 miles east from Japan, and at a northern latitude of 37½ degrees. This letter seems to have made a favourable impression upon the Governor-General and to have induced him to send out an expedition in search of the wonderful island. Consequently, on July 2, 1639, two vessels called the *Engel* and the *Gracht* left Batavia under Commander Matthys Quast. Abel Janssoon Tasman was subordinated to him as commander of the *Gracht*. According to their orders they had to continue their voyage at a northern latitude of 37½ degrees till they reached the continent of America, if at a distance of about 400 miles eastward from Japan they did not find the island. It was not found, although at that latitude they pursued their course to a distance of 600 geographical miles from Japan.

A second expedition was sent out to the same purpose in the beginning of the year 1643, with the two ships *Castricum* and *Breskens;* Commander Maerten Gerritsoon Vries and Under-Commander Hendrik Cornelis Schaep. But this time also the expedition failed. On the east coast of Nippon the two vessels were separated by a hurricane. The *Breskens*, after having lost her captain and nine men, who were captured by the Japanese in going on shore, sailed eastward and was obliged to return on account of sickness among the crew, after reaching a distance of about 480 miles from Japan. The *Castricum* also had to return without any result after having sailed 450 geographical miles from Japan.

Between the 37th and 38th degrees of northern latitude no island certainly could be discovered, but if Matthys Quast had seen his way to execute the orders which he had received, and to continue his voyage till he had reached the continent of America, he would have reached the land somewhere in the neighborhood of San Francisco. The establishment of a colony in those regions would not have been beyond the reach of the Dutch East India Company. It could have been supervised from Batavia just as well as Manila was from Mexico. The East India Company had ill-luck in its search for gold mines."—Letter of REINIER D. VERBEEK, dated Haarlem, Holland, Jan. 14, 1897, and printed in the New York Mining and Engineering Journal, April 17, 1897.

1579. Drake visits Alta California and suspects the presence of gold; one of his officers writing: "There is no part of the earth here wherein there is not a reasonable quantity of gold or silver." Says Edmund Randolph in Min. Res., 1867, p. 271: "The natives mistook them (Drake's officers) for gods, worshipped them and offered sacrifices to them." Had the Indians been aware of the golden resources at their own command and made them known to Drake, it seems not improbable that the worshipped and the worshippers would have changed places.

1584. Expedition of Da Gali.

1595. Expedition of Cermiñon.

1602. Expedition of Biscaino.

1640. The government opens the Testaloma mine in the Real of Santa Ana, or San Antonia, Lower California, and erects smelting works, the ruins of which were visited in 1769 by the European astronomers, who observed the transit of Venus in Lower California.

1642. Mission San José founded by Pacheco.

1650. During the 17th century numerous placers and quartz veins were discovered and reported by the military guards stationed at or near the southern Missions; but the padres discouraged their exploration on the grounds: first, that it would necessitate mine-slavery; second, that it would invite none but depraved and abandoned men, who would surely introduce vice and wickedness among the free Indians, whom the padres were endeavouring with much success to civilize; third, that it would excite avarice and the worst of passions and occasion turbulence, disorder and peonage for debt. The padres had seen enough of these consequences in Mexico, and were confident that they would again follow any permission which might be given to private adventurers to work the mines. Forbes.

1700. Manuel Osio, who had accumulated a fortune from the pearl fishery, opens mines in the vicinity of San Antonia, where there is still to be seen the ruins of the smelting works he erected. San Antonia is an inland town in Lower California, about eighty miles south of La Paz. Forbes.

1721. Capt. George Shelvocke, R. N., visited the coasts of California this year, and observing that the soil near Puerto Seguro contained certain glittering particles, he washed it for gold, but owing to inexpertness, failed to extract the metal. He carried away some of the gravel for future assay. This gravel was lost with his ship on the coast of China. Capt. Shelvocke's voyage was published in London, in 1726.

1750. About the middle of the 18th century the interdict hitherto placed on the mines of California by the missionary priests was partly removed, and numerous adventurers from Mexico explored the country, particularly Lower California, for the precious metals. Forbes.

1768. The Count de Galvez, Governor of the province of California, orders the mines near San Antonia to be reopened and worked on Government account. A decree to this effect was promulgated in the latter part of this year. The names of the mining districts were: 1, El Triumfo; 2, El Oro; 3, El Valle Perdido; 4, El Rosario; 5, El Tale; 6, Las Gallinas, and 7, Los Juntos. Forbes.

1769. The European astronomers who landed upon the coast of Lower California to observe the transit of Venus, in July, 1769, noticed the abandoned smelting works of the Testaloma mine.

1769. Mission San Diego, Alta California, founded July 16th.

1771. Mission San Gabriel, Alta California, founded Sept. 8th.

1775. A gold placer, now known as the Carga Muchacha mine, on the Colorado river, near Fort Yuma, was discovered this year. Henry G. Hanks, in Mineral Resources, 1884, p. 547.

1776. Mission San Juan Capistrano founded November 1st. All of these three Missions are in the present county of Los Angeles. There were twenty-one Missions (altogether) in California.

1780. Filipe de Neve, Governor of California, reports to the viceroy of New Spain that the quartz mines have been abandoned, because quicksilver is too expensive and scarce, and that the miners have gone to the placers where mercury can be dispensed with, because the gold is coarse, each man winning from one to two ounces a day. (Forbes.) The locality of these placers is not known, but it is suspected that they were between Mt. San Bernardino and the Colorado river.

1786–9. Antonio de Alcedo, in his Geography, mentions several large nuggets of gold which were found in California previous to this date.

1797. Mission San Fernando Rey, Alta California, founded Sept. 8th. It was possibly about this time that the placers of the San Fernando Valley were discovered. "The padres in charge of the Mission discouraged the digging of gold as having a demoralising tendency upon their flocks." Mineral Resources, 1872, p. 108.

1814. A silver mine was discovered and worked this year in Salinas Valley, Alta California. This mine is still worked by the Hartnell family. Information from J. A. Forbes, of San Francisco. In 1802, a silver vein was worked in the Olizal or Alisal district, Gabalan Mountains, Monterey county. (De Groot.) It proved unprofitable. Mineral Resources, 1867, p. 13.

1814. About this time a silver mine was worked in what is now Ventura county, Alta California. A tunnel was run in about 300 or 400 feet, and the sides "coyoted," or irregularly stoped. This mine was re-discovered in 1886, together with ruins of old smelting works and a silver brick weighing about a pound. Ventura "Free Press," December 9, 1886.

1815. Wm. Phillips notices gold in California. Calvert's "Gold Rocks of Britain," p. 10.

1815. The Sierra Buttes gold quartz mines in Alta California worked this year, the quartz being crushed in arastras. Henry Janin, in "Mineral Resources," 1870, p. 60.

18—. In the early part of the 19th century silver mines were worked at Cacachilos, Las Virgenes, La Buena Mujer, and La Truichera, in Lower California. All of these locations were abandoned during the Revolution. Forbes.

18—. At Soreta (Lower California, northern part), a man named Garraisal worked a rich copper-mine called "El Sauce." There were other copper mines worked in Lower California at this period. Forbes.

1820. Rancho San Bernardino founded as a branch of San Gabriel Mission, Alta California.

1822. Mexico throws off Spanish rule. Taxes from the Missions exacted by the revolutionary government. Priests' salaries stopped. "Pious funds" declared forfeited to the government. Decline of the Missions.

1823. The mines of San Ydelfonso and Pascal, in the northern part Lower California, worked by J. Maritorena, and those of San Pedro and San Jacinta worked by Feliz Caballera, a missionary priest of Santa Tomás. Forbes.

1824. Gen. M. G. Vallejo, who came to California in 1810, says that in 1824, while he was on a military expedition to the King and Kern rivers, he found a Russian living between these rivers who was well supplied with mining tools and implements, and who had long been mining for gold in that vicinity. Gen. Vallejo himself for many years from this time remitted gold-dust to the authorities in Mexico, but whether from the Russian's mines or elsewhere is not stated. Mineral Resources, 1884, p. 547.

1824. Capt. Walter Comstock, a whaler in the employ of Grinnell, Minturn & Co., of New York, says that he brought gold from California in 1824, and that nearly every vessel which put into the harbour of San Francisco brought away samples of gold. Mineral Resources, 1884, p. 547.

1825. In the museum of the Guadalupe Friars at Zacatecas, Mexico, there was formerly a gold nugget about the size and shape of a pigeon's egg, which had been fetched from California by one of the friars. Mineral Resources, 1884, p. 547.

1825. Gold and silver mines at St. Ysidoro, 35 miles east of San Diego, Alta California, worked this year by a Mexican miner from Guanaxuato. Notes to Wyld's Map of California, London, 1849, p. 14.

1825. J. S. Smith, a trapper employed by the American Fur Company, crossed the Plains this year, followed the river now called the Humboldt, got through the Sierras near what is now known as the old Emigrant Road at the head of the Truckee, reached the Sacramento Valley, and striking south-west, passed through San José to San Diego. On his return, he crossed the Sierras near Walker's Pass, then to Mono Lake, and then eastward to Great Salt Lake. "On this portion of his route," probably meaning from Walker's Pass to Mono Lake, "he found placer gold in quantities and brought much of it with him to the encampment on Green River." Thos. Sprague in "United States Mineral Resources," 1867, p. 305.

1828. Gold diggings were reopened this year at San Ysidoro, San Bernardino county, California. Mineral Resources, 1884, p. 547.

1833. Legal abolition of the Missions. Renewal of mining and exploration. The Yaqui and Mayo Indians hunted down as mine-slaves. Indian outbreaks.

1834. In the latter part of 1833, or the beginning of 1834, J. P. L. Leese brought from Taos, New Mexico, to Los Angeles, Alta California, a considerable quantity of grain gold, the product of the New Mexican placers.

1834. About the same time one Palacios brought into Los Angeles, from Guaymas, several bricks of silver and some placer gold, both having been produced in Mexico. Two lots of this placer gold aggregated about $10,000 in value. A portion of this was exported by the merchants of Los Angeles, chiefly to Boston; a portion was sold to tne jewelers in and about Los Angeles, while another portion was cast into counterfeit Columbian doublons. J. J. Warner, of Los Angeles, in San Francisco "Bulletin," Sept. 7, 1881.

1836. A silver mine "east of St. Inez," Santa Barbara county, Alta California, was discovered previous to this date; "but it has been abandoned." London Penny Encyc., 1836, cited in Appleton's American Encyc., article "California."

1837. In this year one of the padres of the recently abolished Missions endeavoured, through Mr. Young Anderson, to obtain English capital to work the mines in San Fernando Valley. Owing to the remoteness of the mines and the recent unfortunate experience of English capitalists in Mexico, this endeavour failed.

1838. A man, whose testimony in the Bower's poisoning case, 1886, was refused credence, and who was afterwards arrested upon a charge of theft at San Diego, but who, nevertheless, may have told the truth in this case, informed the "Commercial" newspaper that the placers of Santa Feliciana were worked by one Francisco Garcia of Los Angeles, in 1838.

1838. The Mexican government commences to grant patents for lands where placer gold had been found in San Fernando Valley.

1840. A gold quartz mine at Francisquito, near the Mission and valley of San Fernando, was worked this year by a Frenchman named Carlos Baric. The croppings of the vein extended sixteen miles along the ravine. M. Baric produced from his mine an ounce of gold per day. (Notes from Wyld's Map of California, London, 1849, pp. 14, 17, 30.) The mines at Cahuenga, same Mission, had not been worked for want of mercury. Ibid., p. 31.

1840. Don Juan Bandini, of San Diego, worked some copper mines at All Saint's Bay, and others at San Antonio, near the Mission Guadaloupe, about seventy miles below San Diego. These were abandoned in 1844 owing to the troubles during the administration of Governor Manuel Micheltoreno. Forbes.

1841. In the early part of this year Don Andres Castillero, a native of Mexico, a man of scientific attainments and mineralogical knowledge, travelling from Los Angeles to Santa Barbara, saw and gathered up, near the rancho of Las Virgenes, some mineral specimens, which he exhibited in Santa Barbara, and said that generally, if not invariably, placer gold existed wherever this class of pebbles were found.

In the month of June, 1841, two vaqueros (herdsmen) of a neighbouring ranch, while riding over the ranch of San Francisquito, dismounted from their horses by the side of a rivulet to give them a breathing spell, and seeing a bed of wild onions they engaged in gathering some of them. While so doing, one of them, by name Francisco Lopez, who had been present and saw the pebbles which Castillero had said was an indication of gold placers, noticed some of them here and said to his companion: "Look at this; there is gold here, for I heard Don Andres Castillero say that there was gold to be found where-ever these little stones exist"; and immediately scooping up a handful of the sand and gravel which had been loosened by pulling up the onions, he rubbed it with his other hand, and sure enough he found in his handful of sand a grain of gold.

On their return to Santa Barbara these men took with them a few dollars worth of gold which they had obtained from the gravel. The news of the discovery of gold soon spread from Santa Barbara, and people from San Diego to San Luis Obispo hurried to the newly-discovered placers. Although few or none of the native Californians had any practical knowledge of gold-washing, there were, at the time of this discovery, quite a number of natives of Sonora and other parts of Mexico scattered over Alta California, and especially in the southern part of it, who had worked in placers. A large part of the extensive country drained by the Santa Clara river (Ventura county) was now prospected, and gold obtained in many places. During the first two rainy seasons following the discovery, some hundreds of people were profitably engaged in mining, and gold was brought into Los Angeles by miners and sold there every year, from 1841 to the discovery of the richer and broader gold-fields of Central California in 1848, which last soon caused the abandonment of the less productive placers of San Francisquito. The late Abel Stearns, then a merchant of Los Angeles, bought a good deal of Francisquito gold. He sent a small part of it to the mint of Philadelphia. The original receipt or certificate of its deposit was some years ago donated (I believe) to the California Pioneers' Society, at whose rooms, I presume, it can be seen. J. J. Warner, of Los Angles, in San Francisco "Bulletin," Sept. 7, 1881.

1841. Mr. James D. Dana, the mineralogist of Com. Wilkes' Expedition, which visited the Coast this year, published a book in 1842, and mentioned in it that gold had been found in the Sacramento Valley.

1842, April 4. Date of the first mining "location" which appears in the public records of Alta California. This "location" was made by Lopez, Cotta, and Bormudes, and was for a gold placer in San Fernando Valley. The original application and "location" is now in the United States Surveyor General's office, San Francisco. Forbes.

1846. Mr. Thomas O. Larkin, the U. S. Consul at Monterey, reported that gold, silver and other mineral deposits were to be found "all over California." "Mineral Resources," 1867, p. 14.

1847. Com. Sloat communicated an article to "Hunt's Merchants' Magazine," of New York, in which the presence of gold in California was indicated.

1847. John (afterwards General) Bidwell, at that time the agent of Capt. Sutter, and who, many years later, lived at Chico, having visited the placers of San Fernando Valley, and observed the manner of working them, and suspecting the presence of gold farther north, searched for it in the foot-hills on the line of the Cosumnes River, but without success.

1847. Further discoveries of gold near Los Angeles. (Wyld, p. 17.) These were probably made by the Mormons, who emigrated from Utah through the Cajon Pass, to San Bernardino, near which place

(Lytle Creek) some Mormons worked a placer mine for many years, so late as 1890.

1848, January 18. (The Pioneer Society of New York now keep this anniversary on January 24th.) James Marshall, who had also previously suspected the presence of gold in the northern foot-hills, found a large nugget of that metal in the tail-race of Capt. Sutter's mill on the American river. This was the GREAT DISCOVERY which opened a new æra in the history of the precious metals, money, prices and industry throughout the world.

One of the earliest-written reports of Marshall's discovery is found in Dr. J. Trywhitt Brooks's "Four Months Among the Gold Finders"; a book published in London in 1849, and long since out of print. Brooks visited Sutter's Fort in May, 1848, and Sutter told him the story of the discovery. Sutter's version as given by Brooks was that he was sitting in his room at the fort one afternoon, when Marshall, whom he supposed to be at the mill, forty miles up the American River, suddenly burst in upon him. Marshall was so wildly excited that Sutter, suspecting that he was crazy, looked to see whether his rifle was within reach. Marshall declared that he had made a discovery that would give them both "millions and millions of dollars." Then he drew his sack and poured out a handful of nuggets on the table. Sutter, when he had tested the metal and found that it was gold, became almost as excited as Marshall. He eagerly asked if the workmen at the mill knew of the discovery. Marshall declared that he had not spoken to a single person about it. They both agreed to keep it secret. Next day Sutter and Marshall arrived at the saw-mill. The day after their arrival they prospected the bars of the river and the channels of some of the dry creeks, and found gold in all.

"On our return to the mill," says Sutter, "we were astonished by the work people coming up to us in a body and showing us some flakes of gold similar to those we had ourselves procured. Marshall tried to laugh the matter off with them and to persuade them that what they had found was only some shining mineral of trifling value; but one of the Indians, who had worked at a gold mine in the neighbourhood of La Paz, Lower California, cried out, 'Oro! Oro!' and the secret was out"

CHAPTER XXXI.

AUSTRALASIA.

Gold discovered in 1851—Rush to the diggings—Ticket-of-leave men—Statistics of the production—The discovery due, not to scientific prevision, but to chance—Nuggets—Prices—Licenses—Taxes—The chimerical republic of 1854—Wanton and cruel destruction of native life—Cost of production—Chronology.

PLACER gold was discovered on the Macquarie River, New South Wales, in February, 1851. By the first of June following there were upwards of 1,000 men in the diggings. On the 10th of June gold was discovered on a tributary of the River Loddon, Victoria; on the 20th of July at Mount Alexander, Victoria; on the 8th of August at Mount Buninyong, Victoria; and on the 8th of September at Ballarat, Victoria. In the month of October upwards of 7,000 miners were at work. By the end of the year the number of persons in all the placers of Victoria was from 15,000 to 17,000, and some $6,500,000 worth of gold (mint value) had been taken out. In 1852 the number of miners had swollen to something like 150,000 (including about 9,000 ticket-of-leave convicts), when, notwithstanding the extraordinary production of that year, it was found that on the average gold-digging was unprofitable; and many persons left the workings to engage in other pursuits. Nevertheless, their places were soon filled by new adventurers.

Although gold-fields were subsequently discovered in several other of the Australasian colonies, both the total production and the miners employed declined from the year 1856, until December, 1879, when the former scarcely exceeded $28,000,000 a year, and the latter numbered 37,553 men, of whom 28,443 were Europeans and 9,110 Chinese. In 1886 the mines reached their lowest point of productiveness, and for many years they continued with little increase, until 1894, when they produced about $40,000,000 a year, a sum which, in the course of the six years following, was doubled; so that at the present time Australasia is throwing upon the mints of the world no less than $80,000,000 in gold.

The following Table shows the production of gold in all the Australasian Colonies from the discovery of the placers in 1851 to the present time:

ANNUAL PRODUCTION OF GOLD IN AUSTRALASIA.

Sums in Troy ounces fine, worth $20.67 per ounce.

Year.	Ounces.	Year.	Ounces.	Year.	Ounces.	Year.	Ounces.	Year.	Ounces.
1851	328,457	1861	2,479,298	1871	2,234,298	1881	1,470,325	1891	1,519,059
52	2,856,863	62	2,559,150	72	2,034,432	82	1,422,289	92	1,652,440
53	3,028,778	63	2,566,230	73	1,878,895	83	1,313,513	93	1,726,438
54	2,259,882	64	2,247,454	74	1,617,227	84	1,368,371	94	2,060,069
55	2,751,403	65	2,305,928	75	1,537,664	85	1,327,478	95	2,167,117
56	2,979,276	66	2,366,871	76	1,356,220	86	1,275,956	96	2,185,676
57	2,775,271	67	2,264,672	77	1,217,194	87	1,326,480	97	2,695,562
58	2,664,838	68	2,372,446	78	1,411,677	88	1,338,836	98	3,263,313
59	2,470,570	69	2,217,812	79	1,397,500	89	1,600,319	99	4,104,330
1860	2,403,647	1870	1,923,726	1880	1,459,069	1890	1,460,911	1900	3,840,826

REMARKS.—Queensland is not included in the annual returns until 1878. Previous to that year the total produce of Queensland, not separable into years, is given by the Royal Mint at 2,435,163 fine ounces, equal to about fifty million dollars.

In the United States Consular Reports for May, 1901, the following figures are given for the auriferous production of Australasia: 1890, 1,587,947 oz.; 1893, 1,876,563 oz.; 1894, 2,239,205 oz.; 1896, 2,375,735 oz.; 1897, 2,929,959 oz.; 1898, 3,547,079 oz.; 1899, 4,461,105 oz.; 1900 (a retrogression), 4,174,811 oz. It is presumed that these quantities relate to crude gold. Those in the Table relate to fine gold. The great increase since 1896 is attributed to the discoveries in Western Australia. The proportion yielded by vein, quartz, or reef mining was, in 1851, nil; 1853, about one-eighth; 1870, about three-sevenths (Withers); at the present time about three-fourths.

The proportions contributed by the various colonies in the year 1900 were as follows, the quantities being in ounces crude: New South Wales, 345,000; New Zealand, 371,993; Queensland, 951,065; South Australia, 29,397; Tasmania, 89,000; Victoria, 807,407; Western Australia, 1,580,949; total, 4,174,811.

It has been frequently claimed that the discovery of gold was owing to the observations of Sir Roderick Murchison, in 1844, concerning the similarity between the geological formation of the Australian and Ural mountains. The plain fact is that gold was discovered and that the discovery was made public and notorious, before Murchison published his observations. Not only this, but Mr. Phillips communicated the fact to Sir R. Murchison in 1841. Murchison also held that gold is confined to Lower Silurian strata, into which it did not make its appearance until just before the time of the Drift. This theory has been controverted by Mr. Hartt in his "Gold of Nova Scotia of pre-carboniferous age," who says: "As the gold of Nova Scotia was probably introduced into, or assumed its present form in, the quartz veins at the time of the metamorphism of the Silurian rocks, which metamorphism was pre-carboniferous, I have doubted the correctness of this theory (of Murchison and others). The occurrence of gold in the carboniferous rocks at Corbitt's Mill shows that it (the theory) is not to be applied to the province of Nova Scotia." Says Mr. Whitney: "According to this eminent authority (Murchison) gold in paying quantities is exclusively confined to the Palæozoic rocks. . . . It was also a favourite dictum of this geologist that auriferous quartz-veins are a superficial phenomenon, and that mines of this metal would not hold in depth as persistently as those of other metals. The discoveries of the California Survey, in regard to the age of the gold-bearing formation of that State, have entirely refuted the (until 1864) generally accepted theory of the exclusively Palæozoic age of rocks of this kind, although this fact was not admitted by Murchison in his latest publications." Cited in Lock, 755, 780.

From these statements it would appear that Murchison made no discovery, with relation to gold-bearing rocks, which proved to be of any practical value, and that in respect of this matter he was a greatly overrated man, fond of notoriety and praise, which he was willing to accept as his due, even when it rightfully belonged to others. Murchison was also mistaken in regard to the depth of gold deposits. Vast quantities of this metal have been taken from quartz mines in California, Nevada and other American States, at depths of between two and three thousand feet, and from mines in Australia at still greater depths, for example, the Lansell, which is now (1901) down 3,350 feet and going lower.

In regard to the early discoveries of gold in Australasia, a dispatch of the Lientenant-Governor of the Colony to the Secretary of State, dated Sept. 2, 1840, enclosed a report from Count Strzelecki, stating that he had discovered auriferous pyrites in the vale of Clwydd, in 1839.[1] The Rev. W. B. Clarke announced his discovery of the metal in the same colony in 1841. The natives knew of it; and made their arrow-heads of gold. The white shepherds found gold; but prudently refrained from searching for more.[2] In a despatch, dated June 11, 1851, from Sir C. A. Fitzroy, Governor of New South Wales, to the Earl Grey, it is stated that a prospector, named Smith, had discovered gold in 1849. Yet the great discoveries of 1851 were not owing to any of these observations, but to the prospecting of a California

[1] The Count afterwards stated that he was urged by the Governor to keep the discovery secret, for fear it would impair the discipline of the 45,000 convicts on the island. "Thirty Years in New South Wales and Victoria," quoted by Phillips on "Mining." The secret transpired; but it was hushed up.

[2] Some of the shepherds had picked up pieces of gold in Victoria, but they wisely paid no further attention to the matter. Phillips, 103.

miner named Hargreaves. Years afterward the colonial governments recognized him as the principal discoverer of gold by rewarding him with the sum of £10,000.

Australia was not a solitude when Hargreaves visited it. The aborigines had roamed its mountains and encamped upon its auriferous river banks from time immemorial, without troubling themselves about its gold. It had been a penal colony of Great Britain since 1788, and at the period of the discovery of gold it contained a population partly composed of free colonists engaged in agriculture and sheep-farming, whose settlements stretched in every direction; who, before the gold discoveries, had been engaged, among other industries, in mining copper; and who, therefore, had a motive to further prospect the country for metals. Hargreaves was no more clever than the rest; he was simply in luck.

Said Mr. Patterson, in the Westminster Review, 1883: The story of the Australasian gold-fields is in many respects similar to that of the Californian. The gold discoveries were at first suppressed by government, "fearing lest a gold mania and gambling spirit would without any adequate return divert the population from its course of steady industry;" a fear which was justified by events, for in the course of a few months half the male population of Victoria had left their legitimate occupations and gone hot-footed in search of the precious metal. Workshops stood idle, business places were closed, ships lay empty at the wharves, trade was at a standstill, business was allowed to drift where it would; there was but one thing thought of, and that was gold, until the number engaged in this new industry in Victoria amounted to 100,000 persons. This great rush lasted, however, only a few years; the surface diggings soon became exhausted; individual labour was rendered unprofitable and capital and machinery became necessary; meanwhile, the prices of food and other necessaries rose enormously, flour was sold as high as £44 a ton in 1855, and a cabbage cost 5s. Bricks rose from 30s. the thousand to £18, and all other commodities in proportion; but these wants were soon supplied by importation; the famine prices declined, and eventually the supply became greater than the demand, and many merchants and shopkeepers were ruined, one firm losing £90,000 in a twelvemonth.

In the Australasian gold-fields, however, the prizes were tremendous. One nugget found at Ballarat weighed 2,195 ounces, and was sold at Melbourne for £9,325. Many others were but little inferior to it, and every adventurer worked on, hoping that he might be the lucky finder of some such great prize! Now that the fever has ceased, what are the results? An enormous increase in the population, wealth, industry, and commercial prosperity of every land wherein gold is found in paying quantities is certain to follow the discovery, but only after many have suffered untold hardships, privations, sickness, and death. It would seem as though Nature had spread out her gold-field as a tempting bait to the human race, in the same manner as carniverous plants display their deadly leaves, covered with luscious fluid to tempt the unwary insect to destruction.

When the great discovery took place it was claimed by the authorities that all gold deposits belonged of right to the Crown; but Hargreaves, between whom and the Royal Prerogative rolled sixteen thousand miles of ocean, ridiculed such claims as belonging to a past age. As he was a man "strong in counsel and Californian experience," and he voiced the sentiment of a lawless and resolute community, the Australasian authorities deemed it prudent to lower the royal demands to the price of a digger's license, costing thirty shillings a month. The mere doubling of this trifling fee afterwards occasioned a Revolution; and but for the detachment of armed police and troops who stormed the Eureka Stockade, Australia in 1854 would have become an independent republic. It had already declared this desire and had gone so far as to raise a distinctive flag.

At first the fees for mining-licences were paid in gold-dust, there being no money in circulation. "Coin was rare and the digger generally bartered his gold-dust for goods. Change there was none; and reckonings partook of the largeness of view which ignored minute calculations." During the early days of Golden Point, the storekeepers, for example, Burbank, and Colac, issued their promissory notes, payable in money at Melbourne or Geelong "one day after sight." These notes sometimes circulated in the place of money; but as the issuers refused to accept one another's notes, they failed to command a general currency and were discarded. Boxes of matches and potatoes at 3d. each, were used for small change. In 1853 £1 notes were in circulation, but it is not stated who were the issuers. Withers, 49.

In the first edition of the present work attention was called to the cruel and wanton destruction of native life by the Spanish and Portuguese gold-hunters in America and the Orient, as well as by the English in Australasia. An English writer of eminence was quoted, who said of the murdered natives of New South Wales, "They have been wantonly butchered; and some of the christian whites considered it a pastime to go out and shoot them. I questioned a person from Port Stephens. . . . His answer was, 'Oh, we used to shoot them like fun!'" ("Breton," in Bishop Whateley's Essays of Lord Bacon, ed. 1857, p. 339.) Twenty years have since passed. The class of cruelties to which allusion was made were repeated during this interval in South Africa; yet, out of the 150,000 pulpits in Great Britain and

America, countries in which this book is not unknown, not one of them has ever raised its voice against the crimes connected with the search for gold; not even against the betrayal and murder of Lobenguela.

Lest it should be supposed that these infamies belong to a past age, or to inferior races, it becomes necessary to revert to some other doings in Australasia. The time is so late as 1894, when the Coolgardie deposits were discovered and fabulous finds were reported from Ninety-mile Point and elsewhere. Nuggets of 50 and 100 ounces were found in the possession of the natives, who refused to tell whence they were procured. These stories induced prospectors to push into the wilderness, far beyond where white men had ever been before. The early parties had scarcely enough food and water to keep them alive. Some of them were brought back raging with fever and placed in the Coolgardie Hospital. The fate of these desperate men did not deter others from forming a syndicate, who, with heavy packs of provisions, water and rifles, started from the extreme western limit of Coolgardie to penetrate the interior. They set out on December 1st, 1894, and nothing was heard from them for over a month. On January 10th, 1895, Dan Robertson, one of the syndicate, returned from a camp, about 120 miles from Coolgardie. He said the syndicate were returning from a distant point without luck, when Mike Fitzgerald discovered rich outcroppings of gold ore. The men flew to the place and danced around wild with excitement, for the quartz was the richest they had ever seen. Robertson continues: "We found water near by, but our delight was soon changed when one of the men rushed up and said the blacks had stolen some of our provisions. Our joy changed to fury. Some of the boys behaved like demons, and when one of them suggested that we should go to the nearest native camp and take possession of the provisions at the point of our rifles, it seemed right to every one. I led the party, and we came upon the very tribe of blacks which had stolen our provisions. We followed them to their camp and we butchered them like cattle; men, women and children. We forgot that we were men. The slings and arrows of the blacks were nothing to the deadly rifles of the white men. Prayers and entreaties were in vain. The white men showed no mercy, and the blacks were unable to defend themselves." San Francisco dispatch to New York Sun, March 15, 1895.

This butchery is described by one of the actors. What if it were told by one of the sufferers? What if in fact no provisions were stolen, or if they were of trifling moment, or if the real motive of the gruesome tragedy was to plunder the wretched natives of their gold trinkets and arrowheads? Or to disclose the source of these baubles?

These crimes are related with no intent to disparage the English. Similar crimes have been related in this work of the Romans, Spaniards, the Portuguese, the Hollanders, the French and the Americans. They belong neither to races nor religion; they belong to the precious metals and especially to gold; because silver is never found on the surface or in placers. Neither are they related with intent to disparage gold. Essentially they belong neither to one nation nor to one metal; but to all men engaged in the search for hidden treasures. They are a part of the history of the Precious Metals. They serve to illustrate one of the elements in the Cost of their Production. That element is human life, not merely the destruction of a few men, but of entire races: the Phœnicians and Carthaginians in ancient times, the Mexicans and Peruvians in modern times, the Negros to-day, perhaps the Chinese to-morrow. Such crimes are worse than wicked; they are idiotic. They rob us of what is far more valuable than gold: they rob us of the national experience, of the history, the discoveries, the inventions, the dreams, if you will, of the men we have murdered.

CHRONOLOGY OF GOLD IN AUSTRALASIA.

1814. Some convicts of Botany Bay, New South Wales, having found placer gold and made the fact known, they were flogged for the offence and threatened with further punishment unless they kept quiet about the matter. It has been publicly declared that Governor Phillip hanged the first convict who made the discovery, but no better authority for this statement has been found than a Peoria newspaper of March 26, 1897. Writing in 1812–15, Malte-Brun says of New Holland or Australasia: "None of the precious metals have hitherto been seen."

1823. Assistant Surveyor, James M'Brian, found placer gold in several localities and publicly recorded the fact.

1825. A convict found placer gold this year, but was cautioned not to make the discovery public.

1838. McCulloch, whose Geographical Dictionary was completed this year, says: "It may be presumed from the comparatively small amount of old formations in the mountains (this is Murchison's theory) that they are destitute of the precious metals." Yet, he is obliged to add in a foot-note, citing the surveys of Flinders and King: "Gold is found at Timor (a small island near the northern coast of Australia), but the much greater elevation of the Timorean Mountains implies a different composition from that observed in Australia." Further on (under the head of "Timor,") he says: "Gold is found both in grains and large pieces; but the aborigines are said to have a strong aversion to search for it, and once massacred a party of Dutch sent inland to collect the metal."

1839. Count Strezlecki discovered placer gold and informed the Colonial Government of the fact, when he was requested to preserve silence on the subject, for fear it would cause an outbreak of the convicts at Botany Bay.

1841. Rev. W. B. Clarke found gold in the Macquarie and other river valleys of New South Wales, and in 1844 showed it to Members of the Legislature. At a subsequent date, when the discoveries of gold had become numerous and known to the world, Dr. Clarke was rewarded by the government of Australia as one of the prominent discoverers of a metal which had made the fortunes of the colony.

1844. In the early part of this year Mr. A. Tolmein found placer gold in Australia.

1844. Sir R. Murchison predicted that gold would only be found the Silurian formations of Australasia. Although this "prediction" was made after the presence of gold in Australasia was well known, it has no value, because gold is found in totally different formations.

1848. Some gold nuggets found in the alluvium were taken to Victoria and there exhibited in jewelers' windows.

1849. Rev. Dr. Clarke, who had discovered gold in 1841, wrote a letter giving his reasons why he did not deem it prudent at that time to make the discovery public. He looked upon gold-mining as only fitted to make slaves of the natives, and to this he was opposed.

1849. Placer gold was discovered by a shepherd on the Pyrenees mountains of Australia.

1849. Thomas Chapman found a 16-oz. gold nugget, and for fear the authorities would deprive him of it, he fled to Sydney. His discovery was there made public, but was minimized by the authorities.

1849. The discovery of gold in California causes the departure of some three hundred men from Australia. Landed property and stocks of merchandise in Sydney fall in value by reason of the diminished demand which followed this emigration.

1849. Some convicts having been discovered in the act of digging for gold, they were stopped by the authorities.

1849. A number of quartz mines were discovered in this year.

1851. A man named Austin found a nugget of gold worth thirty-five pounds sterling in Australia.

1851, Feb. 12. E. H. Hargreaves, a California miner, discovered alluvium gold in New South Wales, at a place called Summer Hill Creek, an affluent of the Macquerie river. This constituted the great gold opening of Australasia. After this discovery had attracted a numerous and hardy population to Australasia and had thus secured for it advantages which without such discoveries may not have fallen to this remote part of the world for centuries, the colonial authorities rewarded Hargreaves with a present of £10,000 sterling, an action whose generosity strongly contrasts with the indifference and neglect shown by the American government toward Tom Marshall and Capt. Sutter, of California.

Hargreaves' discovery was soon followed by others. In 1853 Australasia rose to be as great a producer of gold as California. At the present time, including New Zealand and Tasmania, it produces five times as much.

CHAPTER XXXII.

BRITISH COLUMBIA.

Movement of California placer miners to the northward—Discoveries in the Fraser River valley—The Cariboo district—The Kootenay district—The Cassiar district—Halcyon period—The tundras of the Stickeen—Dispersion of prospecting parties on the confines of Alaska—Beginning of quartz-mining—Statistics of production.

THE discovery of gold in the placers of California soon imparted to the miners a rough knowledge of geology, practically sufficient for their calling. It was observed that gold was to be sought west of the Mountains, and that "pay-dirt," or gravel containing enough gold to invite mining, was chiefly to be found in the river valleys, in benches or bars, where glacial and fluvial activity had already accomplished the work of breaking down the quartz veins of the mountains, extracting the gold from its matrix and depositing it in the sand and gravel of the valleys, from which it could be readily extracted by the rude processes of digging and washing. These observations were enough for placer miners; so that when their claims ceased to yield, often after lingering long enough over them to lose all they had ever taken out of them, they took up other claims, and in default of these, abandoned the district and went prospecting farther north. It was these prospecting parties who discovered the Fraser River district in 1856, the Cariboo in 1860, and the Cassiar in 1867.

However, gold had already been discovered in Queen Charlotte Islands in 1849 or 1850. Two years afterward, the Hudson's Bay Company sent a party of miners to drive out the Indians and explore the diggings; yet the enterprise failed of any important results. The Indians were duly shot down and the Islands pacified, but the gravel did not pay, and the vicinity was soon afterwards abandoned. In 1853 Captain McLennan's prospecting party discovered gold on the Similkameen River; but nothing came of it: the gravel was too poor.

The great discovery took place in 1856, when a rich find in the Fraser River valley was reported to the Governor of the Colony. In consequence of this announcement a rush occurred to the diggings:

and this was practically the beginning of mining in British Columbia and the Northwestern Territory. In 1857 further and richer discoveries were made in the Fraser River valley, especially on the Thompson River, near Nicommen, from which district the first quantity of gold was brought to the refiners. In 1858 gold was discovered on the Skagit River; 1859, on the Okanagan; 1860, on Harvey's Creek, in the celebrated Cariboo district; and in 1863 on Wild Horse Creek, in the equally celebrated Kootenay district. This was the halcyon period of mining in British Columbia, 1860–63; when these districts were worked out, the produce, which had reached nearly four million dollars a year, began to fall off.

In 1867 gold was discovered in the tundras of the Cassiar district, on the Stickeen, Stekin, or Francis River, lat. 54 to 56 N. It was probably this discovery that led to the subsequent prospecting of the Upper Yukon Valley, on the Klondike and other affluents, in Alaska. The Cassiar district first became productive in 1874, when it yielded a million dollars; in 1875 another million; in 1876 and 1877 each half a million, when the produce rapidly fell to $150,000 in 1883. In the following year the placers were regarded as exhausted. (Min. Res., 1883 and 1884.) As the outlet to this district is in American Alaska (mouth of the Stickeen River), the produce is not always credited to British Columbia. Sometimes it is credited to both countries.

Most of the gold produced in British Columbia is sent for coinage to the Mint at San Francisco, which very obligingly does the work for nothing. After being coined into American eagles, the gold is usually shipped to London and there recoined into British sovereigns.

Mining and prospecting parties had now gradually moved northward from California and Oregon through British Columbia to the Stickeen. Behind them were the partially worked out and abandoned placers of the West Coast; before them the frozen tundras of Alaska, their riches as yet hidden and unknown. The prospect was too forbidding. Most of the men gave up altogether; many died from the privations they had suffered; some went to quartz mining; others drifted into less dangerous and hazardous pursuits than prospecting for gold. After the Cassiar district was exploited the produce of British Columbia fell away to half a million dollars a year. In 1895–6 quartz mining began, when the produce rose to about one and a-half millions a year. Large capitals were now invested in the mines (quartz) and machinery began to take the place of hand labour. It is yet to be seen if these ventures will prove remunerative. The following table traces the development of the product:

PRODUCTION OF GOLD IN BRITISH COLUMBIA.

(*Sums in millions of dollars and tenths.*)

Year.	Product.	Year.	Product.	Year.	Product.	Year.	Product.
1858	0.7	1868	3.4	1878	1.3	1888	0.6
59	1.6	69	1.8	79	1.3	89	.6
1860	2.2	1870	1.3	1880	1.0	1890	.5
61	2.7	71	1.8	81	1.0	91	.4
62	2.7	72	1.6	82	1.0	92	.4
63	3.9	73	1.3	83	.8	93	.4
64	3.7	74	1.8	84	.7	94	.5
65	3.5	75	2.5	85	.7	95	1.3
66	2.7	76	1.8	86	.9	96	1.8
1867	2.5	1877	1.6	1887	.7	1897	2.7
Totals,	26.2		18.9		9.4		9.3

The total product for the years shown in the table, 1858–97 inclusive, was $ 800,000; to the close of the 19th century it may be roughly estimated at $70,000,000. As before stated, the increase since 1895 has been due to quartz or lode mining, which practically began in that year, all the previous product having been derived from the placers.

According to the report of the United States Consul at Victoria, dated March 16, 1901, concerning the mineral production of British Columbia during the year 1900, the number of miners employed underground in the province was 2,426, and above ground 1,305, total, 3,731. If these were all gold miners, the entire produce of the mines was insufficient to pay them $3 a day—the ordinary wages in mining camps on the northwestern coast—to say nothing of the capital invested in prospecting, purchasing, and developing the mines, machinery, supplies, licenses, taxes and other outlays. In a word, if the Consul's vague returns are correctly understood, every dollar produced from the mines of British Columbia must have cost at least two dollars. British Columbia is not alone in this respect: the same conclusion applies to all gold (or silver) mining countries employing free labour. Mining for the precious metals with free labour does not pay; first because the labourer is working against an enormous accumulation of the precious metals which were obtained by conquest, and slavery; second, he is working against the practical slavery which is still tolerated in Africa and India; and third, because he is working against a scale of prices which has for its foundation not only these circumstances but also the credit of the financial world in the shape of bank-notes and its power to multiply their activity by means of telegraphs and railways.

In 1900 the government of the Dominion of Canada, in which British Columbia is included, disallowed or repealed the act which forbade all except British subjects from locating placer claims. It established rules under which all "free miners"—by which term it is presumed is meant any person not under legal restraint or not a contract labourer—may take up such claims. Roughly speaking, discoverers' claims are limited to 300 feet and all others to 100 feet in length, along a stream, bank, bar or bench. Dry diggings are limited to 100 feet square. The charge for an individual miner's certificate (this is equivalent to a permit to mine) is $5; for a stock company of $100,000 capital or less, $50; for other companies, $100; for a Crown grant, $5; for recording a certificate or a claim, or an affidavit, or generally speaking, any paper, the fee is $2.50; and when of unusual length, 30 cents per folio additional. These charges are not unreasonable. They will scarcely more than cover the expense of maintaining that supervision and regulation of the mining lands which the interests of the government and the miners alike demand. No tax appears to be placed upon the product, but of course the government reserves the right to impose such a tax in future, should it deem fit. These regulations, which are confined to British Columbia and do not extend to the Northwestern Territory, can be consulted at greater length in the United States Consular Reports for 1901, No. 247.

CHAPTER XXXIII.

THE MYSORE AND TRANSVAAL MINES.

Practical enslavement of the Mysore miners—"Contract labourers"—Their wretched condition—Their attempts to escape—Wages—Discovery of the mines due to wealth of the neighbouring temples—Annual produce since 1884—Mine tenures—Royalties—The Transvaal mines—The war—Disorganization of the native labourers—Closure of the mines—Royalties and exactions imposed by the Boers—New impost proposed to the British government—Rhodesia—Madagascar.

THE practical enslavement of the Indian ryot and the literal enslavement of the gold miners of Malabar, which is a part of Mysore, are alluded to in other parts of this work. (See Index.) To the evidences there presented it may be added that in 1831 Lieut. Nicolson of the British army reported that at Nilambar, in Mysore, "the mines were worked by Korumba slaves, who were subjected to horrible cruelties if the gold they found was deficient in quantity." (Lock, 336.) In 1865 Mr. Brough Smyth, a British (Australian) engineer, reported that the industry of the Korumbas had so covered many parts of Mysore with tailings that they resembled abandoned Australian washings.[1] Finally, Lock (340) said that the gold washers employed on the alluvions of Mysore only earnt *four annas* (6d.) a day, which may be taken as an approximate measure of the wages paid in the Colar mines of Mysore presently to be mentioned.

What is here intended to be brought into view is the system of employing contract labourers, Indians, Chinese, Malays, Kanakas and Africans, which prevails to-day in nearly all the British dependencies, from Mysore to Honduras and from Borneo to Johannesburg. A recent writer in a London periodical, describing a British tobacco estate in Borneo, represents the manager as being constantly employed in devising means to prevent the escape of the "coolies," who seize every opportunity to regain their freedom, each escape "meaning a loss of about £8 to the estate." The proprietors of the Colar gold

[1] Gold in Sanscrit is *suvarna* and *hemma;* in Kanada it is *honna* and *chinna;* in Tamil it is *pon*. The names of rivers, hills, towns, villages and districts in many parts of Mysore, as well as the rest of India, proclaim its auriferous character.

mines take the greatest pains to prevent this system of slavery, for such in effect it is, from being known in London, for fear of arousing public opinion against its continuance; but enough of it has transpired to justify the conclusion that these properties are worked by natives, who are practically bought from headmen for a premium and forcibly condemned to work in the mines for a pittance which is scarcely sufficient to keep them alive and cover their nakedness. Such is their wretched condition that the year before last (1898) the output of the mines was diminished to the extent of 6,000 ounces, because owing to their poverty and inadequate fare the bubonic plague broke out among them and rendered many of them unable to work.

According to the London Mining Journal the present cost per foot of driving a gallery in the Wynaad mines of Mysore—this means blasting out and removing to the surface $7 \times 4=28$ cubic feet of rock—with "native and Eurasian labour," is only 4s. or $1. It is against this system of peonage that the free labourers of California, Australia and British Columbia have to contend in producing gold at £4 4s. 11½ or $20.67 per ounce fine for the London, Philadelphia, and San Francisco Mints. These circumstances are mentioned not by way of complaint, or invidious comparison, but merely to exhibit the extremely diverse conditions under which gold is produced in various parts of the world at the present time.

Mysore at the period of the British Conquest consisted of the whole of India south of the river Chrishna. At the present time the term Mysore (often used synonymously with Madras) is attached to a somewhat more limited area; but as we shall mainly have to deal with the mines of the Colar or Kolar district it will not be necessary to enter further into topographical details. These mines—already alluded to in another part of this work—have this peculiarity: they are not recent discoveries, but very ancient mines reopened in recent years; and reopened not because of any new developments, but "rather in consequence of the extensive native workings that abounded and the evidences of the existence of the precious metal that were to be found, particularly in temples." (Mint Report, 1899, page 276.) The plain fact was this: that the neighbouring temples ultimately got the bulk of the gold which the ancient miners had produced; and it was the presence of this gold in the temples that put the modern prospectors upon the trail of its source. In common with other Indian mines, they appear to have been worked superficially by the Hindus, and in after ages carried down several hundred feet by the Moslems. Ancient tools and utensils have been found in some of them: in the

Ponaar mines of Mysore a number of human skeletons were discovered, implying that the miners had perished in them, probably by a "cave." The reopening occurred about 1870, and although previous to 1884 some small amounts of gold were produced, that year is the first one which marks the production of a sufficient sum of the precious metals to merit historical attention. The following Table shows the produce from 1884 to the present time:

PRODUCTION OF GOLD IN BRITISH INDIA.

The following returns practieally include only the quartz mines in the Colar district of the Mysore, these furnishing 99¾ per cent. of all. The ounces are of British standard gold, eleven-twelfths fine.

Year.	Ounces.	Value.	Year.	Ounces.	Value.	Year.	Ounces.	Value.
1884	1,166	$ 22,095	1890	108,855	$2,062,802	1896	329,400	$6,130,547
85	6,312	119,612	91	132,420	2,510,159	97	389,779	7,247,241
86	16,452	311,765	92	163,985	3,041,818	98	417,124	7,781,524
87	15,652	296,605	93	106,900	3,814,914	99	447,971	8,517,458
88	35,219	667,400	94	211,770	3,881,319	1900	500,000	9,475,000
1889	80,000	1,516,000	1895	251,978	4,656,243	1901	—	—

In round figures the produce of the Mysore mines since 1884 has been about $70,000,000, and according to the Reports issued to the public by the owners of these properties, there is a strong probability that the present rate of production, which is about $10,000,000 a year, will continue for several years to come. The mines are leased by the Indian government to the proprietors upon "liberal" terms. "It is not proposed to levy any royalty or other tax on the industry for the present (1882), because it is deemed most important to attract capital to the gold-fields." The government, however, reserves the right to impose a royalty whenever it deems fit.

We now turn to the mines of South Africa.

The auriferous produce of the Transvaal is fully set forth in a previous part of this work. It reached its achme with 3,831,075 ounces in 1898, fell to 3,500,000 ounces in 1899, and to (probably about) 625,000 ounces in 1900. The last figure is, however, mere conjecture. The war between Great Britain and the South African republics has liberated the negros, the mines are idle and there are no returns of production. If, as it has been contended by the mine-owners, (shareholders) and their apologists, these mines were not worked by forced labour, why is it that they are idle? The Boers left the mines, machinery and structures uninjured;[2] the properties have been in the

[2] Upon examining the mines after the British were driven out of Johannesburg, the Transvaal government found that secret excavations and workings had been made which were not recorded in the plans of the mines, and that many of them were in

undisturbed possession of the British shareholders for a whole year; yet there is substantially no product. There is but one reason for this: the negros have fled to the Veldt, beyond the reach of the avarice, cruelty and hypocrisy which had chained them to the banket lodes of Witwatersrand.

When there are millions of profits on the one side and nothing except mercy or humanity on the other, there will be no end of denials from the beneficiaries of slavery. Therefore, a word or two of additional evidence on this subject may be necessary to justify the attitude of the present work.

According to the U. S. Consular Report, No. 234, the Transvaal mines in 1898 produced gold to the value of $78,361,000, of which amount $28,858,000 were divided as profit and $49,503,000 devoted to expenditures; the mines employing 92,806 hands. In this Consular report the value of the produce is overstated, while the number of hands employed is understated. The real value of the produce appears to have been about 61 millions, while the number of hands employed was about 120,000. However, to save dispute, the Consular figures will be accepted in this place without objection. Of the amount devoted to expenditures fully two-thirds, say 33 millions, must have been expended upon machinery, supplies, dynamite, mercury, cyanide, fuel for the mills, taxes or royalties,[3] legal and other fees, office-expenses, and the travelling expenses of officials, engineers, surveyors, etc. This would leave about 16 millions for labour. Included under this head were several £10,000 a year salaries to engineers or managers and numerous £1,000 a year salaries to minor officials, beside still more numerous salaries of £300 a year to overseers, millmen, cyaniders, shaftmen, mechanics, etc., all of whom were Europeans (whitemen) of exceptional ability, experience and courage. The salaries of the white staff of the Witwatersrand mines may be roughly estimated at about one-third of the fund devoted to labour. This would leave about 10 or 11 million dollars for the 92,000 negro miners; an average of about $120 a year for each negro, say 33 cents a day: about enough to pay for, say, two drinks of the execrable gin with

bad order and needed repairs. The Boers then repaired and worked them. "The government mines (the ten mines chosen to be worked by the Transvaal authorities), are undamaged and they can resume work at any time. As regards the mines not worked in the interests of the (Transvaal) government no damage has been done to them." J. Klimke, ex-State Mining Engineer of the Transvaal, in the London "South African Review."

[3] The royalty imposed by the Boer government was 5 per cent. Sir David Balfour now proposes to substitute a tax of 10 per cent. on profits and to abolish the dynamite monopoly. London Corr. New York Times, June 16, 1901.

which they are supplied.[4] If this is not slavery; if the fact that the subjects of this cunning system were bought from the head men of their tribes at so much each and forcibly "emigrated" against their will into the subterranean depths of the South African mines, is not slavery, then the term has lost all that significance which it had in the eloquent days of Lord Brougham and the British Anti-Slavery Society.

Returning to the quantitative aspect of the Rand mines, it should be borne in mind that although their produce has hitherto constituted 98 per cent. of the entire auriferous produce of South Africa, this may not continue to be the case. There are other auriferous regions in South Africa, and there are millions of negros yet to be kidnapped, chained to work and paid a conscience-soothing pittance, in order that the sordid but pious shareholders in these enterprises may make peace with their chosen gods.

The mines of Rhodesia first began to produce gold in 1899, when the returns were about 65,000 ounces. Madagascar, under the French, has produced for many years, and still produces, about 3,000 ounces of gold per annum. Some other items may bring the entire auriferous produce of South Africa, outside of the Rand, to 70,000 ounces per annum. It is needless to say that this gold is all of it produced by the same means, that is to say, negro-slavery, disguised under the euphemism of "contract-labour."[5]

The closer one looks into the details of mining for the precious metals the more inexplicable appears the policy of the United States in demonetising silver. Of this metal she was the greatest producer, and had she insisted upon retaining it for full legal tender coins, she might soon have become the centre of the world's system of finance. As a producer of gold she is severely handicapped. Take California for instance. This State embraces what is beyond all question the greatest gold-mining area in the known world. The Mother Lode stretches nearly the whole length of the State from north to south, while the country below it is covered with placer deposits which contain more proved and accessible gold than is known to exist in all the dominions of Great Britain, including India, Africa and Australasia. The facilities for mining are unrivalled; a permanent and secure government; equal laws; a reasonably fair administration of justice;

[4] When, during the war, the Boers worked the mines, they reduced the negro wages to one pound sterling (five dollars) a month, or less than 17 cents a day, and fined the mine managers ten pounds sterling for every infraction of this rule.

[5] An attempt was recently made to introduce a similar system among the negros employed in some of the cotton plantations of South Carolina, but it was frustrated by the American Courts, who fined the parties and declared the contracts invalid. New York Times, June 15, 1901.

abundance and cheapness of food and other supplies; mining machinery of the best types made close at hand; railways, telegraphs, refineries, mints, etc.; abundant water at high levels; entire absence of royalties or taxes of any kind; and a perfect climate. Yet California only produces 15 millions a year, while British India produces 10 millions, the Transvaal 60 millions, and Australasia 80 millions. Why? Simply because California is handicapped. She has no Indians, no coolies, no Zulus, no contract labourers. Her people are absolutely free; and in the present state of civilization and under present circumstances gold cannot be prospected, produced and acquired by the economical efforts of freemen for $20.67 the ounce, which is the Mint price. This conclusion stands upon so firm a foundation, it is supported by such an overwhelming mass of evidence, that it is useless to dispute it. The minimum wages of common miners in California are $2 per diem; in most camps, $2.50; in some, $3; in Nevada (Comstock Lode), $4. The proprietors of mines which pay these wages cannot compete with contract-labour at 15 to 50 cents a day. They are handicapped, so that only those mines which are most favourably situated in respect of ample deposites of ore (this practically means "low grade"), a ready command of mechanical power and water, together with cheap transportation and other advantages, are attractive enough to invite the miner and capitalist.

CHAPTER XXXIV.

ALASKA AND THE KLONDIKE.

Russian occupation of Alaska—The Fur Company—Their opposition to its being prospected for gold—Darrehan's report—Russia resolves to sell the Territory—Its purchase by the United States—Inroad of prospectors—The Cassiar district on the Stickeen—Establishment of Territorial government in Alaska—Schiefflin's expedition—Three sets of prospecting parties—The great discovery on Bonanza Creek—Rush of miners to the Klondike—Character of the deposites—The tundras—Climate—Cost of provisions—Wages—Summer diggings—Taxes on production—Produce, 100 millions—Nome—Tanana—The outlook for the future.

DURING the Russian occupation of Alaska, the territory was practically given over to the Russian American Fur Company, who governed it with power of life and death. Their interest consisted in keeping their agents and servants busy in killing seals and securing seal skins. No heed was paid to any other resources which the country might possess; and every obstacle was interposed to mine prospectors. During Gov. Baranoff's administration, a Russian prospector found gold in the mountains near Sitka, but he was ordered either to give up prospecting or take a hundred lashes. In 1855, the last year of the Czar Nicolas, a mining engineer, named Darrehan, was commissioned by the government at St. Petersburg to report upon the mineral resources of Alaska. This officer went through Siberia to the sea, crossed to Sitka, and with a corps of men spent two years in "prospecting Alaska." During these two years he was under the influence of a notoriously corrupt and unscrupulous trading company. In 1858 Darrehan returned to St. Petersburg with a written condemnation of Alaska. This condemnation resulted in its sale. The Fur Company succeeded in preserving its monopoly, but at the cost to Russia of a territory which has since proved to be worth what was paid for it many times over. The sale of Alaska to the United States for seven and a-half million dollars was made in 1867, down to which time no survey had been made of the interior. But now came the prospectors. In 1871 a prospector named Doyle discovered gold at Silver Bay, south of Sitka; in 1872 Doyle and Mahoney discovered a quartz ledge on Round Mountain, and here, in the same year, Nicolas Haley fired the first blast in Alaska and took out of it $80 in free gold.

In 1874 the placers of the Cassiar district were opened and a million dollars taken out; in 1875 another million was obtained; in 1876 and 1877 each a-half of a million. The number of men, including Indians, at work in 1874 was 2,000; and in the following year about 1,500; in 1879 there were 1,800 men; by the year 1883 the placers were practically exhausted and only yielded $135,000, although 1,000 men were working in them. (Min. Res., 1883, p. 23.) In 1884 the miners pulled out and started north. Next to the Cassiar district the earliest placers worked in Alaska are said to have been those of Sum Dum Bay, opened in 1876.[1]

In 1879–80, Prof. John Muir, the State Geologist of California, made a casual survey of the Alaskan Coast, finding gold at many places between Sitka and the Arctic Ocean; but in this respect the prospectors had long anticipated him. Indeed, as a rule, it will be found that the professional man only appears in time to confirm what the prospectors have already demonstrated. In 1880 Joseph Juneau, prospecting along the shores of Gastineau Channel discovered Gold Creek, where he found rich placers and quartz ledges. The "city" of Harrisburg, afterwards called Rockwell, eventually took Juneau's name. In 1882 its population was only 30; in 1898 this had grown to 3,000. The town is situated near the gold quartz mines on Douglas Island, the two principal ones of which, in 1898, produced a million dollars from ores that only averaged $2.40 (about 10s.) per ton. These two mines work 1,500 stamps and employ 427 men at $2.00 to $3.50 per day and board. Min. Res., 1898, p. 55.

The merit of discovering the auriferous character of the Alaskan tundras belongs rather to the British than the American prospector. From the moment that Alaska was acquired by the United States, the country was searched for gold by prospectors from California. These men never thought of looking under the moss and "muck." Their pannings were all made in exposed gravel or in the sand-bars of receded rivers. They found gold, but not in such startling quan-

[1] The Cassiar district is in the valley of the Stickeen River, which rises in the British Northwest Territory and flows through American Alaska to the Sound. The gold is produced within British lines, but finds its shipping port or market within American lines, hence it figures in the production both of British Columbia and American Alaska. As in respect of this district, the writer, like the Mint Bureau, has relied on the Consular returns, and as these do not always distinguish the produce by mines, he fears that the produce of the Cassiar district has been wholly or partly duplicated, both in the Mint Reports and herein. However, it is relatively unimportant and comparatively antiquated.

tities as to cause a "rush." Among the early locations where pay-dirt was found in placers were Douglas Island and the Creeks which drain the mainland valleys into the Lynn Canal, or Sound. In 1879, diggings which paid from $5 to $20 a day were opened on Gold Creek. In the same year prospecting parties were fully equipped at Sitka and sent out in all directions to explore the land. By the year 1882 a number of gold mines were discovered near the coast at various points between the Aleutian Peninsula and Behring Strait. The output of this year was estimated by Mr. Valentine, who enjoyed excellent opportunities for forming a correct opinion concerning the produce of Alaska, at so large a sum as $250,000. In 1883 the output was estimated by Capt. James Carroll at $400,000, and by R. D. Willoughby at $450,000. Min. Res., 1883, p. 20.

In 1884 an Act of Congress provided a Territorial government for Alaska, but as the officials from Washington did not arrive out until after the close of the mining season, the Act did not have any influence upon mining until the following year. The estimated produce for 1884 was only $200,000. Down to this time the only title to property in the Territory was force. So long as men were thus insecure in their possessions there was little incentive to prospect or locate, and none to improve. "Take what you can find; take it anyhow and at once, or never," was the only practical rule for the prospector. But from the organization of the American Territorial Government, the establishment of the Civil Law and the employment of officials and policemen to enforce it, Alaska dates a new birth. No sooner were these great events consumated, no sooner did it become safe for the prospector to disclose by recording his "find," than the riches of the Yukon Valley began to be talked about and advertised. Mining news travels far and fast. In a few months' time it reached Arizona, where lived Ned Schiefflin, popularly known as "J. C.," who had sold his interest in the Toughnut Mine of Tombstone City, for a million dollars. Schiefflin hurried to San Francisco, bought a steam yacht, filled it with supplies and men and started at once for the Yukon. The poorer prospectors, unable to purchase steam vessels, or employ miners, shouldered their kits and painfully struggled through the Chilkat and Chilcoot Passes. Schiefflin, although an experienced prospector and a brave man, lost eighteen months time and found little beyond some quartz ledges and small placers in the Ramparts, the former being of no present value. Leaving a dozen or more men to work the placers, Schiefflin sold his expensive outfit and returned to Arizona. The poor miners from Eastern Alaska, after enduring many hardships, at last reached the Upper Yukon, where they became the pioneers of the land and were among the first to share the rich discoveries which were made a few years later.

Meanwhile, the route by the Chilkat Pass, which had been opened by prospectors in 1879, was surveyed to Copper River by General Miles in 1882, and beyond Copper River in 1883. The visible emblems of authority exhibited by the soldiers employed in these surveys did much to assure the adventurers in this lonely and remote country that their interests would be protected by a firm government.

By the year 1895 three sets of prospecting parties were converging upon the spot where the great discoveries were subsequently made. Was it professional instinct, or Indian talk, or the disclosure of mining secrets by the prospectors, or mere chance, that guided the footsteps of these sleuth-hounds? First, there were a thousand or more men from the Cassiar district, slowly making their way along the coast, and incidently picking off the croppings from the mountain ranges of the Northwest Territory. As for testing the tundras, this was too arduous and expensive a task for isolated and ill-provisioned miners. Prospecting a tundra means the preliminary loss of two seasons and an amount of work that involves association and some capital. Second, there were the prospectors of the Chilkat and Chilcoot Passes, who were creeping northward along the 141st degree of west longitude, now on the British side, now on the American, looking for gold in the river valleys and never suspecting that all the while it laid beneath them, concealed by the moss and the muck of the tundras. Third, to say nothing of other parties, there were the Schiefflin adventurers in the placers of the Lower Yukon, who had preferred to seek their fortunes in Alaska rather than return with their leader to Arizona.

In 1895, a number of men from these various sources settled upon diggings which paid them a bare living, at places not far from the future centre of auriferous activity. There were camps at Forty-Mile, Sixty-Mile, Circle City, and other places in the Yukon Valley. At Sixty-Mile, Joseph Ladue had a saw-mill, which was employed in cutting boards for miners' sluice-boxes and cabins, for which boards he got $150 per thousand feet. His mill was a centre of supplies not only for boards, but also for news. When a miner needed credit for boards he was obliged to disclose his resources; and these seldom consisted of more than his discoveries. In the summer of 1896, a miner who had recorded a claim on the Klondike with the British Mining Commissioner, applied to Ladue for boards. The prudent miller personally inspected the man's claim before he granted the credit. Then he

GEN. NELSON A. MILES
(To face p. 438)

recorded a claim for himself on the Klondike, and at once removed his little mill from Sixty-Mile to a spot which he selected at the confluence of the Klondike and the Yukon, fifty-five miles east of the boundary line. That spot, located September 1st, 1896, is now called Dawson City.

Pay-dirt on Bonanza Creek, an affluent of the Klondike, was discovered by Robert Henderson, an old prospector, one of the much abused "has-beens," during the early summer of 1896. Shortly afterwards, rich gravel was also found on the same Creek by George W. Cormack, to whom Henderson had imparted his find. The result of these discoveries was the filing by Cormack, on August 19th, 1896, of Discovery Claim No. 1, on Bonanza Creek. Then the news flew. By September 1st, 200 prospectors were on this Creek and its tributaries, trying the ground and staking claims; while 500 or 600 others were in the tents of Dawson City, bent upon prospecting for themselves. On November 3rd, a big strike was made on Bonanza Creek, $25 being taken from seven pans (about 150 pounds) of gravel. This was soon afterwards eclipsed by a strike on El Dorado Creek, where $18 to the pan were taken. The phenomenal strike was in Claim No. 30, on El Dorado Creek, which yielded in some places $800 to $1,000 per pan, and averaged $70 to the square foot. The current wage for miners (anybody) was $1.50 per hour, or $15 per day. Yet even at this rate but few men were to be had, each man preferring to prospect for himself: the claims commanding a premium of from $1,000 to $5,000 and some of them very much more. In January, 1897, "lays" (half of the produce, the proprietor paying all the expenses of the operation, such as tools, fuel, timber, etc.,) were accepted by the miners on Bonanza and El Dorado Creeks, in lieu of wages. During the winter, sinking, drifting and hoisting the gravel was carried on vigorously. On May 7th, 1897, the long expected thaw commenced and water became accessible; when the gravel dumps were shovelled into the sluice-boxes and washing was begun. In four months' time, the extreme limit of the season, 600 men took out three million dollars in gold; equal to $5,000 per man, most of them being proprietors and "lay" men. Not over 100 men worked on wages. Claims were now selling at $5,000 to $30,000. By October 1st, 1897, claims to the number of 1,266 were located on 18 different creeks, and the miners looked forward with assurance to a very profitable season in 1898.

The ground in the Klondike district is covered with about a foot of moss, in some places a foot and a-half, which of course has to be removed before operations can be commenced. Beneath the moss are from two to twenty feet of muck, peat and worthless mould. This also has to be removed and disposed of, away from the ground intended to be mined. This last is the gold-bearing gravel. The ground is frozen solid to a depth of about eight feet. To penetrate this frozen ground, with pick and shovel, it has first to be thawed, by building great fires upon it. The heat thaws the earth to the depth of about two feet, when the fires are extinguished, the dirt is removed, and the operation repeated until the gravel is penetrated or bed-rock reached. Meanwhile, the excavations have to be supported by timbers. When the stratification ceases to be frozen, drifting can be carried on and the gravel stoped more rapidly. As fast as the gravel is stoped, it is hoisted to the surface and thrown on dumps to await the spring thaw, when it is washed in pans or sluice-boxes, usually the latter, and the precious metal is extracted.

The shortness, heat and glare of the summer season and the extravagant prices of provisions, transportation, etc., render the toil of the miner exhausting and unprofitable. It takes all that a man can earn during the mining season to support him through the year. The summer often lasts but seventy days, and seldom more than ninety, after which time the water freezes, gold-washing ceases and the production of gold is arrested until the next year. From December 1st to February 1st the daylight only lasts about six hours per diem, while the temperature falls to 40, 50, and even to 77 degrees below zero. During the summer there is seldom any rain, the sun is in sight twenty hours out of the twenty-four, and the temperature in June and July often rises to 125 degrees Fahrenheit; yet the great heat, which exhausts and sometimes prostrates the miners, only thaws the soil to the extent of three or four feet, in some places only one foot.

In spite of these difficulties and hardships the proprietors, on September 23rd, 1897, endeavoured to reduce wages from $1.50 to $1 per hour; but the men succeeded in defeating the movement. The claim of the proprietors was that at the higher rate the produce would not pay; and this in many cases was true; but until miners became more plentiful, the proprietors either had to pay the rate demanded, or let their properties lie idle and lose the chance of a big strike, or else of selling their claims to outsiders, at prices, which for developed claims of good promise, now ranged from $10,000 to $50,000 per claim. In December, 1897, upwards of 3,000 men were in the Klondike mines, while their output, as events proved in 1898, only amounted to about ten million dollars, an average of about $3,333 per man.

In the deep diggings, (the tundras), the bed-rock, close to which the best gold lies, can only be reached in winter, because below the frozen soil the water from the rivers freely percolates the gravelly detritus which contains the gold. Perhaps the introduction of Poetsch's process may remedy this difficulty. Meanwhile, many miners unwilling to risk the severe winters of Alaska, or else to escape the high prices to which provisions rise during this season, limit themselves to "summer diggings," in other words, to ordinary placer mines, of which great numbers in small patches are to be found in the river valleys.

In American Alaska there are no taxes upon mining; no licenses or permits are required, and no fees are exacted, except the trifling one of $2.50 (10s.) for recording a location. In British Alaska, beside the fee for license $15, continuances $15 a year, and high recording fees, the government exacts a tax or royalty upon production of ten per cent.; and if the produce is over $500 per week, 20 per cent., ad valorem on the surplus. A possible modification of this royalty is mentioned further on. The amount of royalty paid in 1898 was $700,000, which bespeaks a production of less than seven millions; but as the tax was largely evaded, the produce has been estimated at twelve millions. In fact, during that year more than eleven millions from Alaska are known to have reached the Mints and Assay offices of the United States, beside what was carried by the miners to other places. Mint Reports, 1896 and 1898: Consular Report, No. 243.

GOLD PRODUCE OF THE KLONDIKE AND ALASKA.

Table showing the estimated produce of gold in the Klondike region of the British Northwest Territory and in the United States Territory of Alaska, respectively, since 1894. Sums in American gold dollars.

Year.	Klondike.	Amer. Alaska.	Year.	Klondike.	Amer. Alaska.
1894	—	$1,000,000	1898	$12,000,000	$ 2,500,000
1895	—	1,500,000	1899	16,000,000	6,000,000
1896	$ 500,000	2,000,000	1900	18,000,000	8,000,000
1897	2,500,000	2,500,000	1901	20,000,000	10,000,000

REMARKS.—The Klondike estimate for 1901, from the U. S. Consular Report No. 243, is evidently based on the assumption that that district will produce more gold this year than it did in 1900. Some estimates goes as high as $25,000,000. This is opposed to the belief of several experienced prospectors, who are confident that the Klondike has reached its achme. Nevertheless, the estimate has been allowed to stand; first, because it rests on respectable authority; and next, because even if mistaken with respect to the Klondike, the estimate is likely to be made good from the produce of other districts in the British Northwest Territory, some of which promise to yield large returns.

The returns from American Alaska embrace the quartz mines on Douglas Island and vicinity. The estimate for 1899 includes three millions for the Nome district (popularly credited with nine millions for that year). The estimate for 1900, in Consular Report No. 240, is three millions. To this amount five millions have been added for Nome and other districts, of which the Consul (at Victoria) does not appear to have heard. For 1901, the estimate includes the expected returns from Nome, Tanana Valley, and other new districts.

From this Table it will be observed that Alaska has already produced over one hundred million dollars in gold; and as only a beginning has been made in prospecting its vast area, it is, perhaps, not altogether too hazardous to predict that it may yet become the principal producer of gold among all the American States or Territories.

In his report, dated May 24, 1900, the American Consul at Dawson City, thus apportions the output for that year: El Dorado Creek, $3,746,200; Bonanza, $3,216,490; Dominion, including tributaries and hillsides, $2,352,010; Sulphur, $1,456,720; Hunker, $1,211,100; Gold Run, $1,037,050; Gold Hill, $749,100; Cheechuco Hill, $712,-300; Fox Gulch and Oro Fino, $702,000; fourteen other (mentioned) principal, besides numberless smaller creeks and benches (not named), $1,815,030; total, $18,000,000. These mines employ 5,280 men at $1 per hour. The Consul represents the population of the Yukon district (the Klondike) as having diminished since the previous year by two thousand. Hundreds of prospectors have left and others are leaving for the new diggings in the Koyukuk and other districts of American Alaska, where there are no royalties or taxes to pay. In a later report he says that the miners who worked for "laymen" have not gained the ordinary wage. When the year's wash-up took place it turned out that the cost of working the claims was more than half of the output, and, as under later contracts, the layman undertook to defray part of the cost of working, this circumstance curtailed his proportion and diminished the fund out of which his workmen were to be paid. As the latter had agreed beforehand not to hold the claims for any deficit in wages, they had no remedy for their grievance. Many of them have started for the new diggings at Tanana in American Alaska, where there is 20 cent dirt (20 cents to the pan of gravel) with only 2½ feet to bed-rock, which is far more profitable than $1 dirt with 25 feet to bedrock. The Tanana district is of immense extent and will probably entice away a large proportion of the miners from the Klondike. At Dawson City the prices of provisions are falling. Potatoes formerly at $1 a pound are now down to 20 cents, while beef and mutton have fallen to 40 cents. On the other hand, wood for fuel had gone up to $25 a cord. The land in Dawson City has been assessed by the Town Council for taxable purposes at a million dollars; the improvements at a million and a-half; and "the volume of business," by which is probably meant the sales of merchandise (and not including land, mining claims, or bullion), at nine millions. On September 17th the same official reported that all deep claims (tundras) were now worked with boiler, engine and pump; that wages had fallen to $4 a day; that Dawson City was still growing in importance and wealth; that of the output for 1900 about 12 to 13 millions had already been shipped; that from three to four million dollars of British capital had been invested in the mines; that the license and record fees, and especially the royalty of "10 per cent. on the gross output of any claim (yielding) over $5,000 per year," had caused great dissatisfaction;[2] that the forests of Alaska were being rapidly destroyed by the miners, who expressed no regrets on the subject; and that a new auriferous district had been discovered on the Chandelar River, which flows into the Yukon below the Fort of that name. Every thought was on to-day; none on to-morrow. Sauve qui peut!

The fall of wages to $4 a day can only mean that the summer season being practically over and the general clean-up nearly completed, there remained in and about Dawson City a large number of unemployed men, who were willing to accept what was barely sufficient to live upon, until the following month of May. It may safely be held that no gold has yet been produced in Alaska or the Klondike at a lower wage than $10 and much at $15 to $20 a day. In some instances $40 dollars a day were paid.[3] Among the newer districts of Alaska is Cape Nome, with headquarters at Nome, originally Anvil City, on Anvil Creek, whose auriferous character was discovered in 1898. As the district is readily accessible from the sea the discovery occasioned a rush, which was accelerated by the further discovery that the beach itself at Cape Nome was auriferous. In 1899 five thousand men were at Cape Nome ready to undergo the hardships and privations of a prospector's life in the Arctic Circle. The diggings on Anvil Creek and Nakkilla Gulch which runs into it, also on Snow Creek and other places, yielded over three million dollars. The beach diggings, though

[2] This appears to be a modification of the royalty previously mentioned in the text; and the "official" reports, from which both the statements are quoted, are not always correct and sometimes need very careful editing. The first statement, (Mint Report, 1896, p. 285,) taxes all mines yielding up to $500 a week, 10 per cent. on the produce, and over that, 20 per cent. on the surplus. The second statement (Consular Report No 243) taxes all mines yielding over $5,000 a year, 10 per cent., but says nothing about a 20 per cent. tax. These statements may be only two different ways of saying the same thing, namely, that the mines are taxed 10 per cent. on an output of $500 a week, and 20 per cent. on the surplus output beyond $500 a week.

[3] M. de Foville, quoting the Department of Labour at Washington, says that between July 15th, 1897, and July 15th, 1898, 40,000 men reached the Klondike, while 20,000 others started, but failed to arrive. He estimates the cost of their outfit at $500 each, total, 30 millions, and 15 millions more spent in "the creation of shipping and commercial establishments," and concludes that "for every dollar extracted from the sands of the extreme North more than a dollar has been expended." Mint Report, 1898, p. 193. This is doubtless true; yet gold seeking "opens" a new country and distributes the industrial population. This is worth something.

they yielded some gold, were soon abandoned for the more alluring locations on the Creeks. In 1900 the rush increased to 20,000 men; and upwards of 30,000 claims were staked. The scene of the principal finds was on Daniels Creek; also in the Blue Stone district at Port Clarence, the outlet to Grantley Harbour; in the Kougarok district, 60 miles north of Nome; and on Harris, Quartz, and Garfield Creeks. The Kougarok River is an affluent of the Kooseticam which flows into a salt lake, 25 or 30 miles from Grantley Harbour. Here there were 1,500 miners. Topkok, 50 miles from Nome, was a discovery of this year. So was Gold Run, a tributary of Blue Stone River, which enters Grantley Harbour. A thousand men were preparing to dig here when the season of 1901 opened. In Behring City and Teller City, which are eight miles apart, town lots were selling for $1,500. The price of boards at Nome was $250 per thousand feet. "Coal at $75 a ton, and each man must be his own coal cart; shingles, $10 a bunch; flour, $5 a sack; milk, fifty cents a can in case lots; sugar, thirty-five cents a pound; rice, twenty-five cents; butter, seventy-five cents; bacon, thirty-five cents; potatoes, $20 per cwt.; eggs, $2 a doz.; coal oil, $1 a gallon; syrup, $2.50 a gallon, and Klondike strawberries (beans) eight cents a pound. The fare from Nome to San Francisco or Seattle is $100 dollars; so that it is cheaper to go away for the winter than to stay. There is the usual boom in real estate, the choicest properties on Front street having changed hands several times. The last sale was the El Dorado Building, a frame 50 by 75 feet, on a lot 50 by 150, the consideration being $22,000. The sum of $90,000 cash was paid for four mining claims in different localities. The ruling price of labour is one dollar an hour. Longshoremen get twenty dollars a day and overtime, but it is worth fifty dollars, as they work in the water, and, in temperature, the water is distinctly Alaskan."

The jewelers at Nome were doing a thriving trade buying gold-dust and nuggets at $16 an ounce, and selling diamonds at four times their cost at San Francisco. The few women who venture into this region are described as being "loaded" with brilliants. The bench-claims above Anvil Creek yielded fifty cents to the pan; some yielded a dollar. A bench-claim at the head of Nakkilla Gulch, "upon which four men were working with two rockers, using water hauled by horses from Anvil Creek, produced in three days 690 ounces of gold, worth over $13,000. On the third day the men rocked out $5,400. On Garfield Creek some of the gravel went $40 to the pan. A claim purchased for $100 yielded in a few days $4,000. Others turned out $4,000, $6,000 and even $12,000. Some claims near Council City have turned out $25,000. A woman who conducted an eating-house at Checkers Town, at the mouth of the Kougarok, was covered with diamonds purchased at Nome." All these camps were crowded with gamblers, a sure sign of their productiveness. The first pan from Gold Run yielded $10. Pans of $2 and $5 were common. Some yielded $16 to $26. Claims were selling for $10,000. Men with rockers on their backs were labouriously dragging themselves over the moss-covered tundras (these have not been touched as yet) to find easy diggings on the Creeks. The output of the Nome district, which embraces all these diggings, is given at five million dollars for the year 1900. The men are described as "unwashed, unshaven, hair uncut, covered with vermin, clothes dirty and ragged, toes out of shoes, soles worn through," their eyes rolling about in every direction searching for gold and regardless of all else. Some have died of privation or disappointment, others have succumbed to the pistol or dirk in contests over "jumped" claims, all are crazed with the auri sacra fames. Tappen Adney, in Collier's Weekly, Jan. 5, 1901, and other press correspondents.

The mineral laws of the United States are stigmatised by Mr. Adney as "outrageous," because they enable a man to take up several claims and to tie them up for a year, without doing anything to develop them; yet preventing others from doing so. The jumping of claims is common and this often ends in bloodshed. As yet the Territorial government is weak in executive officers, but this will soon right itself, as it has done in the other mining districts of Western America; and lawless acts will be repressed. There is a large reserve fund of common sense among the American people and this promptly makes itself felt in new communities. The future of Alaska is full of promise.

CHAPTER XXXV.

PRODUCTION, CONSUMPTION AND STOCKS OF METAL.

Production of the precious metals since the Discovery of America—Dollars, tons and car-loads—Table—Proportions obtained through conquest, slavery and free-mining—These circumstances destroy the theory that the value of these metals is due to the cost of their production—Admission by Cobden and others that it is due to quantity, or demand-and supply—Production compared with consumption—Large proportion consumed in the arts, or lost, worn-out, or destroyed—Misleading statistics of the Mint Bureau—Their pernicious influence—Exaggerated stocks of India, China and Siam—Taxation of the product affords another proof that value is not due to cost of production—World's Stock on Hand—Difficulty of forming correct estimates—Increase of paper money—Cheque system—The coinages afford no reliable guide—Neither do the import and export accounts—Small amount of gold circulating in America—Table of Stock on Hand in various countries in 1879 and 1899—Bank reserves—Proportions of coins, notes and cheques employed in the United States—Their respective ratios of activity—Time, an element of price.

SINCE the discovery of America there has been brought into the commercial world 9,466 million dollars worth of gold, and 10,189 million dollars worth of silver; valuing the latter at the long-time coinage ratio of 16 for 1. Perhaps the immensity of these sums might be better realized by the generality if they were reduced to tons and carloads. Roughly speaking, the gold would weigh 19,000 tons and the silver 340,000 tons. It would require 35,800 freight cars, each laden with ten tons, to transport the lot. If these cars were each twenty-five feet long, they would extend in an unbroken line to a distance of eight hundred and ninety-five thousand feet, or very nearly one hundred and seventy miles. The following table shows the steps of this production from first to last. It is divided into irregular periods coinciding with the years in which comprehensive estimates of the Stock on Hand in the commercial world were made by reliable and painstaking authorities; a subject that will be treated further on.[1]

[1] The reader who may desire to trace the annual or decennial output of the precious metals since the Discovery of America will fine it tabulated at length in the first edition of this work, Chapter XXII.

PRODUCTION OF GOLD AND SILVER SINCE THE DISCOVERY OF AMERICA.

Sums in Millions of American Dollars.

Period.	Years.	Gold.	Silver.	Total.	Cumulative.
1492 to 1545	54	92	33	125	125
1546 to 1600	55	150	630	780	905
1601 to 1650	50	138	612	750	1,655
Plunder of Asia	(100)	172	343	515	2,170
1651 to 1675	25	70	305	375	2,545
1676 to 1700	25	117	298	415	2,960
1701 to 1725	25	315	270	585	3,545
1726 to 1750	25	575	350	925	4,470
1751 to 1776	26	270	530	800	5,270
Plunder of Asia	(100)	—	—	?	?
1777 to 1808	32	320	980	1,300	6,570
1809 to 1828	20	172	463	635	7,205
1829 to 1838	10	125	220	345	7,550
1839 to 1850	12	452	373	825	8.375
1851 to 1860	10	1,420	405	1,825	10,200
1861 to 1870	10	1,135	490	1,625	11,825
1871 to 1876	6	582	413	995	12,820
1877 to 1883	7	748	697	1,445	14,265
1884 to 1893	10	1,210	1,515	2,725	16,990
1894 to 1896	3	582	633	1,215	18,205
1897 to 1899	3	820	630	1,450	19,655
Totals.	508	9,466	10,189	19,655	

The Table is to read as follows, taking the last line as an example: During the three years, 1897 to 1899 inclusive, there were produced throughout the world 820 million dollars worth of gold and (at coining value) 630 million dollars worth of silver, together 1,450 millions, which, when added to the previous output, makes 19,655 millions produced altogether.

We have next to deal with the Consumption and Residue of these metals, or the Stock on Hand in the world employed as money, after the Consumption is deducted from the Output.

Out of the 6,500 and odd millions acquired by the European world down to the period of the Spanish-American Revolutions of 1810, less than 500 millions were obtained through commerce; the remaining 6,000 millions were the fruits of conquest, plunder and slavery.[2] The cost of their production cannot be estimated in dollars, but in human life, in blood, in tears, and in the agony of immolated and expiring races. Whether any considerable portion of the coins minted before 1810 are still in circulation, or not, has no bearing upon the discussion. Even were they all lost or destroyed, it would remain true that their value affected the value of every ounce of metal which has been

[2] Danson says 5,850 millions; Humboldt, 5,986; Del Mar, 6,192; Tooke, 6,700; Jacob, 6,803; Soetbeer, 7,834 millions. Danson's and Humboldt's estimates are confined to America; the other estimates include the acquisitions from conquest and commerce.

produced since 1810. It follows that the present value of gold, that is to say, its value in commodities, is derived in part from its value during the lengthy period when it was all acquired through conquest and slavery. This result is deducible from the principle of Quantity, or supply and demand, and the practice of free coinage.

Between 1810 and the present time there have been produced from the mines, chiefly those of America, Russia, Australasia, British India and South Africa, about 13,000 million dollars, that is to say, in a single century twice as much as in the three centuries previously. Of this amount, about 4,500 millions were obtained by means of slave, serf, peon, or "contract" labour; while the remainder was mainly the product of free labour, chiefly in North America and Australasia, a small proportion having been derived from commerce with Asia and Africa. The product of Asia, outside of British India (the Mysore mines) is a negligible quantity, because it has nearly all been consumed in Asia, and in addition thereto, 4,000 millions of western metal.

Taking the two periods together, the general results are as follows: from the Discovery of America to the present time the European world has acquired 19,500 and odd millions, of which 1,000 millions were obtained by conquest, 9,500 millions by slavery, and 9,000 millions chiefly by free mining labour.

It is submitted that these circumstances entirely controvert and destroy the theory, so often proclaimed by writers but imperfectly acquainted with mining and the conditions under which the bulk of the precious metals have been produced. That theory is that the value of gold is due to the cost of its production. Who will undertake to compute what it cost in money to plunder the 1,000 millions of gold and silver acquired from Asia by conquest? How is it possible to compute in money what it cost to procure 9,500 millions by slavery? Are the lives of the twenty millions of aborigines, whom the Spaniards and Portuguese destroyed in America, by condemning them to mine-slavery, reducible to dollars and cents? Is there a price of blood, anguish, and despair? To-day the mines sf Nertschinsk and others in Russia are worked by convicts and serfs; the mines of Mysore are worked by native Indians practically reduced to a condition of slavery; the mines of Witwatersrand, in South Africa, so long as they were worked at all, were worked by 100,000 negros, involuntary labourers, contract-labourers, naked African labourers, bought from their chiefs at so much per head and thrust into the subterranean caverns of Johannesburg to win gold for distant shareholders in London and Paris. Who will pretend to assert that the cost of committing these

Crimes against Mankind is reflected in the purchasing power of gold? We know well that it is not; that the value of gold, as of all other things, is not the result of cost, but of supply and demand, wholly irrespective of cost. The very same class of men who uphold the delusive theory that value is due to cost of production, repudiate it, whenever it fails to support their interests. For example, when the Californian and Australian mines threatened Europe with a great and sudden increment of gold, a representative of this class, to wit, Richard Cobden, wrote in the Preface to his translation of Chevalier's work: "There *must* be a fall in the value of gold, in consequence of its greatly increased quantity." Chevalier himself expressed the same opinion in almost the same words. Here it is QUANTITY, or demand and supply, that affects value; not cost of production; and this view of the subject was so quickly and widely recognized to be true, that Holland and Belgium demonetised gold, France was on the point of following suit, and England gravely listened to Maclaren and other doctrinaires and alarmists, who, as a remedy for the expected fall of gold, advised a return to corn-rents and barter! A similar renunciation of the theory that value is due to cost of production, took place a quarter of a century later, when it was pretended that the Comstock Lode threatened to "deluge" the world with silver. The deluge meant *quantity* and no matter what it cost to produce, it was the threatened quantity that worked the demonetisation of silver and not the cost of its production, which was and still remains unknowable.

PRODUCTION, CONSUMPTION, AND LOSS OF THE PRECIOUS METALS.

Table showing the cumulative Supplies of the Precious Metals to the European world, the Stock on Hand,[1] *the cumulative Consumption and the Proportion Consumed in the Arts or by wear, tear and loss, or exported to Asia;*[2] *compared with the Supplies, at various periods. Sums in millions of American dollars.*

Date.	Cumulative Supplies to date.	Stock of Prec. Met. at date.	Cumulative Consumption and Export to Asia.	Proportion per cent. of Consumption to Supplies.
1675	2,545	1,250	1,295	50
1700	2,960	1,485	1,475	50
1776	5,270	1,375	3,895	74
1808	6,570	1,900	4,695	71
1828	7,205	1,565	5,640	78
1838	7,550	1,350	6,200	82
1850	8,375	2,000	6,375	76
1860	10,200	2,800	7,400	72
1870	11,825	3,600	8,225	70
1876	12,820	3,700	9,120	71
1883	14,265	[3] 4,000	10,265	72
1893	16,990	3,700	13,290	78
1896	18,205	[4] 4,500	13,705	75
1899	19,655	5,890	13,775	70

NOTES 1 to 4 belonging to the above Table will be found on the next page.

The Table is to be read as follows, taking the last line as an example: Down to the year 1899 there had been produced 19,655 million dollars worth of gold and silver, of which 13,775 millions had been consumed in the arts, or else exported to Asia, that is to say, about 70 per cent., or more than two-thirds of the whole; leaving 5,890 millions on hand in the commercial world in the form of coins and bullion. The whole of the exports to Asia may also justly be regarded as having been consumed in the arts. The gilding of domes, pillars, statues and pictures; the fabrication of solid images, shrines, fonts, and vessels of gold and silver for the temples; the manufacture of bangles, torques, chains, rings and other articles of jewelry; the gilding and silvering of vestments, books, lacquer-ware and crockery; the filling of decayed teeth with gold or silver; besides numerous other arts, consume immense quantities of the precious metals in the Orient. What is not thus consumed is, for safe-keeping, buried in the earth, with a secrecy that commonly defeats all attempts to discover it after the owner is dead. Practically, it is as non-existent as it was before it was mined. The general result is that two-thirds of the entire output of the precious metals is consumed in the arts; leaving but one-third on hand in the form of money.

With regard to the stocks of metallic money, we regret to be again obliged to refer to the defective and misleading statistics of the Mint Bureau; but this is unavoidable. The duty of the Mint Bureau relates to coinage, not to collecting the statistics of foreign moneys, nor the establishment of a school of Political Economy. Private enterprise and individual opinion, however eminent, have but little chance of success in competing with a school that publishes hundreds of thousands of doctrinal works and distributes them broadcast throughout the world, without price. This system discourages both authors and students; for who will pay to learn the truth, when he can obtain a plausible fiction for the asking? The system is as bad as the illiterate Chinese Calendar, which the Celestial government fabricates and forces upon its people, to the exclusion of all scientific

NOTES belonging to Table on page 449.

[1] The estimates of the Stocks on Hand at the earlier dates given, are those of King, Humboldt, Jacob, Tooke, Newmarch, McCulloch, and other reliable authorities. The four last estimates are those of the present writer. [2] Nearly all the precious metals exported to Asia, except those plundered by the European powers, have been used up in the arts in Asia or buried in hoards; and are therefore practically lost. [3] Soetbeer computes the European Stock on Hand in 1886 at 21 million marks, which is equal to about 5,000 million dollars, crediting Germany with 2,636 millions and the United States with 3,756 million marks; too much in both cases. See his "Materialien," 1886, p. 77, quoted in United States Mint Report, 1886, p. 326. [4] Erroneously estimated in "The Science of Money," ed. 1896, at 800, instead of 900, million pounds sterling.

information on the subject. Moreover, so long as the American Mint Bureau obtains the support of Congress in this matter, will that body deprive itself of all information concerning the production, consumption and stock of the precious metals, but such as squares with the narrow views of the Mint Director.

For example, that officer asserts that India possesses a stock of 588½ millions; China, 750 millions; and Siam, 213½ million dollars, together, 1,532 million dollars, in gold and silver, chiefly silver money; whereas, the highest local and technical authorities and most painstaking estimates on the subject, do not credit these countries with more than 295 millions, a difference of no less than 1,237 millions! In 1893 Mr F. C. Harrison, the Accountant-General of Madras, and Sir David Balfour, both estimated the metallic money of British India at 115 crores. In 1898 Mr. Harrison estimated the money of India at 120 crores, equal in value to about 240 million gold dollars. In China, there is practically no gold or silver money, except in the Imperial treasury and at the seaports, for the purpose of trade with foreigners. A liberal estimate of these stocks is 35 million dollars, and of this amount some portion has recently been plundered by the invading allies. Siam, with an indigent population of five millions, is credited by the Mint Bureau with 213½ millions in gold and silver money! This would be more than $200 for every family in the kingdom: it would exceed the metallic resources of London, which conducts the exchanges of the commercial world; a claim too preposterous for argument. The observance of the Buddhic religion;[3] the frequent employment of rice for money-payments; the prevalence of slavery and barter; the petty dimensions of its commerce and revenues; and the low scale of prices in Siam, combine to warrant the opinion that five millions, instead of 213½ millions, will fully cover the circulation of gold and silver in that country. A somewhat similar condition of affairs in China, where, however, the population is numerous and commerce active, has saved it from the fate of Poland. When the Germans were asked the motive of their "punitive" expeditions into the heart of China they replied that they were looking for the 750 millions which the American Mint Reports asserted were to be found in that Empire. They searched for these millions with rifle, sword and dynamite; they outraged, tortured and slew 50,000 people; they tore down the houses of the living and dug into the graves of the dead; but they got little more than a few trinkets and furs. The currency of China consists of overvalued chuen, or "cash,"

[3] See Buddhic interdict of the precious metals in next note.

practically irredeemable notes cast in copper; and when the Germans, French and other foreign invaders of China became fully satisfied of this fact, they declared that the war was over, and withdrew their predatory bands. Had China possessed the millions of silver assigned to them by our propagandist Mint Bureau, there can be little doubt that the invading armies would have remained long enough to destroy the Empire and appropriate its treasures. Prudent, indeed, were the ordinances of Buddha which forbade his followers from using the precieus metals;[4] and the Chinese have only themselves to blame if they have failed to fully appreciate their significance.

If the proofs already advanced, showing that the value of the precious metals has nothing whatever to do with the cost of producing them, should fail to convince the followers of the erroneous theory so commonly entertained on this subject; the Taxation of the Product, a topic invariably omitted from view by the promulgators of this theory, should of itself be sufficient to suggest doubts of its correctness. All of the Spanish and Portuguese acquisitions of the precious metals, down to the Spanish-American revolutions, yielded to their respective governments in royalties and other exactions, about one-fourth. The royalty during the most productive period was a Quinto, or twenty per cent., but beside the royalty there were exacted other imposts, fees and dues, which often brought the proportion to a third, and sometimes to more; so that although the Quinto was in less productive periods reduced to ten and at last to five per cent., it may fairly be estimated that the Spanish and Portuguese governments obtained about a fourth of the entire product. A still greater proportion of governmental share is to be extended to the 1,300 millions obtained from the mines of Russia and Siberia, whether before or after the Spanish-American revolutions. Of the 450 millions yielded by the Transvaal mines down to the period of the war, it has been stated that the Boer government, by means of royalties, imposts and monopolies, exacted as much as ten per cent. Practically, the only output of the mines which has altogether, or nearly altogether, escaped the payment of royalties, seigniorages, or other onerous mint charges, has been that of America since the Spanish-American revolution, and that of Australasia. Together this output may be roughly estimated at 10,000 millions.

The general outcome is that about one-fourth of one-half of the

[4] This ordinance is mentioned in Appleton's Encyclopedia, 1859, vol. IV, p. 68. Mr. Ball, in his "Geology of India," noticed the aversion of the Buddhists of Laddak to search for gold or silver. The Chinese do not permit gold-mining and greatly restrict the mining of silver.

entire product of the precious metals, since the Discovery of America, has been appropriated by governments in the form of Taxation. The cost of production theorists are quite silent on this subject: they neither charge this governmental sequestration (some 2,500 million dollars) to cost of production, nor deduct it from the produce. To do so would fail to make the result square with their dogma; and so they either omit it entirely from view or else refer to it in some relation not connected with the value, or the assumed source of the value, of the precious metals.

From this topic we turn to a consideration of the Stocks on Hand of gold and silver.

It is impossible to make any estimate of the coins and bullion employed, or ready to be coined and employed, for money, to which objections cannot be made. Some writers suppose that the general stock of metallic money has always a tendency to increase; therefore they object to the validity of any table which fails to show such increase. This theory does not agree with fact. In many states notes or paper-moneys are rapidly usurping the place of coins. This is notably the case in the Roman, European, or Occidental world. The general progress of the note-system is shown by the fact that in the early part of the 19th century notes formed about one-fourth of the entire circulation of the Occidental world. At the present time notes form rather more than one-half. The substitution of notes for coins is also to be observed in British India. In China the use of small cheques is very common. On the other hand, the cheque system, except in very large transactions, is not in vogue in France, nor in some other European States. In England and the United States cheques are largely substituted for coins. Other writers suppose that increased production and great coinages of the precious metals imply an increased circulation of coins. This also is an erroneous theory. More than two-thirds of the entire product of the precious metals are used up in the arts, or else devoted to making good the wear, tear and loss of coins. The coinage, so much relied upon by other writers, is in fact not a reliable guide to the circulation. The system of gratuitous coinage, pursued by the leading states of the world, destroys all traces of production and all indications of the circulation. This system permits the miner, assayer, or bullion broker, to sell any quantity of gold (in the United States not less than $100, in England not less than $50,000 in any one lot) to the Mints, at a fixed price. This price in the United States is $20.67 per Troy ounce fine. The ounce is then coined into precisely $20.67; so that the depositor of bullion is paid for it in coins containing exactly the same quantity of fine gold that he deposited; less a nominal charge for refining. The alloys of copper and silver mingled with the gold in the coins, to harden them, are furnished by the government for nothing. As by this system not a grain of bullion is lost by the depositor, and there is practically no penalty for mutilating or melting coins, the same bullion can be, and often is, deposited and coined over and over again, now in one country, now in another, without loss to the depositor. Immense sums of Australian gold bullion and coins are annually deposited and gratuitously coined in the San Francisco Mint. With the proceeds of these deposits, which are received in American gold coins, bills of exchange on London are purchased. These bills will purchase an equal sum of gold in London, and this gold can be deposited and gratuitously coined in the London Mint. Upon a turn of exchange, these coins (English sovereigns) can be transported to Philadelphia and there recoined into American eagles. Under these circumstances the coinages are largely in excess of production. The Mint Bureau makes it a practice to ask the depositors if their metal is native or foreign; and the replies are gravely entered in the official records and published in the Mint Reports, as facts beyond question. They are about as reliable as the Census returns. If the depositor has any motive for concealment—as very often he has—he will melt his bullion and have it refined and stamped by a native assayer. It then becomes impossible to discover its origin; for all refined gold is alike. Still other theorists believe that an increase of bank reserves indicates an increased circulation of coins. In fact, it may be quite the reverse. The import and export account, so much relied upon by others, is perhaps the most fallacious of all indications concerning the stocks of gold coins and bullion retained in various countries. The dealers in exchange would not be able to make a living if the public were permitted to know how much gold they shipped to and fro. What they disclose is what they choose to disclose: what they conceal is often far more important. Immense sums pass in and out of the country, both from Mexico, or to Canada and elsewhere, without appearing in the American records. But they sometimes appear in other records, and these can be got at if necessary. The amounts carried in and out by travellers, emigrants, seamen and others, are in the aggregate also immense. These sums do not appear in the Commerce and Navigation accounts, they are not "officially" known to the Mint Bureau, which, therefore, entirely ignores them and goes on heaping up hundreds of fictitious millions of gold in the hands of the public; whereas, in fact, except in the sparsely populated mining states of the Pacific Coast, there is hardly any gold at all in circulation in America. A gold piece offered to a tradesman in Boston, New York, or Chicago, would excite his unaffected astonishment.[5]

With regard to the Commerce and Navigation accounts, anyone who will take the trouble to compare the Exports from the United States to Canada, as published by the American government, with the Imports into Canada from the United States, for the same period of time, will perceive at once how defective and unreliable are the former. These two accounts should tally, if not as to value, certainly as to quantities. In point of fact, they do not tally in either respect; and the diserepancies amount to a very large proportion of the entire trade. Still greater discrepancies appear when a like comparison is made in the trade accounts between the United States and other countries. But as such last comparison, to be fair, should run through many years, and be subject to many different allowances, it is not offered as evidence in connection with the present subject.

The objections mentioned are only a few of those which have been made to published estimates of circulating coins and bullion. Their name is legion. In this, as in other technical matters, the reader must be content to accept the judgment of those, who by experience, study and frequent tests and comparisons, venture to form and publish a deliberate opinion on the matter; for at best, with the prevailing systems of free-mining, gratuitous-coinage and the unrestrained melting of gold, all conclusions on the subject are but opinions. The only approximately certain factor in such computations is the visible gold, that which is reported to be on deposit in public treasuries and banks. Even this is not certain, for what is called gold in the reserves is sometimes only gold certificates; the gold being deposited elsewhere and therefore counted twice. With these explanations and reservations we now venture to submit a Table of the coins and bullion employed, or ready to be coined and employed, for Money, in the various countries of the world, at dates nearest to 1879 and 1899. This includes the sums in public depositories and bank reserves, and excludes plate, gold-leaf and jewelry.

[5] The method of the Mint Bureau is to accept the inflated production statistics and deduct from such production the declared exports, less imports and the declared amount consumed in the arts, and then to assume that the remainder must be in the country, in the form of coin or bullion awaiting coinage! This is the method of a child. The truth is that, except the Visible Stock (in Treasuries and in Banks), and a few millions circulating in the mining states, there is hardly any gold at all in the United States. Everybody in America is, and has been, well aware of these facts; that is to say, everybody except the Mint Directors of the last thirty or forty years.

STOCKS OF METALLIC MONEY IN VARIOUS COUNTRIES.

Table showing the Population and Stock on Hand in public depositories and in circulation, of Gold and Silver coins and Bar Bullion, in the Roman and Oriental Worlds, at dates nearest to 1879 and 1899, respectively. Population in millions; sums of coins and bullion in millions of dollars. In the Roman world the silver is computed at the mint value; in the Oriental world at the market value.

	1879		1899			
Countries.	Pop.	Gold and Silver.	Pop.	Gold.	Silver.	Total.
France,	37	1020	39	1100	500	1600
United Kingdom,	34	520	40	400	100	500
German Empire,	44	400	52	500	200	700
Russia in Europe,	74	250	130	750	250	1000
Russia in Asia,	8	25				
United States,	49	250	75	600	517	1117
Mexico,	10	60	13	3	97	100
Other Independent States in America,[1]	36	50	41	14	40	54
Canada,	4	30	5	20	20	40
Spain,[2]	17	170	19	35	35	70
Portugal,	4	55	5	5	18	23
Austro-Hungary,	37	80	46	120	100	220
Italy,	27	40	32	50	40	90
Netherlands,	4	60	5	30	50	80
Belgium,	6	85	7	30	40	70
Switzerland,	3	10	3	10	8	18
Greece,	2	10	2	1	5	6
Sweden,	4	10	5	13	7	20
Norway,	2	5	2	5	2	7
Denmark,	2	5	2	15	5	20
European Colonies,[3]	10	50	16	40	40	80
Egypt,	7	20	9	15	10	25
Turkey in Europe,[4]	5	25	5	?	?	25
Roumania, Servia and Bulgaria,	11		11	?	?	
Turkey in Asia, and Tripoli in Africa,	17	25	17	?	?	25
Total Roman,	454	3255	581	—	—	5890
India,[5]	295	250	297	5	250	255
China,[6]	120	40	120	—	35	35
Japan,[7]	40	50	45	25	30	55
Siam,[8]	5	5	5	1	4	5
Other Oriental,[9]	75	100	75	—	—	100
Total Oriental,[10]	535	445	542	—	—	450

NOTES 1 to 10 belonging to the above Table will be found on the next page.

The intelligent reader does not need to be reminded that this Table does not represent the stocks of Money in the countries named, but only such portion of the money as consists of gold and silver Coins and bullion. Owing to the gratuitous coinage for gold, a policy which, (following the example of Spain in 1608 and England in 1666), has been enacted into law in the principal commercial states, these stocks are continually shifting, the metal being transported from one country to another, at the dictate of stock-speculators, bullion-dealers and bill-drawers. Consequently the estimates are, and only can be, roughly approximative. The national Measure of Value in this boasted age of civilization is in fact a thing of india-rubber.

In combining these stocks of metal with the other moneys employed in the States mentioned, with the view of roughly determining the entire currency or Measure of Value, in each State (a matter that ought not to be, but is perforce, left to conjecture), several important considerations are to be held in view. A portion of these metallic stocks—usually a fourth—should be deducted from the circulation as metallic money reserved to redeem promissory paper issues. In some States the machinery of the financial department is so unskilfully devised that a considerable portion, sometimes nearly all, of the metallic stock remains permanently in the Treasury, without any lawful means of putting it into circulation. These hoards should also be deducted from the circulation: because they do not circulate. In other states, for example, Russia, Germany and France, very large metallic stocks are purposely and permanently kept out of circulation, being dedicated as War Funds, to be used only in cases of emergency. In some States the cheque system so largely supplants the use of money, that to compute the latter without taking the former into consideration, would lead to gross misapprehension. Such is the case in England and the United States. On the other hand, to regard cheques as money, by adding together the sum of both, as

NOTES belonging to Table on page 456.

[1] The "Other Independent States in America" include Central America, with a population in the year 1899 of about three millions, and South America, with a population of about thirty-eight millions. [2] In Spain, gold coins command 27 to 28 per cent. premium, and are not in circulation. Consular Report, No 243, p. 474. [3] The "European Colonies" include the principal West India Islands, population about three millions; Cape Colony, two millions; South African Republics, two millions; Straits Settlements, four millions; and Australasia, five millions. [4] "Turkey in Europe" includes Roumelia, Bosnia, and Herzegovina. These states, together with Roumania, Servia and Bulgaria, total population about 16 millions, were all lumped under "Turkey in Europe" in the estimates of 1879. [5] The stock of coins in British India amounts to 120 crores, or 1,200 million silver rupees, nominally about 528 million dollars, but worth in gold dollars less than half that sum. The sum of five crores is allowed for the native states. Consult Harrison's estimate, in the "History of Monetary Systems," English edition, p. 19. [6] For population of China, see The Cambridge Encyclopedia, vol. I. It is therein shown from numerous authorities and from the censuses of China, from the rice tribute and other indications, that the commonly accepted numbers of the population are the result of Chinese misrepresentations to the foreign ministers. [7] For money of Japan, see "Money and Civilization," chapter XX. Both gold and silver command a premium in the paper-notes, which form the bulk of the circulation. Consular Report, No. 249, p. 271. [8] For money of Siam, see remarks in the text. [9] "Other Oriental" includes Persia, Afghanistan, Turkistan, French Indo-China, Anam, Camboge, Cochin-China, Tonkin, Corea, the Straits Settlements and the Philippines. [10] The Asiatic countries are now for the first time included in a tabulation of the world's stock of the precious metals.

has been done by Sir John Lubbock and others, would be equally misleading. Researches on this subject have been made by several writers, but by none with entire satisfaction, the data being defective and the researches recondite and perplexing. Including cheques, bills of exchange, promissory notes, bonds, book-accounts and all other instruments or devices of credit employed in payments, as money, or as substitutes for money, the efficiency of Credits compared to Money is, in the United States, as one is to twenty. In other words, in computing the Money of that country, a sum equal to one-twentieth of all the Credits outstanding at a normal given time must be added to the current Notes and Coins, after deducting from the latter the Reserves and Treasury Hoards.[6]

The relative efficiency of Money and Credits is due to their respective ratios of activity, or frequency of use and re-use. A coin or note, say to the value of $100, may make, say, a thousand payments in the course of a year, thus aggregating payments and effecting exchanges to the extent of $100,000, per annum. A cheque for $100 will usually make but one payment, after which it is cancelled or destroyed. It is true that other cheques may be issued; but so may be other notes. In an analysis, the enquirer is only concerned with the efficiency of one specimen of each kind of money or money-substitute; not of an unknown number. As to such number it is evident that it can not be illimitable, but must be restricted by the quantity of actual Money in circulation and by the condition of Credit. In short, no attempt to determine the monetary circulation of a country, and its influence upon prices, can be successful without taking into consideration the varying velocities of different kinds of money, or substitutes for money. The precious metals are the product of Nature, whilst money is the product of Law; one may be studied statically, the other must be studied because it operates, dynamically. Time is thus seen to be an element of Price. It is therefore also an element of Value, when Value is expressed in the measurer called Money.

[6] Consult the author's "Science of Money," chapter XIV, for a more complete exposition of the respective velocities of Money and Credit in the exchanges of the United States and Great Britain.

CONCLUSION.

Mining policy of the United States—Historical review—Free-mining demands the resumption of national authority over the mines—This will lead to more correct returns of production—Recapitulation—Evolution of the money metals—Legal creation of money—Substitution of Exchange for Barter—Introduction of paper-notes—Evolution of money apart from metals—Increase of its mobility or frequency of use and re-use—Time as an element of price—The increased production of gold and the increased velocity of money has more than made good the demonetisation of silver.

A RECENT writer on the mining policy of the United States observes that: "There is, perhaps, no part of the public land policy of the United States which reflects so little credit upon the government as the management and disposal of that portion of the public domain containing mineral deposits."[1] By the Act of 1785 there was ordered to be reserved of the public lands "one-third part of all gold, silver, lead and copper mines, to be sold or otherwise disposed of as Congress shall hereafter direct." This fund was intended for the maintenance of public schools. But before tracing the further progress of American mining legislation, it will, perhaps, be better to turn back for a moment and note the status of the mineral lands acquired by the purchase of the province of Louisiana. In 1712 the French Crown granted this province to Antoine Crozat, together with full control of all the mines he might discover in it, reserving to itself one-fifth of the gold and silver and one-tenth of all other minerals produced. In 1717 these grants were transferred to the Mississippi Company, of Paris; in 1722 the Company failed; and in 1731 the grants reverted to the Crown. In 1762 the province of Louisiana was acquired by Spain; in 1763 all of it east of the Mississippi River was ceded to Great Britain, and in 1783 this portion fell to the United States. The western and southern portion remained in the possession of Spain until 1800, when it was ceded to France, who, in 1803, sold it to the United States. It was therefore not until 1803 that the Act of 1785 became operative in the Trans-Mississippi.[2]

[1] Mr. Geo. R. Virtue, on "The Public Ownership of Mineral Lands in the United States," in the Chicago Journal of Political Economy, March, 1895.

[2] Del Mar's "History of Money in America," chap. x.

From that date, or rather from 1807 to 1847, the American government not only exercised the right to tax the mineral produce of the public lands, it undertook the entire management of the mines. This was done for 15 years by the Treasury Department, and for 26 years by the War Department. No gold or silver mines were known at that period in the Trans-Mississippi, so the government devoted all its attention to lead and copper. The lead mines were leased to individuals upon a rental varying from 10 to 26 per cent. of the produce, sometimes $3 to $5 per thousand pounds of mineral raised. After 1821 the rental was uniformly 10 per cent. of the produce, and the lessee was bound to employ a certain number of men for a stated portion of the year. (Virtue.)

In 1829 the lead lands of Missouri were thrown open to private purchase, while those of the Dubuque region continued to be managed by the government. In 1829, owing to a fall in the price of lead, the rental was reduced in this region from 10 to 6 per cent. of the produce. By this time the numbers, wealth and influence of the miners, smelters, and others engaged in this industry had become so great, while the Federal government was so weak, that the authority of the latter was openly disregarded, and by the middle of the year 1835 the entire receipts from rentals ceased. The miners declared that the Act of 1829 relating to Missouri, extended to the Dubuque region, and in 1834, at a sale of "public lands" in the Dubuque region, although the mineral lands were expressly reserved, the miners bought public lands, and while some seized mineral lands under the pretext that they were public lands, others, who held mineral lands under leases, refused to pay the rentals; an unlawful act in which they were encouraged by the Senator from Missouri and the Governor of Illinois. The legal contests to which this condition of affairs gave rise were carried to the Supreme Court and decided against the miners and in favour of the government, who resumed its management of the mines. In spite of this, the miners continued to preserve their disobedient attitude until 1846, when the government, wearied with the contest, favoured the passage of an Act of Congress, which in that year threw open the mineral lands to private purchase. This completed the triumph of the miners.[3]

In March, 1847, the copper lands in the Chippeway tract, which had been leased since 1842 by the Secretary of War, were likewise thrown open to purchase, and the system of leasing the lead and copper mines by the Federal government virtually came to an end.

[3] Mr. James C. Welling, in his Essay on the "States-rights Conflict over the Public Lands," in the Am. Hist. Ass'n Papers, vol. III, pp. 167–183, cited by Virtue.

When the discoveries of gold were made in California, Mr. Ewing, Secretary of the Interior, favoured leasing the placers and selling the quartz mines, reserving as to the latter a share of the produce. In 1850, Mr. Ewing, having meanwhile gone to the Senate, proposed that all the mines should be obliged to sell their gold to the government at $16 an ounce, the coinage value being $20.67. This plan was opposed by Senators Benton and Fremont, his son-in-law; so that nothing was done beyond the passage of an Act charging a nominal fee for the right to mine. The President in his Annual Message, said he was at first inclined to favour the system of leasing; but bearing in mind the experience of the government with the lead and copper mines, he had concluded to recommend the sale of the California mineral lands, under certain restrictions. These views were carried out by the Acts of 1866, 1870, and 1872, and so the matter has remained to the present day.

Under these circumstances there was nothing to stop the silver miner from becoming a gold miner, and in fact he has become one. The legislation of 1873 drove him to seek the yellow metal instead of the white, and with so much energy has he prosecuted the search, that he now produces more gold (in value) than he previously produced of gold and silver together. Not only this, but the discovery of new gold placers and low grade quartz deposites in Colorado, California, Alaska and other States and Territories, has greatly augmented the number of miners and vastly increased the produce of gold: not a dollar of which, though nearly all of it is produced from the public lands, yields any revenue to the government.

Worse than all is the uncertainty of the produce. In the absence of a statistical tax the officers of the government are reduced to the necessity of "estimating" the produce, and when it is remembered that these estimates find their way into the sober accounts of the Secretary of the Treasury, who publishes a monthly official statement of the gold which is supposed to be in the country, forming a part of the "circulation," the necessity of obtaining correct information on the subject appears more and more urgent.

There is yet another consideration to be borne in mind in this connection. The acquisition of the precious metals by means of conquest is virtually over. There are no States possessing any large quantities of the precious metals, which are so weak as to be exposed to rapine. Gunpowder and arms of precision have imparted strength enough to the weakest to resist the invader, while the substitution of paper money has removed the chief object of pillaging expeditions.

Slave mining is also virtually defunct. The systems of contract-labour, which still lingers in the British dependencies, cannot much longer continue to defy that detestation of slavery and resolution, to stamp out which has long characterised the public policy of Great Britain. The facts in the case need only to be brought home to the British public to procure the abolition of the system; and such a consumation cannot long be delayed.

A necessary step towards the preservation of free mining will be the more complete resumption of national authority over the mines, and this will necessarily lead to obligatory and more reliable returns of production.

When slave-mining and contract-labour are entirely swept away and the precious metals needed for the world's arts and coinage are obtained altogether from free labour, it will not only be regarded a necessary measure to determine with precision the exact produce of the mines, but also how far, with respect to the necessities of State and the taxation of other industries, such produce may be required to contribute to the support of government.

Something toward a consideration of these subjects is to be discerned in the pending movement to organize a Department of Mines and Mining in the government of the United States. The International Mining Congress, which met this year at Boisé City, in Idaho, placed the matter of erecting such a Department very fairly before the country. Since the year 1866 Congress has appropriated over a million dollars for the collection of mining statistics and the publication of mineral reports. We have seen how little it has to show for this vast expenditure. The erection of a governmental Mining Department can scarcely fail to direct such expenditures into a more profitable channel.

Glancing backward over the forty centuries which cover the history of the precious metals, a process of evolution is to be observed in their employment as money, only the more palpable features of which have as yet arrested general attention.

The selection of gold from amongst a variety of commodities, as the one best fitted for a common medium of barter—the addition of copper, of bronze, of silver—their coinage—the legal creation of money—the development of Barter into Exchange—the local acquisition of the precious metals by conquest and slavery—the social and political revolutions which ushered in the 19th century—the general introduction of paper money—the abolition of slavery—the establishment of free-mining for the precious metals—the renunciation of full legal tender copper money by Sweden, Russia and other States—the general demonetization of silver—and the establishment of mixed systems of money, composed partly of paper-notes and partly of full legal tender gold coins with subsidiary coins of silver and copper—these are among the steps in the evolution of Money, of which but few intelligent persons can be ignorant.

But behind all this there has been an evolution of Money far more important than that of its substance, the evolution of its mobility: its growth from a statical to a dynamical mechanism, and the concomitant growth in the methods of exchange. So far as we are aware, this phase of money was first noticed by Locke in 1691, and by Necker in 1784; it was alluded to by Thornton and Mill; the subject was treated at length and the velocity of money computed by the present writer some forty years ago, and recomputed at frequent intervals since that time;[4] whilst lately it has received additional light from the researches of Mr. Fawcett and M. des Essars.

This Evolution of Money has waited upon the establishment of peace and justice, the development of intercourse, the application of steam and electricity to the means of transportation and communication; the introduction of the cheque and clearing-house systems; the establishment of trading corporations; the rise of stock exchanges; and numerous other agencies. The general result is that in progressive States, a dollar now performs, in the same interval of time, several fold the duty in facilitating exchanges that it was previously capable of performing; so that TIME, whose influence upon Price was until recently imperceptible, now clearly and unmistakenly enters into its composition.

For the sake of illustrating this view, let it be supposed that steam vessels and railway trains were suddenly reduced to one-half their present speed, and that the telegraphic service was suspended all over the world—in short, that the means of intercourse and communication were reduced to their condition some forty or fifty years ago. Is it not evident that the number of dollars at present employed to effect the daily exchanges of commerce would be insufficient for the purpose, that for the same number of exchanges more dollars would be needed, and that were more not forthcoming there would ensue a stagnation of trade and a fall of prices, until the reduced number and amount of exchanges, or lower prices, accommodated themselves to the sum of the slower moving currency ?

Look at France, where the cheque system has not been fully de-

[4] Del Mar's "Science of Money," chap. XIV.

veloped and observe how much more money per capita she requires and is obliged to retain in circulation, in order to effect her exchanges, than does England, a neighbouring State, with a similar level of prices, an extended trade and very many more exchanges. Observe the lower level of prices that prevail in Spain, in Italy, in Turkey, where intercommunication is but partially developed, or in India, where it is scarcely developed at all.

These illustrations serve to indicate what financial science has long observed and pointed out: that Money, no matter of what substance it is made, is everywhere growing in effectivenesss, because it is increasing in velocity, or in frequency of use and re-use.

This, then, is the reply to all pessimistic reflections concerning the Demonetisation of Silver: it was a mistake, but one that time and the harmonies of political economy have rectified. It would be a still greater mistake to undo it: for those who lost by it would not be the same persons who would gain by remonetisation. There was a time and that but recently, when it was worth while to save the silver mines, the silver States and the silver interests of our country. It is now too late. The silver miners have become gold miners. The silver States are now gold States; and for weal or for woe the country is unmistakenly committed to the policy of basing its exchanges upon a monetary system whose metallic basis is substantially gold coins only.

HYDRAULIC MINING IN NEVADA COUNTY, CALIFORNIA, 1854
(Judge Sawyer in the foreground)

INDEX.

ERRATA.

Page.	Line.	
13	37	After "Getæ," insert a parenthesis ")."
35	39	For "$8," read "$16."
39	36	For "Agatharchides," read "Agatharcides."
59	17	For "ming," read "mining."
95	42	For "Report," read "Resources."
132	23	For "Alkahem," read "Alhakem."
135	24	For "upon any," read "of any."
136	37	For "sinking," read "shrinking."
137	34	After "Metals" insert "first edition."
138	14	For "alcade," read "alcalde." Same on lines 16 and 26.
152	17	For "Donna," read "Doña."
153	25	For "from first to last," read "during the years."
164	4	For "Ovieda," read "Oviedo."
172	2	For "spoilation," read "spoliation."
173	29	For "indifferance," read "indifference."
180	38	For "lives 30," read "lives of 20."
185	14	For "conquisador," read "conquistador."
186	3	For "Sir A. G." read "Sir H. G."
186	18	For "Yutacan," read "Yucatan."
195	11	Strike out the first "of."
197	18	For "Cassamarca," read "Casamarca." Same on lines 27 and 35.
198	2	For "is a mere matter of," read "was subject to."
198	13	For "Cassamarca," read "Casamarca."
198	24	For "Valverdo," read "Valverde."
228	28	For "soft," read "light."
237	10	Insert the note-reference figure "23" at end of paragraph.
242	35	For "eat," read "ate."
250	39	For "octava," read "oitava." Same as to "octavas."
273	7	For "then," read "them."
286	16	For "$692,976," read "$1,692,976."
291	18	For "plentitude," read "plenitude."
324	21	For "suf-" read "suffered."
340	39	For "is," read "are."
345	15	For "its," read "the."
374	20	For "disadventages," read "disadvantages."
380	10	For "Gagarin," read "Gargarin."
380	25	For "contain," read "contained in value."
385	36	For "*Rhinocerus tichorinus*," read "*Rhinoceros tichorhinus*.
388	16	For "1896," read "1899."
388	18	For "40," read "50."
393	11	For "raised," read "revised."
416	5	For "chimerical," read "ephemeral."
423	40	Before "the," insert "in."
427	9	Year 1893, for "4," read "5."
427	16	For "$ 800,000," read "$63,800,000."
427	35	For "labourer," read "miner."
431	31	For "achme," read "acme."
438	16	For "incidently," read "incidentally."
443	31	For "and," read "but."
448	33	For "sf," read "of."
452	9	For "precieus," read "precious."

APPENDIX A.

THE HALCYON AGES OF GREECE.

Causes of Greek intellectual pre-eminence—Opinion of Mr. Clodd—Dr. Francis Lieber—Real cause obscured by false chronology—This was the discovery of Iron—Ies Chrishna, Tubal-Cain, Eric-theus, and other mythological inventors of iron—It was first used for weapons—Afterwards for mining and agriculture—These considerations connect the systematic opening of Laurium, the increase of Money during the Solonic æra, and the rise of Greek intellect—Eminent Greeks of this period—Vast yield of Laurium—Assumption of mining and coinage prerogatives by the State—Achme of Laurium, tempo Themistocles—Increase of Money—Rise of Prices—Stimulus to trade and invention—Rewards of genius—First Halcyon Age—Decline of Laurium, tempo Xenophon—Fall of prices—Arrest of development—Interval of inertia—Opening of mines and military conquests of Philip—Plunder of Asia by Alexander—Increase of money—Rise of prices—Revival of Greek invention and enterprise—Eminent men of the period—Return of the precious metals to Asia—Decline of prices and industrial stagnation in Greece—Internecine conflicts—Termination of the Second Halcyon Age—Conquest of Greece by the Romans.

In his recent very able work called the "Pioneers of Evolution," Mr. Edward Clodd intimates that the cause of Greek predominance in the ancient world of thought and invention was that they had no Bible. "That old Greek habit of asking questions, of seeking to reach the reason of things, gave the greatest impulse to scientific inquiry." This line of thought seems superficial. Did not other peoples ask questions; did not others seek to reach the reason of things; did not the Greeks themselves possess a Bible in the works of Homer, so that down to the days of Strabo these works were regarded with the same profound veneration that we now accord to the writings ascribed to Moses and the Hebrew prophets? Did the Greeks not believe in oracles, and submit themselves to their guidance; finally, did not the Greek intellect, once so predominant in the word of thought, suddenly collapse after the æra of Alexander, so that never since has it evinced any superiority? In fact, Greek predominance only existed at two intervals. The first one extended from about the time of Periander or of Solon, to the Peloponnesian War; the second, from that of Philip or of Alexander, until the Roman subjection of Achaia. Neither before nor after those periods did the Greek intellect exhibit any marked superiority. The causes of such superiority must therefore reside in some circumstance or circumstances which were transient and peculiar to those periods, and not in the possession or absence of a body of sacred writings; because this last-named circumstance is not of a transient, but of a permanent and continuous character.

In his article on the "Antique," published in the Encyclopedia Americana, Dr. Francis Lieber attributes Greek predominance to the following causes: First, "A religion, which saw in the gods, ideal men; which raised men to the rank of gods; and personified every quality in its multitude of gods and demi-gods." Second, "The number of small states" into which Greece was divided. Third, "The joint celebration of the Olympic games in all of them." And Fourth, "The inventive and finely tempered spirit of the people, their happy views of life, the mildness and beauty of the climate, and the fine marble which the country afforded in abundance." Reasoning of this sort is of little worth. It does not explain why Greek predominance began and ended at certain well-defined periods and afterwards wholly ceased, nor does it connect the deification of heroes or the personification of attributes with the capacity to excel in philosophy, mathematics and the fine arts. As to the influence of the Olympic games or of the possession of marble quarries upon the intellectual and inventive faculties of a nation, one can only wonder that so learned a man as Dr. Lieber should have regarded them as worthy of the slightest consideration.

The principal obstacle that prevents us from utilizing the experience of the past by tracing the causes and effects of well-marked sociological phenomena is a false chronology, which insists upon according place to Adam, Abraham, Moses and other "men raised to the rank of gods," and assigns them to an impossible antiquity. For example, we are told in the Word of God that Tubal-Cain first worked in iron. A statement upon such authority cannot be doubted, but when we are asked to believe that Tubal-Cain was an antediluvian who lived some two or three thousand years before our æra, we know from hundreds of other circumstances either that this is false, or else that Tubal kept the process of making iron to himself, which the Bible itself does not permit us to believe. The general result of employing pious chronologies is to put the false mark of a pious origin upon all sociological phenomena. This result is exhibited by both of the authorities cited. Mr. Clodd attributes Greek predominance to the absence of a Bible. Dr. Lieber attributes it to an anthropomorphic religion. But suppose it can be shown that such predominance followed closely upon the invention of iron, what would become of these pious theories?

The process of making iron was invented somewhere and by somebody. When it was invented it must have occasioned a revolution in every department of human activity and mentality; in arms, in art, in the mode of living, in the relations of social life, and even in philosophy and religion. It was far more important than the long subsequent invention of gunpowder, which is credited with the most far-reaching consequences in all these respects, for example, with the conquest of America, the exploitation of its mines, the Halcyon Age of Europe, and the Protestant Reformation. Can it be doubted that the invention of iron had similar influences? The writer has elsewhere indicated that Ies Chrisna, Tubal-Cain, Eric-theus, Bacchus, and other so-called inventors of iron are mere astrological myths; that in fact iron was unknown to Western Asia, Egypt, or Greece before the period assigned to Homer; and that it did not come into practical use until after the Messenian revolt, which Pausanias explicitly connects with the earliest use of iron weapons in the Levantine world. [1]

The first use to which civilized man is induced to put a new invention is military, because excellence in the art of war is essential to the security of the State. The second use is economical, because profit is essential to progress and happiness. It can scarcely be doubted that iron was originally used in battle, or that it was afterwards used for mining and agriculture. These considerations connect the invention of iron with the systematic opening of the great Laurium mines, the rise of prices which occurred during the Solonic period and the sudden emergence from obscurity of that intellectual host whose works have shed an immortal lustre upon the annals of the Greek "Colonies" and States.[2] Here follows a partial list of them:

EMINENT GREEKS OF THE SOLONIC PERIOD: B. C. 600—400.

	Born.	Fl.	Died.
Periander of Corinth, Patron of Learning . . .	665	625	585
Arion of Lesbos, Poet, ...	—	625	—
Draco of Athens, Lawgiver, ...	—	621	—
Alcæus of Lesbos, Lyric poet, ...	—	600	—
Bius, ...	—	600	—
Chilo of Sparta, one of the Seven Wise Men of Greece, .	—	600	—
Myson of Sparta, one of the Seven Wise Men, . ,	—	600	—
Ilychis, ...	—	600	—
Stesichorus, Poet, ...	632	600	552
Cadmus of Miletus, ...	—	600	—
Anacharsis of Scythia, one of the Seven Wise Men, . .	—	594	—
Solon of Athens, Lawgiver, ...	638	592	558

[1] The Cambridge Monthly Encyclopedia for January, 1899, article on "The Early History of Iron."

[2] The so-called Greek Colonies in Asia are mentioned before Greece itself, because modern archæological research has shown that it is more probable that Greece was populated by the Colonies, than that the Colonies were populated from Greece.

EMINENT GREEKS OF THE SOLONIC PERIOD: B. C. 600—400.—*Continued.*

	Born.	Fl.	Died.
Sappho of Lesbos, Poetess, "Hymn to Venus,"	—	592	—
Thales of Miletus, Astronomer and Philosopher,	636	584	546
Anaximander of Miletus, Astronomer, "Sun-dials,"	610	578	546
Æsop of Phrygia, Fabulist,	—	570	—
Depœnus of Athens, Sculptor,	—	568	—
Scyllis of Athens, Sculptor,	—	568	—
Susarion of Athens, Comedian,	—	562	—
Dolon of Athens, Comedian,	—	562	—
Pisistratus of Athens, (Poems of Homer,)	612	561	527
Anaximenes of Miletus, Philosopher,	—	556	504
Scylax of Caria, Geographer and Math., "Periplus,"	—	550	—
Æsculapius of Croton, Physician,	580	548	—
Œnopides of Chios, Astronomer,	—	542	—
Callimachus, Architect, "Corinthian order,"	—	540	—
Phocylides of Miletus, Philosopher and Poet,	—	535	—
Simonides of Cos, Lyric Poet,	557	535	467
Thespis of Athens, Tragedian,	—	535	—
Pythagoras, Astronomer and Philosopher,	569	533	470
Theognis of Megara, Elegaic Poet,	570	530	490
Hipparchus, son of Pisistratus of Athens,	560	530	514
Xenophanes, Philosopher and Poet, [3]	—	520	—
Heraclitus of Ephesus, Philosopher,	—	513	—
Parmenides of Elis, Astron., "Sphericity of the Earth,"	—	505	—
Aphrodisius, Astronomer,	—	500	—
Anacreon of Teos, Lyric Poet,	533	500	478
Diogenes of Apollonia, disciple of Anaximenes,	—	500	—
Aristides of Athens, Statesman,	—	500	468
Pindar of Thebes, Lyric Poet,	522	482	442
Æschylus, Tragic Poet,	525	480	436
Myron of Athens, Sculptor,	—	480	—
Harpalus, Astronomer,	—	480	—
Themistocles of Athens, Statesman,	514	476	449
Anaxagoras of Clazomenæ and Athens,	500	464	428
Cimon of Athens, Soldier,	—	462	449
Zeno of Elia, disciple of Parmenides,	488	460	—
Pericles of Athens, Statesman,	—	460	429
Phidias, Sculptor,	490	460	432
Sophocles of Athens, Tragic Poet,	495	450	405
Bacchylides of Cos, Lyric Poet,	—	450	—
Antimachus of Colophon, Poet and Musician,	—	450	—
Philolaus of Thebes, Lawgiver,	—	450	—
Herodotus of Halicarnassus, Historian,	484	445	408
Epicharmus, Earliest Greek Comic Poet,	—	445	—
Empedocles of Sicily, Philosopher,	490	444	—
Euripides of Salamis, Tragic Poet,	480	442	406
Leucippus of Abdera, Astronomer,	480	440	—
Socrates of Athens, Philosopher,	468	435	399
Meton of Athens, Astronomer,	—	432	—
Eupolis of Athens, Comic Poet,	—	428	—
Alcibiades of Athens, Statesman,	450	411	405

[3] Xenophanes remarked that "the Ethiopians represent their gods black and flat-nosed, while the Thracians give theirs red hair and blue eyes." From this observation it appears that in the ancient time men used to fashion their gods after themselves. It also appears that so long ago as the epoch of Xenophanes, the Getæ, or Goths, of Thracia possessed the same red hair and blue eyes which distinguish them to-day, whether in the Baltic provinces, in the lowlands of Germany, in Scandinavia, or in the British Isles.

EMINENT GREEKS OF THE SOLONIC PERIOD: B. C. 600—400.—*Continued.*

	Born.	Fl.	Died.
Democritus of Abdera, Philosopher,	470	409	361
Apollodorus of Athens, Painter,	—	408	—
Thucydides, Historian, "Peloponnesian War,"	471	404	401
Lysander, Spartan General and Admiral,	440	404	395
Ctesias of Cnidus, Historian,	—	400	—
Euclid of Megara, Philosopher, disciple of Socrates,	—	400	—
Plato of Athens, Philosopher and Lawgiver,	427	400	347
Aristophanes of Athens, Comedies,	444	400	380
Archytas of Tartentum, Mathematician and Philosopher,	—	400	—
Hippocrates of Cos, Father of Medicine,	460	400	357
Xenophon of Athens, Soldier and Historian,	444	400	359
Aristippus of Cyrene, Philosopher, disciple of Socrates,	—	—	—
Antisthenes of Phrygia, Philosopher, "Unity of God,".	—	400	—
Thrasybullus of Athens, Soldier,	—	400	389
Epaminondas of Thebes, Soldier,	—	400	362
Scopas of Ephesus, Sculptor,	—	375	—
Eudoxus of Cnidus, Astronomer,	—	366	—

Here are nearly a hundred of the most glorious names in the annals of Ilium, names which stand for excellence in every departnent of human activity: philosophy, religion, statesmanship, science, art, and arms. These heroes of antiquity all flourished within a period of two centuries, namely, from 600 to 400 B. C. We have now to enquire what it is that occurred during this interval to create, encourage, and sustain such a galaxy of stars. That the mines of Greece were first opened by an Asiatic race whom we vaguely call Pelasgii, or Stone-workers, is susceptible of proof from existing evidences, for the remains of their works are still to be seen in many places. Some of these are in the Sunium Peninsular, a fact which indicates that the Laurium mines were originally opened by these people. But their deeper and more-systematic working awaited the invention of Iron and the employment of iron tools. This last event could scarcely have occurred earlier than the seventh century; for it was at this period that the Greeks got control of Laurium and still later when the mines were thrown open by the Athenian government to be worked subject to the control and regulation of the State. At that period gold was valued in the coinages of India at 1 for 5 of silver. By raising the value of gold in its coinages to 1 for 10 of silver, the Athenian government got its silver from the farmers of the mines at one-half the price in gold coins for which it exchanged its silver with India for gold, thus making cent per cent profit on the entire product of Laurium. This district reached its achme in the time of Themistocles, about 475 B. C., at which period it annually yielded from 360 to 480 quintals of pure silver, equal in value at that time to from 36 to 48 quintals of gold; that is to say, over a million dollars of the present coinage. From the first employment of iron tools in these mines to the æra of Themistocles was an interval of about two centuries, during which, Laurium must have annually thrown into the Greek mints between 100 and 200 quintals of silver. Among the numerous effects of this steady accession to the monetary circulation was the resumption by the Greek states of their control and regulation of the mines, which previously had fallen into private (chiefly ecclesiastical) hands; and the control and regulation of the coinage, which had been usurped in a similar manner. The Athenian coins were struck from pure silver, without alloy, and bore the device of an owl. Their weight and purity was so rigidly and uniformly maintained that they commanded for centuries the circulation of the Mediterranean and western Asiatic world. Egypt, Arabia, Ethiopia, Scythia, and even Persia and Hither India gave them currency.

But the principal result of the large and steady production of silver from Laurium and its exchange for gold with India, was an enormous increase of money in Attica and the surrounding Greek states and a prodigious yet not simultaneous rise of prices, which, though imperfectly understood, has been faithfully chronicled by the labourious Boeckh, in his "Political Economy of Athens." The period of this rise of prices agrees so perfectly with that of the State control of the silver mines and currency that no doubt can be entertained of their connection. It also synchronizes with the rise of Greek intellect. "Until this period," (the Solonic,) says the Rev. Henry F. Clinton, in his Fasti Hellenicæ, I, viii, "the Athenians displayed no signs of that intellectual superiority which they were destined to assume." A rise of prices means greatly increased rewards for exertion; and when such rewards are to be obtained, there never fails to appear a class of men to claim them and to establish their claims by works. A steady rise of prices offers freedom to the slave, competency to the freeman, riches to the ingenious, enjoyment and security to the wealthier classes, and opportunity to all. Such was the condition of the Greek States during the Solonic days. Unfortunately it was not to last forever.

About the middle of the fifth century B. C. the mines of Laurium were worked to the limits then practicable. A generation later they yielded so poor a return, (at the rate fixed upon silver by the Attic mints,) that they were gradually deserted. Xenophon, who wrote his "Revenues of Athens" about B. C. 400, pathetically alludes to their abandonment and couples it with the low state of the public treasury, which he proposes to remedy by reopening the mines with 10,000 slaves, to be worked on public account. The proposal proves that, whilst he was a famous soldier, he was a poor miner and a worse financier.[4]

[4] President Grant's ignorance of the fact that the silver dollar had been demonetized during his own administration affords another instance of the incompatibility of soldiering and finance.

The epoch of Xenophon saw the end, for a time, of Greek predominance. The mines were practically exhausted, the rise of prices ceased, the industry of Greece stagnated, and the Greek intellect became petrified. In spite of Mr. Clodd's theory of Enquiry, in spite of Dr. Lieber's theory of Anthropomorphism, Olympic games, and Marble quarries, the light of Greece was extinguished, to be relighted only when Alexander plundered the oriental world.

It cannot be too often repeated nor too strenuously asserted that its function as a measure of value gives to Money a social and political bearing which transends in importance all other economic influences and which renders it an element of historical consideration that never can be safely left out of view.

Turning now from the Solonic to the Alexandrian age, we begin, in like manner, with a list of those eminent Greeks whose names stand for excellence in the various pursuits with which they were connected.

EMINENT GREEKS OF THE ALEXANDRIAN PERIOD: B. C. 350–150.

	Born.	Fl.	Died.
Theopompus of Chios, Historian,	—	354	—
Æschines of Athens, Orator,	389	350	314
Ephorus of Cumæ, (Etolia,) Historian,	—	350	—
Demosthenes of Athens, Orator,	385	348	322
Philip, son of Amyntas, King of Macedon,	382	340	336
Callipus of Athens, Astronomer,	—	330	—
Apelles of Cos, Painter,	336	327	333
Alexander the Great, King of Macedon,	—	330	—
Callisthenes of Olynthus, Historian,	—	328	—
Aristotle of Stagyra, Morals, Philosophy, Political Economy,	384	325	322
Euclid of Alexandria, Geometrician, "Elements,"	—	320	—
Theophrastus, Philosopher, "De Lapidus,"	370	320	304
Pytheas of Massilia, Geographer and Navigator,	—	320	—
Zeno of Citium, Stoic Philosopher,	362	320	263
Megasthenes, Traveller and Ambassador to India,	—	312	—
Menander of Athens, Comic Drama,	342	306	291
Aristarchus of Samos, Astonomer and Philosopher,	—	280	—
Bion of Smyrna, Bucolic Poet,	—	280	—
Epicurus, Attic Philosopher,	342	280	270
Theocritus of Sicily, Idyls,	—	273	—
Aratus of Cicilia, Astronomer, "Phenomena,"	—	272	—
Ephorus of Ephesus, Painter,	—	265	—
Agesandros of Antioch, Sculptor, "Venus di Milo,"	—	265	—
Lycophron of Chalcis, Tragic Poet, "Cassandra,"	—	260	—
Archimedes of Syracuse, Mathematician,	287	250	212
Moschus of Syracuse, Idyls,	—	250	—
Callimachus of Alexandria, Epigrams,	—	250	240
Cleanthes of Assos in Troas, Stoic Philosopher,	330	250	240
Apollonius of Perga, "Conic Sections,".	—	240	—
Eratosthenes, Astronomer, Philosopher,	274	240	194
Apollonius Rhodius, "Argonautics," (Poem,)	—	200	—
Alexander, the Isian Orator,	—	200	—
Attalus Rhodius, Mathematician,	—	173	—
Polybius of Megalopolis, "History of Rome,"	206	165	124
Hipparchus of Nicæa, Astronomer, "Eclipses"; "Lat. and Long.",	—	160	125
Apollodorus of Athens, Mythologist, "Biblotheca,"	—	150	—
Nicander of Colophon, Physician and Poet, "Theriaca."	—	—	—

It will be observed that the eminent persons of the Alexandrian period are not half so numerous as those of the Solonic period, although the latter is over two centuries earlier. This is due to that general destruction of the later Greek literature, which the pretentions to divinity of Alexander, Seleueus, Ptolemy, and several of their successors rendered necessary. The same thing may be observed of other ancient literatures: everything was destroyed which did not harmonize with the worship of emperors.[5]

However few in number, these names represent the acme of human attainment in the moral and political sciences, in philosophy, mathematics, astronomy, and art. For accuracy of observation and profundity of thought, Aristole, for ages and ages, stood unrivalled. The "Elements" of Euclid are still employed as a text book in our schools; the Venus di Milo of Agesandros is the finest piece of statuary in the world. That so many great men should have appeared in a single country and within a period of two centuries, to be followed by twenty other centuries of degeneracy and retrogression, bespeaks a cause peculiar to the time, the place, and the circumstances. We venture to assert that no other cause than the Rise of Prices which followed the campaigns of Philip and his son, Alexander the Great, will satisfactorily account for this second Halcyon Age of Greece. To say nothing of Philip's mining revivals and military conquests, Alexander, within the space of four years following the battle of Issus, seized upon the treasures of Egypt, Persia, Bactria, and Hither India, and coined them into those beautiful pieces which are yet so abundant as to be more common than the coins of Washington, which were struck twenty centuries later. A rise of prices immediately followed this great increase of money, and the dormant energies of the Levantine States were once more set in motion. Notwithstanding the disordered condition in which the Macedonian empire was left by the premature death of Alexander; notwithstanding the division of his empire into several kingdoms; they all enjoyed a degree of prosperity which the countries they comprised had not known for several centuries. Egypt under the Ptolemies, Syria under the Seleucidæ, Greece during the Achæan League, and Bactria under Diodotus and his successors, were both rich and progressive. The rewards of talent and industry were ample and stimulating; direct intercourse with India threw open new stores of wisdom, new avenues of adventure, and new means of profit. Unfortunately this commerce was neither beneficial nor lasting. There is a fatality in the Indian trade which has attended every nation that has enjoyed it. Perhaps it is

[5] "Middle Ages Revisited," Appendix F.

the influence of Indian thought; more likely it is the proneness and capacity of India to absorb the precious metals from the West, a subject that eighteen centuries ago attracted the attention and elicited the laments of Pliny the Elder and that still engages the attention of economists and statesmen.[6]

The effects of Alexander's conquests were more permanent than the profits of the Indian trade. They rendered money plentiful; prices rose; industry was stimulated; genius was rewarded. But conquest is not production. The plunder of Asia gradually found its way back to Asia; the rise of prices turned to a fall; industry gave place to internecine wars, and these paved the way to ruin. Within two centuries after Alexander returned to Babylon laden with the spoils of Persia and India, the Greek States were under the heel of Rome and their second and last Halcyon Age was over.

[6] So late as the year 1899, the Boer War and the stoppage of gold supplies from South Africa, compelled England to postpone its policy of introducing a gold currency into India and to order home the gold that had been sent out in pursuance of such policy.

[7] After having remained closed for more than two thousand years, the Laurium mines during the decade A. D. 1870-80, yielded over two million dollars

DEL MAR'S WORKS

AND OTHERS.

PUBLISHED BY

THE CAMBRIDGE ENCYCLOPEDIA CO.

No. 240 WEST TWENTY-THIRD STREET,

NEW YORK.

For expedition, address Post Office Box 160, M. S., New York.

REMIT BY POST-OFFICE ORDER OR DRAFT ON NEW YORK.

ROME: THE MOTHER OF STATES.

From The Boston Public Library Bulletin, July, 1901.

THE MIDDLE AGES REVISITED; or the Roman Government and Religion from Augustus to the Fall of Constantinople. By Alex. Del Mar; 8vo, pp. 400. The Cambridge Encyclopedia Publishing Co., 240 West Twenty-Third Street, New York. Net, $3.

THE author of this work, formerly a Bureau officer of the United States Treasury, delegate to Russia, etc., is rapidly rising into public esteem as an historical writer. His preparation for this difficult eminence was a ripe scholarship and 15 years of close study in the British Museum and Bibliotheque Nationale, during which time he issued several monographs on classical literature, Roman history, archæology, ancient manuscripts and coins; all of which obtained immediate recognition in England and France as works of the latest and most complete research. His first appearance as an historian of that theme of themes, the Roman Empire, was in "Ancient Britain in the Light of Modern Archæological Discoveries," of which work the British critical press said: "This is a boundless store of information neglected by ourselves and garnered by a scholarly American. He reconstructs Roman Britain, a country full of busy cities, seaports and industrial centres, connected by fine highways, of majestic temples and villas, and of a splendidly organized commerce." His second work in the same role is the one before us, "The Middle Ages Revisited."

We defer our opinion of this work until after some review of its contents; a task which, owing to its immense scope, its brevity of style and the grandeur of the theme, is sufficient to tax all the resources of condensation. Perhaps this may best be accomplished by placing ourselves as it were somewhat in the attitude of the author.

In describing the Roman government and religion and their relation to the states of the modern world, it will scarcely fail to appear that the Constitution of the Empire, the Christianization of its institutes and the position of the Medieval Empire and the provinces, until the latter became independent kingdoms, is the key to all modern history; that it has its practical importance and conveys its lessons for the future. In weighing the evidences which throw light upon these subjects, the author is compelled to trace the ancient systems of mythology and religion. It is evident that he would gladly have avoided a subject of so much contention; but this was found impracticable. Society is to some extent the product of religious belief. To appreciate the spirit of the laws under which we live and must act, it becomes necessary to follow the evolution of religious systems. Says the author: "We have entered the arcana of the Sacred College not to profane its mysteries, but to fill our pitchers at its holy fount."

When civil strife had so much exhausted the Romans that they were unable to prevent the overthrow of their republican institutes, or resist the erection of a Pagan Hierarchy, they accepted from their tyrants a form of religion so impious and degrading as to speedily disgust the better classes of citizens and turn them against a government in whose support they had formerly taken an active and prominent part.

"Cæsar claims to be a god," cried Cicero. "He has his temples, steeples, priests and choristers," and the orator sealed his indignation with his blood. This feeling found popular echo in distant provinces like Judea and Britain, where it occasioned those frequent insurrections which distinguished the first century of our æra. The religion which fomented these insurrections was the worship of Cæsar as the Supreme Being. Though it led to Cæsar's assassination by a party of Roman patricians, he was supplanted by Augustus, who, after his conquest of the Roman world, adopted precisely the same impious pretensions. Were not Ptolemy, Antonius, Sextus Pompeius, Deiotaurus and many other sovereigns, who were destroyed by Augustus, worshipped by their subjects as gods; and could Augustus be less of a god who had subdued them all, who had extended the Roman Empire from Gades to India and from Britain to the extremities of the known world? In the reign of Trajan, the careful Tacitus could afford to write: "The reverence due to the (ancient) gods was no longer exclusive. Augustus claimed equal worship. Temples were consecrated and images erected to him; a mortal man was worshipped; and priests and pontiffs were appointed to pay him impious homage." But there was a dread interval of nearly a century when to have written as much would have cost the historian his life, subjected his relatives to banishment and confiscated his and their patrimonies.

Our author shows upon a body of evidence drawn largely from contemporaneous inscriptions, coins and customs, that it was upon this pivot, the worship of the Cæsars, that turned the history of Rome for centuries; because even after the impious belief was rejected by the educated classes, it was cherished by the vulgar. Yet only the faintest allusions to it will be found in our standard works of reference. In Mr. Del Mar's work it is brought into relief. It is then perceived that the true grandeur of Christianity and the moral lessons of its conquest over Paganism have been hidden from the light by a false history of the Roman religion and its development. "No greater struggle was ever fought and none so belittled by petty conceits and fables. Not only this, but if the edifice by which the aims of civilization are supported, continues to be poised upon the flimsy foundations which the medieval monks constructed, it is exposed to the risk of being injured by the attacks which modern criticism and satire may make upon these childish and vulnerable elements."

Passing from the religion to the civil institutes of the Roman Empire, the author challenges the accepted origin and spirit of the Feudal System. The views sf Robertson, Hallam, Guizot, Buckle, Bishop Stubbs and others, are examined with a justice and acumen that belong to the highest order of historical criticism. Their attribution of Feudalism to a barbarian origin, their fixing it upon the basis of military service, their treatment of beneficium and commendatio, are scattered into thin air. Feudal systems have been found in India, Japan, Egypt and Mexico, countries which had nothing in common with the institutions of medieval Europe, except their hierarchical governments. Feudalism is even to be discerned in the early days of the Roman Empire, in the charters of Julius and Augustus, in the laws of Diocletian and Justinian, in the land tenures and customs whose roots were buried in the Sacred College of Paganism. We will not divest our author of the interest with which he has invested this problem by anticipating its solution. We recommend its treatment as the best specimen of historical writing which has appeared since the publication of Gibbon's immortal work.

The institutes of the Roman Empire; the rise of Christianity; the christianization of these institutes; the rise of the Medieval Empire; the Lost Treaty of Seltz (between Charlemagne and Nicephorus, defining their respective boundaries, powers and prerogatives); the Constitution of the Medieval (German) Empire; the fall of the Roman (Byzantine) Empire in 1204; the Guelf and Ghibelline Wars; and the legal and actual position of the Roman provinces during these changes, are told with a force of diction, an elegance of style and a wealth of illustration, which leaves nothing to be desired by the reader. The work is a revelation. It proves that the archæological finds of the past half century have placed at our command a store of learning which only needs scholarship, mental digestion and charm of style to render it of absorbing interest and practical value to the reading world. These are the materials which our author brings to his great task. The scaffolding of the work is hid from sight; one sees only the perfected edifice, in which there are no awkward joints, no evidences of patching, no tiresome digressions, no second-hand evidences, no unnecessary foot-notes. A perfect grasp and critical sifting of original evidences; a ripe judgment in the selection and arrangement of materials; a modest but complete mastery of his subject; thorough assimilation of its elements; and a practised hand in wielding the pen. Such are the impressions which the work conveys; a work which we venture to say must place its author upon a very high literary pedestal.

There is but one fault we have to find with it. Its title, in full, is "The Middle Ages Revisited; or the Roman Government and Religion and their Relations to Britain," and we are bound to say the work is faithful to the title. But why only Britain? Why not "And their Relations to Modern States?" The author shows very conclusively that Britain, long after the time, when, according to received history, it was an independant monarchy, was in fact merely a province of the hierarchy, governed variously at Treves, Aix la Chapelle, or Rome, according as medieval Emperor or Pope maintained the paramountship of the slowly dying Empire of Cæsar. This position was as true of France as of Britain. Why not then have embraced France in those chapters on the "Earliest exercise of certain regalian rights," "The Birth of the independent Monarchy," etc., which close this memorable volume? Mr. Del Mar's earlier works have been translated into French, and have a wide reading in France. Has he not, in this instance, unwittingly cut himself off from a friendly market?

Our Public Libraries will peculiarly appreciate Mr. Del Mar's work. It is printed in bold type (old style, ten-point, leaded, with eight-point notes), on clear stout paper and copiously indexed. One of its chief features for the Librarian is the Bibliography, which takes up 14 pages of eight-point type and includes a number of rare works, of which only a student in the great Libraries of Europe would be likely to have any knowledge. To such works, the author attaches a brief descriptive notice, which will be useful to book collectors not having access to the originals; and to all of them he appends the shelf number of the British Museum Library; in order to save the student the trouble of searching its immense catalogue, in itself a Library, we believe, of several hundred volumes.

DEL MAR'S WORKS.

Alexander Del Mar was born in the city of New York, and educated as a Civil and Mining Engineer. In 1862 he published "Gold Money and Paper Money," and in 1865 "Essays on the Treasury." In that year he was appointed Director of the Bureau of Statistics, at that time a Board of Trade, with executive functions, among others the Supervision of the Commissioners of Mines, Commerce, Railways, Immigration, etc. In 1866 he was appointed the American delegate to the International Congress which met at Turin, Italy, and in 1868 delegate to the Hague. In 1868 he was nominated by Mr. Seymour's friends, and in 1872 by Mr. Greeley's friends, for Secretary of the Treasury. In the same year he represented the United States at the International Congress in St. Petersburg, Russia.

In 1876 he was appointed Mining Commissioner to the United States Monetary Commission; 1878, Clerk to the Committee on Naval Expenditures, House of Representatives; 1880, he published his "History of the Precious Metals;" 1881, "A History of Money in Ancient States;" 1885, "Money and Civilization, or a History of Money in Modern States;" 1885, "The Science of Money," (original edition); 1889, "The Science of Money," 2d ed.; 1899, "A History of Monetary Crimes;" 1900, "The Science of Money," 3d ed.; 1900, "A History of Money in America;" 1901, "A History of Monetary Systems," 3d ed.; and "A History of the Precious Metals," 2d ed.; besides several historical works and archæological treatises of great interest, all of which have been reviewed with the highest commendations by English, French and American critics.

For the past twenty years, Mr. Del Mar has given his whole leisure time to original research in the great libraries and coin collections of Europe. His future works, both of which are well advanced toward completion, will be "The Romance of the Precious Metals;" and "The Politics of Money."

The price of these works, in fine cloth, will be $3.00 per volume; half-morocco, $4.00.

ANCIENT BRITAIN—(*Continued.*)

Mr. Del Mar turns the light of research upon the dark period antecedent to the Norman Conquest with such effect that much of what has been credited without demur must now be revised or cast aside. Mr. Del Mar has drawn his inspiration from a very wide field, and in every instance his views appear to be well founded. The chapters on the Roman House of Commons, the Prerogative of Money, and those dealing with the Gothic power, government and language, and the pretended Bretwealdas of the Heptarchy, are among the most important in the book, which will of necessity be read by all who seek the truth in history, and those not content to accept the stereotyped conclusions which have so long held sway.—MANCHESTER COURIER.

Revolutionises the early history of the country.—LONDON MORNING POST.

He reconstructs Roman Britain, a country full of busy cities, seaports and industrial centres, connected by fine highways; of majestic temples, villas and other architecture in the style of the Mother Country; of a splendidly organized commerce; in short, bearing the same relation to Rome as a British Colony to-day bears to Great Britain.—LIVERPOOL MERCURY.

The History of Money in Ancient States, comprising China, Japan, India, Ariana, Bactria, Caubul, Afghanistan, Egypt, Persia, Assyria, Babylon, Palestine, Aboriginal Europe, Greece, Greek Colonies, Carthage, Etruria, and Rome, from the Earliest Periods to the Dark Ages, by Alex. Del Mar; 8vo, pp. 400; cloth, $3.

In this volume Mr. Alex. Del Mar sets forth the view that "the essence of money is limitation, and that coins and notes are alike symbols of money."—LONDON TIMES.

It would be impossible to impugn the learning and industry which a work such as Mr. Del Mar's undoubtedly displays.—GLASGOW HERALD.

A singularly fascinating work for all readers, . . . which never fails to arouse and sustain our interest. . . . A brilliantly-written work upon a theme of vital importance and perennial interest.—MANCHESTER COURIER.

Evinces industry, patience, and research of no ordinary kind. . . . It is an invaluable service towards the elucidation of a problem on which men's minds are much exercised at the present day to bring together so vast a body of evidence on the world's experience on this subject.—EDINBURGH SCOTSMAN.

The value of the work is such that there can be little doubt that it will come to be regarded as a standard authority on the subject of which it treats. . . . Mr. Del Mar remains severely impartial in drawing deductions from his historical researches.—BOMBAY GAZETTE.

Money and Civilization; or a History of the Monetary Laws and Systems of various States since the Dark Ages and their influence upon Civilization, by Alex. Del Mar; 8vo, pp. 475; cloth, $3. (*Edition sold out.*)

As an authority on monetary systems this work deserves to rank high. It is, in fact, an encyclopædia on the subject, and no one who is making a study of this important matter can afford to be without it.—NEWPORT HERALD.

Del Mar's magnificent "History of Money and Civilization" is the crowning work of a lifetime. It embraces a complete and impartial collection of materials relating to the monetary experience of the various States of the world, bringing together the most ancient and the most modern data. Nobody who proposes to write or talk upon money can afford to be without it.—PARIS STENTOR, ORGANE COMMERCIAL, INDUSTRIAL, AGRICOLE, FINANCIER, ET D' ASSURANCES.

It is with peculiar pleasure we welcome a new work by Mr. A. Del Mar, particularly when it deals with solid historic facts.—LONDON BANKER'S MAGAZINE.

PUBLISHED BY THE CAMBRIDGE ENCYCLOPEDIA CO.

A History of Monetary Systems; or a Record of Actual Experiments in Money made by various States of the Modern World, as drawn from their statutes, customs, treaties, mining regulations, jurisprudence, history, archæology, coins, nummulary systems, and other sources of information. By Alex. Del Mar; cloth, pp. 450; $2.

This is an American reprint of the London edition, which is now out of print.

Apart from any views which an author may propound, as derived from his study of the facts on which these are based, it is always a grateful task to recognize ability, industry and learning in the collection and statement of the facts themselves. In Mr. Del. Mar's present work these qualities are conspicuously present, for none can open the book without seeing in every page evidences of the most laborious inquiry on fields of research and learning wide as the world itself. Every age and country, from Egypt, centuries prior to Abraham's sojourn, the Brahminical records of India, the parchments of Greece, Judæa, Rome and Arabia, down to the exploits of the Argentine Republic in our own day, have been explored and made to furnish their quota to the great museum of the facts of monetary history. . . . All being treated with a wealth of reference and allusion, with a breadth of knowledge and minuteness of detail which can leave little to be desired. . . . Treating of a subject nearly as old as the hills, it is as interesting as a novel and immensely more profitable. . . . Mr. Del Mar's book will continue a monument of his learning and industry wherever monetary science is studied; deeply interesting and valuable to advocates of every shade of opinion or school of finance. He fairly follows the promise of his preface, "that he has not laid his historical works under contribution to support theories," but honestly "seeks, by analysing the various experiments that have been made with this subtle instrument (money), to derive from them whatever light they may be able to throw upon the questions that vex us to-day."—NORTH BRITISH ECONOMIST.

Will no doubt constitute, for many years to come, the standard work on the world's systems of interchange. It is impossible to speak in too high terms of praise of the erudition displayed by the author, without one trace of anything like pedantry. The book is as interesting as a good novel, and vastly more entertaining. A most valuable and interesting contribution to the history of economics; voluminous in its contribution; exhaustive in detail.—LONDON JOURNAL OF THE INSTITUTE OF BANKERS.

Mr. Del Mar's reputation as a statistician and as a writer on monetary subjects makes this work a necessity at the present time to every person who wishes to know or wishes to teach the lessons of experience.—NEW YORK WORLD.

Mr. Del Mar's rank as a student and writer is high. He writes with a view of imparting information, not of establishing a theory. . . . Full of exact facts eminently pertinent to the discussions now in progress.—CHICAGO INTER-OCEAN.

A mass of information which can scarcely fail to be of service in future monetary discussions.—LONDON FINANCIAL TIMES.

A valuable work containing much careful research and few theories.—LONDON BOOKMAN.

Destined to throw a flood of light upon the questions whic' are vexing civilization today.—PEORIA JOURNAL.

The latest solid work in which the gold and silver question is receiving common-sense treatment.—BROOKLYN EAGLE.

A remarkable store of facts regarding ancient and modern systems of money. A storehouse of curious information, a work of remarkable industry, which only a man with a real enthusiasm for his subject would have undertaken.—NORTH BRITISH MAIL.

Alex. Del Mar is recognized as one of the most able and well-equipped students of the monetary problem. His extensive investigations, his skill in sifting and grasping the enormous mass of material gathered in a lifetime of unwearied research, his originality and independence of judgment, have made his works authority.—CLEVELAND PLAINDEALER.

In the present volume, Mr. Del Mar has been as unprejudiced as he has shown himself to be in all of his former writings, his expressed desire being simply to examine and profit by the experience of the past. . . . To statesmen and financiers this work will commend itself for the fulness with which it treats the various systems of the past and present.—NEW YORK TIMES.

A HISTORY OF MONETARY SYSTEMS—(Cont'd.)

After passing through the crucible of Mr. Del Mar's vigorous mind, this work has emerged without bearing any trace of partisan bias. . . . A deeply interesting book.—BOMBAY GAZETTE.

Mr. Del Mar writes without prejudice, without bias concerning any system of money. . . . This honest, unprejudiced spirit of inquiry is one of the most commendable features of a valuable and erudite work.—CAPE ARGUS.

A valuable work.—HAMBURG BORSENHALLE.

Those bewildered by the mass of literature issued on all hands supporting this or that currency scheme cannot do better than obtain Mr. Del Mar's work. We welcome it with peculiar pleasure.—BANKER'S MAGAZINE.

Mr. Del Mar displays a commendable grasp of his subject, and enables his readers to see many episodes of the past from a new point of view.—LONDON MORNING POST.

Heine was said to have achieved an almost impossible literary feat when he was witty in German. Mr. Del Mar has completely succeeded in what may be, without doubt, considered yet more difficult in the writing way: he has given us a book having both charm and value upon the subject of money. Charm there indubitably is, as indeed there always will be, when a deep student of the ancient world gives us the result of his studies in the form of "unassuming expression of knowledge that has been perfectly assimilated," as Hamerton once said.—LONDON SENATE.

Able, thorough, and interesting.—MADRID ECONOMIST.

The most complete collection of historical materials on the subject of money.—PARIS STENTOR.

Peculiarly worthy of attention from students of money problems.—LONDON FREE REVIEW.

A valuable and weighty volume on the history of currency in the various countries of the world.—LONDON PALL MALL GAZETTE.

The literature of monetary science and history is undoubtedly enriched by this able and exhaustive work.—THE SCOTSMAN.

Those interested in the subject will do well to study this informing volume.—LIVERPOOL MERCURY.

We quite agree with the author that money is something more than the mere metal or paper of which it is composed.—MANCHESTER GUARDIAN.

We follow Mr. Del Mar in his attack on the extravagent waste of public money, both in this country and the United States, in coining gratis gold which bullion dealers melt down again periodically into bullion.—LONDON DAILY CHRONICLE.

As an authority on monetary systems this work deserves to rank high. It is, in fact, an encyclopædia on the subject, and no one who is making a study of this important matter can afford to be without it.—NEW YORK HERALD.

History of Money in America; from the Spanish Conquest to the Foundation of the American Constitution, by Alex. Del Mar; 8vo, pp. 200; cloth, $1.50.

La admirable obra "History of Money in America" debida a ese portento de erudicion, quiza el mas fecundo y original escritor economista de nuestros tiempos.—REVISTA ECONOMICA.

Mr. Del Mar, the author of the "History of Money in America," probably deserves to be called the most enthusiastic and indefatigable of writers on the theory and history of money. This latest work from his pen forms no exception to the rule. It has earned a place in the library of every student.—PHIL'A POLITICAL AND SOCIAL REVIEW.

Del Mar's "History of Money in America" should be as familiar as household words. How can we discuss a subject intelligently unless we have consulted the chief authority?—SPIRIT OF THE TIMES.

The passage in which the difference between "barter" and "purchase-and-sale" is shown, and its bearing upon the theories of money which dominate the present mint laws of nations, is the most brilliant chapter in the whole range of political economy. Every student of money will have to study this exposition.

PUBLISHED BY THE CAMBRIDGE ENCYCLOPEDIA CO.

History of Money in the Netherlands; by Alex. Del Mar; pamphlet; 8vo, pp. 32; 50 cents.

"Replete with historical information of the highest value."

"Though written by an American, it has become authority in Holland."

History of Money in China; by Alex. Del Mar. Illustrated with fac-similes of the most rare and curious coins. Pamphlet, 8vo, pp. 32; 50 cents.

"Indispensable to a correct knowledge of the monetary situation in the Philippines and to our future relations with the Orient."

The Science of Money; or The Principles deducible from its History, ancient and modern, by Alex. Del Mar; third edition; 8vo. pp. 226; cloth, $1.

"The Science of Money," by Alexander Del Mar, M. E., formerly Director of the Bureau of Statistics of the United States, Mining Commissioner to the United States Monetary Commission of 1876 and author of several books concerning the precious metals, published in large type and handsome dress by the Cambridge Press of New York, is a contribution to this large and intricate subject that demands the closest study. It is written in a style of powerful simplicity and the originality of many of the observations it contains on modern society as affected by monetary mechanisms are of profound interest and convincing value. Taken altogether, Mr. Del Mar's book is one that no publicist and no thinker can afford to be without. It pays a large premium on perusal.—NEW YORK PRESS.

Of great value because of its wealth of financial facts.—BOSTON DAILY ADVERTISER.

The "Science of Money" should be carefully read by all who desire to form clear ideas upon the subject.—FINANCIAL NEWS.

Terse and vigorous and comes directly to his meaning without circumlocution.—MINING WORLD.

A work which has had for ten years considerable circulation and prestige as an authority on monetary questions.—THE REVIEW OF REVIEWS.

"The Science of Money" is a work written for the student and to take a place among the classics of economic literature.—THE PHIL'A AMERICAN.

An able treatise . . . whose conclusions will obtain general acquiesence. The work will be read with interest by those concerned in economical and monetary transactions.—MORNING POST.

The chapter on the "Precession of Prices" has been pronounced by a leading critical authority in England as "the finest effort of analytic faculty in the whole range of economic literature."—THE JOURNAL, PEORIA, ILL.

The argument is brief, but singularly clear and forcible; the work is one of marked ability and strong interest. It may be read with profit by every business man, legislator and student—in a word, by everybody.—CHICAGO TIMES.

This is a new edition of a valuable work by a writer who has deeply and conscientiously studied the subject on which he writes. The book contains much readable and valuable matter.—THE BOSTON COMMONWEALTH.

Thoughtful persons who have already made some study of monetary questions will find this book worth reading. Considering the vast amount of literature now being turned out on the subject of finance, this is saying a good deal.—BUFFALO EXPRESS.

It possesses considerable value to students of economics. . . . Its author has written several other able works treating of financial matters—notably, a History of the Precious Metals and of Monetary Systems in various States. . . . The work is of real value to students of economics.—THE CALL, SAN FRANCISCO, CAL.

Those who desire to master the monetary question should read Del Mar's "Science of Money" once, twice, thrice—aye, even four times, not because it lacks ludicity or comprehensiveness, but because it is a mind-opener.—ELPHINSTONE V. A. MAITLAND IN THE LONDON COUNTRY GENTLEMAN.

The merit of the book as a comprehensive treatise cannot be denied.—LONDON JOURNAL OF THE STATISTICAL SOCIETY.

PUBLISHED BY THE CAMBRIDGE ENCYCLOPEDIA CO.

A History of the Precious Metals; from the Earliest times to the Present; by Alex. Del Mar. Second edition, complete in one volume; pp. 500, 8vo. Cloth and gold, $3. This is not a recension of the First edition (London, 1880), but an entirely new work constructed on an improved plan by the same author. The following are from the Press Notices of the First edition.

Abounds with vivid description and practical knowledge.—LONDON ATHENAEUM.

Replete with information; evinces much care and study.—LONDON ACADEMY.

Shows the most conspicuous advance beyond his predecessors.—LONDON SAT. REVIEW.

A work of great weight and elegance of style.—LONDON ECONOMIST.

No such able and exhaustive work since that of William Jacob.—LONDON STATIST.

Years ago Mr. Del Mar gave to the public "A History of the Precious Metals" which has since become a standard work on the subject. In that work was traced the adventures of the Phœnicians, Romans, Spaniards, Californians and Australians—those Argonauts who variously, from the dawn of history to the present time, have led the search for the precious metals. Beside discovery, the author showed that these metals had been largely obtained through conquest and slavery. The Persian, Greek, Roman, Egyptian, Spanish and British conquests of mining countries, and the Roman and Spanish systems of mine-slavery, were next delineated, and finally the conditions of free mining were examined. Upon a general review of the entire subject, the author found ample reason to agree with the celebrated dictum of Montesquieu, that, upon the whole, gold and silver, when obtained by free mining and paid labour, had cost to the world a far greater price than they represented in the exchanges. Montesquieu arrived at this conclusion by instinct; Mr. Del Mar reached it through historical research.—SACRAMENTO RECORD-UNION.

Mr. D. has thoroughly surmounted every obstacle, and furnished a volume replete with information. Every line will be read.—LONDON MINING JOURNAL.

A complete text-book on the subject.—LONDON MONEY.

Based on independent research.—LONDON DAILY NEWS.

Of the highest scientific value, yet readable as a novel.—NEW YORK ECONOMIST.

The peculiar advantages which Mr. Del Mar has enjoyed have alone enabled a work of such magnitude and historical value to be completed in a single lifetime. He has travelled into every country whose monetary system he has delineated; he has examined the official records consulted the local histories and numismatic monuments, surveyed the gold and silver mines and traced the mining rivers, in each of them separately; and by the light of these researches he has been enabled to construe many obscure passages in the texts of Cæsar, Pliny, Plutarch, and other ancient writers. Mr. Del Mar's works are valuable contributions to the rather small amount of literature that has been published on such important subjects, and they are the more useful because they are devoted to the setting forth of facts, and not to the upholding of any particular theories.—LONDON FINANCIAL NEWS.

Next to its versatility and vast scope, the absolute impartiality of the work is perhaps its most striking feature. It advocates no theories, it indulges in no deductive arguments, it is strictly historical, and it is history of the highest order, fit to rank with the works of Gibbon, Robertson, Alison, Macaulay, Niebuhr, and Mommsen.—BULLIONIST.

All of the above works will be furnished bound in

HALF-MOROCCO AND GOLD

FOR ONE DOLLAR EXTRA.

Remit by P. O. Money Order or draft on New York to

THE CAMBRIDGE ENCYCLOPEDIA CO.,

240 WEST TWENTY-THIRD ST.,

Address, P. O. Box 160, M. S., NEW YORK.

(The P. O. Box Address ensures more prompt delivery.)

Printed in Great Britain
by Amazon.co.uk, Ltd.,
Marston Gate.